Joy Tivy

Biogeography

A Study of Plants in the Ecosphere

Third Edition

Pearson Education Limited
Edinburgh Gate, Harlow
Essex CM20 2JE, England
and Associated Companies, throughout the world.

First published 1993
Reprinted 1995, 1996 (twice), 1998 and 1999

ISBN 0–582–08035–5

British Library Cataloguing in Publication Data
A CIP record for this book is available from the British Library

Library of Congress Cataloging-in-Publication Data
A CIP record for this book is available from the Library of Congress

Disc conversion in 10/12 Ehrhardt by 8
Printed in Malaysia, PP

Contents

Preface to the first edition

Biogeography is the 'Cinderella' of geography. It is still a relatively neglected and underdeveloped field of study at both school and university, despite the considerable lip-service that has been paid to its significance. Many have drawn attention to its value in providing (within the framework of the early formulated and recently 'rediscovered' ecosystem concept) an integrated holistic approach to environmental studies. Some have accorded it the distinction of being the 'vital' link between physical and human geography. Others have enthusiastically advocated (without necessarily demonstrating) its potential as a means of unifying a subject whose peripheral organs often seem to be growing at a rate greater than that of its body! That biogeography is slowly gaining greater recognition is unquestionable. However, there are still all too few British or American universities where, in general physical geography courses, it receives or attracts the same attention as either climatology or geomorphology – and fewer still where it can be pursued at a more advanced Honours level. In the latter cases instruction is often relegated to departments of biology, botany, zoology or pedology. Those who go so far as to profess biogeography are 'rare', indeed some would say 'odd' species. English textbooks on the subject, written by geographers, can be counted on one hand. A large proportion of its most elementary data must be distilled from non-geographical and often highly specialised and technical biological sources. And developments in the latter have long outstripped those in biogeography.

There are good reasons if not necessarily valid excuses for this state of affairs. First, because of the range of phenomena with which the biographer is faced, his field of study is less amenable to isolation and systematisation than other branches of physical geography, such as climatology, hydrology or geomorphology. Biogeography is not easy to define or delimit precisely; and the distinction between it and the closely related subject of ecology often tends to be more a function of scale and emphasis that of content or method. Second, the majority of teachers and students of geography, whether at school or university, know much less about the nature and scope of biogeography than about practically any other aspect of their subject. Opportunities to combine a study of biology and geography are limited. Unfamiliarity (exacerbated by a high degree of urbanisation) with even the most common plants has tended to create a psychological barrier which makes the average geography student chary of involving him- or herself too deeply in a field which requires the acquisition of a whole lot of new and apparently highly 'technical' terms (among which Latin names seem to be the most daunting!) and concepts. And the continuing emphasis in many, particularly school, texts on 'explanatory' descriptions of the vegetation or soils of the world, often completely divorced from any concrete terms of reference, has done little to overcome this 'built-in' resistance and to stimulate interest in the subject.

This book is the product of a long period of 'experimentation' in the teaching of biogeography at various levels to university students, many of whom had no previous training in biology, or for that matter in any of the physical sciences. Its aim is twofold. The first is to summarise and explain (in a way which it is hoped will be comprehensive and palatable to teachers and students at sixth form and university levels) those biological processes and concepts which the author considers basic to the understanding of the principal characteristics of, and the complex interrelationships within, the 'organic world' or 'biosphere'. In the interests of clarity an attempt has been made to cut a path through the terminology jungle of ecology and to be selective, indeed as sparing as possible, in the use of uncommon and/or exotic plant and animal names. The second is to bring together the bases and aspects of biogeography which at present can only be obtained from a great number of widely scattered sources and, in doing so, to bridge the gap between the earlier traditional zonal study of vegetation and soil and modern developments, particularly in plant ecology. The theme is that of organic resources, and the reciprocal relationship between these and man. Throughout, the emphasis is on plants, as the primary food producers which form the essential link between man and his physical environment. The first half of the book is concerned with a systematic analysis of the effect of environmental (ecological) factors – climate, soil, biological competition, animals and man on the functioning, evolution and adaptation, and distribution of plants. The second half deals with the nature of vegetation and a consideration of the principal characteristics of the structure and function of the major types of ecosystems – marine, forest, grassland and desert – and the potentialities and problems of their particular organic resources for use by man through time.

It is hoped that this book will go some way to answering the perennial *cri de cœur* of the aspiring student of biogeography for a basic textbook which would provide a starting-point and a guide to the highly ramified highways and byways of a vast and complex field. In its compilation the author has drawn on and selected from a wide variety of sources. However, it was decided in the interest of readability to omit numbered references in the text. The lists of references are not, nor were they intended to be, comprehensive. They do, however, include all those sources on which each chapter has been based, and they were selected as those most valuable and relevant to further elucidation and a deeper study of the subjects under consideration.

Finally the author would like to acknowledge her indebtedness to those generations of geography students at the Universities of Edinburgh and Glasgow whose lively interest and constructive criticism in the classroom and extreme fortitude in the field were a continuing source of inspiration. Without them, and the constant encouragement of long-suffering friends and colleagues, this book would never have been written.

Preface to the second edition

In the decade since the first edition of this book was published biogeography has 'come of age', and what David Watts has called 'The new biogeography' has emerged to take its place as a generally accepted and reasonably respectable member of the geographical academic establishment – Cinderella has gone to the ball! Development in Britain has been particularly marked along three closely related fronts. One is in the teaching of the subject. The number and diversity of courses offered at the higher educational levels have multiplied. There are now few university or college geography departments which do not include biogeography as an integral element in their basic introductory and common physical geography courses. The number and range of option courses in the field have proliferated at an almost exponential rate. And at least one member of staff professing biogeography is now a *sine qua non* of any self-respecting university geography department. Undoubtedly, as Watts notes, the growth in the scope and range of the subject during the 1970s has been encouraged by the concurrent increase in joint Honours degrees, the development of new multidisciplinary courses in the life and environmental sciences, and in resource management, to all of which biogeography has contributed an often important integrative element. In addition, teaching at introductory levels in universities and colleges has been greatly facilitated and stimulated by the increase in numbers of young second- and third-generation biogeographers coming into the profession; and not least by the publication of several basic, and happily complementary, textbooks of which those by Cruickshank, Curtis and Trudgill, Pears, Simmons and Watts, are important milestones.

Unfortunately, in spite of these trends in higher education, the feedback into school examination syllabuses has not yet had any really marked impact. A recent report by the Geographical Association on 'Biogeography in the Sixth Form' deplores the fact that its treatment at school level is, if not neglected entirely, variable; it is still too descriptive of zonal world types of soil and vegetation and is not yet sufficiently ecological in its approach; while its integrative potential is rarely appreciated or developed by those who design examination board syllabuses.

A second, major factor in the recent development of biogeography has been the simultaneous and successful establishment, in 1974, of the *Journal of Biogeography* and the Biogeography Study Group within the Institute of British Geographers. Both have opened up venues for publication and discussion within biogeography, as well as channels of communication with related disciplines and with research-funding organisations. More importantly, they have provided two long-needed and invaluable foci for the subject.

Third, the current vitality of biogeography is reflected in the growth in research and in the increasing number of publications appearing in all the major British and American geographical journals. It is particularly interesting to note how the range of research problems has expanded from the early emphasis on the historical development of former 'natural' or remaining 'semi-natural' vegetation to virtually all the formal and functional aspects of ecosystems and, more particularly, to the role of man in current biological and ecological processes. Watts has identified five trends in biogeographical research in the 1970s as reflected largely in the contributions to the *Journal of Biogeography*. These include: the soil vegetation (–man) complex, within which the savanna problem has maintained its early momentum; the relationship between major vegetation types and animal species; the distribution of species; man–ecosystem relationships with increasing emphasis on impact studies; and Quaternary and Holocene ecosystem change and development. In fact, historical biogeography has retained and strengthened its initial dominance, and has played a major part in the vitality of Quaternary studies not only in Britain but in Europe, Africa and Australia. A large proportion of the teaching and research in biogeography in Britain is still in this field; and the palynologists and soil scientists probably form the largest and certainly the two most coherent subsets within the Biogeography Study Group.

The academic 'take-off' of biogeography in Britain in the 1970s is partly a function of its own feedback mechanisms. It has, however, also been influenced by related developments outside the field, mainly, though not exclusively, in biology. Among the most important of these has been the fast-rising flood of quantitative data from a world-wide range of ecosystems generated by the International Biological Programme (IBP) which was initiated in the early 1960s. Another has been the interest in and concern for ecosystem conservation and management which had begun to manifest itself in the immediate post-Second World War decades. This found expression in a spate of new journals and publications, such as the *Journal of Applied Ecology*, *Biological Conservation*, *Applied Biology*, *Journal of Environmental Management*, to mention but a few, which were launched in the late 1960s or during the 1970s. They have provided stimulating platforms for multidisciplinary contributions to the man–ecosystem debate. It is not surprising that the most significant conceptual trends in ecology in the past decade have been closely related to, and have undoubtedly been influenced by, these developments. In the light of the new data now available, there has been a critical reassessment of the concept of ecosystem stability and its implications for those of diversity and those of diversity and fragility; and a revival of interest in McArthur's concept of island biogeography, in view of its relevance for population and ecosystem fragility and conservation.

Finally, the concern, initially generated in the 1950s and 1960s, about the productivity of the biosphere in face of a spiralling world population and the rapid development of resource-oriented outdoor recreation found expression in the past decade in the rapid growth of two areas of study within the field of biogeography. One has been the ecological impact of recreation to which biogeographers have made a not insignificant contribution; and many are members of the very active multidisciplinary Recreation Ecology Research Group (of the British Ecology Society). The other more recent growth area has been that of energy efficiencies in man-managed ecosystems, and the ecological and economic implications of the increase in energy-subsidised systems of intensive agriculture, forestry and fishing.

This second revised edition of *Biogeography* has endeavoured, within the inevitable economic constraints, to take the most significant of these developments within and related to biogeography during the past decade into account. To this end four main changes have been effected.

1. Bibliographies have been updated in such a way as to combine the most important new material relevant to the particular chapter with the initial authoritative sources from which the new ideas have evolved;
2. Quantitative data available from publications of IBP work have been incorporated in new tables and diagrams wherever possible and appropriate; in most cases they have served to illustrate and reinforce rather than radically alter the basic concepts propounded in the first edition;
3. Two areas in which modification and/or elaboration in the light of recent developments have been required are that of ecosystem stability in relation to the climatic concept; and recreation as an ecological factor;
4. A new chapter has been added on 'The soil ecosystem' which it is hoped will make good deficiencies in respect of, particularly, soil formation which was not dealt with in the previous edition.

Otherwise the philosophical underpinning and basic aims stated in the Preface to the First Edition remain substantially the same.

Preface to the third edition

It is now over 25 years since *Biogeography* was first conceived and written in response to a growing demand at that time for an introductory text to this newly emergent field of study in geography. Biogeography, with its focus on the spatial variation of forms, composition and processes within the biosphere, is now firmly and universally established not only as a discipline in its own right but as an integral part of geography. Its continued development since the mid-1970s has been stimulated by concurrent advances in biology – particularly in ecophysiology and population dynamics, by a revival of interest in historical biogeography and the role of the environment in organic evolution and adaptation, and, not least, by the growing concern about the escalating impact of human activities on the biosphere.

The revised second edition of *Biogeography* allowed neither the time or scope to do adequate justice to the flood of empirical and theoretical work already emerging by the end of the 1970s; nor did it reflect the changes in the author's interests, her approaches to and ideas about the subject since the publication of the first edition. She was, therefore, particularly pleased on the eve of retirement, to accept the publisher's suggestion that the third edition should be rewritten rather than merely revised.

However, this edition has not only been rewritten (though the astute critic may detect a few scattered 'palaeorelicts' of the original text), it has also been re-expanded and reorganised. As far as possible the material has been updated to incorporate new data and concepts as well as the more important trends in terminological fashions. More emphasis has been given to the spatial variations in the biosphere (particularly to the animal component) and to the impact of natural and anthropogenic disturbance on it. While most of the figures and tables are new, some have been retained in the conviction that their illustrative value is not necessarily a function of date of publication. The bibliography is the result of a process of successive selection from a universe that becomes increasingly difficult to encompass. It is hoped, however, that the result achieves a reasonable balance between the classic and more recent sources and contains a sufficient number of important basic references on which the student who wishes to delve into a particular aspect of biogeography can draw.

Accordingly the text, after an introduction to the development of biogeography, has been organised in three parts.

Part I. The biosphere – presents in seven chapters, a systematic analysis of the temporal and spatial variations in the components and processes in the biosphere, dealing more fully with the biomass characteristics, primary biological productivity and biological cycling than in the previous editions.

Part II. Ecosystems – in which seven chapters are devoted to a consideration of the origins, formal and functional characteristics of, and human impacts on, the major (forest, grassland, arid) ecosystems and to distinctive island, mountain and aquatic ecosystems.

Part III. Biotic resources – discusses, in the final five chapters, the nature and impact of natural and human disturbance on the biosphere with respect to problems of ecosystem fragility and resilience; management and sustained yield; and the conservation of organic resources.

The aim of this third edition is to produce a more systematic and structured textbook than previously which covers much the same field as before and, it is hoped, retains the original theme of the spatial variation in human–organic resource interactions. It is also hoped that it will provide undergraduate students of geography, environmental science and related disciplines with an overview of the wide-ranging scope and the important inter- and intradisciplinary implications of biogeography, as well as a springboard from which he/she may be stimulated to pursue particular aspects or problems in greater depth. As in previous editions, every effort has been made to summarise and explain those biological processes and concepts which are basic to an understanding of the complex interrelationships within the biosphere in a way that will be palatable and comprehensible to those readers with a limited or no formal training in the natural sciences.

Any attempt to distil the essence of as wide and complex a field as biogeography within such a narrow compass must inevitably be a very personal, and to that extent subjective, interpretation of the author's corpus of knowledge and understanding at a particular point in time which should be reflected in the significance of the ideas rather than in the volume of data presented. The author alone must take full responsibility for any errors of fact, misinterpretations of others' work and for all other shortcomings contained herein.

The author would like to reiterate her indebtedness to the generations of geography students at the Universities of Edinburgh and Glasgow who were the initial stimulus behind the first edition of this book and to the support and encouragement of friends, colleagues and not least my publisher without which *Biogeography* would not have been rewritten. Finally, she would like to thank Betty Johnstone and Tilly Wright whose word-processing skills and indefatigable good humour in face of an unfamiliar terminology and an often marginally decipherable manuscript ensured the expeditious production of copy; and to the cartographer, Bill Thomson, for his meticulous attention to style and accuracy in in drawing and/or redrawing all diagrams; and to James McCormack for assistance checking references and typescripts. In addition, she is indebted to the Leverhulme Trust for a generous Emeritus Award towards the costs of production.

Acknowledgements

We are grateful to the following for permission to reproduce copyright material:

Academic Press Ltd. for Table 6.6 (Weigert & Owen, 1971); American Association for the Advancement of Science and the author, Dr. E. P. Odum for Table 16.1 (Odum, 1969) Copyright 1969 American Association for the Advancement of Science; American Chemical Society for Fig. 18.1 (Likens, 1976) Copyright 1976 American Chemical Society; The American Geographical Society for Fig. 3.3 (Thornthwaite, 1948); American Institute of Biological Sciences and the author, Dr. H. A. Mooney for Fig. 6.1a (Mooney & Gulman, 1982). Copyright 1982 by the American Institute of Biological Sciences; American Society of Agromony Inc., Crop Science Society of America Inc., and the Soil Science Society of America Inc. for Fig. 19.5 (Smith & Hill, 1975); Annual Reviews Inc. and the author, Dr. I. Noy-Meir for Figs 12.2 & 12.4 (Noy-Meir, 1973) © 1973 Annual Reviews Inc.; Edward Arnold (Publishers) Ltd. for Table 2.1 (Stace, 1989); Edward Arnold (Publishers) Ltd. and the author, Dr. W. O. Pruitt Jr for Fig. 10.2 (Pruitt, 1978). © William O. Pruitt, Jr; Association of American Geographers for Fig. 7.7 (Gersmehl, 1976); Blackwell Scientific Publications Ltd. for Figs 5.5 (Grubb & Whittaker, 1989), 7.3 (Stout *et al.*,1976) 8.2, 8.3b (Brown & Southwood, 1987) 15.2 (Moss, 1980), 15.5 (Barnes & Mann, 1980), 19.3 (Krebs, 1985 redrawn from Beverton, 1962) & Tables 3.2, 3.3 (Bannister, 1976), 6.4 (Humphries, 1979), 7.3 (Coleman, 1976), 11.1 (Sims *et al.*, 1978); Blackwell Scientific Publications Ltd. and the respective authors for Figs 5.1b (Watkinson, 1986), 8.1 (Miles, 1987); the authors, Dr. J. H. Brown & Dr. A. Kodric-Brown for Fig. 13.1b (Brown & Kodric-Brown, 1977); Butterworth Heinemann for Fig. 15.4 (Tait, 1972); Cambridge University Press for Fig. 9.3b (Huttel, 1978) & Tables 10.6 (White *et al.*, 1981), 11.2, 11.3 (Coupland, 1979); Cambridge University Press and the respective authors for Figs. 9.5 (Oldeman, 1978), 10.3b, 10.4 (Rydén, 1981) & Tables 6.8 (Cooper, 1975), 7.8 (Sprent, 1987), 10.3, 10.5 (Wielgolaski *et al.*, 1981), 10.4 (Bliss, 1981), 12.1 (Le Houérou, 1979), 19.3 (Pimentel, 1979); the author, S. Carlquist for Table 13.3 (Carlquist, 1974);

Chapman & Hall for Figs 5.1a & 5.2b (Putnam & Wratten, 1984); Chapman & Hall and the respective authors for Figs 4.6 (Meyers & Giller, 1988), 4.10 (Lynch, 1988) & Tables 13.1 (Gorman, 1979), 20.2 (Gilbert, 1989); Crop Science Society of America Inc. for Fig. 6.2 (Pearce *et al.*, 1967); Ecological Society of America for part Table 5.3 (Evans & Dahl, 1955); Elsevier Applied Science Publishers Ltd. for Table 16.3 (Crawford & Liddle, 1977); Elsevier Science Publishers and the respective authors for Figs 12.1b (Shmida, 1985), 12.3 (Evanari, 1985a), 14.2 (Flohn, 1969) & Table 12.2 (Evanari, 1985b); Verlag Eugen Ulmer for Fig. 20.1 (Sukopp & Werner, 1986); Food and Agriculture Organization of the United Nations for Table 6.1 (FAO, 1978); Gauthier-Villars for Fig. 12.5 (West & Skujins, 1977); Harper Collins Publishers Inc. for Figs 3.7 (Pianka, 1988), 19.3 (Krebs, 1985) Copyright © 1985 & 1988 by Harper & Row Publishers Inc.; Hodder & Stoughton Ltd. and the author, Prof. A. D. Boney for Table 15.2 (Boney, 1989), © 1989 A. D. Boney; the author, Dr. H. B. N. Hynes and Fig. 18.3 (Hynes, 1963); the author, Dr. V. Jensen for Table 7.1 (Jensen, 1974); Kluwer Academic Publishers for Fig. 6.1b (Sestak, 1985) & Tables 6.5 (Heal & MacLean, 1975), 7.6 (Reichle *et al.*, 1975), 20.1 (Sukopp & Werner, 1983); Longman Group UK Ltd. for Figs 4.1 & 4.11 (Good, 1974), 7.2 (Richards, 1974) & Table 12.4 (Louw & Seely, 1982); Longman Group UK Ltd. and the author, Dr. K. A. Longman for Fig. 9.3a (Longman & Jenk, 1987) © 1987 K. A. Longman & J. Jenk; the author, Dr. D. S. McCluskey for Table 15.3 (McCluskey, 1981); McGraw-Hill Inc. for Fig. 5.6 (Trewartha, 1954); Macmillan Publishing Company for Fig. 3.4 & Table 3.4 (Brady, 1974) from *The Nature & Property of Soils*, 8th Edition by Nyle C. Brady. Copyright © 1974 by Macmillan Publishing Company & Fig. 6.3 & Table 6.2 (Whittaker, 1975) from *Communities & Ecosystems* 2nd Edition by Robert H. Whittaker. Copyright © 1975 by Robert H. Whittaker; S. A. Masson for Table 7.4 (Duchafour, 1965); Methuen & Co. for Fig. 7.4 (Pitty, 1979) & Table 14.3 (Wardle, 1974); Methuen & Co. and the author, Dr. P.J. Webber for Table 14.4 (Weber, 1974); National Academy Press for Figs 7.6 (Duvigneaud & de Smet, 1970), 15.3 (Bunt, 1975) & Tables 6.3 (Petrusewicz & Grodzinski, 1975), 6.7 (Crisp, 1975); Oxford University Press for Tables 2.6–2.8 (Lovelock, 1979), 13.2 (Williamson, 1981); Pergamon Press PLC for Fig. 2.6 (Jantsch, 1980) Copyright 1980 Pergamon Press PLC; Routledge for Fig. 14.1 (Ives & Barry, 1974); Royal Swedish Academy of Sciences for Table 7.7 (Rosswall, 1976); Scientific American Inc. for Figs 2.5 (Cloud & Gibor, 1970), 18.2 (Likens *et al.*, 1979) & Tables 2.4 (Delwiche, 1974), 17.2 (Mangelsdorf, 1953) Copyright © 1953, 1970, 1974 & 1979 Scientific American Inc. All rights reserved; Sigma Xi, The Scientific Research Society for Table 12.3 (Hadley, 1972); Sinauer Associates Inc. and the author, Dr. S. L. Pimm for Table 16.2 (Pimm, 1986); Springer-Verlag for Fig. 18.4 (Moore, 1967) & Tables 3.6 (May, 1982), 10.1 (Davis, 1981); Springer-Verlag and the respective authors for Figs. 9.4 (Hallé *et al.*, 1978) 10.1 (Kimmins & Wein, 1986), & Tables 10.2 (van Cleve *et al.*, 1986), 11.4 (Owen-Smith, 1982), 19.2 (Pimentel *et al.*, 1990); the author, Prof. J Tivy for Figs 2.3,

2.4 & 19.4 (Tivy & O'Hare, 1981); UNESCO for Fig. 12.1a (Meigs, 1953) & Table 7.9 (Ellenberg 1971b) © UNESCO 1953 & 1971; The University of Chicago Press and the author, Dr. R. E. Ricklefs for Table 13.4 (Ricklefs & Cox, 1972) Copyright © The University of Chicago Press for Fig. 13.2 & Table 13.4 (Ricklefs & Cox, 1972) © 1972 by The University of Chicago. All rights reserved. University of Washington Press for Figs 3.6 (Shimwell, 1972), 19.2 (Allen, 1980); The University of Wisconsin Press for Fig. 18.5 (Rudd, 1964) © 1964 The Board of Regents of the University of Wisconsin System; John Wiley & Sons Ltd., for Fig. 5.3 (Watkinson, 1981) & Tables 18.2, 18.4 (Treshaw & Anderson, 1989) Copyright © 1981 & 1989 Wiley & Sons Ltd.; John Wiley & Sons Ltd. and the respective authors for Figs 2.2 (Ramade, 1984) & Tables 18.1, 18.5 (Dix, 1981), 20.3 (Hounsome, 1979) Copyright © 1979, 1981 & 1984 John Wiley & Sons Ltd.; John Wiley & Sons Inc. for Figs 4.2, 13.1a (Pielou, 1979) Copyright © 1979 John Wiley & Sons Inc.; Yale University Press for Fig. 4.7 (Stehli, 1968) Copyright © Yale University Press.

Whilst every effort has been made to trace the owners of copyright material, in a few cases this has proved impossible and we take this opportunity to offer our apologies to any copyright holders whose rights we may have unwittingly infringed.

Chapter

1

Introduction

Biogeography, as the term indicates, is both a biological and a geographical science. Its 'field of study' is the biologically inhabited part of the lithosphere, atmosphere and hydrosphere – or, as it has become known – the *biosphere*. Its subject-matter covers the multitudinous forms of plant and animal life which inhabit this relatively shallow but densely populated zone, as well as the complex biological processes which control their activities. The approach to, and aim of, the subject is geographical in so far as it is primarily concerned with the distribution (together with the causes and implications thereof) of organisms and biological processes. However, although this 'field of study' is shared by, and is common to, both biology and geography it is not the exclusive preserve of either of these two sciences. By its very character, biogeography is situated at, and overlaps the boundaries of, a great number of other disciplines. The geologist, climatologist, pedologist, geomorphologist as well as the botanist, zoologist, geneticist and geographer (to mention but a few) all 'cultivate' or 'crop' as the case may be, particular parts of this very large and varied field; and in doing so they are, to a greater or lesser extent, essential to, as well as being dependent on, an understanding of biogeography. As a result, the approach to – or concept of – biogeography is in large measure determined by the training, interest and objectives of the particular student.

The geographer's interest tends to focus on (to be organised around) the spatial variation of two basic characteristics or processes, rather than on any particular components of the biosphere. The first is the intimate interrelationship between the organic and inorganic elements of the earth's environment; the character of the biosphere is primarily a product of the continual interaction or interchange between the lithosphere and the atmosphere. The second is the reciprocal relationship between humans and the biosphere. On the one hand the latter provides the vital link between man and his physical environment; and despite the advances of modern science and technology man is still, whether he realises or likes it, completely dependent on the biosphere for his food. On the other hand, because of an ability, greater than that possessed by any other form of life, to exploit organic resources he is not only an integral part of the biosphere but is now the ecologically dominant organism in it. The significance of biogeography is particularly well expressed by the ecologist M.G. Lemée (1967: 4) in *Précis de Biogégraphie*, when he states:

> C'est [la biogégraphie] aussi une science géographique, car elle tend à établir les rapports avec des peuplements végétaux et animaux avec les autres grandes phénomènes géomorphologie, sols, activités humaines, pour atteindre à une vue synthétique des aspects de la surface du Globe. Pour le géographe la connaissance de la partie vivante du paysage intervient comme un élément de première importance de ce complex, car liée aux autres éléments par d'étroites relations mutuelles, elle constitue un indicateur très sensible des caractères du milieu géographique.

However, the definition and, hence, the delimitation of the scope of biogeography varies dependent on whether or not the aim or point of view has been primarily biological or geographical and on the role accorded to man in the study of the biosphere. Until relatively recently ecologists tended to regard humans as an important but 'unnatural' elements in the environment of plants and animals. The geographer's rather greater concentration on, and appreciation of, the nature and extent of man's role as an ecological factor have tended to distinguish his approach and contribution to biogeography up to date. Some would perhaps subscribe to the view of the late Margaret Anderson (1951) that the study of the biological relations between man, *considered as an animal* (author's italics), and the whole of his animate and inanimate environment is the essence of biogeography.

In both biological and geographical literature attention has been primarily (though not exclusively) concerned with the study of plant rather than animal geography. There are various reasons for this. Greater mobility combined with the small size and an elusive habit of life of the majority of animal species make a study of their distribution more difficult than that of plants. Also, until fairly recently, zoogeography had not developed to anything like the same extent as plant geography. Other considerations have, however, been responsible for the geographer's preoccupation with plant life. Plants – vegetable matter – both living and decaying, comprise the greatest bulk of the total world biomass (i.e. volume of living material) both above and below the ground surface. Not only is the animal biomass small in comparison, but most of it is composed of micro-organisms the majority of which live in the soil. Hence plants are the most conspicuous components of the biosphere and, *en masse*, form a major landscape element. Further, plants are more directly dependent upon, and affected by, their physical habitat than animals. In comparison to the latter, plants are relatively immobile or at least lacking effective means of independent locomotion. As a result they provide a better index (outward and visible expression of the total environment – physical, biological and human) of the site they occupy; and this is no less true of the isolated tree in the city square than of the cornfield or the tropical jungle! Also, because of their greater biomass, plants exert a greater influence on the character of the atmosphere and soil they occupy. They not only modify the physical habitat, but in doing so they create a particular biological environment which would not otherwise exist. Finally plants provide the primary source of food energy for all other living organisms – including man. They are basic to the geography of animals and constitute the most important of man's resources.

Biogeography is firmly rooted in the biological sciences on whose data, concepts and methods the geographer must draw and whose developments have inevitably influenced his particular interest in, and approach to, the biosphere. Biogeography originated – as did so many of the other closely related but now highly specialised and discrete disciplines of botany, zoology, geology, climatology, etc. – in the early and more catholic field of natural history or what are sometimes called the earth sciences. A growing curiosity about the nature of the earth became organised around the description and classification of natural phenomena. And by the latter half of the eighteenth century, the Swedish botanist, Carl von Linné (Carolus Linnaeus 1707–78), had laid the foundations of modern biological taxonomy and nomenclature. The late eighteenth century and the nineteenth century constitute the era of such great explorer-naturalists as Alexander von Humboldt, Edward Forbes, Joseph Hooker, Louis Agassiz, Alfred Wallace and Charles Darwin. As early as 1804, von Humboldt, often described as the father of plant geography, had started to publish volumes recording his observations on plants and other environmental data collected during the course of his extensive travels in South America. Exploration, motivated by a combination of economic and scientific incentives, led to the accumulation of a growing body of factual data. In addition, far-reaching voyages early drew attention not only to the extent of biological diversity, but to striking variations and anomalies in the distribution of different types of plants and animals. And from the search for the causes of these variations there emerged two interrelated concepts which revolutionised the study of natural history and initiated the modern development of the biological sciences. One was that of the adaptation of organisms to their physical environment, the other that of the process of natural selection of those best fitted or adapted to survive in a given habitat. Together these formed the basis of the Darwinian theory of evolution and the origin of species now powerfully supported by modern genetics (Dawkins 1988).

Initially dependent upon the data collected and the concepts formulated by the early naturalists, the subsequent development of biogeography has been distinguished by two distinct, though not completely exclusive lines of investigation – one primarily *taxonomic* (or *systematic*) the other *ecological*. The first is characteristic of plant (or phyto-) geography (and its counterpart animal or zoogeography). Although frequently employed in a wider sense, plant geography is, strictly speaking, an accepted branch of botany. It is the botanist's study of the distribution of different types of plants or plant taxa (families, genera, species, etc.) and originated in the early inventories or floras compiled from different parts of the world. Its object is the analysis, and explanation, of the geographical range of particular taxa or floras as a means to a fuller understanding of their origin, evolution and dispersion. And it should not be forgotten that in so far as the distribution of a taxon cannot be fully understood without a knowledge of its ecology the rigid distinction sometimes made between systematic and ecological biogeography is, to say the least, misleading. Plant geography has made and continues to make important contributions to the elucidation and assessment of the relative importance of the factors which determine floristic distributions; and one of the most important has undoubtedly been the light that it has thrown on the effect of past events on present distributions, including the time of origin of a taxon and the environmental events (e.g. continental drift and climatic change) that have taken place during the course of its evolution.

The geographer's approach to biogeography, however, has to a large extent been influenced and determined more by ecological than taxonomic concepts. Indeed the late Marion Newbigin (1948), the early doyenne of English biogeographers, regarded its scope and aims as virtually synonymous with those of ecology. The term 'ecology' (from Gr. *oikos* = home) was originally coined by zoologists at the end of the nineteenth century to define the study of the reciprocal relations between organisms and their environment. But it was not the ecological approach alone which influenced the development of biogeography: the understanding of the distribution of taxa (of plant geography) was also dependent on a knowledge of the interrelationships of particular species to their environment or, as it became known, *autecology*. It was, however, the study of the distribution of vegetation rather than floras which became the geographer's prime concern and basic to his concept of biogeography. The nature of vegetation (the sum total of all the plants) is dependent on the relative proportions of the various species, on the particular assemblage, group or community of plants occupying a given area. Vegetation geography (or ecological plant geography as it has sometimes been termed) is more or less synonymous with that branch of ecology referred to as *synecology*, i.e. the study of (plant) communities in relation to their environment: and the development of biogeography has been influenced to a lesser or greater extent, over time, by ecological studies of the four main attributes of plant communities – their composition, function, structure and development.

The study of vegetation geography, however, developed nearly a century before the recognition and acceptance of ecology as an academically respectable branch of the biological sciences. In fact, it emerged at the same time and for the same reasons as did plant geography; indeed, initially there was little distinction between the two subjects. Descriptions of the very obvious variations in the world's vegetation cover (which must have made such an impression on the early globe-trotting naturalists) were an integral part of the inventories of natural phenomena that were being accumulated. At an early stage they focused attention on the very obvious significance of climate as an important ecological variable. This, on the one hand, resulted in the concept of *life-zones* as originally defined by C.H. Merriam in 1894. These were, primarily, thermal climatic zones each distinguished by a characteristic flora and fauna. They were later to find quantitative expression in the growing season as a means of defining either vegetation or agricultural limits. On the other hand, von Humboldt's successors – Grisebach (1877–78), de Candolle (1885) and Drude (1897) – had, during the course of the nineteenth century, been putting increasing emphasis on the variation, particularly in the form and structure of the major types of variation, from one part of the world to another. Attempts to explain the distribution of vegetation and its morphological features in terms of adaptation to environmental, and more especially climatic conditions culminated in the publication of A.W.F. Schimper's classic survey of world vegetation, *Plant Geography on a Physiological Basis*.

The English translation of this major work appeared in 1903, and it soon became the main reference source on which many subsequent descriptions and explanations of world vegetation were based. The, by now, presumed causal relationships between climate and vegetation distributions were beginning to have a profound influence on the study of all the natural sciences. It is not without significance that one of the earliest classifications of world climates was proposed by a biologist, Köppen (1918); it was, in fact, an attempt to establish the climatic parameters which coincided with the boundaries between major types of vegetation. Further, vegetation–climate relationships profoundly influenced geographical thinking of the time. They were the basis of the major natural regions of the world set out by A.J. Herbertson in 1905. Indeed the assumption that climate was the major or master ecological factor early became, and to a marked extent has remained, an unchallenged tenet in both ecological and geographical studies.

Until the turn of the century concepts of vegetation were essentially static. The gradual growth of ecological work, however, began to direct attention towards the significance of other factors in the determination of the nature of vegetation. The importance of the time element gained increasing recognition. The pioneer work of the American ecologists, inspired originally by Cowles (1901) on the development of vegetation on a sand-dune complex in Michigan, is a landmark in the history of ecology. It demonstrated the process of succession in the establishment of a vegetation cover and the essentially dynamic nature of both the physical habitat and its associated biological communities. The generic approach to the study of vegetation was firmly established by the subsequent writings of Clements (1916), one of Cowles's most distinguished pupils. He finally elaborated the concept of the climax – expressed in the form of the dominant plants and determined (in his opinion) by climate – as the ultimate terminal stage in vegetation development. These trends in ecology closely paralleled the contemporaneous, but largely independent, developments in geomorphology and pedology, which had been influenced by the Darwinian theory of organic evolution. The concept of landforms as the product of processes operating over a period of time had already gained acceptance among geologists. The *cycle of erosion* as reflected in the varying stages in the development of landforms from youth through maturity to old age, and culminating in the peneplain, was initially propounded by William Morris Davis in 1899.

Concurrently the importance of climate and vegetation as factors in soil formation were attracting attention in Russia. This resulted in the concept of the *zonal soil* and the genetic classification of major soil groups whose profiles reflected, predominantly, the overriding influence of the climatic regime and associated type of vegetation under which they had developed (Glinka 1927). Attempts, particularly in America, to apply the Davisian concept of stage to soil studies, led to the equation of the zonal soil with the mature soil – the final terminal phase in soil formation when the resulting profile has attained a state of dynamic equilibrium under the prevailing climatic conditions. Theoretically the attainment of soil maturity would be accompanied by a similar stage in the development of associated landforms and the final establishment of a climax vegetation. The latter would be dominated by that form of plant life best adapted to (and hence representing the most complete expression of) the prevailing climatic conditions. And in contrast to geomorphology, pedology and climatology, biogeographical studies have been dominated to a much greater extent and for a considerably longer period by the zonal approach and the concept of the climatic climax.

While the American–British school of ecology, under the influence of Frederick Clements and A.G. Tansley respectively, was pursuing studies of the habitat and status of vegetation communities, European botanists were more concerned with the floristic composition (the sociological attributes) of plant communities together with inter- and intra-specific associations and relationships. The detailed and precise data demanded by these studies stimulated the development of more quantitative, and hence standardised, methods of sampling and describing vegetation. The quadrat (a rectangular area of a size appropriate to the sampling of a particular type of vegetation cover) became the unit area of study, the species list provided the basic data. From the 1930s onwards the Zurich–Montpellier school of ecology exerted (particularly after the publication of J. Braun-Blanquet's textbook (1932) on plant sociology) a major influence on phyto-sociological concepts and methods. These were centred on the recognition and description of *associations* – stands of vegetation characterised by the possession of common sociological features distinctive enough for them to be grouped together as a *community type*. The association became the basic unit in the classification of vegetation and the definition of the association a long-standing source of contention among ecologists. And in the immediate post-Second World War years ecological studies focused on empirical studies of plant communities and problems of vegetation taxonomy. These studies served to highlight the role factors other than regional climate (e.g. soil, local and microclimate, biotic interrelationships and anthropogenic factors such as livestock, grazing and fire) in determining the variable nature of semi-natural vegetation cover, at regional and local scales, and to undermine the Clementsian monoclimax theory.

In recent decades, however, the emphasis in ecological studies has shifted not only from extensive to intensive study but from vegetation taxonomy to detailed work on ecological relationships and processes. These trends were influenced, on the one hand, by the development of controlled field and laboratory experiments designed to test empirically based hypotheses; of new and more efficient methods of analysing data; and on the other, the recognition of the key role of energetics in ecological relationships. However, despite major technical developments, the nature and interaction of ecological variables are such that not all biological phenomena are amenable to the same degree of either objectivity, accuracy or standardisation of measurement as can be achieved in the physical sciences. The 1970s and 1980s, in particular, have seen the rapidly increasing use of the following:

1. Statistical analyses designed to assess the significance and 'margin of error' of the results of ecological investigations;

2. Mechanical aids in the collection, storage and processing of data, which have been accompanied by a tremendous increase in the volume of data available;
3. Methods other than the traditional hierarchical classification of vegetation as a means of analysing causal relationships between species or between species and their environment.

Of these the most important is that of ordination; this is an attempt (by those who maintain that the variation in vegetation is continuous) to arrange sampled stands of vegetation in sequences or an order which can be related to either environmental or community variations.

At the same time there was a growing interest in the analysis and quantification of the process of energy flow and nutrient cycling on which biotic–abiotic relationships in the biosphere were dependent. The significance of man as a universal and long-established ecological variable is, at last, being fully appreciated by biologists. 'Over-cropping', exacerbated by direct and indirect habitat modification (not least in recent years by the pollution of air, water and soil) has sparked off biological chain reactions which man has, as yet, been unable to control completely. The resulting imbalance has manifested itself in drastic fluctuations of populations, particularly of animals and pathogenic organisms. Increasing population combined with technical developments continue to intensify pressures on the biosphere and to increase the seriousness of these problems. The need to maintain both the diversity and amount of organic production and, where possible, to increase food production is becoming even more urgent.

The solution of such problems and the efficient management or organic resources is dependent on a better understanding of ecological functions and processes and of the nature of ecological interrelationships. And this need has undoubtedly stimulated a revival of interest in the ecosystem as a functioning unit. Neither the term nor the concept are new. The term ecosystem was originally coined by A.G. Tansley in the 1930s to express the sum total of organisms and their physical habitat and the concept has long been the basis of ecological studies. Until quite recently, however, concentration and specialisation on particular components or aspects of the biosphere have tended to divert attention from the necessity of relating the former to the whole interacting system. Population studies by zoologists such as Charles Elton (1958) and V.C. Wynne-Edwards (1962) combined with quantitative studies of energy flow and nutrient circulation in the biosphere helped to reinstate the ecosystem as a fundamental integrating concept in the study of animal communities.

Studies by Odum (1971) and his co-workers of aquatic communities revived and built on the earlier concepts of energy flow and conversion in the biosphere propounded by Lindemann (1942) and of the ecosystem as defined by a set of interrelated inter-reacting organisms and their habitat by Tansley (1954). And the publication of the first edition of Odum's textbook *Fundamentals of Ecology* (1971), which explored the concept of the ecosystem, marked an important turning point in both ecological and biogeographical studies. It demonstrated the value of the concept in providing a much more satisfactory conceptual framework within which the biotic (plants and animals), and abiotic (soil and atmosphere) could be related and their complex interactions analysed. It is not limited by scale and can be applied to any ecological system from the whole biosphere (or ecosphere) to a rock pool. Further, it provides standardised bases in terms of energy equivalence and nutrient status by which a comparison of structurally dissimilar organic communities or physical environments can be made. Finally, humans can be viewed – as they should be – as an integral element in the ecosystem, as the now most important ecological factor rather than as an unnatural biological accident.

The ecosystem paradigm, by providing the means of analysing animal communities on the basis of their feeding relationships (or trophic levels), also stimulated the whole field of animal ecology. Since the publication of Elton's *The Pattern of Animal Communities* (1966), this field has expanded at an ever-increasing rate to the extent that it has become the basis for many of the more recent developments in both applied and theoretical ecology. The latter are related particularly to the population dynamics of individual and interrelated (predator/prey) species. The early emphasis on micro-organisms and invertebrates (especially insects) reflected on the one hand their abundances and short life cycle which facilitated the generation of both field and laboratory data on which the mathematical modelling of population growth, variation, balance and control could be based; and on the other, the growing need to control pests and parasites.

From the 1960s onwards ecology began to change from a static descriptive to a dynamic and experimental science which according to Krebs (1985) was developing in two main areas. The first is towards the quantitative and experimental analysis of population dynamics and interactive processes, communities and ecosystems. The second is in the already rapidly developing fields of evolutionary ecology and ecophysiology. The late MacArthur and Wilson (1967) did much to awaken interest in this approach which is concerned with the changing relationship between organisms and environment through time as a result of speciation and natural selection (or what Krebs calls 'ecology in action') of those best fitted by reason of adaptive physiological and/or morphological strategies to exploit the available environmental resources at a particular time. Contemporaneously with the evolutionary approach there has been a growing recognition of the significance of both natural and anthropogenic disturbance in organic regeneration and in evolutionary trends; and of the reciprocal interrelationships between disturbance and ecosystem fragmentation. This recent concern with the spatial characteristics of ecosystems has found expression in conservation and in the rediscovered field of *landscape ecology*, or the interaction between disturbance and the spatial arrangement of ecosystem fragments.

The geographer's, as distinct from the ecologist's, approach to biogeography early became and indeed tended to remain until relatively recently synonymous with the study of the relationships between types of world vegetation, climatic regimes and major soil types. The description and explanation of the natural or potential climatic climax vegetation (Eyre 1968; Küchler 1949) became an end in itself. And the zonal study of vegetation dominated most school geography syllabuses in Britain up to the end of the 1970s. As a result biogeography failed to keep pace with the immediate post-Second World War advances in plant ecology.

In the 1950s and 1960s biogeographical research was primarily focused on large-scale vegetation mapping and Quaternary ecology.

The former saw the publication of a large number of local habitat studies in which the methodological problems of classifying vegetation at varying scales (Frenkel and Harrison 1974; Pears 1968) and assessments of the status of the vegetation in relation to burning and grazing were central. The latter early attained and has since the mid- to late 1950s maintained a dominant position in biogeographical research. To a considerable extent this was a function of the dominance of geomorphology in most schools or sub-departments of geography. And with the rapid growth in Quaternary studies and the associated techniques of fossil pollen analysis and carbon-14 dating of fossil organic material biogeographical teaching and research became strongly oriented towards the palynological reconstruction of pre-glacial environmental characteristics and change. And from an early concentration on the formerly glaciated areas of Europe and North America, biogeographers have more recently made important contributions to a clearer understanding of Quaternary environmental changes in the non-glaciated tropics and subtropics (Flenley 1979). And despite the subsequent developments of new fields of enquiry historical biogeography still ranks high in the publication lists and is still the major research activity of over half the members of the Biogeography Research Group of the Institute of British Geographers.

Biogeographers have also made significant advances on the effect of human activities on the distribution of species as a result of the accidental or deliberate introductions from one area to another, and also on the problems of the origins and dissemination of domesticated species of plants and animals. And the ecological and economic consequences in the lesser developed countries and in formerly biogeographically isolated areas merit even greater attention than they have received to date.

From the late 1960s, the study of biogeography was undoubtedly invigorated by the holistic process-oriented approach of the ecosystem concept and the quantitative data on global variations in energy flows and nutrient circulation that resulted from the world-wide ecosystem studies initiated under the International Biological Programme. This was reflected in the series of basic textbooks on biogeography written by geographers (Tivy 1971; Watts 1971; Pears 1985; Simmons 1974, 1979; Furley and Newey 1982) that became available in the 1970s. They all developed, to a greater or lesser extent, the role of humans in and the impact of human activities on the resource values of ecosystems. This was, perhaps most evident in the emergence of the soil – as a distinct sub-field within biogeography (Cruickshank 1972; Curtis *et al.* 1976; Pitty 1979) – and in the increase in the contributions geographers started and continue to make to a deeper understanding of the origins, the changing distributions, the formal and functional characteristics of, and the effect of, human activities on the major world ecosystems.

Increasing awareness of, and concern, about the escalating exploitation and pollution of the biosphere has, particularly since the mid-1970s, witnessed a rapid diversification of lines of biogeography enquiry into the ecological impacts of different types of land and water use and of air and water pollution; the characteristics of man-managed ecosystems and problems of sustained yield; the relative fragility of ecosystems and the problems of organic conservation – a heady embarrassment of riches. That biogeography has a key role to play in the field of environmental sciences has become increasingly obvious. However, the increase in the range of biogeographical problems and in the new techniques of remote sensing, computerised mapping and geographic information systems has far exceeded that in the number of research-active biogeographers. The breadth of its field has always been at the same time both a potential weakness and the strength of biogeography. And at present to a greater extent than ever before it can lead either to excessive dilution and loss of identity or to the development of another integrative paradigm – such as that of landscape ecology.

PART I THE BIOSPHERE

Chapter

2

The biosphere

The biosphere or 'organic world' is that biologically active part of the earth at the interface between the atmosphere, the lithosphere and the hydrosphere. It is a relatively shallow zone which extends at its maximum a few hundreds of metres above the land surface and below the oceans, and less than a metre below the earth's substratum. Despite its limited extent the biosphere is densely populated, teeming with myriad forms of life of infinite variety and complexity. They range from plants and animals of microscopic size and simple one-celled structures to those of large size and complex morphology. A conservative estimate might put the total number of known species of plants at *c.* 500 000, of animals at *c.* 1 million (Table 2.1). The most obvious components of the biosphere – the large terrestrial plants and animals – represent a very small proportion of the total biotic population. A very high proportion are of microscopic size and live in the soil or

Table 2.1 Estimated numbers of species in main plant and animal groups (from Stace 1989)

Seed plants (conifers and flowering)	240 000
Ferns	12 000
Mosses	23 000
Algae (eucaryotic)	17 000
Fungi	20 000
Lichens	16 000
Blue-green algae	500
Bacteria	3 000
Protozoa	30 000
Invertebrate animals (non-chordate)	1 000 000
Vertebrate animals (chordate)	50 000
Total	1 512 000

in water. It has, for instance, been estimated that there may be up to 100 000 algae, 16 million moulds and other fungi and several million bacteria in 1 g of rich topsoil and that the average number of micro-organisms living under an acre of tropical forest in Panama is *c.* 100 000 million.

The upper limit of life in the atmosphere is *c.* 6000–6500 m OD (Fig. 2.1) above which insufficient oxygen, low air density and temperature together with a

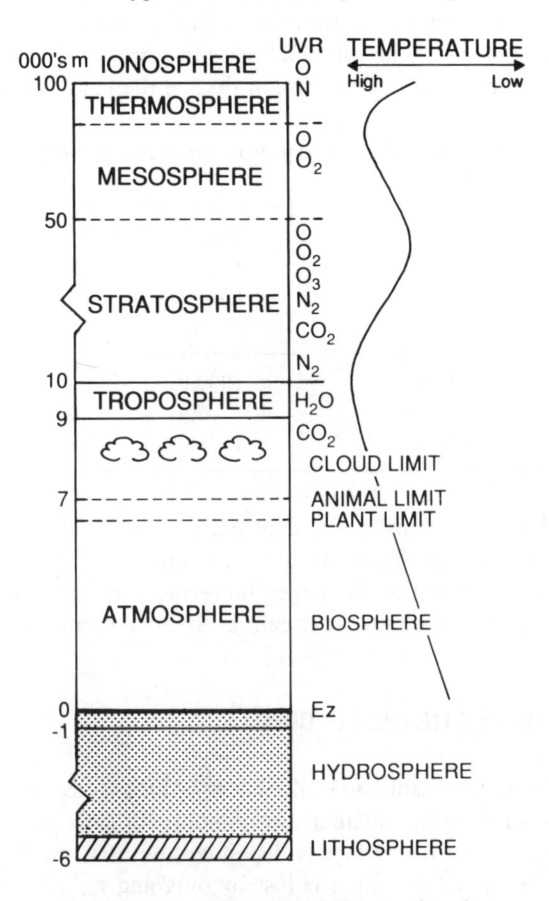

Fig. 2.1 Spatial range of biosphere in relation to variations in altitude, gaseous content and temperature of earth's atmosphere. Ez = enphotic zone

high intensity of ultraviolet radiation (UVR) are inimical for practically all organisms. Some aeroplankton, in the form of dormant and inactive spores, can be carried up to the parabiospheric zone to an altitude of *c.* 9000 m OD Only a third of the lithosphere, the mineral substratum of which the earth is composed, is exposed as land (10% of which is covered by the polar ice-caps) varying in altitude and relief and whose weathered rocks provide the parent material for the active (i.e. biologically occupied and modified) soil layer. The highest density of organisms, however, is concentrated in the surface vegetation layer (which varies from a few centimetres to *c.* 100 m) and in the upper part of the associated soil from a few centimetres to *c.* 1 m deep. Some land and sea birds can range higher. Two-thirds of the biosphere is covered by water (the hydrosphere) in which organisms, particularly plants, are concentrated in the well-lit upper euphotic zone the depth of which varies dependent upon the clarity and mobility of the water. With increasing depth, oxygen deficiency, low temperatures and high pressure become limiting for all but a few very specialised forms of animal life.

Although the hydrosphere occupies such a high proportion of the total surface of the biosphere its *biomass* (i.e. the total mass in terms of dry weight or volume) of living matter is only a minute fraction of that on the land (Table 2.2). Further,

Table 2.2 Area, phytobiomass and mean net primary productivity (NPP) of terrestrial and marine ecosystems (from Whittaker 1975)

Ecosystem	*Area* (10^6 km^2)	*Total phytobiomass (dry matter)* (10^9 t)	*Mean NPP (dry matter)* ($g\ m^{-2}\ year^{-1}$)
Terrestrial	149.0	1837.0	773.0
Marine	361.0	3.9	152.0
Total	510.0	1841.0	333.0

in the oceans the plant biomass is composed mainly of microscopic one-celled organisms and is less than that of the animal biomass. In contrast, the terrestrial plant biomass is dominated by the larger flowering and cone-bearing forms, the land animals accounting for only 3 per cent of the total biomass.

ENERGY FLOW IN THE BIOSPHERE

All but a very insignificant amount of the energy which 'activates' the biosphere is supplied by incoming solar radiation (insolation). As indicated in Fig. 2.2, the amount reaching the earth's surface is dependent on how much is intercepted at very high altitudes and how much is lost by outgoing radiation from the lower atmosphere. Above *c.* 25 km light energy is used in the dissociation of molecular oxygen (O_2) into atomic oxygen (O) – some of which recombines with O_2 to form ozone (O_3) – some to convert O_3 back into O_2 and O. This process removes most of the short-wave UVR, exposure to which can be detrimental to many organisms. In the lower atmosphere most of the long-wave infra-red radiation and some of the shorter long-wave radiation (i.e. the white visible light) is absorbed and/or scattered by water vapour and H_2O molecules. Some is intercepted and scattered by suspended particulate matter and water droplets in clouds, mist, snow, etc. and radiated back to space or to the earth's surface. The amount of direct insolation reaching the surface varies with latitude and season. The intensity of light decreases away from the tropics as the angle of the incident light becomes more oblique and the thickness of the atmosphere traversed increases. During

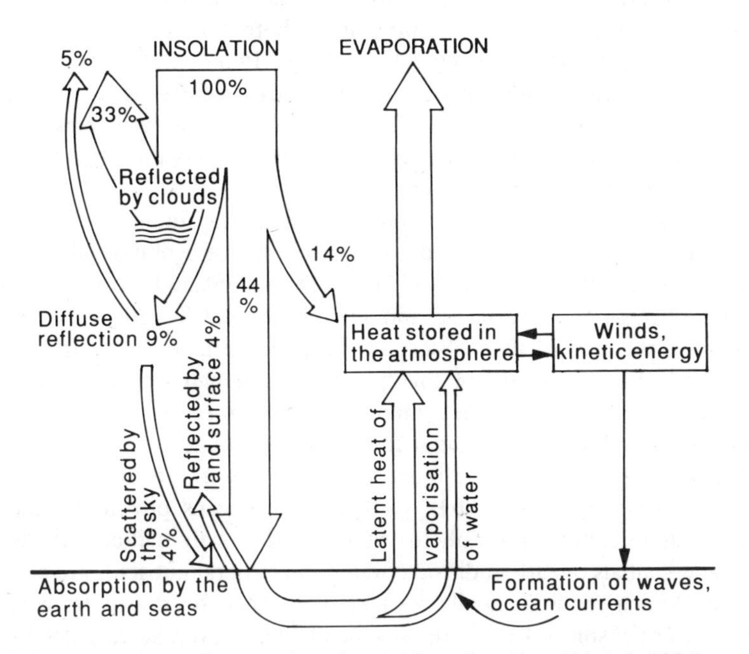

Fig. 2.2 Energy input to and output from the earth's surface (from Ramade 1984)

the northern and southern summers, however, the length of daylight all but compensates for the lower light intensity characteristic of high latitudes.

The direct radiation reaching the earth's surface is not only unevenly distributed but is of low energy density – rarely exceeding 1 kW m^{-2} except in the most favourable conditions. About 30 per cent is reflected back into space, the actual loss or *albedo* being determined by the colour and smoothness of the receiving surface (i.e. bare land, vegetation, water, ice or snow). Just over 20 per cent is used to evaporate water from water bodies, from precipatation and from the surface of plants and animals and hence to maintain the earth's *hydrological cycle*, that continuous circulation of water between the land and water and the atmosphere. Forty-five per cent is expended in heating the land, water and atmosphere, thereby providing the kinetic energy that powers the circulation of the atmosphere and of the oceans.

Of the remaining energy, less than 1 per cent of the light provides that on which all but an insignificant amount of the life on earth depends. About 50 per cent is of a wavelength which can be used by green plants in the biochemical process of *photosynthesis* whereby the inorganic elements of carbon, hydrogen and oxygen derived from the atmosphere are combined to produce complex inorganic carbohydrates – the building blocks for all living matter. This is a biochemical process which has not yet been replicated outside the chloroplas-containing plant cells. Of the light energy absorbed by the green plant, about a sixth is used in this process; the remainder is converted into the chemical or potential food energy of the plant tissues. This provides the energy, on the one hand, for plant metabolism and, on the other, for all other organisms. Eventually this food energy is released as heat in the reverse process of respiration, in both plants and animals, during which oxygen is used and carbon dioxide produced. Photosynthesis constitutes the basic difference between most plants and all animals. As a result, the nutrition of green plants is *autotrophic* (self-feeding), that of animals is *heterotrophic* (diverse feeding). The former are the primary biological producers – the latter include the secondary producers and decomposers.

The necessity for plants to carry out photosynthesis under a wide range of environmental conditions has played a major role in the evolution of the diverse forms of plant life that exist today. Plants obtain light and carbon dioxide by absorption through their surfaces, and their efficiency in this respect will be dependent on the area of their absorbing surface in relation to their total bulk. Fossil records leave little doubt that plant life originated in the oceans; the remarkable similarity of chemical composition between sea-water and that of the living cell has prompted one biologist to compare land and freshwater organisms to packets of sea-water that have found ways of living in a wide variety of non-marine environments (Bates 1964). Certainly the oceans provide a much less extreme or variable medium for photosynthesis and growth than does the land. In the sea the essentials of carbon dioxide, water and mineral nutrients are combined in one enveloping medium, while on land they must be drawn from the two separate sources of soil and air. In the sea the plant is entirely surrounded by water which is much richer in dissolved carbon dioxide than the atmosphere. It is, therefore, not altogether surprising that the plant life of the oceans is the most prolific form of life on earth. It consists, however, almost entirely of the lower and more primitive (in the evolutionary sense) types of plants. The bulk is composed of microscopic unicellular organisms – the plant plankton – which float in the upper light-absorbent layers of the sea. Their minute size ensures a very large absorbing surface in relation to their volume. Given the efficiency of this simple form for photosynthesis and the remarkable uniformity of the environment, the absence of a greater variety of plant life in the sea is perhaps understandable.

On the other hand, the evolution of land plants required successful adaptation not only to a different but to a much more variable set of environmental conditions. The amount of carbon dioxide in the atmosphere remains relatively constant at 0.03 per cent – much less than that dissolved in the sea. Plants growing on the land are subjected to much greater variations of temperature and moisture from place to place and from season to season, than in the oceans. There is, in addition, a fundamental separation of the atmospheric environment, from which plants obtain light and carbon dioxide, and the soil from which water and mineral nutrients must be drawn. It is necessary that most land plants be fixed in one spot and have the means of extending into and extracting water from the sunless area of the soil. The land plant must also be of a form or growth habit that enables it to obtain sufficient light in competition with other plants and at the same time have as large a photosynthetic area in relation to volume as possible. The elaboration of the external form by branching and the subdivision of leaves has provided a means of maintaining maximum photosynthesising surface with increasing size. While in the sea large size would not be of any great advantage, on land, competition for light among immobile plants would obviously give those with greater size, and particularly height, an advantage.

Height, however, creates problems associated with the passage of water and mineral nutrients from the soil to the photosynthesising organs and with the mechanical support of longer stems. Also, the necessity for the absorbing cells in the leaves to be protected from desiccation by an impermeable skin or cuticle, while at the same time allowing carbon dioxide entry through minute openings (i.e. stomata) inevitably exposes the land plant to the loss of water by evaporation. The problems of attaining maximum photosynthesis on the land have therefore created those of maintaining an adequate water supply in the plant. The variety and complexities of form which have developed in the higher cone-bearing (gymnosperm) and seed-bearing (angiosperm) plants have enabled them to cope most successfuly with the problems of attaining maximum efficiency of photosynthesis in competition with other plants under a diverse range of physical conditions and have enabled them to become the dominant plants in the terrestrial vegetation cover today.

The total amount of energy fixed by plants is the gross primary production (GPP). Some of the energy is used in plant metabolism, the remainder, the *net primary production* (NPP), is 'stored' in the plant tissues. This available food energy is passed from one group of animals to another, i.e. herbivores → carnivores → predators, along what is called a *food chain* (Fig. 2.3). Some passes

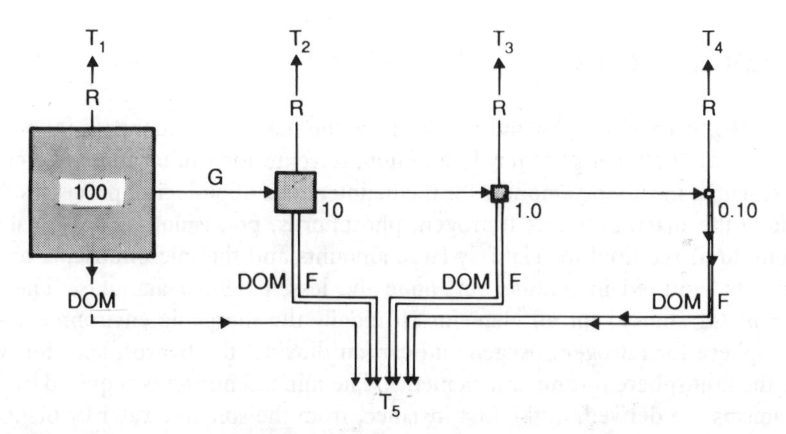

Fig. 2.3 Main types of food chain and trophic levels involving living and dead organic matter in the ecosystem. T = Trophic level; F = faeces; G = grazing route (from Tivy and O'Hare 1981)

along the *grazing route*; more passes either directly or indirectly (as faeces and dead organisms) into the *detrital or decomposing route*. The relative importance of these two routes varies from one part of the bisophere to another. The grazing route is more important in the sea; the latter is more important on land where only a relatively small proportion of the NPP is consumed by herbivores. Also, while the food chain provides a model of the way in which energy passes from plants to animals, the actual feeding relationships among organisms are rarely (except in simplified agricultural systems) so simple. One type of plant may provide food for several different types of animals; carnivores usually hunt a variety of prey and many animals derive their food from more than one *trophic level.* The result is a complex food web in which the trophic levels are not always as clearly distinguishable as in the single food chain.

In the passage from one trophic level to another there is a fairly rapid loss of energy since each time one organism eats another it uses some of the food energy for its own metabolism, with the remainder stored in its body tissues. The model of energy flow along the food chain was originally based on Lindemann's (1942) 'rule of the tenth' which envisaged a 90 per cent transformation of energy to heat in respiration and excretion at each trophic stage. Since the amount of primary biological production determined the absolute amount of animal life that can be sustained, the number of links in food chains or of trophic levels was thought to be limited to four or five. However, the assumption that the amount and *conversion efficiency* of energy alone determine the number of trophic levels has been questioned by many biologists. Nevertheless, there is little doubt that the *photosynthetic efficiency* (i.e. the percentage of the light energy absorbed by land and marine plants converted into food energy) of the biosphere is remarkably low – an estimated mean annual of 0.1–0.2 per cent. This value, however, obscures wide variations as a result of differences in the physical condition of the biosphere and the large area occupied by deserts and ice-caps where no plants exist.

BIOLOGICAL CYCLING

Hydrogen, carbon, oxygen and nitrogen are the main constituents (*c.* 90 per cent dry weight) of all living matter. In addition, thirty to forty other mineral elements are essential in varying amounts for the maintenance of biological processes. They include the macronutrients (nitrogen, phosphorus, potassium, sulphur, calcium, magnesium) required in relatively large amounts and the micronutrients or trace elements required in minute but, none the less, essential amounts. The main *reservoir* (or source) for all elements is initially the inorganic environment – the atmosphere for nitrogen, oxygen and carbon dioxide, the hydrosphere for water, and the lithosphere for mineral elements. The mineral nutrients required by living organisms are derived, in the first instance, from the soil and water by plants and stored for varying periods of time in what is often called the *labile* or *available pool* of living or dead organic matter (DOM). These inorganic elements are eventually released in the process of organic decomposition and reabsorbed either directly or indirectly by growing plants (Fig. 2.4). In the course of this biological cycle some of the nutrients in the dead organic matter can be 'lost' by erosion and leaching from land and surface waters to deep-water sediments beyond the influence of upwelling ocean currents. Recovery from this nutrient sink can only be effected by major geological events. In conditions unaffected by human activities the assumption has been made that such losses are small and are naturally compensated by inputs from volcanic explosions or the weathering of

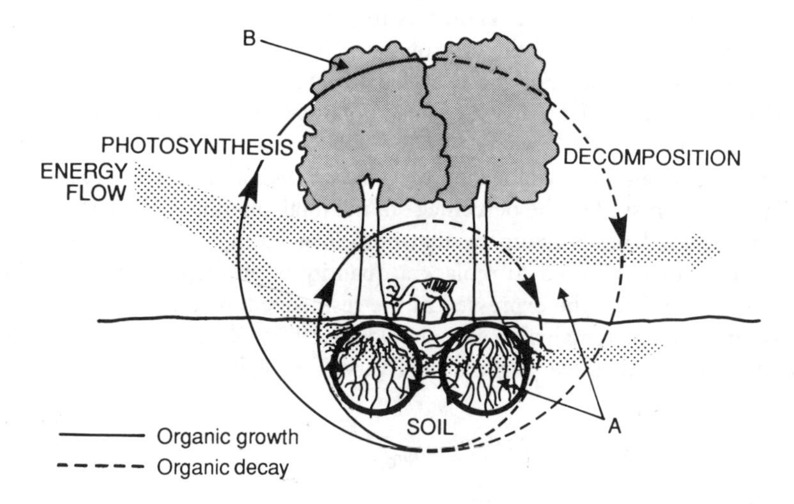

Fig. 2.4 The biological cycle. A = abiotic substances of low chemical energy; B = biotic substances of high chemical or potential energy (from Tivy and O'Hare 1981)

rocks. However, accelerated soil erosion consequent on human disturbance of the terrestrial vegetation cover and the underlying soil has resulted in an increasing loss of mineral nutrients to rivers, lakes, ground-water and eventually to the oceans. Some may be used and recycled by aquatic organisms; much, however, joins the steady drain to the deep-water sink.

While biological elements are frequently categorised in terms of their main reservoirs (i.e. gaseous, mineral or liquid) and analysed individually, this is a pedagogic expediency which tends to obscure reality. As Deevy (1970) noted, water, carbon, nitrogen and sulphur in the form of carbon dioxide, methane, nitrogen, ammonia, hydrogen sulphide or sulphur dioxide are mobile in both liquid and gaseous form and must be cycled together along with water if carbohydrate formation is to be maintained. As a result, the biosphere is both a temporary sink and a source (or reservoir) for these elements. However, while carbon dioxide can be reduced by the green plant, nitrogen and sulphur reduction are dependent on the activities of anaerobic micro-organisms. Nitrogen and sulphur plus phosphorus are essential for protein formation. In contrast to the other two elements, phosphorus follows the hydrological cycle only partially, i.e. from the lithosphere to the hydrosphere and the ocean is its only sink. Finally, for all other elements which are soluble but not volatile the only route by which they can be cycled is through DOM; their reservoir is the lithosphere, their sink the oceans.

RATE OF CYCLING

Those elements involved in biological cycling vary not only in their availability, the quantities in which they are cycled and the routes by which they circulate, but also in the rate by which they move from one sort of pool to another. And the amount of any substance (per unit, volume or area) that moves in a given unit of time from one pool to another is known as its *flux rate*. This is dependent on a number of factors, particularly the size of the pool and the efficacy with which substances become mobile. For those cycled through organisms (the biological cycle) and temporarily stored in living or dead organic matter the flux rate tends to be relatively rapid.

The flux rate is also determined by the *turnover rate* and *residence time*. The turnover rate is expressed as the percentage of the total amount of an element in a pool that is released or taken up by that pool in a given length of time. The turnover time is that required to replace a quantity of an element equal to the amount in the pool. It can be expressed as the residence time or the time a given amount of an element remains in a pool (Odum 1983). The turnover times of elements that are held in the organic pool vary (Fig. 2.5). Oxygen and carbon

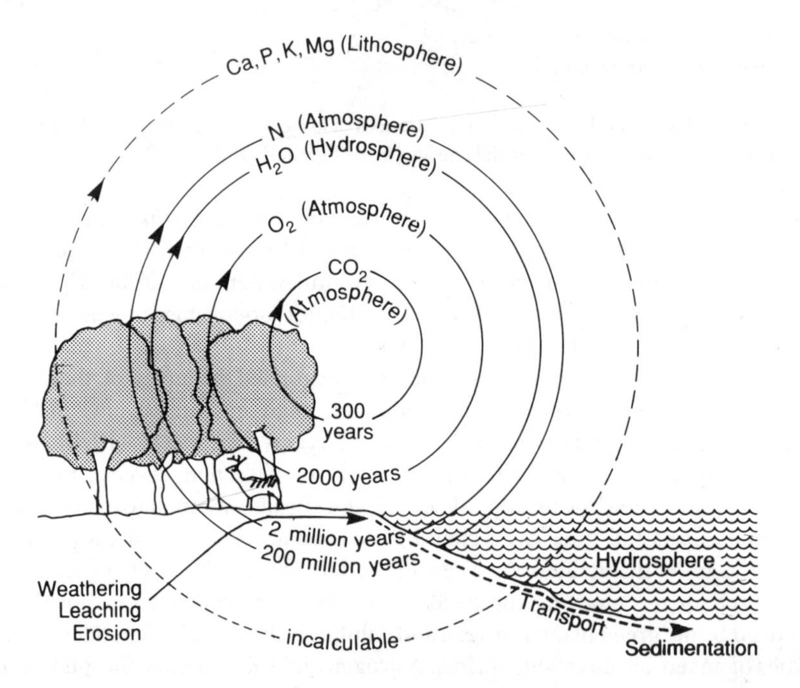

Fig. 2.5 The geochemical cycle showing rate of flux of chemical elements between the lithosphere, the hydrosphere and the atmosphere (from Tivy and O'Hare 1981, adapted from Cloud and Gibor 1970)

dioxide circulate relatively rapidly between organisms and the environment in the basic processes of photosynthesis and respiration. Water and nitrogen have more complex local routes from the atmosphere via the soil to plants, and hence their turnover time is more protracted. Mineral elements are released more slowly in the process of rock weathering and soil formation; moreover, these elements are lost in terms of human time-scales from the cycle by erosion to deep-water sediments. The residence time of elements in organic pools depends on the life span of the organisms, which may vary from a few hours (micro-organisms) to several hundred years (trees) and on the nature and rate of decomposition. Some may be stored in peat for thousands and, in coal and oil deposits, for millions of years.

THE ECOSYSTEM CONCEPT

Energy flow and biological (nutrient) cycling link the organic and inorganic components of the biosphere in such a way as to form an ecological system or *ecosystem*. While ecologists early appreciated the nature of biological interrelations, the term was first coined by Tansley (1935) to express the idea of a group of interdependent, interacting organisms and their habitat (or home) and later refined by Evans (1956: 1227) on the basis of the functional characteristics as the 'circulation, transformation and accumulation of energy and matter' by the action of organisms and, as a result, the ability of the ecosystem to regulate and, within limits, to perpetuate itself. *Ecoregulation* (Odum 1983) is mainly a function of two processes – ecosystem development or *succession* and *homeostatis*. In the former it is assumed that organisms modify the atmosphere and substratum they occupy. In doing so one group may make the habitat more suitable for some and less suitable for other plants and animals. As a result one group of organisms will be succeeded by another more diverse and complex and with a larger biomass than the preceding. Theoretically, the final stage in this development is reached when the maximum use of environmental resources available has been attained. This has been variously called the climax stage or the mature stage of the ecosystem. It is assumed that at this stage the ecosystem can maintain itself in a steady state provided variations in the external or internal conditions do not exceed the limits to which the system can adjust.

The ecosystem as such is dimensionally undefined (Waring 1989) and can be applied and studied at a scale from a rock pool to the tropical rain forest or to the entire biosphere or ecosphere (Cole 1958). Ecosystems can be characterised by one or more of their formal or functional attributes dependent on the objects of study. On the basis of the associated physical environment the major contrast is between aquatic and terrestrial ecosystems; and within these environments ecosystems can be defined in terms of the climatic regimes (tropical, temperate, Arctic) or substrata (saline, acid, alkaline) in which they occur. In so far as plants dominate the biomass of terrestrial ecosystems and the vegetation is the most obvious (visible) component of an ecosystem, classification on the basis of the dominant vegetation (forest, woodland, grassland, etc.) is common, or in the case of the world's biomes the vegetation and animals associated with major climatic areas (e.g. rain forest, Mediterranean, tundra, Boreal forest). Ecosystems can also be characterised by inherent attributes such as stage of development, by biomass, productivity, etc.

No ecosystem exists independently of the others; the biosphere is, in fact, composed of a mesh of interlocking systems, each one being part of another larger one and containing smaller subsystems, dependent on the scale of study. No ecosystem is completely closed. All are dependent on the continuous input of solar radiation. Losses of material from one ecosystem will eventually become the gains for another in the biosphere. The only closed system in this sense is the biosphere (or ecosphere) itself, for which reason the term 'space-ship earth' become a popular ecoterm.

GEOCHEMICAL CYCLING

The biological cycle is part of and is dependent on the more extensive geochemical cycling of elements between the atmosphere, the lithosphere and the hydrosphere – the three reservoirs from which the biosphere draws its basic requirements.

THE ATMOSPHERE

The oxygen cycle

Oxygen occurs in three forms in the atmosphere: dissociated atomic oxygen (O), molecular oxygen (O_2) and ozone (O_3). Molecular oxygen, essential for organic respiration, accounts for about a fifth of the lower atmosphere. Ozone is present only in small amounts. It is an unstable, easily dissociated element which forms as a result of the combination of O_2 and O at some 30–40 km above the earth's surface with a maximum concentration at *c.* 20–30 km. Together with O_2 it is responsible for the absorption of *c.* 1–3 per cent of the short-wave UVR which is the main source of energy for atmospheric circulation above 30 km. In addition, ozone shields the biosphere from the otherwise damaging effects of UVR.

Most atmospheric oxygen is thought to be primarily of biological origin produced by the breakdown of water during photosynthesis; other sources include atmospheric carbon dioxide, water and mineral oxides. The exchange of oxygen between living organisms and the atmosphere is, however, only part of a more complex global cycle. Molecular oxygen is the most abundant element in the lithosphere, comprising almost half of the total atoms in the earth's mineral crust. It is a particularly reactive gas which combines chemically with a great number of other elements including hydrogen, carbon, nitrogen, sulphur, phosphorus, etc. The affinity of oxygen for other molecules is expressed as the *redox potential* (O : R) of the environment. It is dependent on the ratio of oxidising to reducing atoms per atom of mineral with which it combines. The higher the redox potential the more oxygen can be used, the lower the ratio, the less can be used and some can be released. In addition, there is a continuous interchange of oxygen between the atmosphere and water bodies – the latter, particularly the oceans, serving to maintain the oxygen balance in the atmosphere. It is not known to what extent the oxygen cycle has been affected by the increased use of oxygen in the burning of wood, in the combustion of fossil fuels and as a result of an increased demand for oxygen created by increased amounts of human sewage and animal waste and of increased organic growth as a result of cultural nutrient enrichment (eutrophication) of water bodies.

The carbon cycle

Carbon derived from carbon dioxide is fundamental for the process of photosynthesis. Atmospheric carbon dioxide also exercises an important climatic function in absorbing incoming infra-red light waves and by blocking outgoing radiation of heat from the earth's surface and the lower atmosphere. It thereby dampens diurnal and seasonal temperature ranges that might otherwise be more extreme. There is a fairly rapid and continuous exchange of carbon between the atmosphere and the biosphere. Of that taken up by green plants some is returned in the concurrent process of respiration; some is temporarily stored in plant tissues (e.g. wood) and DOM before being released by organic decomposition. One of the most important natural sources of carbon dioxide results from a combination of oxygen with methane (CH_4), the latter produced during the anaerobic decay of organic matter in the soil.

There is a slower exchange of carbon between the atmosphere and the oceans. This takes place most rapidly in the upper 100 m above the colder deep water. Carbon is also added to ocean water in the form of organic and inorganic materials (particularly carbonates and bicarbonates) leached from the land surface. Inorganic carbon is absorbed from the water by photosynthesising phytoplankton. While some is immediately recycled in decomposition, a much higher proportion is lost from the surface water as DOM which sinks slowly to the sea floor. The organic carbon, together with carbonate and bicarbonates derived from marine shells and terrestrial leaching, make the sea-floor sediments a major global sink for carbon. Dependent on the depth of the ocean floor, this carbon may not be brought into circulation with the atmosphere for thousands of years (Lewis 1989). The oceans then function both as an available pool and as a sink for carbon. They contain some fifty or more times carbon than the atmosphere and hence regulate the flux of carbon to and from the atmosphere as well as buffering potential changes to the small atmosphere pool of carbon dioxide (0.003%) in the atmosphere.

However, since the rise of industrialisation in the mid-nineteenth century with the combustion of fossil fuels, and increasing deforestation, the input of carbon dioxide into the atmosphere has been increasing. In 1978 Woodwell estimated that, as a result of human activities, the content had increased by 290–*c.* 330 ppm, 25 per cent of which had been added in the last decade of this period; and, at these rates, the carbon dioxide content could double by the year 2020. Continuous carbon dioxide monitoring since 1958 (when the first station was established in Mauna Loa, Hawaii) and remote sensing by Landsat imagery have allowed a more accurate quantification of current trends. Measurements now indicate an increase in atmospheric carbon dioxide, estimated at about 8–10 ppm per annum in 1959–88. This has, however, been accompanied by annual and seasonal variations. The latter are particularly marked on the land in the northern hemisphere, where the increase of carbon dioxide in winter can vary from 0.5 to 1.5 ppm consequent upon seasonal variations in photosynthesis in mid-latitude deciduous forests. During the growing season more carbon dioxide is used up than is returned by respiration or combustion, or by diffusion from the oceans; at 30° N it has been estimated that there is a net decrease of *c.* 0.3 per cent (Bolin and Cook 1983; Bolin *et al.* 1986) consequent on cessation of growth by deciduous trees. The carbon dioxide content of the atmosphere also varies daily and vertically within the terrestrial vegetation cover with low daytime values in tree canopies and high night-time values at ground level. In order to identify and accurately quantify long-term changes in the concentration of atmospheric carbon dioxide, more accurate information is needed about the flux of carbon between the atmosphere, the hydrosphere and the biosphere and, particularly, the inputs to and outputs from the oceans.

The nitrogen cycle

The cycling of nitrogen is more complex than either that of oxygen or carbon. Although the proportion in the atmosphere is very high, it exists as an inert gas which only a few organisms can utilise directly. It also has an unusually high number of oxidation levels or valencies (Table 2.3). Some nitrogen compounds are

Table 2.3 Nitrogen valencies

N-ions	*Chemical formula*	*Valency*
Nitrate	(NO_3^-)	+5
Nitrite	(NO_2^-)	+3
Nitroxyl	[NHO]	(not stable)
Nitrogen gas	(N_2)	0
Hydroxylamine	$(HONH_2)$	−1
Ammonium	(NH^{+}_4)	−3

produced naturally as a result of a reaction with atmospheric oxygen or hydrogen during exceptionally powerful energy pulses generated by cosmic radiation and lightning (Table 2.4). However, most of the available nitrogen is fixed by certain terrestrial and marine micro-organisms (prokaryotes). Some of these are free-living in soil and water, others exist in a symbiotic relationship with higher plants. The former include the nitrogen-fixing bacteria (aerobic *Azobacter*, anaerobic *Clostridium*), blue-green algae (*Cyanobacter*) and a purple bacterium (*Rhodospirillium*) capable of synthesising ammonia (NH_3). This can either be absorbed

Table 2.4 Estimated global nitrogen budget. 9×10^6 metric tonnes net gain may represent rate at which biologically fixed nitrogen is accumulating in the biosphere in the soil, ground-water, reservoirs, rivers, lakes and ocean deeps (from Delwiche 1974)

	Nitrogen fixation	*Denitrification*
	(10^6 mt (metric tonnes))	
Terrestrial (historic)	30.0	43.0
Legume crops	14.0	—
Marine	10.0	40.0
Atmospheric	7.6	—
Juvenile	0.2	—
		(Sediments 2.0)
Total gain	92.0	Total loss 83.0
Net gain	9.0	

directly by plant roots or oxidised by nitrifying bacteria (*Nitrosomas*) to nitrite (NO_2^-) which is rapidly converted by the *Nitrobacter* into nitrate (NO_3^-). The relative nutritive value of ammonia and nitrate has long been debated. The ammonia cations are easily absorbed by the soil colloidal complex and held until oxidised. They are less immediately available than the nitrate anions which remain in the soil solution and can be rapidly taken up by the growing plant. By the same token, however, the nitrate anions are very susceptible to leaching.

The symbiotic nitrogen fixers include the fungi *Actinomycetes* and the *Rhizobium* bacteria which live in the root nodules of a few, but diverse range of, higher plants. The most significant are the bacteria associated with leguminous plants whose annual fixation rate in nature is about a hundred times that of non-leguminous symbionts. The actinomycete fungi are found in the root nodules of non-leguminous woody perennials – shrubs and trees – belonging to about nine families: Alnus, Ceanothus, Compostoma, Elaergnus, Myrica, Casuarina, Coriarca, Araucaria and Ginko (Torrey 1978). In addition, the blue-green algae can fix nitrogen in association with lichens, mosses and ferns. Together with the blue-green algae these nitrogen fixers allow the continous cultivation and the maintenance of soil fertility levels (albeit low) of the padi rice fields without fertiliser application.

Organic matter, living or dead, is the most important pool of available nitrogen in the biosphere. Organic nitrogen is eventually recycled when plants and animals die and decay. Some organic nitrogen is recycled by the living plants; some is further reduced, particularly in anaerobic environments, by denitrifying bacteria (*Pseudomonas*) to nitrogen which is returned to the atmosphere and maintains the nitrogen balance in the atmosphere. Some is taken out of circulation in the form of organic material (either eroded from land surfaces or produced in the surface layers of water bodies) which sinks to deep sea-floor sediments.

In contrast, human activities have, at an ever-growing rate, been increasing the inputs to the nitrogen cycle by the use of inorganic nitrogen fertilisers on the one hand, and by the emission of nitrogen compounds from agricultural and industrial processes on the other. Delwiche (1974) estimated that industrial fixation of nitrogen is far greater than that by any other means and that it now equals that fixed by all terrestrial ecosystems before the development of modern intensive agriculture. However, it is difficult to assess the extent to which the inputs and outputs have changed because of a lack of precise information about the amounts of nitrogen fixed in the atmosphere and of that lost to ocean-floor sediments.

THE LITHOSPHERE

The sedimentary cycle

The sedimentary cycle involves the weathering of the rocks of which the lithosphere is composed and their erosion, transport and deposition. The type (chemical or mechanical) and rate of weathering are principally a function of climate, that of the erosion and transport of weathered material, of ground surface gradient and vegetation cover. Weathered mineral matter transported by downslope gravity, by rivers or ice, and by wind is deposited either on the land surface or via surface run-off into the oceans. The former provides the soil parent material in which physical and biological processes maintain the chemical breakdown of rock particles and release minerals in an available ionic form, i.e. soluble and capable of being absorbed by plant roots or of being rapidly washed (leached) out of the root zone. The available form of the most important mineral nutrients are the positively charged cations which can be adsorbed and retained by the soil clay–humus (colloidal) complex and the negatively charged ions which occur in the soil water and are hence more susceptible to leaching.

Sulphur and phosphorus

Two elements of particular significance and whose cycles differ in some respects from those of most other minerals are sulphur and phosphorus. Sulphur like nitrogen is mobile in both air as sulphur dioxide (SO_2) and hydrogen sulphide (H_2S) and as sulphide (S^-) and sulphate (SO^-) ions in water. While sulphur is not required in large amounts by organisms it is, nevertheless, essential to the process of protein assimilation and plays a vital role in the availability of phosphorus. Much of the sulphur used in terrestrial ecosystems is volatilised from the oceans or reduced from sulphates in estuarine muds. It is also recycled directly to plants as sulphates from decomposing organic matter or via the atmosphere as sulphur dioxide. The loss of sulphur compounds deposited in lake and sea-bed sediments is increased by the tendency for sulphur to be fixed in insoluble and often potentially toxic sulphide compounds (e.g. iron sulphide, Fe_2S) particularly under anaerobic and acid conditions. The large-scale industrial use of fossil fuels and associated industries, however, has increased the input of sulphur into the atmosphere (by over 90 per cent). Industrial pollution now contributes over a third of the atmospheric sulphur – and together with nitrous oxide – is one of the main constituents of acid rain.

Phosphorus is (along with nitrogen and potassium) one of the 'big three' minerals required in larger amounts than others for biological processes. It is not an atmospheric element, is hardly detectable in the hydrosphere and is relatively scarce in its main reservoir, the lithosphere. In contrast to sulphur it is freely soluble and available only in acid conditions. In neutral to alkaline, and particularly in anaerobic, conditions, it readily forms insoluble compounds with other minerals (e.g. Fe, Ca). However, when iron sulphides are formed in anaerobic sediments, phosphorus becomes available. Phosphorus deficiency is one of the main factors limiting natural primary biological productivity in many parts of the world; and increased loss by erosion and leaching of this scarce substance may well have a greater impact on the long-term productivity of the biosphere than the loss of any other of the mineral nutrients. The relatively limited reserves of phosphorus-rich rocks (including guano deposited by concentrations of sea birds) are being mined to provide inorganic fertiliser, but reserves are probably insufficient to make good losses by leaching and erosion. Although the input of phosphorus by industrial pollutants has increased, this has tended to retard and deplete rather than enrich and speed up the cycle. Leached into water bodies phosphorus enrichment results in an increased rate of production of organic matter, and an increase in oxygen demand such that creates anaerobic conditions and drastically retards decomposition and mineralisation of the accumulating DOM.

THE HYDROSPHERE

The hydrological cycle

The hydrological cycle is the circular route by which water in a liquid, solid or gaseous phase moves continuously between the atmosphere and the earth's surface as a result of the alternating processes of evaporation and precipitation. The oceans are the earth's main reservoir (see Table 2.5) from whose surface water is

Table 2.5 Approximate percentage distribution of earth's water

Oceans	97.0
Ice-caps and glaciers	2.0
Ground and surface water	1.0

evaporated at a rate dependent on the moisture-holding capacity of the contiguous atmosphere. This is a function primarily of the temperature, the water-vapour pressure and the movement of the particular air mass. Much of this water is returned directly by precipitation falling on the oceans or freshwater lakes and rivers which ultimately drain into the sea. The remainder is returned at varying rates and by a variety of routes via three main terrestrial reservoirs: snow and ice – about 2 per cent of the earth's water is held in the permanent ice-caps; ground-water in subsurface aquifers (water-holding sediments); and in the biosphere (in the soil and in organic tissues).

The amount of water which circulates through the terrestrial part of the biosphere is dependent on (1) the amount, intensity and duration of precipitation; (2) the amount and form of the vegetation cover; and (3) the physical nature of the underlying soil. Precipitation is checked (i.e. intercepted) by and evaporated from the vegetation cover. The percentage loss by interception is dependent on the one hand on the density and form of the vegetation cover and on the intensity of the precipitation on the other. Of that which reaches the ground surface some may run off and some may infiltrate the substratum, dependent on the surface slope and the texture of the substrate. The latter also influences the proportion of infiltration water which percolates rapidly below the rooting zone and that which is held within it and hence becomes available for plant growth.

Available precipitation $(Pa) = Pt - (I + RO + If - D)$

then

$I =$ interception $(P - Et)$

where RO is run-off; If the infiltration, D the soil drainage, Et the evapotranspiration, Pa the available precipitation and Pt the total precipitation.

Water reaching the terrestrial surface is returned either to the oceans by slow underground movement from ground water or faster surface runoff to lakes and

Table 2.6 The most important gases in the atmosphere. In dynamic perspective (through-flow per year) O_2 and CO_2 play the major role in recycling of gases of organic origin. Six other gases together have a through-flow *c.* 0.1 per cent of O_2/CO_2 cycle when measured in moles (from Lovelock 1979)

Gas	*Chemical disequilibrium factor*	*Static content (ppm)*	*Mean residence time (years)*	*Through-flow year^{-1}*		*Inorganic share in through-flow*	
				10^{13} moles	*10^9 tonnes*	*Nature*	*Human technology*
N_2	10^{10}	790 000.0	10^6–10^7	1.6	1.0	0.001	—
O_2	Base reference	210 000.0	1000	344	110.0	0.00016	—
CO_2	10	320.0	205.0	354.0	156	0.01	0.10
CH_4 (methane)	10^3	1.5	7.0	6.0	1.0	—	—
H_2	10^{30}	0.5	2.0	4.4	0.09	0.00016	—
N_2O	10^{13}	0.3	10	1.4	0.6	< 0.01	—
CO	10^{30}	0.08	0.3	2.7	0.575	< 0.001	0.20
NH_3	10^0	0.006	0.01	8.8	0.5	—	—
$(CH_2)_n$	Hydrocarbon	0.001	0.003		0.4	—	0.50

rivers or to the atmosphere by the process of evapotranspiration. It is estimated that as much as 20 per cent of the terrestrial precipitation water is circulated by this means, particularly by forests in humid climate zones.

GAIA CONCEPT

Although the biosphere is relatively limited in extent it is so reactive that it has an influence on the atmosphere, lithosphere and hydrosphere inversely proportional to its mass; this has been estimated at an average of 580 mg cm^{-2} of the earth's surface and 580 mg Cm^{-3} of marine water (Deevy 1971). In 1974 Lynn Margulis (an American biologist) and James Lovelock (a British chemist) propounded the hypothesis that the biosphere and the present atmosphere constitute an autopoietic (a self-organising, self-maintaining) system, which has been given the name Gaia – the Greek 'Earth Goddess'. In Lovelock's (1979: 9) words:

> the entire range of living matter on earth from whales to viruses, from oaks to algae could be regarded as constituting a single living entity capable of maintaining the earth's atmosphere to suit its overall needs and endowed with faculties and powers far beyond those of its constituent parts.

Simply expressed, this envisages the earth's atmosphere as a product of organic processes which continue to maintain it at a composition and volume favourable for life.

The Gaia concept is based on a wide range of established empirical evidence and theory about the chemistry of the atmosphere and the evolution of organisms. The earth's atmosphere is unlike that of any other known planet (see Table 2.6); indeed, it is in a state of high chemical non-equilibrium, i.e. there are more gases present than would be expected, or possible, in an equilibrium system with the existing oxygen context (see Table 2.7). It is now considered that the production of this atmosphere has been dependent on the evolution of life during which process an initial reducing, anaerobic environment has been replaced by an aerobic,

Table 2.7 Comparison of chemical composition of present atmosphere of the earth and of a hypothetical equilibrium state (from Lovelock 1979)

	Principal components	*Per cent at present*	*Per cent at chemical equilibrium*
Air	CO_2	0.03	99.0
	N	78.0	0.0
	O_2	21.0	0.0
	Argon	1.0	1.0
Oceans	Water	96.0	63.0
	Salt (NaCl)	3.5	35.0
	$NaNO_3$	Trace	1.7

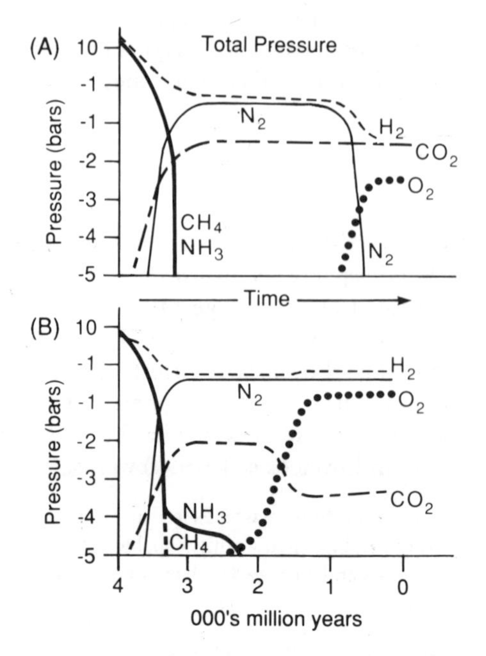

Fig. 2.6 Hypothetical evolution of the gaseous composition of the earth's atmosphere (A) without and (B) with life (from Jantsch 1980)

oxidative one (see Fig. 2.6 and Table 2.8). In the initial absence of oxygen it is believed that the earliest organisms were those adapted to anaerobic conditions (i.e. in water or saturated sediments). They were probably heterotrophic reducers, obtaining carbon, oxygen, nitrogen and sulphur from the reduction of inorganic compounds, and protected from high UVR by the water in which they existed. These organisms then produced oxygen, derived by reduction, in the processes of respiration and decomposition.

It is also assumed that terrestrial life was unable to evolve until the ozone layer in the atmosphere (itself a product of the splitting and recombination of molecular oxygen) was sufficiently well developed to block the incoming UVR light. The

Table 2.8 Comparison of earth's physical environment without life and at present (per cent) (from Lovelock 1979)

	Without life	*At precent*
CO_2	98.0	0.03
N	1.9	79.00
O_2	Trace	21.00
Argon	0.1	1.0
Average surface temperture (°C)	290.0 ± 50.0	13.0
Average pressure (bars)	60.0	1.0

first photosynthetic land organisms for which there are fossil records (*stromalites*) were the blue-green algae (*Cyanobacteria*) which grew on the land surface and were protected from UVR by layers of dead cells. As a result, the oxygen content of the atmosphere started to increase while the carbon dioxide content began to decrease. Much of the initial oxygen produced, however, would have been used up in the oxidation of the earth's mineral substances. A more rapid increase in atmospheric oxygen levels was dependent on the evolution of the *prokaryotes* (simple unicellular organisms with an unenclosed nucleus), some of which were aerobic and autotrophic, i.e. capable of photosynthetic productivity. As organic production increased, at a rate greater than that of decomposition, so the oxygen produced as a by-product of photosynthesis began to accumulate in the atmosphere to its present concentration of 21.0 per cent. This is the safe limit for life. Oxygen is such a reactive substance that a 4 per cent increase would promote spontaneous combustion (Allaby 1986). It also sets a limit to the amount of ammonia (NH_3), methane (CH_4) and other gases that can be present. Nearly all the existing atmospheric oxygen is now used in photosynthesis and recycled fairly rapidly by respiration. However, the fact that it has remained at a constant level for so long suggests that some control mechanism might be operating. The Gaia hypothesis identifies methane (CH_4), together with nitrous oxide (N_2O), as the key two-way regulators of oxygen which buffers the loss of oxygen resulting from the many non-biologic oxidative processes. On the one hand, methane combines easily with oxygen in the upper atmosphere where it is thought to help maintain the stability of the ozone layer. On the other hand, it is oxidised in the lower atmosphere, particularly in tropical areas. Nitrous oxide also plays a part in the regulation of oxygen and ozone.

Concomitant with the increase in biological productivity and the accompanying decrease in atmospheric oxygen, there was a drastic decline in the amount of carbon dioxide and a comparably large increase in nitrogen. As has already been noted, the carbon dioxide balance of the atmosphere is maintained in the short term by simple reactions with oceanic waters. Carbon dioxide and oxygen are in equilibrium with carbonates; and about fifty times as much carbon is held in the oceans as in the atmosphere. The very high proportion of nitrogen in the atmosphere is largely the product of denitrifying bacteria. Before the evolution of nitrogen-fixing bacteria, the denitrifiers returned more nitrogen to the biosphere than was removed in the form of soluble compounds (NH_3, $NHCO_3$, NO_3) produced in electrical storms in the atmosphere. The resulting high proportion of nitrogen serves to dilute the oxygen content of the atmosphere and reduce the risk of spontaneous combustion. It also maintains air pressure which, in its absence, would probably necessitate a reduction in water vapour (an important greenhouse gas).

The oceans are, as in the case of the atmosphere, in a non-equilibrium state which, it is argued, was also produced and is maintained in a constant chemical condition by organic processes. In the first place, the phytoplankton use hydrogen. Previous to the evolution of life, hydrogen was produced mainly by the rapid oxidation of water during submarine volcanic activity. It was, however, rapidly lost to the atmosphere because of its low density. Had this process not been checked the oceans would eventually have disappeared. However, oxygen produced during photosynthesis combines with hydrogen to form water. The heavy water vapour cannot escape since, with increasing altitude, it either condenses or freezes and falls back to earth. Also, the mineral composition of sea-water is much lower than might be expected given chemical equilibrium, and the present salinity is favourable for the maintenance of plant and animal life. Mineral salts are continually being added by runoff from land and by subterranean sea-floor spreading. Removal of excess salt (particularly chloride) by evaporation and precipitation occurs in shallow bays and lagoons where the inflow of sea-water is greater than the outflow of fresh water. Among the areas which contribute most to this process are the large lagoons enclosed by the offshore coral reefs and atolls of tropical areas.

One of the most important stabilisinig feedback processes in the Gaian universe system is that exerted by the organically produced atmosphere on the climatic conditions of the earth today. The temperature of the biosphere must have been maintained within the relatively narrow range of 0–50°C in order to allow the evolution of the organic diversity that now exists. There is ample evidence that during the evolution of the earth, incoming solar radiation gradually increased. Had the concentration of what are now called the *greenhouse gases* remained constant the temperature of the earth's surface would have been high enough to evaporate all the water and would, in addition, have been subject to more extreme daily and seasonal ranges than is the case today. The greenhouse gases are those which absorb outgoing short-wave heat radiation and reradiate it back to earth. One of the most important in this respect is carbon dioxide, together with water vapour and the very small amounts of ammonia (NH_3) and methane (CH_4). It is postulated that the decline in the carbon dioxide content of the atmosphere, as a result of photosynthetic activity, counterbalanced the increase in solar energy and that fluctuations in atmospheric carbon dioxide have been paralleled by the alternation in warmer with colder climatic conditions. Isotopic (O-16,H-1, O-18,H-2) recording of the distribution of erratic rocks suggest that there may have been at least fifteen ice ages dating back to Jurassic and Cretaceous periods. Furthermore, the heat energy radiated from the earth's surface powers the hydrological cycle of evaporation and precipitation, and the atmospheric (wind) and oceanic (currents) circulation on which climatic variations in temperature and precipitation are dependent.

The Gaia hypothesis assumes that the living world (the biosphere) and the global environment form a single interactive system which has, in some instances, been interpreted as an organism with distinct emergent properties; and that Gaia is self-regulating in face of disturbances from within and without. While this is a very controversial hypothesis it has, as Watson (1991) notes, served to focus attention on the nature of the biosphere, the evolution and spatial variation of the diverse forms of life that occupy it and the implications of the increasing human impact on them.

Chapter

3

Ecological variables

The distribution of a species population is a function of its resource requirements for growth and reproduction and of its ability to exploit available environmental resources. If Tansley's (1954: 210) original definition of an ecological factor or variable as 'any substance, force or condition affecting the vegetation directly or indirectly in such a way as to differentiate it from the behaviour and distribution of other vegetation' is accepted, four interacting groups of factors can be distinguished:

1. Abiotic or physical;
2. Biotic (excluding human;
3. Anthropogenic;
4. Historical or evolutionary.

The first two will be considered in this chapter, the third in Chapter 4. The human impact will be considered in the context of particular ecosystems in Part II, and more specifically in Part III.

ABIOTIC FACTORS

Both plants and animals are either directly or indirectly dependent on the physical environment for the resources that sustain life (Table 3.1).

The ability of any species population to maintain and reproduce itself in a particular physical environment will depend on the presence of those conditions necessary for, and the absence of those detrimental to, its growth and development. For every physical parameter there is a minimum level below and a maximum above which an organism's metabolism cannot be sustained. The optimum level is that at which growth and reproductive capacity – the fitness of the organism – will be at a maximum. Any condition less than or in excess of that required to sustain life will limit an organism's distribution.

On a global scale the two most important *limiting factors* are temperature and water, the levels of which vary widely both spatially and temporally.

Table 3.1 Plant and animal resources

Plants	*Animal*
Oxygen	Oxygen
Carbon dioxide	—
Light	(Light)
Water	Water
Mineral nutrients	Nutrition (plant food)
Temperature condition	Temperature condition

WATER

Water is essential for all life. It accounts for a very high percentage of the fresh weight of all organisms. Few plants can survive if their cellular water drops below 10 per cent saturation level and most will die below 50 per cent (Fig. 3.1). All

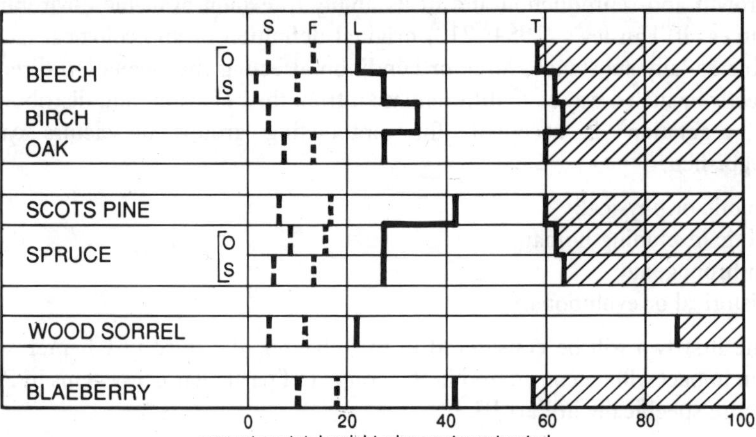

Fig. 3.1 Critical values for water content of leaves of certain types of vegetation in temperate and cold zones. S = water content at which stomatal closure begins; F = water content causing complete stomatal closure; L = lethal water content; T = dry weight as a percentage of the total weight when tissues saturated with water; O = shade leaves; S = sun leaves (from Birot 1965)

organisms take in water from the surrounding environment and lose it from their body surfaces by one means or another. Water loss is the means whereby excess heat and waste are dissipated.

Most land plants obtain water from the substratum via their root systems, though mosses and lichens can absorb moisture directly from the atmosphere. Over 90 per cent of the water absorbed by plants is given off, in the process of transpiration through minute pores (stomata) situated on the surfaces of all green organs or through the cuticle of mosses and lichens and from which it is removed by evaporation. Efficient metabolism requires a constant stream of water be maintained in the plant in order to keep all cells (and particularly the guard cells which control the stomatal apertures) turgid. The water balance in a plant depends on the difference between 'income' and 'expenditure'. If the latter exceeds the former the plant will be subject to stress, the effect of which will be dependent on the severity, i.e. the amount and duration of the deficiency. Thornthwaite (1948) defined drought as occurring when potential evapotranspiration exceeds that available in the soil. The former is a function mainly of the evaporative capacity of the atmosphere, and the latter a function of the infiltration and water-storage capacity of the soil. The rate at which plants absorb water is determined by the temperature and chemical composition of the soil solution. Low temperatures, high acidity, alkalinity and salinity can all reduce the rate of water absorption and, when evapotranspiration demand is high, cause water stress which results in temporary wilting, reduced growth or, if prolonged, death.

Animals lose water from their bodies as perspiration or as waste material (urine and faeces) and, as in the case of plants, terrestrial animals vary in their ability to tolerate water loss. Few can exist for long periods without drinking. Under conditions of high evaporation, water loss may exceed that available for drinking and animals will be subject to water stress. Small animals are particularly vulnerable to stress because of their high surface area to body mass.

TEMPERATURE

The temperature of the physical environment is as important a limiting factor as is water. It controls the rate of biological processes. Van't Hoff's law – that the speed of a chemical reaction doubles with every 10 °C increase of temperature – is applicable, within limits and up to an optimum level, to the metabolic processes of all organisms. Although there are few places in the biosphere where temperatures are too hot or too cold for life, most organisms operate somewhere within the range of 0–50 °C. However, some algae can grow and reproduce at temperatures below zero while others can tolerate those as high as 70–80 °C in hot springs. For each species of plant and animal there is a range of temperatures within which it can exist. The critical limits are set by the minimum threshold temperature below which and the maximum temperature beyond which an organism cannot survive; the optimum temperature is that at which the organism functions most efficiently (Table 3.2). Critical minimum temperatures are higher and maxima are lower during youth. In addition, some species are thermoperiodic in that they require lower night-time or winter temperatures for successful reproduction.

LIGHT

Light varies both temporally and spatially in quality, intensity and duration. However, only in deep water and below the ground surface is there a lack of sufficient illumination for photosynthesis. In terrestrial habitats the quality of the visible white light necessary for this process is not so variable as to be an important ecological factor. However, light is absorbed very rapidly in water; the

Table 3.2 Maximum, minimum and optimum temperatures (°C) for net photosynthesis in various groups of plants (from Bannister 1976)

	Minimum	*Maximum*	*Optimum*
Herbaceous plants			
Tropical (including C_4)	5.0 to 7.0	50–60	35–45
Crop plants (including C_3)	0 to over 0	40–50	20–30
Sun plants	−2.0 to 0.0	40–50	20–30
Shade plants	−2.0 to 0.0	*c.* 40	10–20
Arctic–alpine	−7.0 to −2.0	30–40	10–20
Woody plants			
Tropical trees	0.0 to 5.0	45–50	25–30
Arid shrubs	−5.0 to −1.0	42–55	15–35
Temperate deciduous trees	−3.0 to −1.0	40–45	15–25
Evergreen trees	−5.0 to −3.0	35–42	10–25
Actic/alpine dwarf shrubs	*c.* −3.0	40–45	15–25
Lichens (cold regions)	(−25) −15 to −10	20–30	5–15

red and blue bands are filtered out. Green light is poorly absorbed by chlorophyll, but the red marine algae have supplementary pigments which facilitate the utilisation of this wavelength and hence allow them to live at greater depths than the green algae.

Light intensity and duration are more ecologically significant. For each plant species there is a compensation point, i.e. a minimum light intensity at which, given the requisite temperature and availability of water, photosynthesis commences (Table 3.3). Similarly, there is an optimum intensity (i.e. the saturation point) at which photosynthesis attains a maximum rate. For most, if not all plants, however, the latter falls below that of full daylight. Because of the existing low carbon dioxide content of the atmosphere, species which respond to high light intensities are the sun-loving or heliophytic plants. Those which tolerate, or perform best at, low intensity are the shade plants or sciaphytes. Most trees, many grasses and herbaceous weeds and the majority of arable crop plants are heliophytes. Mosses and ferns are sciaphytic and some cave-dwelling mosses and deep-water algae can survive where the light intensity may be no greater than that of moonlight. There are, however, many plants which are facultative sciaphytes which can grow in dim light but attain their maximum development under high light intensities. In contrast, obligate sciaphytes grow best in shade.

The duration of light varies seasonally and diurnally with latitude and aspect. Photoperiodism, particularly day-length, is an important stimulus in the timing of daily and seasonal organic rhythms. It is the factor which triggers leaf-fall of many temperate deciduous trees as well as the beginning of mating, migration and dormancy in several animal species. Some plant species will not flower until the day-length (the photoperiod) is less than the critical threshold of 12–14 hours, i.e. they are short-day (or long-night) varieties. Others are long-day (or short-night) species requiring more than 12–14 hours day-length for successful reproduction.

Table 3.3 Light compensation and saturation in various groups of plants: lux = candles per square metre; C_3 and C_4 plants refer to the form in which carbon is fixed in photosynthesis: in the first the pathway involves formation of acids with three carbon molecules and in the latter with four (from Bannister 1976)

	Compensation (K-lux)	*Saturation (K-lux)*
Herbaceous plants		
C_4 plants	1.0–3.0	Over 80
Crop plants (C_3)	0.1–2.0	30–80
Sun plants	0.1–2.0	50–80
Shade plants	0.2–0.5	5–10
Deciduous trees		
Sun leaves	1.0–1.5	25–50
Shade leaves	0.3–0.6	10–15
Evergreen trees		
Sun leaves	0.5–1.5	20–50
Shade leaves	0.1–0.3	5–10
Mosses	0.1–2.0	10–20
Algae	—	15–20

THE GROWING SEASON

In order to complete their life cycles successfully all organisms require a period of time during which environmental conditions remain favourable for growth and reproduction. In those parts of the world where low temperatures limit organic activity for part of the year, the thermal growing season is defined as that period (in days or months) when the mean temperature is above a selected minimum threshold (Tivy 1991). In temperate latitudes the latter can vary from 0 to 10 °C dependent on the type of plant or the objectives of study. Those thresholds most commonly used to define the growing season are 6 °C, originally considered to be near to the average for the commencement of growth of temperate cereals, and 0 °C, the limit of the frost-free period. Differences in the temperature range and hence amount of heat received by two locations with a similar length of growing season can be expressed in terms of accumulated temperatures, i.e. the sum of the mean daily or monthly temperatures (i.e. day-degrees) above the selected growth threshold (see Fig 3.3).

The concept of the growing season was originally synonymous with the thermal growing season and was applicable only to temperate arable agriculture. It was based on the assumption that temperature was a more fundamental limiting factor than water which could be overcome by irrigation, while temperature was not susceptible to modification on a similar area. Where water is the most important limiting factor the definition of the growing season is more difficult. Most definitions are based on the concept of aridity, expressed as the difference over a

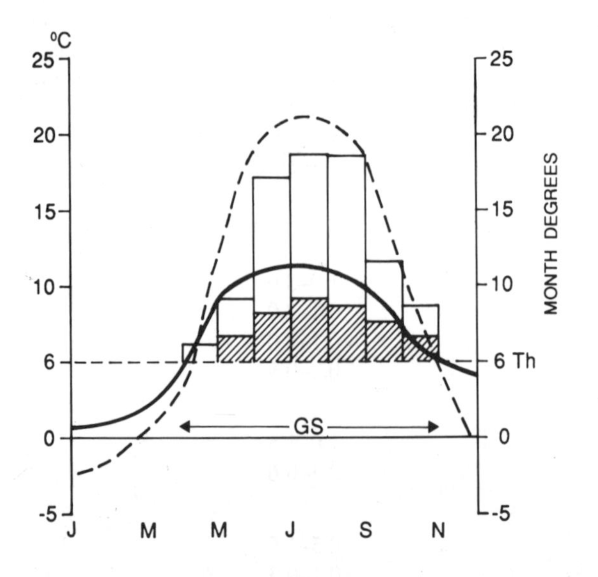

Fig. 3.2 Length of thermal growing season (Th) and monthly accumulated temperatures over mean of 6 °C: continuous curve and shaded bars for Aberdeen (57° N 2° W; altitude 2.4 m) and broken curve and unshaded bars for Chicago (42° N 88° W; altitude 7 m). Total day-degrees: Aberdeen 1095; Chicago 2328; Gs = growing season

given period between precipitation and evaporation or as a ratio between precipitation and temperature (one of the most important factors influencing evaporation rates for which long-term measurements are universally available).

It is, however, more difficult to give a quantitative expression of the humid growing season because of the relative paucity of measurements of either evaporation or evapotranspiration. The concept of potential evapotranspiration (Thornthwaite 1948; Penman 1963) allowed the calculation of temporal variations in the soil-water balance and hence in the length of the period of water deficiency and ecological drought (see Fig. 3.3). Potential evaporation is defined as the water loss that will occur if there is no deficiency of soil water – in other words the water that would be evaporated given the evaporative capacity of the atmosphere. Both authors constructed empirical formulae by which either potential evapotranspiration (or potential transpiration) could be expressed as a function of available climatic parameters. Thornthwaite used mean air temperature and day-length; Penman's set of complex but more comprehensive formulae combined saturation deficit, temperature, wind speed and net radiation.

SOIL

The concept of the growing season is still applied in one form or another as a basis for bioclimatic comparisons on regional and global scales. However, whether a particular species can occupy a given physical habitat does not depend on climatic conditions alone. The climate provides opportunities; the extent to which these are exploited will depend on the physical and chemical composition of the soil. Soil

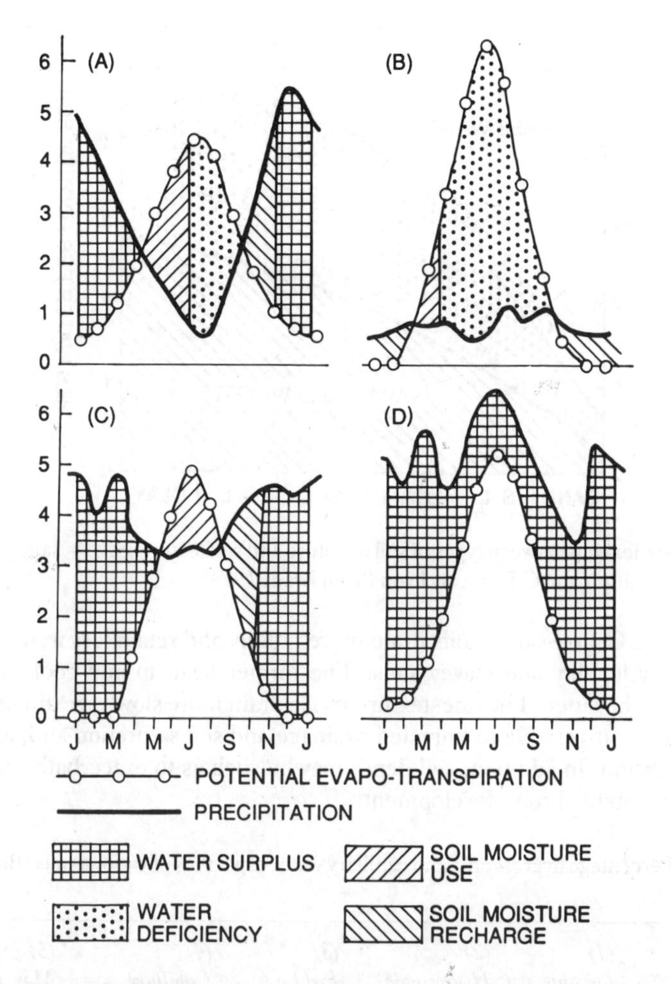

Fig. 3.3 Annual march of precipitation and potential evapotranspiration at: (A) Seattle (Wash.); (B) Grand Junction (Colo.); (C) Bar Harbour (Maine); (D) Brevard (NC) (from Thornthwaite 1948)

climate is more critical for the initiation of growth than that of the atmosphere. The lag between above-ground and below-ground soil temperatures is a function of the depth of the insulating layer of vegetation and dead organic matter (DOM) and of the texture of the substratum. The amount of soil water available to plants is that held by capillary tension after all surplus water has drained away. At this point soil water is at its maximum (i.e. field) capacity, and given suitable environmental conditions photosynthetic rates are highest. Soil-water deficits occur as the water drops below field capacity to the wilting point. Below this level the film of hygroscopic water around soil particles is held with a force that cannot be overcome by plant suction. Figure 3.4 and Table 3.4 show the relative proportions of available and non-available water in soils of varying texture. Water retention increases from coarse- to fine-textured soils and is reflected in variations in the

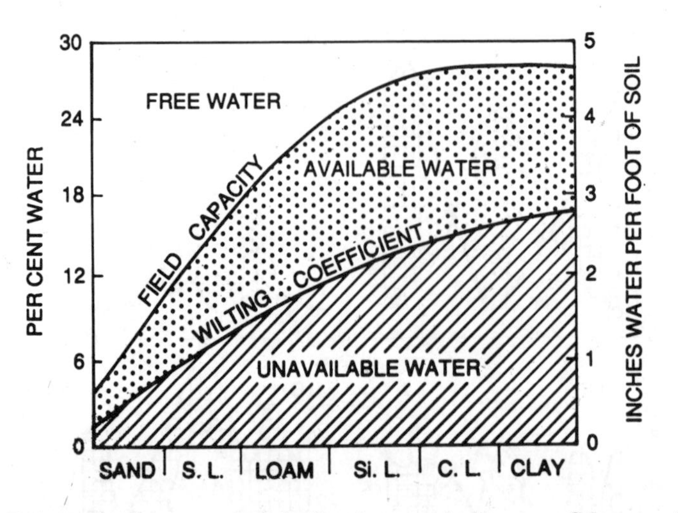

Fig. 3.4 Relationship between types of soil moisture and soil texture. S.L. = sandy loam; Si.L. = silty loam; C.L. = clay loam (from Brady 1974)

specific heat. Coarse sandy soils drain more rapidly and retain less water than fine, less permeable, silty and clayey soils. The former heat up and cool down more rapidly than the latter. The fine-textured soils drain more slowly, retain more water and are more susceptible to impeded drainage and soil saturation and, as a result, to poor aeration. In addition, soil depth may be such as to exacerbate soil drought by limiting potential root development.

Table 3.4 Percentage hygroscopic and capillary water capacities of various soils (Brady 1974)

Soil	*(1) Organic matter*	*(2) Hygroscopic coefficient*	*(3) Field capacity*	*(4) Capillary water (3–2)*	*(5) Max. retentive capacity*
Fine sand	2.13	3.4	7.6	4.2	44.5
Sandy loam	3.01	6.9	15.5	8.6	58.0
Silt loam	3.58	10.4	24.0	13.6	76.5
Silty clay	5.91	16.1	30.4	14.3	87.0

The soil is also the main source of the mineral nutrients essential for the successful growth and development of plants and of animals dependent on plant food. The macronutrients including nitrogen, potassium and phosphorus are required in relatively large amounts; others such as iron, magnesium, sulphur and calcium in lesser quantities. The micronutrients or 'trace elements' (including boron, manganese, molybdenum, copper, zinc, chlorine and, in some cases, sodium, cobalt, aluminium and silicon) are essential in extremely small amounts. They can, however, be toxic at concentrations only slightly above required levels.

Plants vary in their tolerance of highly acid, alkaline or saline soils, all of which conditions affect nutrient availability and the nutrient balance. Also, in highly alkaline, saline and acid soil conditions water uptake by roots is impeded and given high evaporation rates, plants may be subjected to what is called physiological drought. Further deficiency or excess of one or more of the essential or of inessential but potentially toxic elements can depress the vigour and rate of growth or, in extreme conditions, preclude it from a particular habitat. Among the most limiting soil-nutrient conditions are those associated with poor 'barren' substrates where the essential elements are either lacking or are present in critically low amounts or in toxic concentrations. Such conditions are often associated with soils derived from old, highly weathered and leached rocks of Australia, Africa and South America; or in soils derived from calcium-rich rocks (e.g. limestone, dolomite, gypsum) in which the availability of phosphorus and trace minerals such as boron and molybdenum is low and heavy minerals (lead, zinc, copper manganese, etc.) may occur in toxic concentrations. One of the poorest substrata is the rock serpentine (a magnesium iron-silicate) in which a deficiency of calcium, phosphorus and nitrogen is often accompanied by near-toxic levels of chrome and nickel (Proctor and Woodwell 1975). The extent to which soil chemical conditions (e.g. lime-rich, acid, etc.) preclude some species or depress growth so that some species are less able to compete than others is still a debatable ecological question (Salisbury 1929; Grubb 1987).

RANGE OF TOLERANCE

Organisms then vary not only in their essential requirements for successful growth and survival but also in their tolerance of unfavourable (i.e. stressful) environmental conditions. For every abiotic condition there is an optimum level at which an organism attains its maximum performance. With either an increase or a reduction in the quantity or quality of a particular variable there will be a decline in performance to a minimum or maximum tolerance level (see Fig. 3.5). These limits define the species range of tolerance and hence its potential spatial range or distribution.

Tolerance of the physical conditions discussed differs not only between species but between individuals in a species population. It also varies in respect of both physiological processes and of growth phases from youth to maturity. Young organisms usually have a lower tolerance than older ones. Conditions limiting reproduction are critical. Also, an organism's tolerance may be wide for some variables but narrow for others; its potential distribution will be determined by that condition for which it has the narrowest range of tolerance at the most vulnerable stage in its life cycle.

Under natural conditions it is not easy to identify the critical limiting factor for a given species or to quantify the level at which it becomes limiting. In the first place, environmental factors do not operate independently – they interact. The level at which precipitation becomes limiting depends on the prevailing temperature and soil conditions. High wind force can increase the rate of evaporation and intensify the effects of either high or low temperatures. For

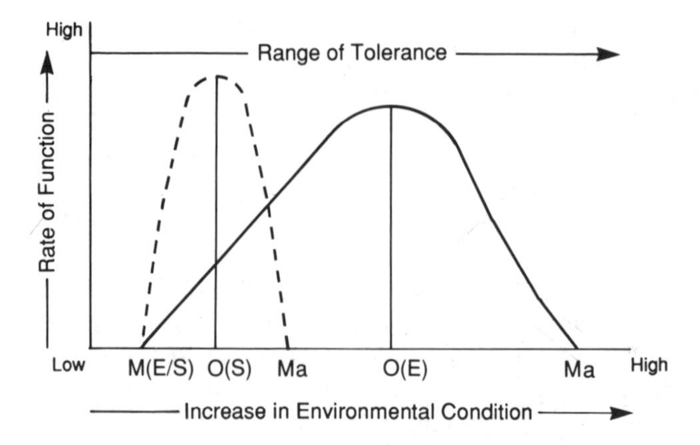

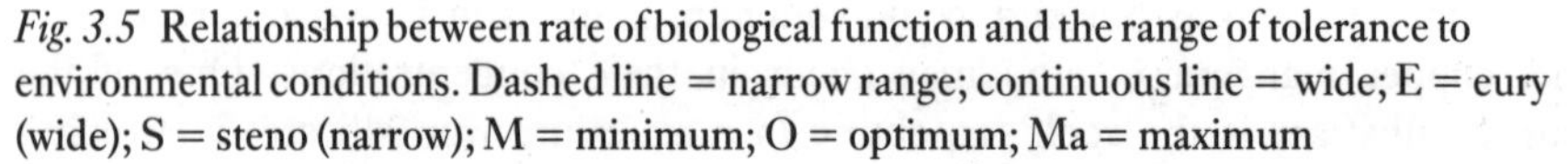

Fig. 3.5 Relationship between rate of biological function and the range of tolerance to environmental conditions. Dashed line = narrow range; continuous line = wide; E = eury (wide); S = steno (narrow); M = minimum; O = optimum; Ma = maximum

instance, it cannot always be ascertained with certainty whether exposure to wind or low temperature is the principal factor limiting tree growth at high altitudes. Also, because of the close interaction of environmental conditions – factor interchange or compensation (i.e. when one factor compensates for the deficiency of another) – can mitigate the effects of limiting factors. Under climatic conditions that would normally limit the growth of a particular species a sheltered site, a south-facing aspect or a sandy soil may result in local temperatures sufficiently high during the growing season to allow survival in an area generally climatically unsuitable. Conversely, in another area, the texture of the soil may be such as to reduce water availability or give rise to oxygen-deficient conditions such that species which might otherwise grow there would be excluded. Finally, an organism's range of tolerance to the total physical condition of the environment may well be less than that for any one factor.

ADAPTIVE STRATEGIES

The distribution of a species is dependent not only on the extent to which the physical environment can satisfy its resource requirements but also on its ability to tolerate periods and levels of environmental stress. Grime (1979) describes environmental stress as a deficiency of light, moisture and mineral nutrients and suboptimal temperatures which limit photosynthetic production, i.e. the external constraints which limit the rate of dry-matter production of all or part of the vegetation.

ANIMAL ADAPTATIONS

Because of their mobility and life histories animals can avoid stressful environmental conditions more easily than plants. Many transmigrate over longer or shorter distances in response to food and water deficiencies. Birds in particular undertake long-distance migrations from cold and cool temperate to tropical and subtropical regions avoiding the winter season, and back again. Herbivores often migrate in response to seasonal and/or spatial variations in climatically determined vegetation growth. Some animals become dormant or semi-dormant and hibernate, while others survive unfavourable periods in a larval or egg stage. Many animals can, to a greater or lesser degree, adapt to the onset of stress by controlling their internal temperatures and hence water loss either by behaviour or because they are homoiotherms with homeostatic control of their body temperatures. Putnam and Wratten (1984) distinguish between conformer (non-homeostatic) and regulator (homeostatic) organisms.

PLANT ADAPTATIONS

One of the earliest and most successful attempts to correlate plant morphology (growth or life forms) and life-history characteristics with environmental conditions was made by Raunkiaer (1934). His classification of life forms, despite an unwieldy terminology, has survived because it is both comprehensive and ecologically sound. Plants are grouped according to the height above the ground surface of their perennating buds. This is a visible indication of the way in which a plant survives periods unfavourable for growth. The original classification of land plants included five primary life-form classes (see Fig. 3.6):

1. (Th) Therophytes; annual plants which, having completed their life cycles, survive periods of cold or drought as seeds or spores;
2. (Cr) cryptophytes: including all those perennial plants whose resting buds are

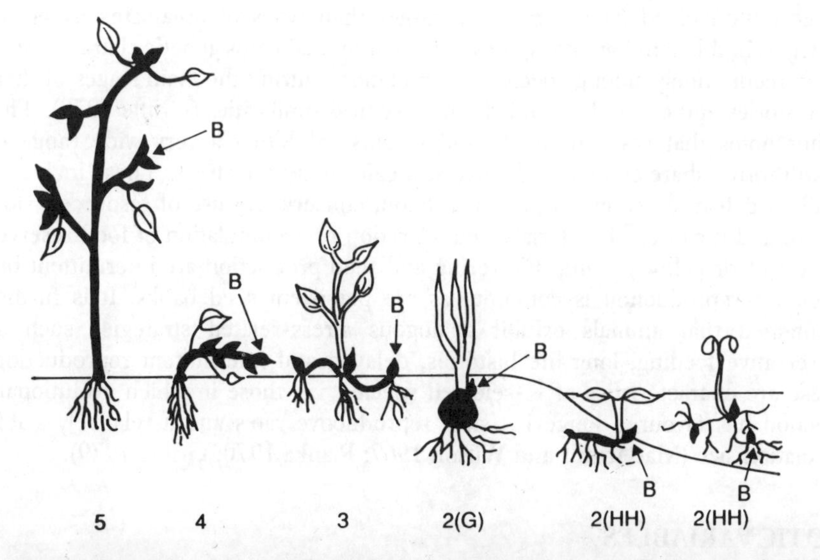

Fig. 3.6 Examples of the main Raunkiaerian perennial life forms: (5) phanaerophyte; (4) chamaephyte; (3) hemicryptophyte; 2(G) = genophyte; 2(HH) = hydrophyte; B = perennating bud (from Shimwell 1972)

below soil or water surfaces; these can in turn be subdivided into (G) Geophytes: perennial, herbaceous plants with underground, food-storage organs such as bulbs, tubers, rhizomes, etc. and (HH) Hydrophytes: marsh and water plants;
3. (H) Hemicryptophytes: mainly herbaceous plants with resting buds located on the soil surface; these include many mosses and lichens as well as plants with a tussock or rosette arrangement of their leaves;
4. (Ch) Chamaephytes: herbaceous or small, low-growing woody plants with resting buds carried on stems up to, but not exceeding, 25 cm above the ground surface;
5. (Ph) Phanaerophytes: woody perennials; trees and shrubs with resting buds on upright perennial stems 25 cm or more above the ground surface.

Raunkiaer further subdivided the final class on the basis of height, duration of leaves and bud protection. Later he included classes comprising such special forms as succulents and epiphytes. There have been many modifications of this system which have involved the subdivision, in greater or lesser detail, of Raunkiaer's primary classes, as well as the inclusion of the micro-organisms of the sea and soil. The main purpose of Raunkiaer's scheme was to analyse the relationship between plant form and climate. The percentage of species in the flora of a particular region belonging to each of the life-form classes formed what he termed the biological spectrum. A comparison of biological spectra from different parts of the world with the normal spectrum (an approximation of the percentage life-form composition of the world's flora) revealed that the former could be correlated with climatic variations. On this basis, Raunkiaer distinguished four main types of phytoclimate (see Table 3.5).

A more recent approach to environmental adaptation is that of Grime's (1979) categorisation of adapative strategies rather than types of organisms. Adaptive strategies in this sense are groupings of similar or analogous genetic characteristics which recur widely among species or populations during the main stages of their life histories and cause them to exhibit ecological similarities (Grime 1979). This author notes that vascular plants and lichens exhibiting a very wide range of growth forms share common adaptive strategies to severe stress. These include a small hard-leaved evergreen perennial habit, conservative use of resources, slow growth and turnover of materials, sequestration or accumulation of food reserves above and/or below ground. Flowering and seed production are intermittent but vegetative reproduction is common, as are persistent seed banks. It is further maintained that animals exhibit analogous stress-related strategies such as conservative feeding, long life histories, delayed and intermittent reproduction. These are characteristics of K-selected species, i.e. those in which evolutionary selection has favoured longevity, slow reproductive rates and a relatively stable population size (MacArthur and Wilson 1967; Pianka 1970; Grime 1979).

BIOTIC VARIABLES

The range of tolerance of an organism to abiotic conditions alone is usually referred to as its physiological or potential range. However, no organism exists in

Table 3.5 Life-form spectra of four major phytoclimates (according to Raunkiaer 1934)

Locality and phytoclimate	*No. of species*	*Percentage life-form classes* *Ph*	*(Ch)*	*H*	*Cr*	*Th*
(Ph) *Phanaerophytic climate:*		(warm, humid, tropics)				
St Thomas and St John	904	61	12	9	4	14
Seychelles	258	61	6	12	5	16
(Th) *Thereophytic climate:*		(tropical and subtropical; arid and semi-arid)				
Death Valley (California)	294	26	7	18	7	42
Argentario (Italy)	866	12	6	29	11	42
(H) *Hemicryptophytic climate:*		(mid-latitude cool and warm regions)				
Altamaha (Georgia)	717	23	4	55	10	8
Denmark	1084	7	3	50	22	18
(Cr) *Cryptophytic:*		(high latitude (tundra) and high altitude)				
Spitzbergen	110	1	22	60	15	2
Alaska	126	0	23	61	15	1
Normal spectrum	1000	46	9	26	6	13

isolation but in company with others of similar or differing type, and its actual or ecological range will also depend on the suitability of the biotic environment. The latter is, on the one hand, a product of the interactions between organisms and their abiotic habitat and, on the other hand, of the interactions between organisms of the same or different species living together. Biotic interactions may be direct or indirect, exclusively or mutually antagonistic, depressive or beneficial. They may preclude or be essential for the presence of one or more species in a given habitat. The principal types of biotic interaction (Krebs 1985) include:

1. Negative interactions:
   (a) Competition: both species suffer;
   (b) Herbivory predation and parasitism: one species benefits at the expense of the other;
2. Positive interactions:
   (a) Commensalism: one species benefits, the other is unaffected;
   (b) Mutualism: both species benefit.

COMPETITION

Competition has been defined in two ways by ecologists (Law and Watkinson 1989). In the first instance it is described as the struggle of two or more organisms to exploit the same limited resource. In the second, it occurs 'when the interaction between two or more individuals or populations adversely affect the growth, fitness and/or population size of the other' (Giller 1984: 22). Of all the biotic interactions competition is thought to exercise the strongest selective pressures on the distribution of species and on the evolutionary adaptation of their populations in such a way as to avoid marked ecological overlap and to allow coexistence in the same habitat. Although the competitive exclusion principle (Gause 1932, 1934) (i.e. when two species are grown in a similar environment one either wholly or partially displaces the other) has been experimentally demonstrated, field evidence is limited and often conflicting because the coexistence of similar species populations appears to be common. The problem is further compounded by the fact that it is difficult to assess the most significant environmental parameter and/or that level at which similarity between species makes coexistence impossible. Field evidence of competition is relatively rare because, it is argued, existing distributions are the result of past competition. Most obvious current competition in plant and animal populations occurs in disturbed habitats, as in the case of crops and weeds in cultivated soils and in areas where recently introduced more competitive species are in the process of successfully displacing the weaker natives. The current displacement of the native red squirrel by the introduced grey in Britain is an often quoted but not universally accepted example. Direct competition is most intense between young individuals of the same species dependent on similar limiting resources of space, light, water and nutrients. The way in which organisms compete and the effects of competition vary between different plants and animals, between different taxonomic groups and particularly between terrestrial and aquatic organisms.

Land plants rooted in a substratum are relatively immobile. They must therefore compete for a fixed space (or site) in the habitat from which they obtain essential resources (Krebs 1985). One of the main forms of competition among green plants is for sunlight. The creation of shade and the consequent reduction in light intensity is the most common way in which one type of plant may suppress or prevent the growth of those not adapted (i.e. obligate sciaphytes) to low-light intensities. In the struggle for light relative success is dependent primarily on size, growth form and habit. Obviously, the tallest plants capable of growing in a given habitat will have a competitive advantage over smaller forms, which in turn will have an advantage over even lower-growing plants. Among plants of similar potential growth height, however, competitive ability will depend on the form, density and duration of their canopies. Deciduous trees such as beech, oak and ash have comparable growth heights; the beech, however, casts a deeper shade in the growing season than the others, while the ash is relatively translucent.

Above ground surface, competition for light dominates the struggle for existence among plants. However, its influences and its final outcome may be determined by competition for nutrients and water drawn from the soil. In this case competition will be most severe between individuals and/or species populations exploiting the same volume of soil. Competitive ability then is dependent on the density and extent (both horizontal and vertical) of the root systems and the rapidity with which these develop. The spatial development of a root system is dependent partly on the physical condition of the soil (depth and texture) and partly on the species.

Competition for light, nutrients and water are closely linked. Hence, it is not easy to distinguish with certainty between the relative effects of root and shoot competition. That between root systems tends to commence sooner in the growth cycle. The more vigorous and nutrient-demanding individuals or species may so deplete the supplies of nutrients and water as to check severely the rate of growth of the less demanding. Reduction of light intensity by shading not only depresses the growth of both roots and shoots but also reduces the plant's ability to compete for water and nutrients.

A plant's competitive capacity tends to be greatest in optimum habitat conditions. In suboptimal conditions – near the limits of its range of tolerance – rate of growth, vigour and reproductive capacity decrease. It becomes more susceptible to depressive competition from those species better adapted to these habitat conditions. Many plants, however, are excluded by competition from their optimum growth habitats. This is revealed in the greater size and vigour that some species attain when they colonise, or are cultivated in, a more favourable site. It is particularly characteristic of many light-demanding herbaceous annuals which normally grow in poor soil and harsh climatic conditions free of competition for light.

It has been suggested that the preference of many plants for soil of a particular mineral composition, nutrient status or pH is the result not so much of a casual relationship as of the relative competitive ability of the plants adapted to these conditions. Salisbury (1929) early noted that while sorrel (*Rumex acetosa*) shows a definite preference for acid soils, deprived of competition it can grow with enhanced vigour on limed rather than non-limed soils. And Tansley (1954) quotes the classic example of the heath bedstraw (*Galium saxatile*) which avoids calcareous soils not because it cannot grow on them but because, under such conditions, its seedlings develop only very slowly and are, as a result, handicapped in competition with other plants that can grow more vigorously. More recently Grubb (1987) has confirmed this edaphic influence.

Competition among plants is mainly indirect. Interference is not so evident except in so far as the varying size of different plants results in competition for sufficient space in which to reach maturity. A more distinctive type of interference is that of allelopathy or the chemical inhibition of one species or individual by anoıther (Whittaker and Feeny 1971). Some plants produce on decay, or as exudates from living roots, harmful or toxic chemical substances which can depress or kill the seedlings of the parent plant as well as those of other species. Certain desert shrubs, prairie forbs and kauri pine trees 'poison' the soil in this way. The wide spacing of individual plants that results is thought to reduce competition for either water and/or nutrients between the allelophatic individuals.

ANIMAL COMPETITION

The process of intra- and inter-specific competition between animals is more complex than that between plants. Not only are animals more mobile but their morphology, physiology and behaviour are more diverse and specialised. And while plants are autotrophic (organisms dependent on a common resource and the physical environment), animals are heterotrophic using a wide variety of organic food resources including plants, animals and dead and decaying organic matter. As a result the distributional range of an animal is directly or indirectly determined by the types of plants available. Also, because animals are adapted to live and move in one or two physical media (air, water, soil) and because they have such a diverse range of activities, the parameters that determine their distribution are more complex than in the case of plants.

Competition among animals, as among plants, is primarily for living space, food and breeding partners, and their competitive ability is dependent on the efficiency with which they can attain these ends. It has been variously correlated with population numbers, rate of reproduction and methods of food acquisition. Interference appears to be the main type of competition between animals and is expressed in relative aggressiveness and territorial behaviours in delimiting and protecting feeding, mating and rearing territories.

Aquatic organisms, however, are thought by many biologists to be the most prevalent exception to the competitive exclusion principle. Hutchinson (1957) noted that phytoplankton use a common resource pool in which nutrients are the principal limiting factor. Different species can coexist because the instability of their environment is such that before the complete displacement of one by the other could occur seasonal changes in the environment would shift the competitive balance. Also, freshwater fish are generally less specialised (Krebs 1985) than terrestrial animals, have a wide range of habitat and feeding conditions and a plastic growth rate that allows them to survive long unfavourable periods.

Other negative or antagonistic interactions which can affect the numbers and distribution of plant and animal species include herbivory, predation (carnivory) and parasitism. Since all involve feeder–food interactions they can be, and indeed often are, all considered as forms of predation. Predators are found in every taxonomic group and vary in size, form and methods of feeding. All tend to depress the abundance of their food (prey or host) by reducing the survival and reproductive rate of the prey organism(s).

HERBIVORY

The most important herbivores include the plant plankton feeders in aquatic habitats, and insects, ungulate mammals and monotremes in terrestrial habitats. Grazing of aquatic photoplankton is effected by mechanisms for filtering the sea-water. Grazing by terrestrial herbivores is dependent on the animals' foraging and eating habits, on the one hand, and on the palability, nutritive value and morphology of plants on the other (Denno and McClure 1983). Some (sheep, rabbits) 'crop' short plants, others (cows) pull up taller ones, while many insect herbivores tear leaves and cut into other organs. Many herbivores, particularly among insects and birds, are selective, specialist feeders; others such as the larger mammalian and monotreme herbivores are generalists, grazing a wide range of plant species (Crawley 1983). Grazing may involve organ damage, reduction of seed and seedling numbers and defoliation. The net effect of one or a combination of these impacts will tend to reduce the rate of growth and of seed production. Plants vary in their susceptibility (or resistance) to grazing, dependent on morphological or chemical attributes. Those which may reduce or inhibit grazing include tough leaves and stems of low nutritive value; low palatability as a result of spines, hairs, etc. or high levels of condensed toxins or other toxins; flattened 'rosette' and creeping forms; protected perennial buds, etc.

Current theories have been much concerned with the evolutionary significance of plant defence mechanisms (Hassel and Anderson 1989; Edwards 1989). A distinction is frequently made between neutral resistance (making plants more difficult to exploit) and defence adaptations (tough leaves, low nutritive value and toxins) though it has proved difficult to make a clear distinction between the two. Further, many of the grazing defence characteristics are also associated with soil-water and mineral deficiencies, ultraviolet light and micro-organisms. Whatever the cause of selective grazing it gives the more resistant plants a competitive advantage over those more susceptible to pressure. As a result some species can be excluded from an otherwise favourable habitat. Also, reduction of competition favours the growth of others that might, in the absence of grazing, have been shaded out by taller forms. While grazers in an undisturbed unmodified terrestrial habitat usually only consume a relatively small proportion of the total plant biomass available, overgrazing can result in the elimination of a species from part or the whole of a habitat.

OVERGRAZING

Overgrazing involves the removal of plant material at a rate greater than it can be regenerated. It is a common hazard in habitats where natural or man-induced disturbance results in a greatly increased number of herbivores. This may be a result of a natural 'population explosion', the introduction of a herbivore into an area free of competition and predators or of high densities of domestic livestock. In the latter case susceptibility or resistance to various types and intensities of grazing depends on the type of animal involved as well as on the palatability and growth form of the species available. Many plants are avoided by domestic animals because they are distasteful or poisonous. The persistence and continued spread of bracken (*Pteridium aquilinum*) on rough pasture in Scotland is undoubtedly aided by the fact that it is slightly poisonous, particularly to young stock. The survival and prevalence of shrubs such as elder (*Sambucus nigra*), gorse (*Ulex* spp.), broom (*Sarothamus scoparius*) and the common weeds, ragwort (*Senecio jacobaea*) and creeping thistle (*Cirsium arvense*), can be attributed to their lack of palatability for one reason or another. Freedom from attack, however, is relative to the availability of other more desirable forms of forage and the degree of selectivity of the animals present. In this respect sheep are notoriously selective of the more palatable and nutritious plants; cows and in particular goats are much less fastidious.

Among those most susceptible to the effects of grazing are often, therefore, the taller, more conspicuous or palatable species. Annual plants and tree seedlings are particularly vulnerable. They may be destroyed completely, their numbers thereby drastically reduced and their regeneration limited or, under extreme circumstances, prevented. Perennial herbaceous plants with underground food-storage organs, such as bulbs, corms, rhizomes, etc. whose buds are well protected from injury or removal, are more resistant to grazing and are capable of making new growth despite the removal of some or all of their vegetative parts. However, a balance must be maintained between the rate of removal of their green photosynthesising material and the storage of food for continued growth.

Grazing reduces the proportion of those plant species less able to withstand certain intensities of grazing than others. It results in a reduction of plant biomass and species richness. The depredations of sheep, goats and, in some instances, pigs, have not only contributed to deforestation but have directly prevented the regeneration of natural woodland in the long-settled areas of Europe, not least in Mediterranean countries. Many plants, while not necessarily eliminated, are reduced in number; the vitality and vigour of the survivors are weakened, competition from other less-affected plants is thereby favoured. Relieved of competition from tree growth, bracken grows with greatly increased vigour. With the disappearance of heather, grasses and other plants can become more abundant. The greater vigour and competitive ability of many species more resistant to or less affected by grazing are further stimulated by concurrent soil changes. Selective grazing, particularly of the more nutritious and demanding plants, extracts from the soil mineral nutrients and potential organic matter which can never be fully replaced by animal excreta alone. The gradual decline in soil fertility which inevitably results contributes to the greater competitive capacity and aggressiveness of poorer, less demanding species. The spread of the tough, wiry, moor mat grass (*Nardus stricta*) and heath rush (*Juncus squarrosus*) over considerable areas of rough grazing in Scotland are indicative of a concomitant decline in grazing quality and soil fertility. Such changes are the outward and visible signs of overgrazing, a condition under which intensity of grazing (either in terms of numbers of animals or duration of grazing) causes a depletion either of a particular species, or of the vegetation cover as a whole, at a rate greater than they can be renewed and maintained by regrowth. Under extreme circumstances overgrazing can result not only in the elimination of certain species and decline of soil fertility but also in the weakening or removal of part or most of the vegetation cover. The exposed, unprotected soil becomes, particularly where slopes are steep or drought occurs, susceptible to accelerated soil erosion.

PREDATORS AND PARASITES

The predators includes those animals which feed on other animals (prey) and parasites which live on other plants or animals (hosts) to the detriment of the prey and host. Both groups contain a tremendous diversity of organisms varying in size, life-span, hunting and feeding characteristics (see Table 3.6). Parasites include micro-organisms (viruses, bacteria, protozoas) and macroparasites (helminths and arthropods). Parasitoids are the alternately parasitic and free-living, hymenopterous or dipterous insects, the larvae of which develop inside a host and eventually kill it.

Under 'natural' conditions a balance is normally achieved between host and parasite. The latter, however, can depress the host's vigour and reduce its competitive ability. Under certain circumstances parasites and pathogens can destroy a species in a particular area. Classic examples are the chestnut blight which all but eliminated the sweet chestnut (*Castenea*) from the eastern USA, the potato blight in Ireland in the 1840s and the phylloxera infestation of the grapevine in France. These were consequent on the accidental introduction of a parasite into an environment in which the host plants did not possess a natural resistance to attack.

Table 3.6 Comparisons between some characteristics of the life histories of microparasites, macroparasites, parasitoids and predators (from May 1982)

Life-history characteristic or other property	*Microparasite*	*Macroparasite*	*Parasitoid*	*Predator*
Ratio of average life-span to that of host or prey	$<<$ 1 (very small)	$<$ 1 (fairly small, $1–10^{-2}$)	1 (usually about equal)	$>$ 1 (usually live longer)
Ratio of body sizes	Much smaller than hosts	Mature stages often of similar size to host	Usually larger than their prey	
Intrinsic growth rate of population	Much faster than hosts	Faster than hosts	Comparable, but usually slightly slower than hosts	Usually slower than prey
Interaction with individual hosts, as observed in natural populations	One host usually supports a number of populations of different species	One host usually supports from a few to many individuals of different species	One host can support one or several individuals of one (or rarely two) species	One prey item can feed one or a few individuals of the same predator species, but many individual prey are required during a predator's life span
Effect of the above interaction on the host	Mildly to fairly deleterious	Variable: not usually too virulent in definitive host; can be very virulent in intermediate host	Eventually fatal	Usually fatal immediately

Table 3.6 (Cont'd)

Life-history characteristic or other property	*Microparasite*	*Macroparasite*	*Parasitoid*	*Predator*
Ratio between numbers of species, at the population level	Many species of parasites recorded from each member of the host population	Many species of parasites recorded from each population of hosts	Most host species support several parasitoid species, both specialist and generalist (but only a proportion of hosts are actually attacked)	Individual species of predators tend to use more than one prey species
Degree of overlap of the ranges of the two species	Occur as diffuse foci thoughout host's range	Occur as diffuse foci throughout host's range	Usually present throughout host's range	Range is usually greater than that of prey
Gentotypes per host of prey	Single or multiple	Multiple	Single or multiple sibships	Single

POSITIVE INTERACTIONS

In contrast to the preceding negative biotic interactions, mutualism and commensalism are the two most important positive biotic interactions. Mutualism (or symbiosis) is the mutually beneficial association of two dissimilar organisms. The lichens (composed of a fungus and alga) are probably the most advanced form. Among the most important is the association of symbiotic nitrogen-fixing bacteria with the root nodules of legumes, the fermenting bacteria in the rumen of ruminant herbivores and the mycorrhizal fungi in the roots of trees and shrubs. In the latter the absence of the fungi can prelude the regeneration of certain plants (*Calluna vulgaris*, *Pinus sylvestris*) on poor acid soils. Commensalism occurs when one of two species benefits from an association while the other is unaffected, as in the case of plants or animals growing on the shells of marine animals or of plants supporting themselves on other plants (epiphytes).

The distribution of many plants is dependent on pollination and/or seed dispersal by animals. The greater bindweed (*Calystegia sepium*) seldom sets seed in Britain because its pollinator, the convolulus hawk-moth, has become so rare. The successful introduction of *Smyina* figs to California and red clover to Australia depended on that of the pollinating wasp in the first instance and the bumble-bee in the second.

NICHE CONCEPT

The presence or absence of a species in a given habitat depends on the availability of a suitable niche. This term was originally used (and often still is) synonymously with either microhabitat or trophic level (Schoener 1989). Hutchinson (1957), however, originally defined a niche in terms of the animal rather than the plant or habitat it occupied. He described it as the multidimensional space (or hypervolume) (Fig. 3.7) – what might be called the 'environmental package' – delimited by the combined range of all the environmental variables (physical,

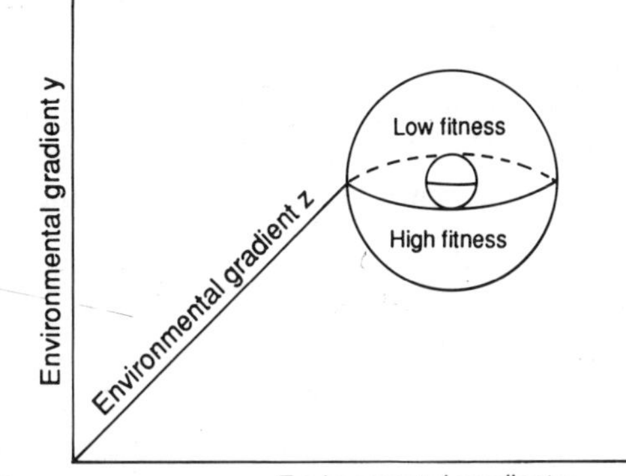

Fig. 3.7 Diagrammatic representation of the niche concept (from Pianka 1988)

chemical and biotic) to which a species is adapted. Crawley (1986: 5) has expressed it more succinctly as 'a multi-dimensional description of a species resource needs, habitat requirements and environmental tolerances'. Some generalist organisms have very wide niche requirements, other more specialist organisms have narrow niche requirements. Hutchinson further distinguished between the fundamental (potential) and the actual niche. The former is a function of species' adaptation to physical (i.e. abiotic) conditions and could be thought of as the physiological or auteocological niche. The latter is the ecological (or more precisely synecological) niche which is an expression of both biotic and physical parameters. As has already been noted, according to the competitive exclusion principle two organisms with identical or very similar environmental requirements cannot occupy exactly the same niche. Although, this model has been demonstrated in the laboratory, field evidence is limited and often conflicting and the coexistence of similar species seems to be the rule rather than the exception. The problem is further compounded by the fact that it is difficult to determine the relative importance of all the operative variables, not all of which are amenable to linear measurements or retain constant values. As a result the applicability of the concept in the field tends to be limited, and the niche is most frequently defined in terms of the correlation of one or two food characteristics with variation in related morphological attribute of the animals concerned (e.g. size of food : length of beak in the case of birds).

The niche concept has not been applied to plant ecology to the same extent as to animal ecology in which area it was initially developed. This is probably because of the basic difference in the nature of the food resources of the two types of organisms. As Crawley (1986) again notes, while it is possible for plant resources to be quantitatively apportioned to competing animal species, this is not so easy in the case of plants whose abiotic resources are light, water and nitrogen. The niche concept, nevertheless, is fundamental to current interpretations of the evolutionary adaptation of organisms to their environmental conditions and, hence, of their range of distribution.

The present distribution of plants and animals, however, cannot be fully understood in the light of current biotic and abiotic processes alone. They may explain why certain species are able to exist in certain habitats but not how they come to be present in one habitat and not in another similar one. For the answer to these questions one must look back in time and from the evidence available attempt to assess the effect of past environmental factors on the evolution and distribution of organisms.

Chapter

4

Historical biogeography

The geographical limits of the area within which a particular population of organisms occurs are known as its range. While its presence suggests that current ecological conditions are (or have very recently been) suitable for a species' successful growth and reproduction, its absence does not necessarily mean the reverse. The actual range of an organism does not necessarily coincide with its potential range and the former can really be fully understood only in the light of past events. The aim of historical biogeography then is to reconstruct the origin, spread, and extinction of taxonomic groups of organisms through time and to explain the influence of past geological and climatic events on present distributions (Myers and Giller 1988).

TYPES OF TAXONOMIC RANGE

Although some taxonomic groups have a very wide range, few are truly *cosmopolitan* in the sense that they are found in every suitable habitat throughout the world. The higher groups (i.e. orders and families) have, in general, a wider range than the lower-level genera and species. Four of the six families of the order Angiospermae (flowering plants) – Compositeae, Gramineae, Leguminoseae and Cyperaceae – are almost cosmopolitan. Species with a comparable range are mainly weeds whose spread has been dependent on human activities. With the exception of bats (*Chiroptera*) and marine mammals only three of the eighty-nine animal families – Old World rats and mice (Muridae), hares and rabbits (Leporidae); dogs and wolves (Canidae) are found on all continents except in Antarctica (Pielou 1979). In contrast, other taxonomic groups have very local, restricted (i.e. *endemic*) ranges, and the smaller the area the lower the taxonomic level (species or varieties) of endemic organisms that will be present. Many more taxa have *discontinuous* or *disjunct* ranges with their population groups occurring in two or more areas so widely separated as to preclude the possibility, under existing conditions, of normal means of disperal. Some large disjunctions at

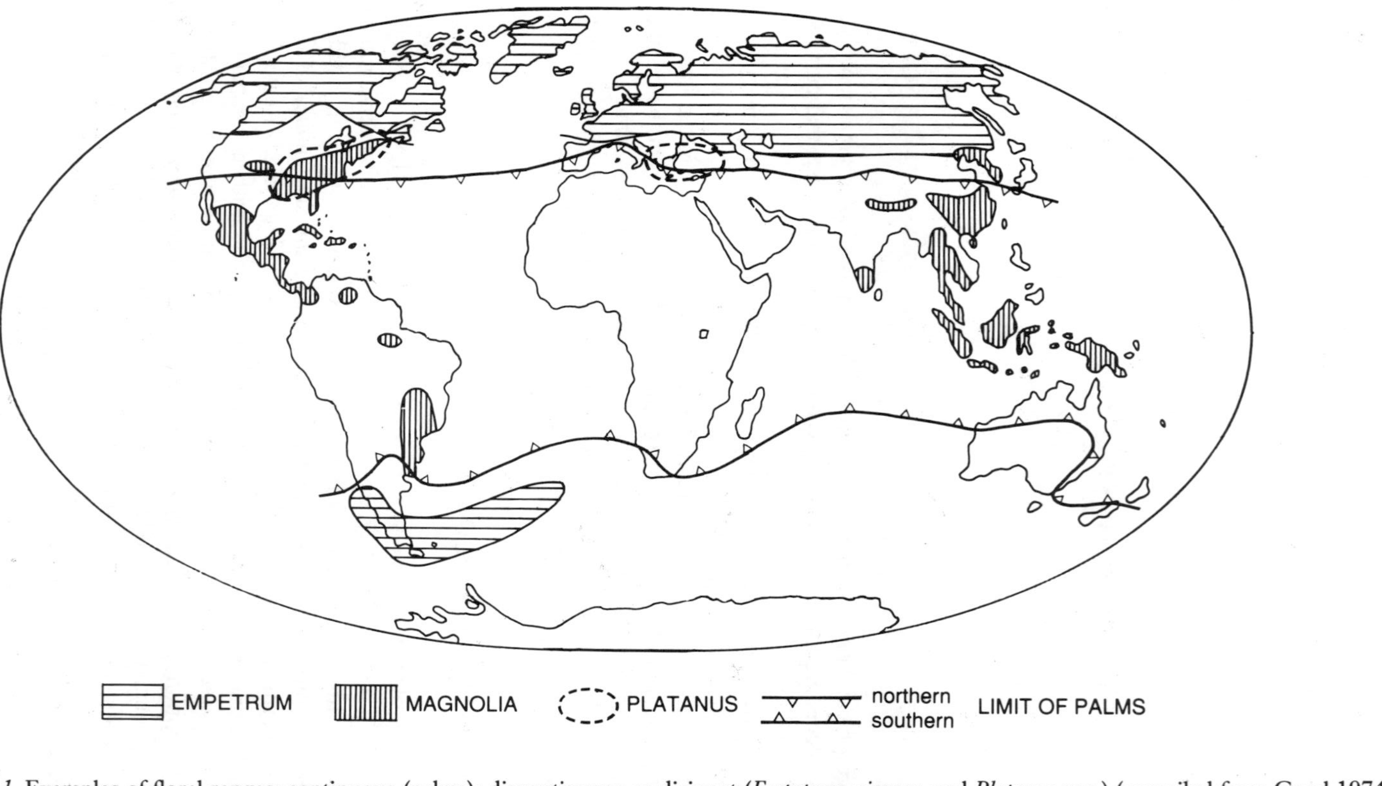

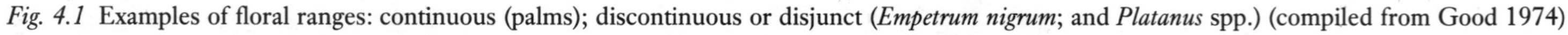
Fig. 4.1 Examples of floral ranges: continuous (palms); discontinuous or disjunct (*Empetrum nigrum*; and *Platanus* spp.) (compiled from Good 1974)

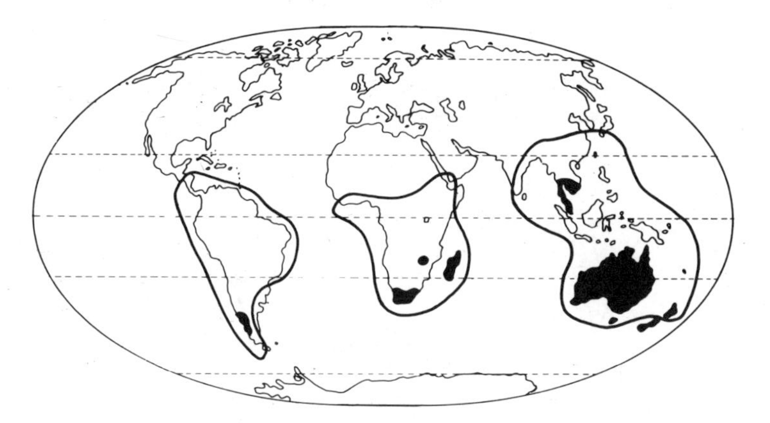

Fig. 4.2 Disjunct ranges of angiosperm families: Restionaceae (black) and Proteaceae (outlined) (Pielou 1979, redrawn from Hutchinson 1926 and Good 1974)

the global scale include taxa with bipolar; amphi-Atlantic; eastern area–North America; and South America–South Africa–Australia (see Figs 4.1 and 4.2). Many more have been identified at varying scales and taxonomic levels (Thorne 1972).

TAXONOMIC REGIONS

Taxonomic spatial units or regions have been distinguished at various scales on the basis of their floral and/or faunal composition. There are, however, no universally agreed rules by which such regions can be classified. Stott (1981) notes the criteria most commonly used:

(a) the percentage of endemic taxa;
(b) the size and importance of the different floras or faunas;
(c) demarcation knots, i.e. the points or lines where there is a marked change in flora or faunal composition which generally coincides with the limits of a large number of taxonomic ranges.

At the global scale terrestrial floral kingdoms are differentiated by means of angiosperm endemism; and faunal realms by endemic mammal families. In Fig. 4.3 Lemée (1967) combines the two in biogeographic regions which are subdivided into subregions and sectors (or districts) that can be identified on the basis of successively lower taxonomic levels. French plant geographers have long distinguished a hierarchy of floral regions: regions, with a high proportion of endemic genera; domains, with a high proportion of endemic species; and sectors and subsectors (or districts) with a high proportion of subspecies (varieties) or ecotypes (see Fig. 4.4).

Because families of marine organisms tend to be more widespread than terrestrial organisms, only two marine faunal realms are normally recognised at the global scale. These are correlated with the two distinctive environmental zones – the continental shelf (epicontinental seas less than 200 m deep), and the ocean

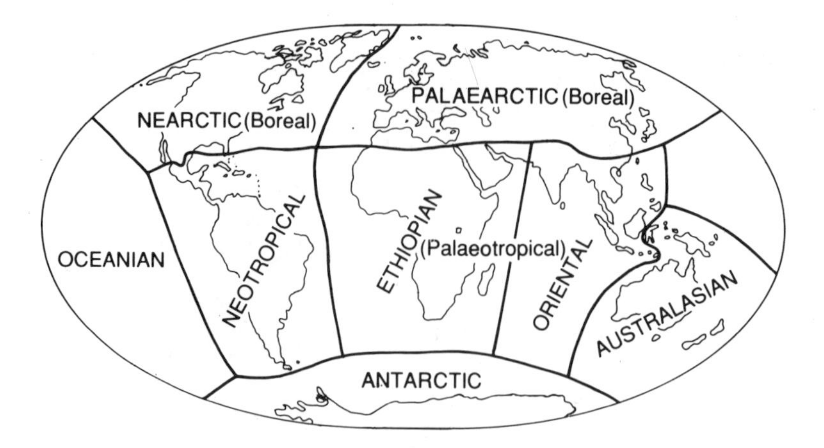

Fig. 4.3 Biogeographic regions of the world. Alternative names of floral as distinct from faunal regions given in brackets

deeps. It is less easy to delimit regions and provinces in the ocean than on the land because of the very much wider range of marine than of terrestrial genera and species.

Floristic and faunistic taxa can also be classified by criteria (or elements) other than that of their geographical range. These include:

1. *Historical elements*: the geological period when taxa were incorporated into the flora or fauna. A distinction is sometimes made between native or indigenous taxa established by 'natural' means and aliens (or exotics) deliberately or accidentally introduced by humans; but it is not always easy to establish the status of a taxon. For instance, many 'weeds' once considered to be aliens in Britain have since been identified by fossil remains as indigenous in late glacial and post-glacial deposits.
2. *Migrational elements*: the location of the existing range of a taxa relative to a given area, i.e. endemic (restricted to the area in question); or extraneous (or foreign) when the range of a taxon extends beyond the area in question. In the latter case the shape and orientation of the range is thought to be indicative of its migrational history. Matthews (1955) classified British plant species according to their 'migrational elements'. The number of endemic species of flowering plants in Britain is relatively small and the present flora is an extension of that in Europe. Species that have a very wide range throughout and even beyond Europe account for more than half the total British flora. The remainder can be classified into a dozen migrational elements. One half of these species have their main centre of their range in the south and south-west Mediterranean and central Europe; the other half in northern and north-western Europe (including Arctic areas) and the alpine zones of the mountains of west, central and south-east Europe (Fig. 4.5a and b).
3. *Genetic elements*: centres of origin (i.e. of evolution) from which taxa migrated. It was originally assumed that the centre of origin of a taxon would coincide with its centre of greatest genetic diversity. High diversity would occur where

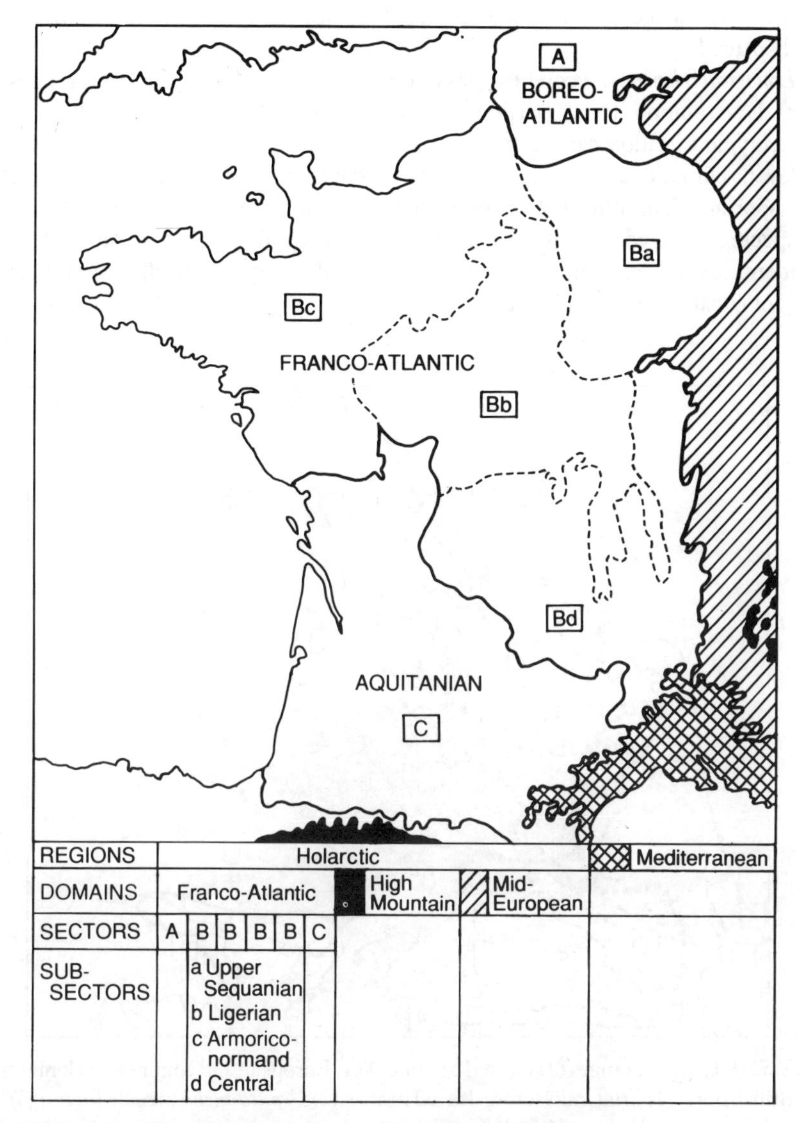

Fig. 4.4 Hierarchy of floral regions in France (from *Atlas de France*)

conditions were most favourable for growth, reproduction and speciation and the number of species would decline in parallel with biotype depletion (Stace 1989). Vavilov's (1949–50) analysis of the centres of origins of crop plants on the basis of species diversity has been subjected to considerable criticism. Nevertheless, the twelve centres of diversity for a large number of unrelated taxa identified by Zeven and Zhukovosky (1975) still contain the main representatives of older forms of the taxa concerned. This rather simple model, however, is only really valid in the case of young, recently evolved taxa. For older taxa there may be two or more recently established secondary centres of diversity, while the original centre has, for one reason or another, been displaced.

4. *Ecological elements*: ecotypes classified on the basis of their preferences for, and adaptation to, a particular environmental habitat or ecological niche. Ecotypic variation can be analysed at any taxonomic level. At the higher family level, climatic adaptation is more marked and hence diagnostically more important than other habitat characteristics. At the generic and particularly the specific level adaptation to a particular ecological niche is significant for the identification of genetically unrelated and morphologically dissimilar but ecologically equivalent taxa.

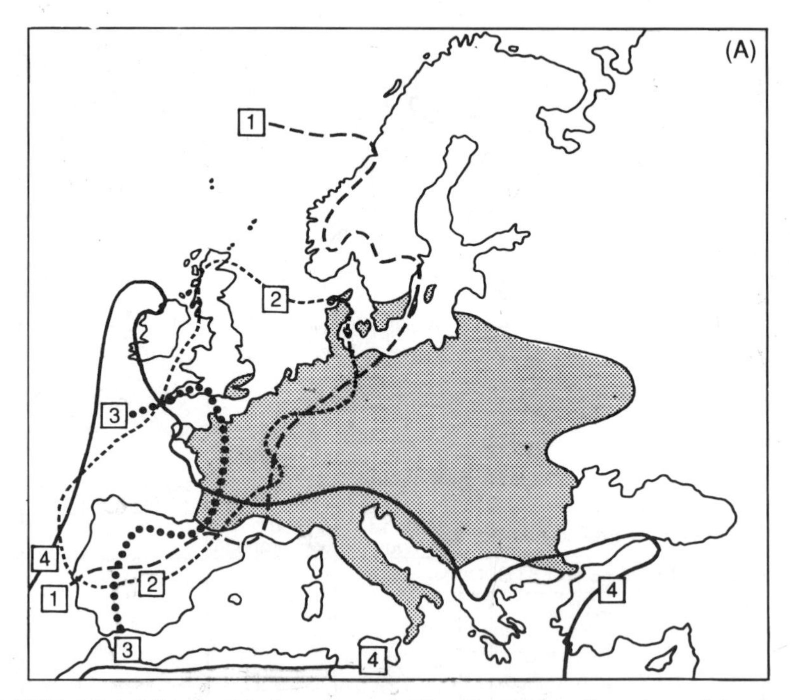

Fig. 4.5(A) Limits of ranges of selected oceanic West European and continental elements in the British flora: (1) *Erica tralix* (cross-leaved heath); (2) *Genista anglia* (needle furze); (3) *Erica ciliaris* (ciliated heath); (4) *Arbutus unedo* (strawberry tree); shaded area = *Carpinus betula* (hornbeam) (from Matthews 1955)

DISTRIBUTION OF TAXA AND TAXANOMIC REGIONS

The present distribution of biotic taxa and of biotic regions is a result of three concurrent and continuing processes:

1. *Evolution*: location, time of origin and rate of organic evolution;

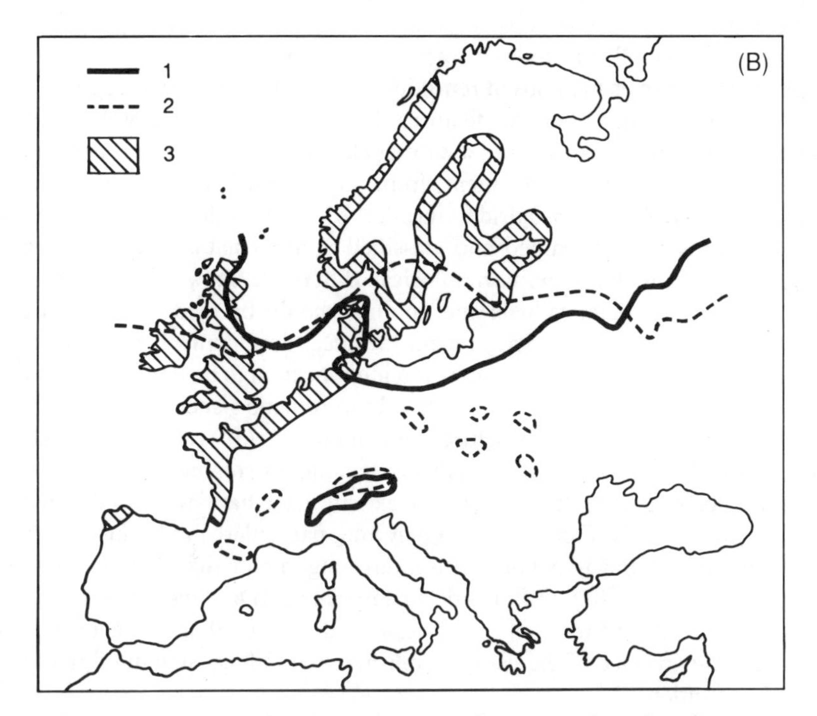

Fig. 4.5(B) General limits of distribution or range of representatives of northern montane and oceanic elements in the British flora. Montane: (1) *Linnaea borealis*; (2) *Salix phylicifolia*; oceanic: (3) *Myrica gale* (compiled from Matthews 1955)

2. *Migration*: direction and rate of dispersal and establishment of organisms;
3. *Environmental evolution*: changes in geological, geomorphological and climatic conditions during the course of organic evolution and migration.

ORGANIC EVOLUTION

As already noted, the basic organic taxonomic unit is the species. The problem of defining a species is one with which biologists have wrestled for a long time without reaching a unanimous solution. For many it is merely a convenient classificatory unit based on all the relevant information available whereby one population of organisms can be distinguished from another. Others accept the concept of the biological species defined by Mayr (1970) as groups of actually or potentially interbreeding populations that are reproductively isolated one from the other. Isolating mechanisms include, on the one hand, differences in times and methods of mating courtship in animals, or pollination in plants, or the production of sterile hybrids; on the other hand, differences in ecological requirements or niche width. Finally, the range of variation within a species population should theoretically be less than that between it and any other species population.

Morphology (anatomy) is still the principal diagnostic character by which individual organisms are described and classified. In this respect plant forms are not so complex or diverse as those of animals and many different types of plants share the same overall organisation into roots, stems, flowers, etc. However, in both plants and animals methods of reproduction and the form of the reproductive organs are the most important features by which organisms were initially organised into what was considered a natural classification which was assumed to reflect the evolutionary development from lower primitive to higher, more advanced organisms. The visual identification of the morphological features by which organisms are still named and classified is of its nature, very subjective. Decisions as to the level of morphological difference at which a distinction between one group of individuals and another should be made have been and continue to be arbitrary. However, recent technical developments in microscopy have enabled more accurate description and identification to be undertaken as well as making available other types of taxonomic information such as: (1) chemical make-up; (2) chromosome structure and number; (3) genetic information (Heywood and Moore 1984). Such analytical techniques combined with the use of high-speed computers to store, sort and classify data have witnessed a recent resurgence in the study of organic taxonomy and, particularly, in its application to genetics and to evolutionary studies. It is interesting to note that chemo-taxonomy (Briggs and Walters 1984) and genetic fingerprinting (Dawkins 1988) by which the ancestry of organisms can be investigated are tending to confirm the evolutionary significance of the early-established hierarchical system of plant and animal classification.

Speciation

Speciation is the evolutionary process whereby new species come into being. Although there continues to be considerable, and often acrimonious, debates about the mechanisms involved in organic evolution, the majority of biologists today accept Darwin's basic hypothesis that species have evolved from a common ancestral stock as a result of natural selection of those whose attributes which best fit them to survive (i.e. to compete for essential but limited resources) in a given habitat. The process of speciation (see Fig. 4.6) can be gradual or abrupt (Briggs and Walters 1984; Barton 1988). The former can result from the extension or contraction of a species' range, resulting in some of an original species population being subject to a change in environmental conditions, followed by either geographical and/or genetic isolation of the surviving variants. One of the most important means of abrupt speciation is *polyploidy* or the natural production of plants containing more than the normal double set of chromosomes. This can be the result of either a doubling of the parents' sets in the offspring (i.e. autoploidy) or hybridisation between individuals from two closely related populations usually after polyploidy (i.e. alloploidy). However produced, polyploid individuals are usually larger, more vigorous and competitive than the diploid parents and have a wider range of tolerance to environmental conditions. They can either displace all or some of the parent populations and/or colonise less favourable habitats. Polyploidy is common in perennial flowering plants and in the older cereal cultivars. It is perhaps significant in this respect that polyploids are particularly

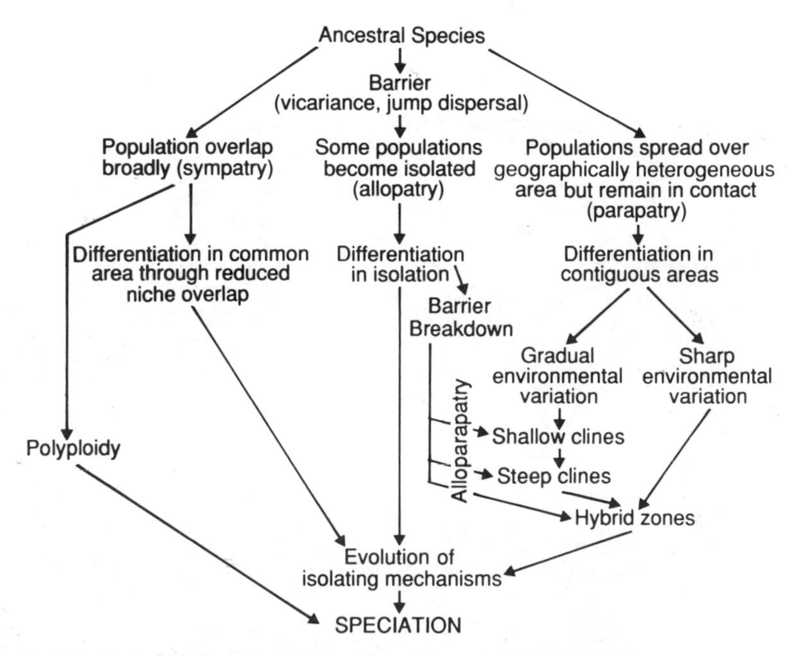

Fig. 4.6 Principal pathways of speciation (from Myers and Giller 1988)

prevalent in cold and arid regions of the world where they have given rise to new species better adapted to extreme climatic conditions than their parents.

Centre of origin

The geographical area in which a species (or any other taxon) originated and from which it has spread is difficult to establish. A number of criteria by which it could be identified have been suggested (Cain 1974), but none have proved universally acceptable. The first and most frequently used is diversity, since the evolutionary centre of origin of a taxon might be expected to possess a greater diversity of forms than in its periphery. Diversity of forms can, however, be a function of variety of habitats. Also, in the course of evolution the original centre may be 'lost' and two or more secondary centres arise as a result of polyploidy, hybridisation and/or adaptive radiation. Diversity combined with a concentration of what are assumed to be more primitive forms of the taxon have been used in the location of the centre of origin of particular genera (Holland 1978) (see Fig. 4.7).

To assess the time of origin of a species and its subsequent rate of evolution and dispersion recourse must be made, in all but the most recently developed species or subspecies, to the fossil record (Hallam 1972, 1977b). This involves the identification of macrofossils and microfossils. The former are the remains of the organs, some preserved as imprints, some in a petrified or organically preserved condition. The latter comprise plant spores and pollen grains preserved in anaerobic sediments and peat deposits. The age of such fossil material must be dated relative to an already established geological (rock strata), geomorphological (lake varves) and/or archaeological (artifact) datum line. In the case of younger

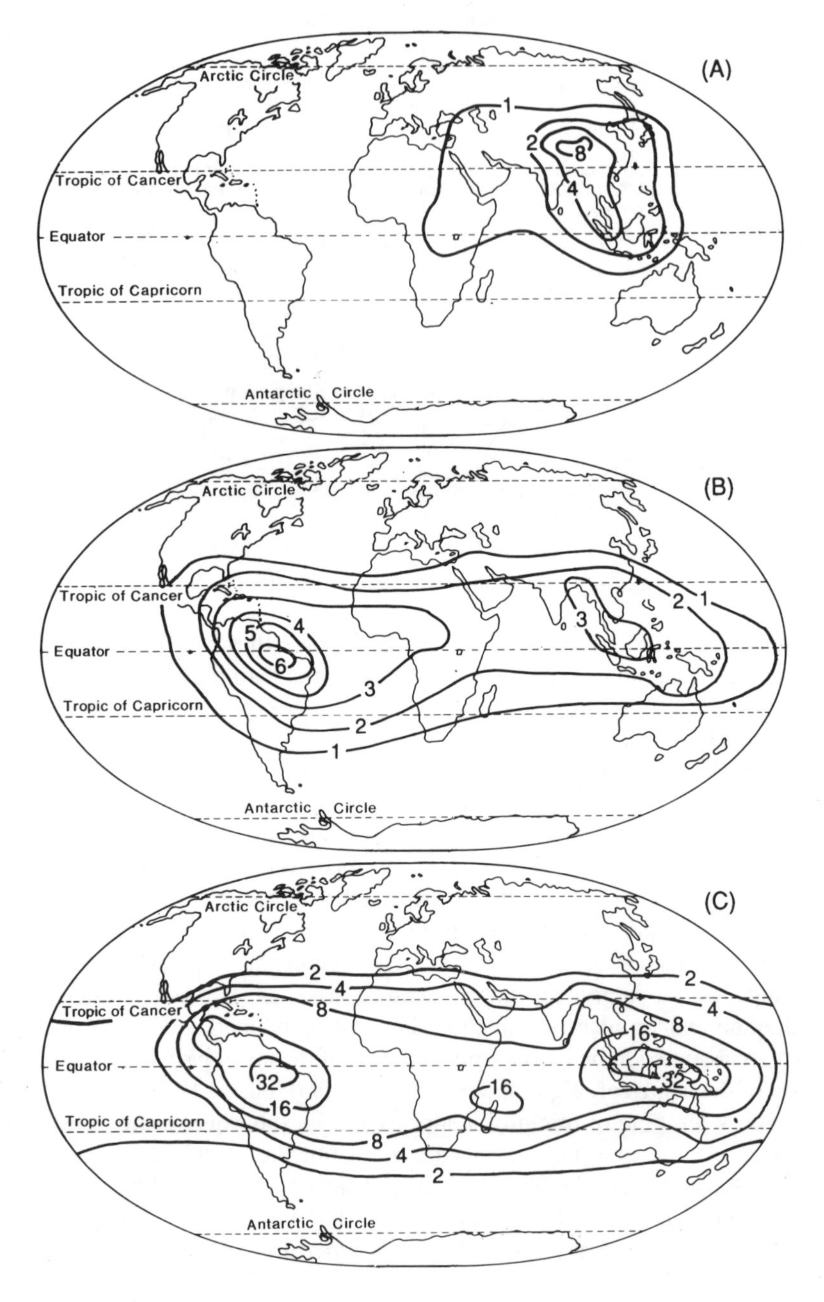

Fig. 4.7 Centres of origin of (A) genera of pheasants (Phasiainanae); (B) species of *Crocidilia*: (C) genera of palms (Palmae); the isopleths join points of equal taxon density (from Pielou 1979, adapted from Stehli 1968)

organic material the known rate of decay of its constituent isotopes of carbon-14 can be used to give a date within an accuracy of a few hundred years. However, despite the technical advances in identification and dating of fossil material, the problem of establishing the generic, taxonomic and ecological relationship between it and existing organisms is difficult and, as in all retrospective methods of study, the record is not spatially or temporally continuous; nor is it amenable to verification.

MIGRATION

The absence of an organism from an otherwise suitable area may be because insufficient time has elapsed since the taxon evolved or the habitat became available. Migration from one area to another is dependent, first, on the dispersion of individuals and, second, on their successful establishment (ecesis) in the new habitat. The generally held opinion is that little variation in the distribution of a species is a result of dispersal capacity. Krebs (1985), however, points out that there is not much empiricial or experimental evidence to support this view. He also notes that common, widespread plant species often have a greater dispersal capacity than those with a more restricted distribution. However, while dispersal capacity may not be an important factor limiting distribution at the local scale, it can be critical at the global scale. The existing world biotic distributions are the result of the dispersal of organisms from their initial centres of origin in the course of which taxa are brought into contact with new habitats and with each other. Migration is an important process in evolution.

Plants and animals vary in their methods and rate of long-distance dispersal, as well as in their ability to cross physical or ecological barriers. They can spread by means of either vegetative (cloning) or of sexual reproduction. The former is slow and has the disadvantage that shoots (clones) from the parent stocks do not possess the high genetic variability of seeds or spores. Hence they lack the potential to adapt to variations in new habitats that may be encountered as they spread. Seeds and spores are generally more mobile and can be rapidly dispersed, often over very great distances. The efficiency of seed and spore dispersal is dependent on a number of factors of which the most important are: the number and size of seeds produced; the age and frequency of seed production by the parent; the means of transport (i.e. wind, water, animals, man).

While animals, in contrast to plants, are mobile, they do not possess such specialised means of dispersal (Elton 1966); and, in any case, their natural power of movement is directed towards finding food, water and mates. Small animals, such as insects, which have larval stages, can be transported easily by wind. However, dispersal of animals is generally slow and motivated by changes in habitat. Large-scale rapid dispersion is more often associated with pressure of population on food resources as in the case of locusts, grouse and lemmings.

The dispersal of an organism from one ecologically suitable habitat to another may, however, be prevented by intervening ecological barriers whose height and/or width may present obstacles to movement. One of the most effective barriers is salt water. Very few land or freshwater plants have seeds adapted to long periods of immersion in salt water and it is also difficult for most terrestrial mammals (other than bats) to cross oceans. In addition high mountain ranges and extensive deserts also present insuperable barriers to the migration of many species. The efficiency (or effectiveness) of any barrier, however, will depend on the scale of the barrier and on the migrational ability of the organism involved. Also, although many organisms can cross barriers, successful migration is dependent on the one hand on the establishment and survival of sufficient individuals to form or initiate a breeding population. In this respect animals require male and female individuals, while many plants are dioecious. On the other hand, successful migration is dependent on the presence of a vacant niche in the new territory or of immigrants with a competitive ability to displace existing organisms.

Variant or allopatric speciation

Species migration or range extension will obviously involve only part of the potential parent gene pool. Because of the particular gene mix of the migrants and the selective pressures of an environment differing from that at the centre of the ancestral range, a new *allopatric* (or *vicarious*) but closely related species can evolve (Barton 1988). In areas which are difficult of access and geographically isolated, the absence of competition provides opportunities for speciation by divergent evolution or adaptive radiation, i.e. the multiplication of species from the same ancestral stock (Diamond 1984). If the variety of available unoccupied habits and niches is large, the initial immigrants may give rise to a large number of species of varying form and habit each exploiting different though sometimes overlapping niches. The classic examples are Darwin's finches on the Galapagos Islands and the honey-creepers on Hawaii (Cox *et al*, 1976). *Adaptive radiation* is also common among flowering plants with a high dispersal capacity. In isolated habitats, beyond the disperal range of trees, relatively small herbaceous plants have 'spawned' a variety of species with shrub and/or tree form. There are five different tree species (4–6 m high) on St Helena which have evolved from different types of sunflower. Similarly lobelias, in the absence of orchid competition, are represented by some 150 species, ranging from trees *c.* 9 m tall to herbaceous plants (less than 1 m). Adaptive radiation is most commonly, though not invariably, exemplified on large oceanic islands sufficiently isolated and stable to ensure the persistence of the species (Cox *et al.* 1973). The significance of islands for adaptive radiation will be considered more fully in Chapter 14.

Conversely, convergent evolution may give rise to unrelated species with similar forms in similar but widely separate habitats.

ENVIRONMENTAL CHANGE

During the long period of organic evolution environmental conditions have undergone changes varying in magnitude, intensity and rate. Environmental change has been an important stimulus to evolution as well as a major factor in the existing distribution of plants and animals. At the global scale the three main types of change involve: (1) the relative location of land and sea; (2) climate; (3) sea-level.

Continental drift

It has now been firmly established that the relative position of existing continental land masses has been changing continuously at the same time as organisms have been evolving. Recent geological studies in plate tectonics have confirmed that the lithosphere is composed of six to ten separate, rigid plates of igneous rocks some 100 km thick composed of one or two layers. The upper layer which underlies the continental masses, including the continental shelf, is made of a granitic-type rock (sial). This rests on a lower layer of denser basalts (sima) called the asthenosphere which is completely submerged. The tectonic plates, on which the continents rest, have moved and continue to move relative to each other as a result of:

(a) sea-floor spreading along and away from mid-oceanic fissures;
(b) convergence and collision of plates with the subduction of one beneath the other with consequent uplift, faulting and folding and the formation of oceanic volcanic islands;
(c) shearing and rotation of plates along their main fissures.

The accompanying movement of the continents, i.e. *continental drift*, has had a profound effect on the present distribution of plants and animals (Flessa 1980). As indicated in Fig. 4.8, over the period of organic evolution the continents have undergone alternating periods of separation and combination to varying extents and degrees. Combination has allowed more extensive migration and mixing of species. Separation has facilitated speciation as a result of the isolation and of vicarious evolution and adaptive radiation. When new land connections were formed land biota converged while marine organisms diverged and vice versa.

Fossil evidence indicates that in the mid-Permian (Palaeozoic) period there was one supercontinent, Pangaea, with a relatively uniform, warm climate and a relatively uniform flora and flora. Some time during the Mesozoic, Pangaea began, as a result of continental drift, to break in two, forming Laurasia (existing North America and Eurasia) and Gondwanaland (existing Antarctica, Australia, New Zealand, South America, Africa, India). The separation of continents into eight land masses would seem to have been most extreme in the late Cretaceous–early Tertiary period when fragmentation and separation was reinforced by high sea-levels. The split-up of Gondwanaland appears to have started earlier than that of Laurasia, consequently continental drift had its most profound effects in the southern hemisphere. While there is general agreement as to the broad outline of events, Pielou (1979) points out that there are many unresolved contentions of biogeographical significance including:

1. The existence of a single continent, Pangaea;
2. The contiguity of the east coast of India with Australia and Antarctica before the break-up of Gondwanaland;

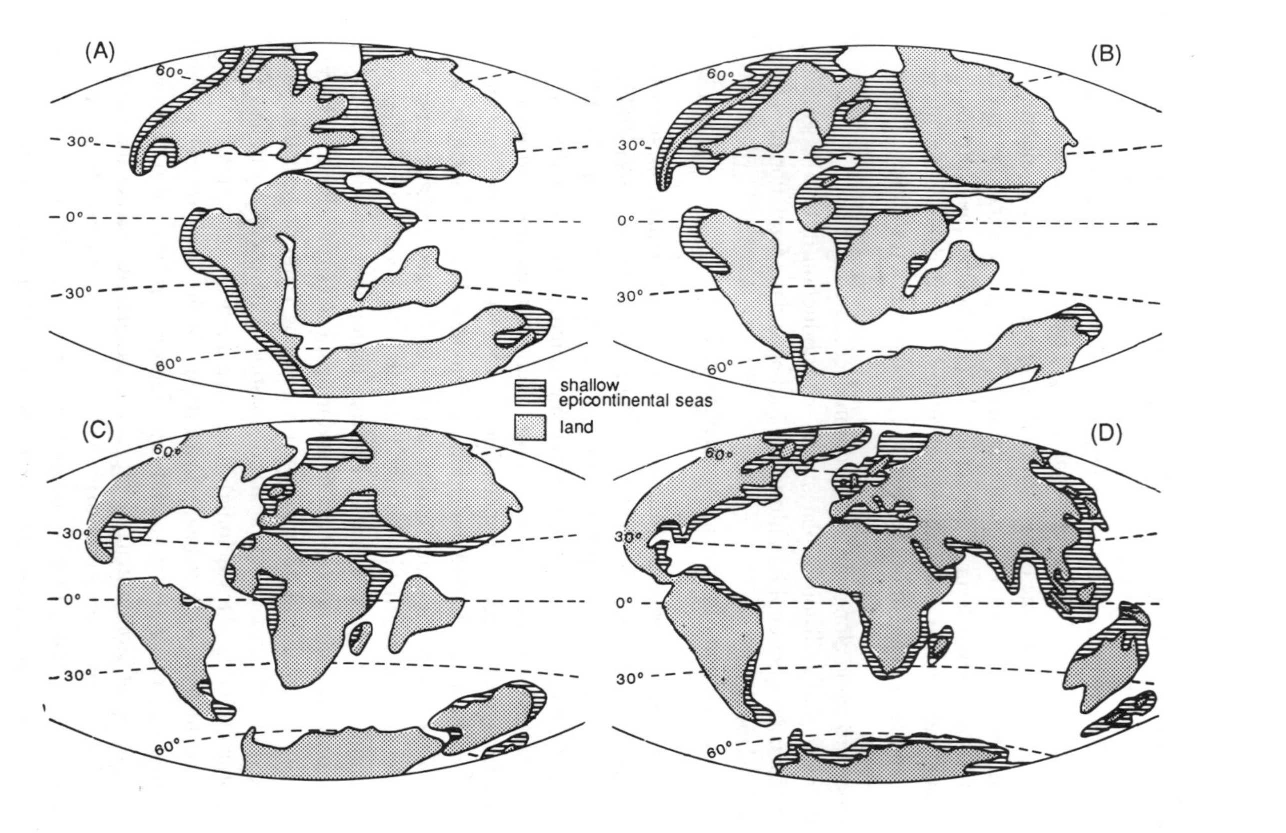

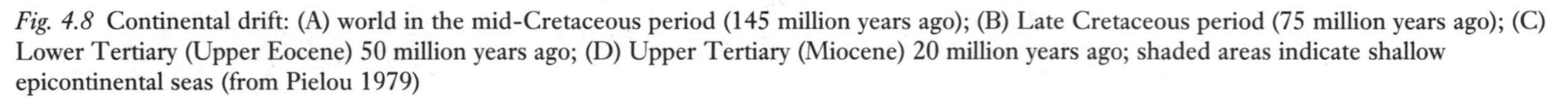
Fig. 4.8 Continental drift: (A) world in the mid-Cretaceous period (145 million years ago); (B) Late Cretaceous period (75 million years ago); (C) Lower Tertiary (Upper Eocene) 50 million years ago; (D) Upper Tertiary (Miocene) 20 million years ago; shaded areas indicate shallow epicontinental seas (from Pielou 1979)

3. The history of the Tethys Sea and its existence as a dispersal route which expanded and shrank in the geosyncline now occupied by the Mediterranean;
4. The history of the Central American isthmus;
5. The junction between west Alaska and Siberia;
6. The origin of the Himalayas as an intracontinental range rather than a plate collision zone.

Continental drift was accompanied by climatic changes consequent on variations in the latitudinal position of the continents: the maritime influences of new intervening oceans and the formation of mountain ranges. As a result differences in temperature between temperate and tropical areas and in rainfall between humid and desert areas replaced the more uniformly warm climate that had persisted up to the early Tertiary period.

Floral distributions

The fossil record indicates that the evolution of plants was initiated earlier than that of animals and that by the early Cretaceous period the flowering plants (angiosperms) had attained their present dominance (i.e. exceeding all other groups in numbers of species and individuals) before the emergence of mammals (Valentine 1972, 1978). Primitive angiosperms had a poor dispersal capacity in comparison to the modern more advanced members of this group. The latter have spread to all parts of the world and their speciation has been relatively rapid. In contrast, the former have been able to persist only in very restricted areas (*refuges* or *refugia*) where habitat and particularly climatic conditions have remained the same as when they evolved (Lynch 1988). These are interpreted as *relics* or *old endemics* (palaeo-endemics) which according to fossil evidence formerly occupied a much wider range than they do today. They include evolutionary relics (Cox *et al.* 1973) such as the cycads (Cycadaceae) and the primitive gymnosperms. The former – the so-called 'tree-ferns' – were among the most important elements of the vegetation in the Mesozoic period. Today, however, they are represented by only 9 families and 100 very rare species with restricted distributions in tropical and subtropical regions. They failed to survive in the northern hemisphere where they were replaced by the angiosperms in warmer and by the 'modern' cold-resistant conifers in the colder circumboreal part of the Holarctic region. Today they are represented by scattered relic families of the Araucariaceae (Chilean pine), the Podocarpaceae (including *Nothofagus* spp. or the southern beech), the Proteaceae and the Winteraceae often in very disjunct areas which are almost exclusively in the southern hemisphere (see Fig. 4.2). These and some angiosperm families have markedly disjunct ranges in South America and South Africa, and in South Africa and Australia which testify to the relatively early fragmentation of the formerly continuous land area of Gondwanaland. The floral affinities between these three areas are further emphasised by the fact that the distribution of six families of angiosperms is restricted to the three southern continents while 700 species are more or less confined to two or three of them. While these affinities are a product of a former contiguity of the land masses, indigenous groups are a result of isolation. This is most strikingly shown in the dominance of the Australian genus, *Eucalyptus* (*c.* 450 species) unrivalled elsewhere in the world; and the diversity of the families Restionaceae, Proteaceae and Myrtaceae (with 630 species of *Acacia*).

Fragmentation of Gondwanaland appears to have commenced earlier than that of Laurasia. As a result, continental drift has had more pervasive effects on both the flora and fauna of the southern hemisphere than of the northern. Even after the North Atlantic opened up, the connection between western Alaska and Siberia by the land bridge (Beringia) across the Bering Strait seems to have persisted for some time. Hence, as well as regional floral differences, there are close relationships between the broad-leaved temperate and coniferous forests of the now disjunct areas of the Holarctic region and many groups have a cosmopolitan, i.e. circumboreal distribution.

Faunal distribution

While the first mammals appeared in the Triassic period, their evolution by adaptive radiation did not take place until the late Cretaceous, after the break-up of Laurasia and Gondwanaland. The emergence of wide oceans imposed a greater restriction on the dispersal of all but the small and/or flying animals than on plants. Migration of the former became dependent on corridors where the distance between the larger masses was not impassible and/or where land bridges or island stepping-stones existed for longer or shorter periods of time. In this respect the later break-up of Laurasia (already noted) and particularly the persistence of the Beringia Corridor between western Alaska and Siberia allowed the movement of animals between Asia and North America and later, as the climate became warmer and the Tethys Sea began to shrink, between Europe and Asia (Cox 1974). The North Atlantic barrier, however, made for an earlier isolation of Europe and eastern North America. As a result, the two Holarctic areas have many groups of endemic animals either because of isolation and vicariant evolution or extinction of a formerly common taxon from one of the continents. The horse and camel, wild pigs, the hedgehog and murid rodents (typical mice and rats) did not become native to Nearctica. The pronghorn antelope, pocket gophers and pocket mice did not reach Holarctica. While the Oriental and Ethopian regions remained in closer contact, the development of the Saharan and Middle Eastern deserts isolated the savannas of South Africa from Europe and northern Asia and resulted in the evolution of the distinctive endemic herbivorous mammalian fauna (giraffe, zebra, gnu, antelope and hippo) and of birds in the southern tropical grasslands.

Continental drift had a more radical effect on animal distributions in the southern than in the northern hemisphere (Cox 1974). This is reflected today in the existence of monotreme and marsupial animals which are exclusive to the South. Of the former, the platypus and the springtail ant-eater are found only in Australia; of the latter there are 170 living species in Australia and 70 in South America. Elsewhere these primitive mammals were displaced by the more advanced marsupials and eventually by advanced placentals. The distinctive nature of the Australian fauna is witness to a long period of isolation – even more marked in New Zealand – during which marsupials, protected from the placentals, survived and experienced adaptive radiation and the convergent evolution of forms ecologically equivalent to the placentals elsewhere: e.g. marsupial mice, rats, squirrels, jerboas, moles, badgers, ant-eaters, wolves. The placental ungulates, however, have no equivalent form in Australia where their place is taken by the kangaroois and wallabies. The only placental mammals are the dingo and rats whose early immigration may well have been closely associated with that of aboriginal man. In contrast, the isolation of South from North America was not so persistent and the periodic connections by island bridges allowed the southward migration of placentals; the only earlier mammals that have survived are primitive placentals (i.e. armadillos, tree-sloths and ant-eaters) and the marsupial possums.

CLIMATIC CHANGE

While the continental drift was accompanied by the zonal differentiation of the world's climate, gradual cooling during the latter part of the Tertiary period culminated in the Pleistocene glaciations (West 1968; Birks and West 1973) – a long period of extreme climatic variability associated with the advance and retreat of the Quaternary ice-sheets and the consequent alternation of cold glacial and warm interglacial periods across the northern hemisphere (Fig. 4.9). At the same

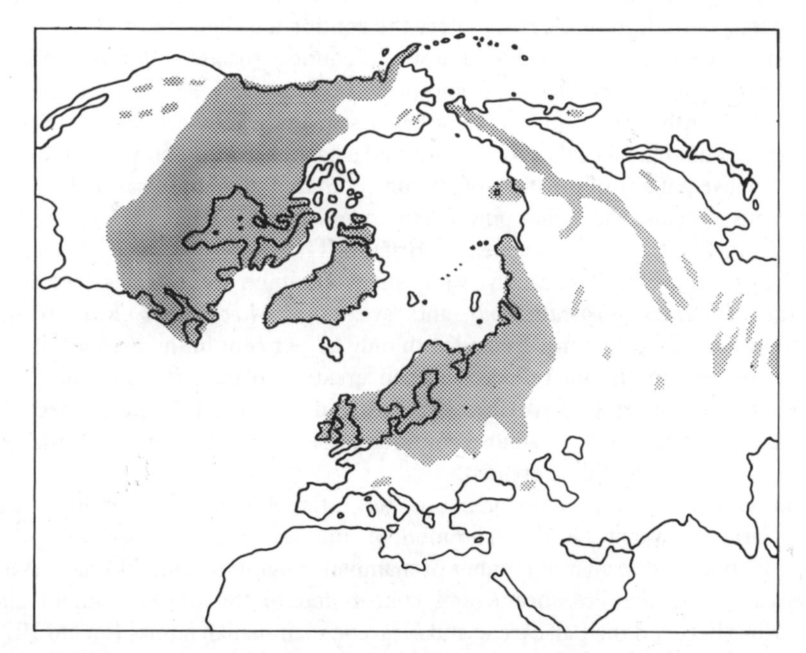

Fig. 4.9 Total area covered by maximum extent of Quaternary (Pleistocene) ice-sheets

time there was a parallel extension of mountain glaciers in the North American Cordillera, the Alpine–Himalayan mountains and other high mountain areas in both temperate and tropical latitudes. Over most, if not all, of the areas covered by ice at varying stages of glaciation, the pre-existing vegetation would probably have been destroyed. The soil and its unconsolidated parent material was either completely stripped away and/or reworked in the processes of glacial deposition. Biotic convergence and mixing resulted in very active hybridisation zones and speciation, particularly in those areas where the retreating ice left unoccupied habitats.

Beyond the ice margins climatic zones and associated vegetation belts were displaced Equatorwards. There is evidence that the existing semi-arid and arid zones experienced more humid conditions and that the Equatorial region, at least in the Amazon Basin, was drier than at present (van der Hammen 1974). Macro- and microfossil (particularly pollen grains) material preserved in late and post-glacial organic deposits suggest that, with the final retreat of the ice-sheets *c.* 10 000 years ago, climatic conditions ameliorated and the existing temperate and Boreal forest vegetation zones were re-established (Roberts 1989). In addition, the advance and retreat of ice were accompanied by eustatic and isostatic changes in sea-level varying from *c.* 160 m lower than at present to *c.* 70 m above OD. This resulted in the alternating emergence and submergence to varying extents of the continental shelf and the consequent opening and closing of important intercontinental and interoceanic migration routes for terrestrial and marine organisms.

Fluctuations in the relative levels of land and sea were universal throughout the Pleistocene period. In those areas where the continental shelf is widest, a lowering of existing sea-levels by only 30–60 m would connect Britain with the Continent. The remains of submerged forests and peat deposits (moorlog) dredged from the bed of the North Sea are indicative of lower sea-levels. The establishment of the English Channel in post-glacial times formed a barrier to migration which resulted in the relative paucity of the biota of Britain compared to that of France. Only four coniferous species, the Scots pine (*Pinus sylvestris*), two *Juniperus* spp. and the yew (*Taxus bacata*) are indigenous to Britain. The Norway spruce (*Picea abies*) failed to reinstate itself naturally, while many common tree species (i.e. horse-chestnut (*Aesculus hippocastanum*) and sycamore (*Acer pseudoplatanus*)) have been introduced by humans. Ireland with only 67 per cent of the number of plant species present in Britain reflects an even greater isolation (Pennington 1969). With a similar lowering of sea-level the East Indies would become connected to Asia and Australia and an almost continuous land bridge between Alaska and Siberia (Beringia) would be created.

The emergence and submergence of the intercontinental land bridges were particularly important for the migration of animals in glacial and post-glacial times. It appears that a large number of mammals migrated from Siberia to North America during the Pleistocene and contributed to the preservation of close similarities between the Palaearctic and Nearctic mammalian faunas (Pielou 1979). Alaskan and Canadian species common or identical to those in Siberia include the snow sheep (*Ovis nivicola*), wolverine (*Gulogulo luscus*), wolf (*Canus lupus*), musk ox (*Ovibos moschatus*), moose (*Alces alcea canadensis*), brown bear (*Urus Arctos middendorfi*), red fox (*Vulpes vulpes*) and arctic hare (*Lepidus timidus*). Among the species which diverged south were the plains bison (*Bis bison*), bighorn sheep (*Ovis canadenisis*), Rocky Mountain goat (*Obea auros montanus*), grizzly bear (*Ursus horriblis*), black bear (*U. Americanus*), coyote (*Vulpes macrotis*) and bobcat (*Lynx rufus*). The most recent mammal to cross this route has been human.

Glacial relicts and refugia

Glacial advance and retreat during the Pleistocene were accompanied by the advance and retreat of biotic zones. The pre-existing flora and fauna were eliminated or replaced except where habitat conditions provided suitable refugia in which the former could persist. The disjunct distribution of similar plant species in both coastal and mountain areas of Britain is thought to be a relic of a formerly more widespread tundra vegetation in the late glacial period. Their present distribution is now restricted to bare, open and frequently unstable habitats in which they have been able to survive the competition of the forest vegetation which was re-established as post-glacial climatic conditions ameliorated. In North America glacial relict flora and fauna have persisted in the so-called 'sky-islands', above the tree-line (*c.* 10 000 m OD) on mountain peaks in Arizona and New Mexico. In the now arid or semi-arid areas of the USA (Nevada) and Africa (Kenya) relict populations of freshwater animals have been found in water-holes – remnants of more extensive water bodies which existed in interglacial pluvial periods.

It has also been postulated that, during the periods of glacial advance and retreat in the northern hemisphere, the tropical rain forest contracted and expanded. During the former periods, the forest became fragmented into isolated 'islands' in a predominantly grassland area (see Fig. 4.10). These islands provided forest refugia for indigenous and in-migrating animals. Alternating forest isolation and convergence stimulated speciation, resulting in the very high degree of diversity characteristic of the present tropical rain forest. This hypothesis first

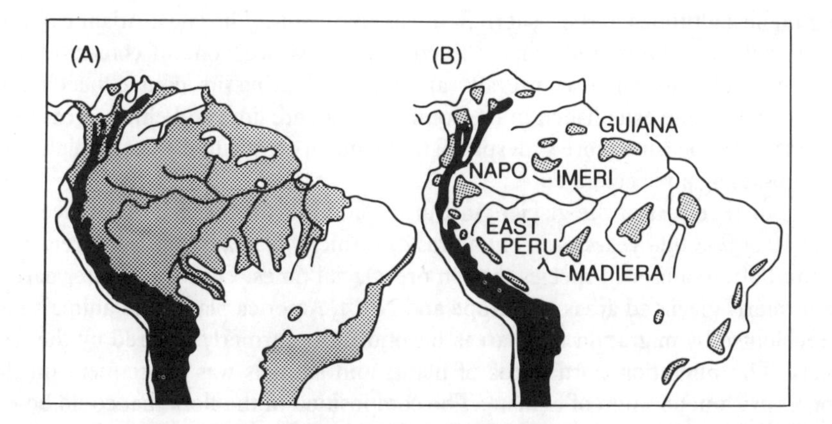

Fig. 4.10 Distribution of tropical rain forest (lowland) in South America at (A) present time and (B) forest refuges during Pleistocene glacial maxima. Continental boundaries have been adjusted to reflect changing ocean levels; Andes (above 1000 m) black. In both maps the general outline of the Amazon drainage is preserved so as to allow comparisons (from Lynch 1988)

applied to bird species, was extended to other animals and latterly to the whole biota. It provided an alternative hypothesis to the long-held contention that the diversity of the tropical rain forest was a function of a long period of environmental stability. However, the refugia hypothesis is difficult to test and has been variously accepted or rejected by biologists (Lynch 1988).

Post-glacial chronology

Identification, dating and analysis of fossil pollen (Godwin 1975; Birks and West 1973; Stott 1981) have allowed a relatively detailed reconstruction of post-glacial biotic changes in the temperate areas of the northern hemisphere. The vegetation chronology which has been established suggests continuous climatic fluctuation of drier continental, alternating with more humid oceanic, phases rather than a gradual amelioration of glacial and periglacial conditions to those of the present.

The early Boreal period, somewhat warmer and drier than at present, is marked by a predominance of pine and hazel in England. This period, sometimes referred to as the xerothermic period, was paralleled in Europe by the northward migration of many characteristically Mediterranean plants and, in North America, by a movement eastwards of prairie grassland species. Post-glacial climatic conditions appear to have attained their optimum in the succeeding Atlantic period when temperatures remained higher than at present but were accompanied by increased humidity. Deciduous forest reached its maximum extension and greatest development in north-western Europe; in Britain elm (*Ulmus*) became more prevalent than at any other stage and lime (*Tilia*) attained its maximum northward range. In France the beech (*Fagus*) markedly increased its area. In cooler areas, such as north-west Britain, higher humidity favoured the increased development and extension of peat-bogs. There is, however, evidence that these became drier and were recolonised by birch and pine in the ensuing somewhat drier sub-Boreal phase (another more continental climatic phase) when pine and birch ascended to their highest altitudes in the eastern Scottish Highlands. The latest Atlantic phase, though subjected to minor climatic fluctuations, has been one of cooler, wetter, oceanic conditions such as prevails at present. During its course beech and hornbeam became established in the south of England, lime and elm decreased in amount, birch became more widespread in the north-west and the accumulation of peat mosses again accelerated.

The final retreat of the glaciers in Europe and North America took place only some 10 000–8 000 years ago. The floras of formerly glaciated areas are relatively new and often poorer in species than in pre-glacial times. Over the greater part of the formerly glaciated areas in Europe and North America plants and animals had to recolonise by migration from areas beyond those formerly covered by the ice-sheets. The migration northwards of plants and animals was consequent on the progressive amelioration of climate. The composition of the flora that could be re-established was, however, dependent on the availability of seed stocks and the existence of routes by which migration could take place. It is noteworthy, for instance, that the vegetation of eastern North America and eastern Asia is much richer in number of plant species than is that of Europe. In the latter area many Tertiary species did not survive the pre- and full-glacial climatic deterioration. Just as the transverse barriers of the Sahara Desert and the Alpine mountain ranges blocked escape routes southwards, similarly in the post-glacial period they cut Europe off from the reservoir of plant species that existed further south. In contrast, no such transverse barriers existed in North America, so many pre-glacial species were able to find refuge further south, particularly to the south-east. Thus the reservoir of possible plant colonists was much richer and eventual migration northward in post-glacial times was unimpeded. Such events help to explain the discontinuous ranges of the redwoods (*Sequoia*), the tulip tree (*Liriodendron*), and the swamp-palm (*Nipa*) (shown in Fig. 4.11) families which were obliterated in the southward plant migrations in Europe, but survived in south-east USA and Asia. They also help to explain why tropical Africa is much less rich in species than similar areas in South America or South-east Asia. The effect of relief on post-glacial migration is also strikingly demonstrated in British Columbia. In terms of present climatic conditions the dominance of coniferous trees and the relative paucity of broad-leaved deciduous and evergeen species is somewhat anomalous. It has been suggested that the longitudinal and very effective barrier of the Western Cordillera cut the area off from the rich plant reservoir of the south-eastern USA. Restocking of the north-west then came from the mountain ranges further south where conifers in particular had been able to survive.

SPECIES EXTINCTION

Millions of species of organisms have become extinct in the course of evolution. And while there is a continuing debate among biologists as to whether organic extinction has a purpose, many would regard it as an inevitable concomitant of speciation. Extinction alone obviously results in a decrease in species diversity. However, it may result in niche release and provide an opportunity for the establishment of new species. Susceptibility to extinction varies. In general, land plants are less vulnerable than animals (Marshall 1988). Their resistance is closely linked to their reproductive abilities which allow them to survive severe reduction of their biomass. Among animals the most vulnerable to extinction are those with a large density and low birth-rate, poor dispersal abilities, restricted range and a specialised diet or habitat and which belong to the highest trophic level.

The causes and processes of extinction are varied. They include competition; the breakdown of predator–prey trophic relationships; prey naïvety as a result of immaturity or the introduction of an unfamiliar predator into a habitat; natural catastrophic events such as volcanic eruptions, magnetic reversals, meteorites and comet showers. While environmental changes have long been considered the main causes of extinction, the effect of human activities at the time of the Pleistocene upheaval has recently been given more emphasis. The mass extinction of the large vertebrate mammals, such as the woolly rhino, mastodon, mammoth, giant beaver and sabre-tooth tiger which were extant in the late Tertiary and early Pleistocene periods has been variously attributed to either climatic change or to overkill by prehistoric man (Bunney 1990). Hunting pressure by man at a time when these

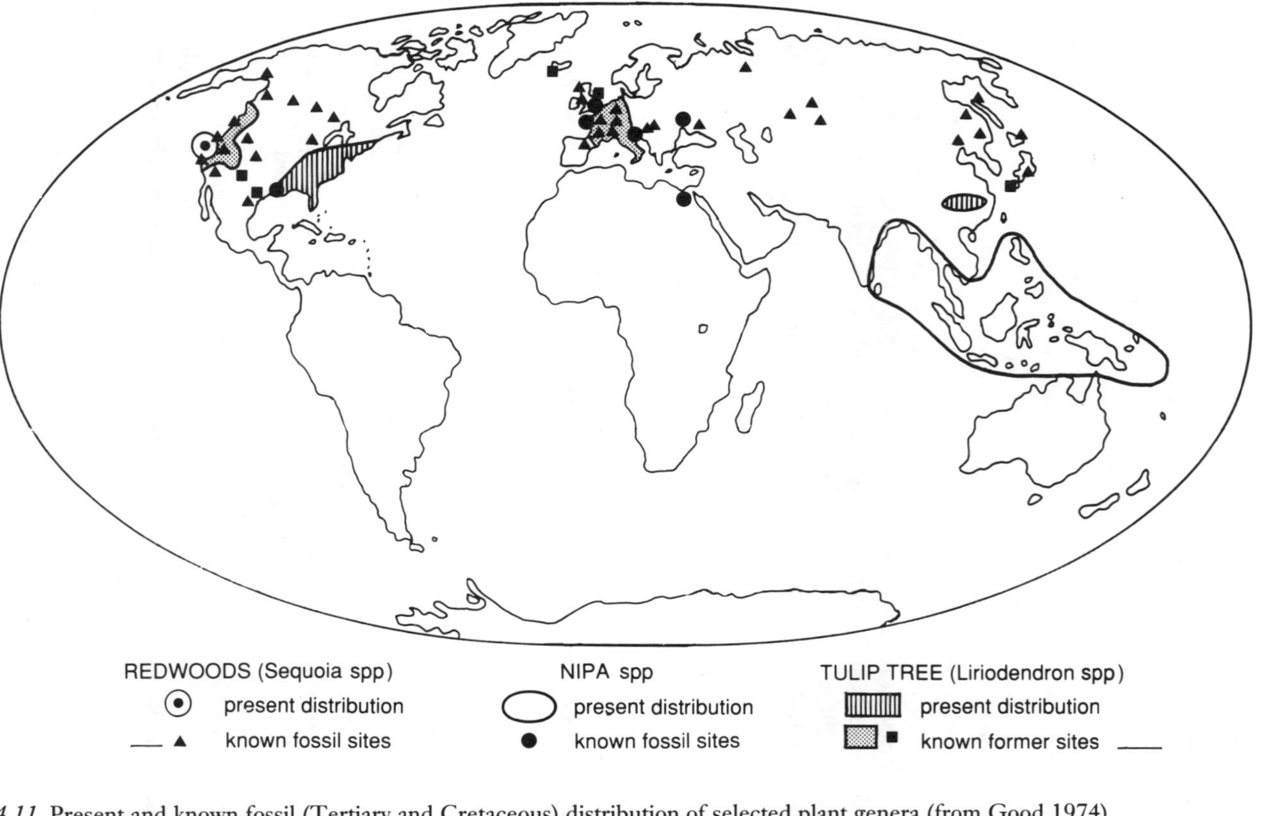

Fig. 4.11 Present and known fossil (Tertiary and Cretaceous) distribution of selected plant genera (from Good 1974)

large animals were becoming more vulnerable due to deteriorating climate and decreasing food supply may, however, be nearer the truth.

THE EFFECT OF HUMANS ON ORGANIC EVOLUTION, DISPERSAL AND MIGRATION

The more recent course of organic evolution and distribution has been no less eventful than that of the more remote geological past. Indeed, it has been more dramatic and spectacular. Within the brief span of some 40 000 years human activity has triggered off a new phase of organic upheaval which is continuing with increasing intensity and during which the rate of evolution has far outstripped that in any comparable period of time in the past. Evolution has been influenced in two interrelated ways. First by human selection and propagation of individual variants in a species population. Second, by the creation of new 'disturbed' or 'hybrid' habitats. In relatively stable, fully occupied habitats the survival of variant offspring is difficult, without segregation from the parents, because of continual back-crossing or introgressive hybridisation. Also, in the absence of a suitable niche in which they would have a competitive advantage it is difficult if not impossible for them to become established and maintain their genetic identity.

Fresh habitats with open and unstable soil conditions are a natural and continual product of those areas subjected to marked erosion or deposition at present. They are characteristic of areas directly affected by river action, by coastal erosion or deposition, or by slumping and sliding on steep mountain slopes. However, the extent and diversity of such naturally disturbed habitats are infinitesimal in comparison to those which have been, and continue to be, produced by human activities. These are the 'hybrid' habitats born of the interaction of human beings and their physical environment; they provide sites in which fertile hybrids and new variants could become established. They are characteristically open, exposed to full sunlight and not infrequently enriched in organic and mineral matter as a result of human activities. The simultaneous evolution occurred not only of species which were adapted to and could tolerate the disturbed habitats created by humans but also of those plants which they selected for their own particular uses. By the conscious or unconscious selection of certain species (and moreover of particular variations and strains within them) best suited to their needs, cultivated crops and domestic livestock were 'created' by the process of domestication.

As well as being an evolutionary catalyst, humans have also become an ever more powerful agent of plant and animal dispersal and distribution. Their effectiveness as agents of dispersal has been gathering momentum particularly during the last 400–500 years with exploration, the growth of trade and the development of more and faster means of intercontinental communication. The latter have bridged the barriers that formerly restricted the range of many species. Many alien plants and animals which in their native homes are only minor or inconspicuous 'citizens', often become much more vigorous, aggressive and abundant in a foreign environment. They may even spread so rapidly and effectively as to compete with the natives. The cumulative effect of human activity has been to favour, either deliberately or accidentally, the evolution and distribution of some types of plants and animals at the expense of others. In Tertiary times, one region or habitat was isolated from the others by physical barriers. These barriers which formerly isolated habitats and allowed an independent evolution of plants and animals have now been broken down. Over considerable land areas a rich variety of different plants and animals has been replaced by the variants of a few specialised species which man deliberately protects or which can tolerate the modified habitat conditions he has created. In the process an increasing number of wild plants and animals have become extinct.

Chapter

5

Biomass characteristics

The species of plants and animals present in a given area constitute its flora and fauna (jointly its biota). However, neither a particular organism nor a species population normally occurs in complete isolation, even in assiduously cultivated habitats. All organisms live in company with a greater or lesser number of others of the same or different species populations and together form a *biomass* or *standing crop*. This is the total amount of living material usually expressed as the mean total of dry weight (tonnes) or dry weight (g or kg) per unit area (m^2). In terrestrial habitats a high proportion of the biomass is composed of plants, i.e. phytobiomass or vegetation. In contrast, in aquatic, particularly marine, habitats the animal biomass exceeds that of the plants. Except in estuaries, algae beds and coral reefs the mean plant biomass per unit area of aquatic habitats is comparable to that of the most extreme terrestrial deserts. However, because of the great extent of the oceans the total marine biomass is equal to that of the land.

BIOMASS VARIATION

The earth's biomass varies with environmental, particularly climatic, conditions, as reflected in the general gradient in plant biomass from forest to grassland tundra and desert. Forest and woodland still account for 90% (of which 70% is tropical forest) of the terrestrial biomass. In contrast, the marine phytobiomass decreases from coastal to open waters, and with increasing depth. In both land and aquatic habitats, and particularly in those areas with a marked variation in the growing season or subject to periodic natural disturbance, the biomass varies temporally. In addition, it can vary both spatially and temporally as a result of direct or indirect human activities. As well as varying in quantity the earth's biomass also differs in the number of constituent species, in species composition and structure, in the abundance (size) of species populations and in the life forms of which it is composed.

SPECIES RICHNESS AND DIVERSITY

Species richness (or poverty) is a function of the number of species present in given area or habitat. While it is often used as a measure of species diversity it fails to take account of the relative abundance (i.e. the heterogeneity) of the species populations present. Assessment of either richness or diversity is complicated by problems of scale and of sampling. The larger the area the greater will be the number of species contained within it. Hence the appropriate minimum size and number of sample areas must be related to the scale at which studies are undertaken. In this respect, Whittaker (1975) distinguishes between *α-diversity* (within habitats) and *β-diversity* (between habitats). The problems of sampling are further complicated by the tendency for some species populations to be highly aggregated while others are more evenly distributed within the total biomass. In addition, species individuals naturally vary in size and it might be expected that the number of young would be greater than that of old individuals and of naturally small-sized species to be greater than those of larger size per unit area. Finally, there is likely to be a range in the number and abundance of species populations in the same area.

A variety of *indices of diversity* have been proposed which attempt, by statistical analyses of species–abundance curves, to weight the relative importance of species according to population size (Krebs 1985; Pielou 1979; Whittaker 1975; Hill 1975) and assess the *equitability* (Putnam and Wratten 1984) of the distribution of species populations within the biomass, as for example:

$$S = \alpha \log_e (1 + N/_2) \tag{1}$$

where S is a number of species and N the number of individuals.

$$H = \mathrm{E}[p_1 \log(p_1)] \tag{2}$$

where $p_1 = n_i/N$ is the relative abundance of the ith species. The latter allows for increasing complexity as the number of species relative to number of individuals increases. However, as Crawley (1986) points out, it matters little which index is used since the two parameters, of species numbers and population abundance, summarise most of the important information on diversity.

Variations in species richness and diversity from one part of the global biomass to another have long been recognised. Comparative studies have documented differences particularly between tropical and temperate biota (with large species numbers in the former but large populations of fewer species in the latter); between disturbed and stable habitats; between mountains and lowland; and between isolated and accessible areas. However, neither the reasons for, nor the ecological significance of, such variations in species composition are fully understood.

The causal factors which have been proposed for the latitudinal diversity gradient, and are applicable to many other gradients, include evolutionary time, environmental conditions, biotic processes and human activity. The diversity of tropical compared with temperate biomass has been attributed to the more constantly favourable climatic conditions over a period of time sufficiently long to allow speciation and dispersal. It is argued that, in the absence of environmental constraints, the high niche potential of the humid tropics has been fully exploited by both plants and animals while the vertical and horizontal variations in microhabitat have undoubtedly contributed to the species richness of herbivores and their predators. Biotic processes such as competition, predation and dispersal may operate to increase or decrease diversity. On the one hand, competition for light as a result of shading between plant species may reduce diversity. On the other hand it may, together with predation, limit the size of animal populations and hence make for a larger number of actual niches and a greater species diversity than might otherwise exist. Indeed, it has been suggested that biotic stability and productivity may be a prerequisite for the development of diversity. However, empirical studies, particularly of plant diversity, indicate that under similar environmental conditions nutrient-deficient and disturbed habitats are often characterised by species richness rather than poverty. However, such apparent correlations do not explain diversity and it is probable that it results from the interaction of several factors whose relative importance may vary between plants and animals and from one habitat to another. As important as the causes of species diversity is its ecological significance and particularly its relation to ecosystem stability which will be considered in Chapter 16.

POPULATION ABUNDANCE

The individual members of a single species constitute its population. This term can be applied to: (1) the total existing population, or (2) a local genetically related group of individuals, i.e. a *deme* or *breeding population*; or (3) those species individuals which occur within a defined area (Watkinson 1986). Species populations vary in size or density, in distribution and in stability. Their numerical abundance at any one moment is determined, in the first instance, by the primary population parameters of birth (B), death (D), immigration (I) and emigration (E), i.e. $P = (B - D) + (E - I)$.

These parameters, however, are not constant. They vary between and within species populations. Differences in the former case are a function of the potential *fecundity* (fertility) (see Figs 5.1(A) and (B)), reproductive rate and longevity (survival rate) of the particular organism. Large animals have, in general, lower fertility rates, slower reproductive rates and live longer than small animals. Differences in the latter case are related to the actual 'fitness' of the individuals (i.e. their ability to reproduce and survive) which may either be genetically and/or phenotypically (i.e. environmentally) determined. All other factors being equal, it might be expected that a species population abundance would be highest where environmental conditions (physical and biotic) were most favourable for reproduction and survival. The primary parameters can also vary temporally. Dependent on variation in resources and/or change in niche width, natural increase ($B > D$) and/or immigration will result in population growth, while natural decrease ($D > B$) and/or emigration will cause a population decline. While the latter can result in the eventual extinction of a species from an area, the former obviously cannot be maintained indefinitely.

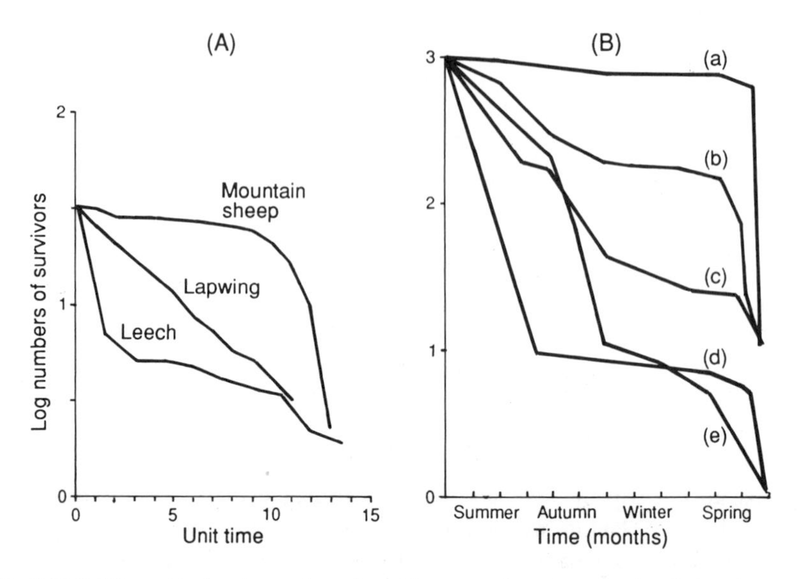

Fig. 5.1(A) Theoretical and actual survivorship curves for selected animals: leech (*Glossiphoma*) (Mann 1957); lapwing (Lack 1943); mountain sheep (Murie 1944) (from Putnam and Wratten 1984). (B) Survivorship curves for natural populations of four winter annual plants from seed production to maturity: (a) *Vulpia fasciculata* (2); (b) *Phlox drummondii* (23); (c) *Bromus rubens* (76); (d) *Minvarita uniflora* (305); (e) *Spergula vernalis*. Average number of seeds, i.e. plant fecundity, is given after species name (*n*) (from Watkinson 1986)

Under conditions unmodified or undisturbed by human activities many species populations attain a level at which they remain relatively constant for long periods of time. Provided there is no drastic environmental disturbance or rapid genetic drift, population numbers will tend to fluctuate around a mean which reflects the mean variations in the physical and biotic conditions. Populations, however, cannot increase indefinitely and a point is eventually reached at which either the rate of growth ceases and the population falls rapidly or growth slows down to a relatively constant rate. In the first instance, the model growth curve is J-shaped or geometrical, in the latter it is logistic or S-shaped (sigmoid) curve (Fig. 5.2A). In undisturbed environments the J-curve is characteristic of organisms with short seasonal cycles of reproduction and discrete generations, as in the case of annual plants, many insects and *lagomorphs*. The S-curve, the traditional albeit simple model, is characteristic of organisms with overlapping generations, longer and slower reproductive cycles and longer life spans as in the case of perennial plants, birds, mammals and some insects.

According to the logistic (S) curve the rate (r) of population growth must eventually be limited by the *carrying capacity* (K) of the environmental resources and numbers would be expected to attain saturation point at or just before this limit is reached. The equations which describe the logistic curve are as follows:

$$\frac{dN}{dt} = \frac{r^m N(1 - N)}{K} \qquad (3)$$

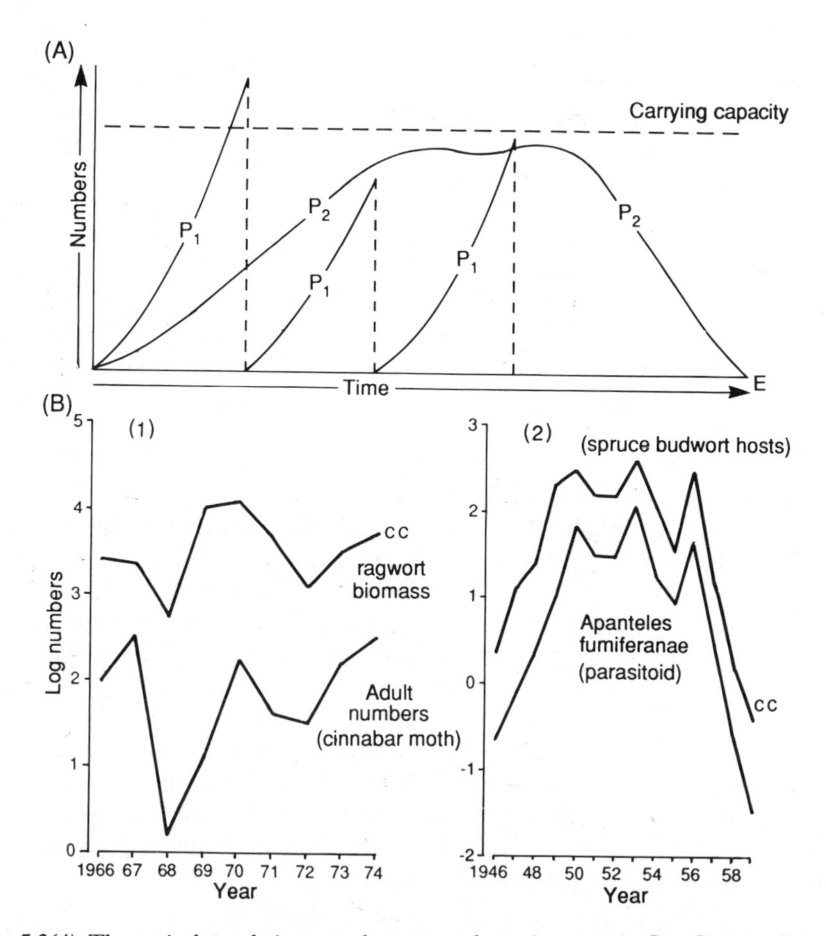

Fig. 5.2(A) Theoretical population growth curves and carrying capacity P_1 = J-curve; P_2 = sigmoid curve; E = extinction. (B) Variations in (1) herbivore and (2) predator numbers with respect to carrying capacity. CC = carrying capacity (adapted from Putnam and Wratten 1984)

where N is the number; t the time and r^m the rate of natural increase under fixed conditions;

$$N = \frac{K}{1 + a \exp(-r^m t)} \quad (4)$$

where K is the carrying capacity and a a constant defining the point at which the curve begins to flatten out.

The logistic curve is based on the assumption that, at any one time, all the organisms involved are demographically identical; that all have an equal probability of dying; that the rate of increase is determined only by the number present. However, given that the rate of increase is proportional to a decrease in the carrying capacity of the environment, the logistic equation needs to be modified to allow for the effect of other demographic parameters. The main weakness of all such logistic models is the difficulty of identifying and expressing the carrying capacity. It remains an essentially theoretical construct which, in reality, is not constant since the environmental variables which determine it are themselves subject to change through time. Hence, it is usually expressed in terms of the relative density of one of the environmental resources, such as in the case of the difference between the density of a predator animal and that of its prey (i.e. food resource). In this case the density of the prey would at any one time be a measure of the carrying capacity for a predator population (see Figs 5.2(A) and (B)). This concept of density is applicable to resources composed of discrete individuals of population of plants and animals. However, there is no comparable measure of the environmental resources on which plant carrying capacity depends. In addition the concept of density is misleading since it is expressed in population size unrelated to unit area or volume.

The *population saturation point* is not absolute but is rather a relatively constant mean value around which species population numbers fluctuate. Indeed, all plant and animal population numbers vary temporally to a greater or lesser degree, at shorter or longer intervals, at slow or rapid rates, with greater or lesser range. Also many organisms exhibit cyclic fluctuations that are not related to the length of their life cycles. Vertebrate cycles of 3–4 years for rodents (of which the arctic lemming and snowshoe hare are the classic examples) and 10 years for lagomorphs have been identified. Other organisms exhibit non-cyclic and irregular fluctations. Although species populations fluctuate, they nevertheless tend to maintain their relative abundance (i.e. relative to those of other species with which they occur in a particular habitat) provided they are not subject to a high level of disturbance. This is reflected in the identification of plants and animals as abundant, common or rare in particular localities.

Population regulation

Until relatively recently most studies of population control were concerned with animals and more specifically with insects and birds. Comparable work on plants followed later, mainly within the last 20 years (Harper 1977). The development of this field of study has been accompanied by varying emphases on the relative importance of *density-dependent* and *density-independent* factors. It is now generally accepted that the fundamental factor in the regulation of species numbers is the density of the species population itself. Rate of population growth and density are inversely correlated, with the former decreasing as numbers increase. Without this density-dependent control the balance in the absolute and relative abundance of populations could not be maintained. Indeed, regulation necessitates that at least one of the primary biological parameters referred to must be density dependent so that as population numbers increase either the birth-rate decreases and/or death-rate increases and/or emigration exceeds immigration (see Fig. 5.3).

The factors which affect variations in the primary population parameters can be either intrinsic or extrinsic or a combination of both. The former result from changes in the environmental resources (i.e. space, microclimate) and in the fitness

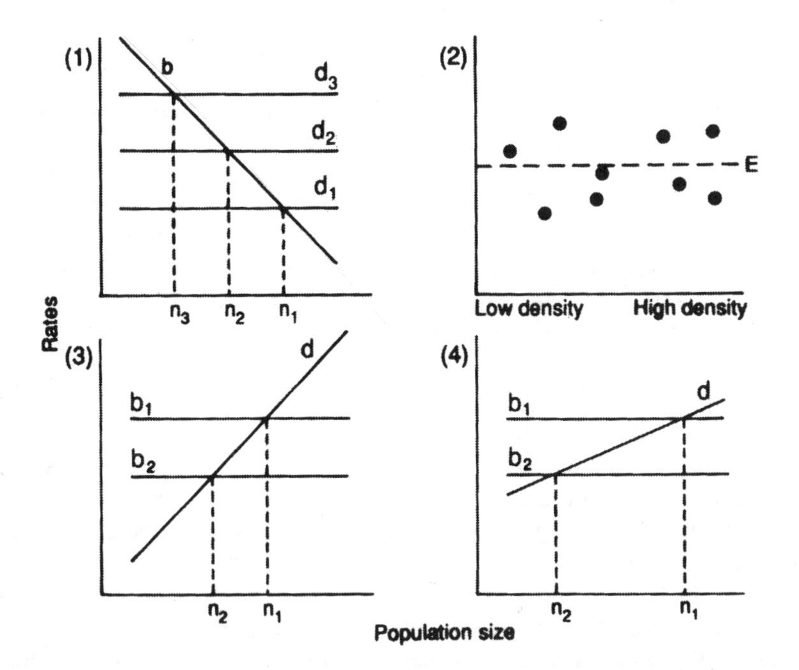

Fig. 5.3 A graphical model showing how density-dependent and density-independent processes interact to determine population size: (1) birth-rate (*b*) density-dependent; the three different levels of density-dependent mortality ($d_1d_2d_3$) lead to different average population sizes ($n_1n_2n_3$); (2) equilibrium population (*E*) exists where mean birth-rate = mean death-rate; in reality average birth- and death-rates vary and population fluctuates around equilibrium; (3) relationship between density-independent birth-rate and population size when form (rate) of density-dependent death-rate varies between populations; (3) high death-rate; (4) lower death-rate (from Watkinson 1981)

of the component individuals to maintain their reproductive and survival rates. Some populations may, to a greater or lesser extent, be self-regulating (or self-thinning) as a result of intrinsic factors operating independently of, or in conjunction with, variations in extrinsic variables. Most vertebrate populations show evidence of self-regulation, often at levels well below that of the carrying capacity expected, given the environmental resources available. Above a given density the reduction in animal territorial areas leads to suboptimal levels of crowding and a decrease in birth-rates and also in death-rates. The former is affected by either physiological and/or behavioural changes, i.e. decrease in fertility, increase in infant mortality, and cannibalism; the last by socially induced stress. Intraspecific competition for space and light which increases with density is a particularly significant regulatory process referred to as 'self-thinning' in plant populations.

Extrinsic factors can be either density dependent or density independent. They include weather and natural enemies such as herbivores (in the case of plants), predators, parasites and disease which either depress reproductive and survival rates directly by lowering the fitness of individuals or indirectly by increasing the susceptibility of organisms to the depressive effects of these regulatory factors consequent upon their reduced fitness. However, the relative importance and operation of density-dependent factors are still far from being fully understood. Many would maintain that while population regulation is a function of a complex of intrinsic and extrinsic factors, the relative importance of one or more 'key' factors can vary both temporally and spatially. Also, the significance of density-dependent factors in the regulation of population cycles is unclear. Both density-dependent extrinsic and intrinsic factors have been proposed to explain some 4-year cycles, but there is no clear evidence for density dependence operating in 3- or 10-year cycles (Putman and Wratten 1984). Further, as Solomon (1969) notes, there are non-regulatory density-independent factors. Many of these are associated with disturbance, the effects of which are not influenced by the density of the population. Weather could be considered in this light. Its effects can be correlated directly with either density or in the case of extreme events be unrelated or indirectly correlated with density as in the case of forest microclimates in which the damping of extreme climatic or weather variations increases with plant density. In the case of extreme events the impact of weather may be unrelated to density.

As a result of a long-term study of the grey partridge in England (Sussex) chick mortality was identified as the key factor in population change (Potts and Aebischer 1989). However, it was revealed that this was not a function of chick density (see Fig. 5.4) but mainly of a shortage of food (insects), as a result of weather in the previous 6 weeks. Density-dependent mortality was associated with

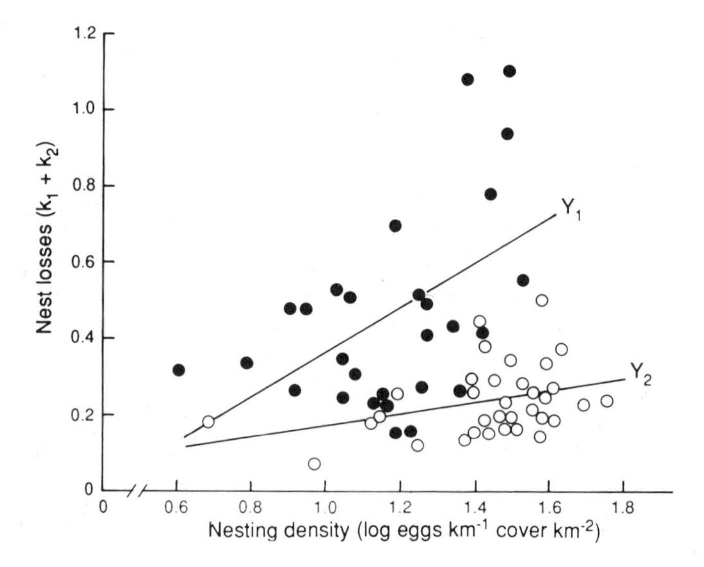

Fig. 5.4 Relationship between nest losses of grey partridges ($k_1 + k_2$) and nest density in Sussex with gamekeepers (open circle) and without (full circle) each point represents one area-year; k_1 = egg losses, k_2 = chick losses; $Y_1 = 0.22 + 0.58 \times$ ($p < 0.01$, SE slope + 0.19) $Y_2 = 0.03 + 0.15 \times$ ($p < 0.05$, SE slope $\pm$ 0.07) (from Potts 1986)

egg and hen predation by crows and foxes respectively. Other losses included density-dependent shooting, emigration and density-independent predation and severe winter weather (see Fig. 5.5). The last can cause a higher proportion of loss from predation at high than low densities. The authors point out that both density-dependent and density-independent factors are essential for population control. Although the importance of the former cannot be denied, the latter will always be the ultimate one because if they continue to cause a decrease the remnant population left by density dependence will eventually become extinct.

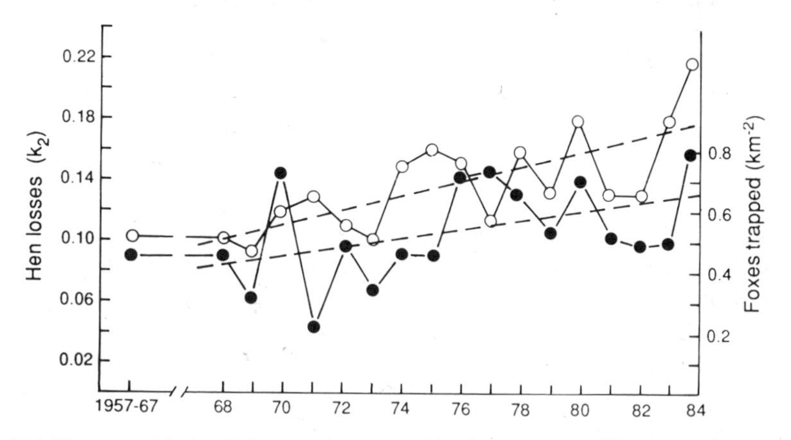

Fig. 5.5 Changes with time in losses of grey partridge hens in part of Sussex study area mostly resulting from predation by foxes (open circle) and in the number of foxes trapped per km^2 in the same area between 1 March and 1 August (a) upward trends (full circle) in the same area. In both, measures are significant ($p < 0.05$) as is the correlation between them ($p < 0.02$). Control of foxes in this area ceased after 1984 (from Grubb and Whittaker 1989)

Plant population control

The study of plant populations developed later than that of animals (Harper 1977) and both empirical and theoretical work has been biased towards short-lived abundant species that tend to grow in almost pure stands (Meijden 1989). These are easier to count than long-lived or rare plants. In contrast to animals, plants are relatively immobile and their distribution is 'patchy'. Carrying capacity is determined by a combination of all the biotic and abiotic components affecting the environment of a species, e.g. other plants, herbivores, pathogens, weather and soil. In view of the variability of the biotic environment – the plants' basic resource for growth – carrying capacity is even more difficult to measure than for animals. However, while the assessment of the relative importance of density-dependent and density-independent factors in population regulation is fraught with the same problems as in animal populations, intraspecific competition is widely accepted as one of the most significant density-dependent factors in the case of common if not of less common (rare) plants. In monocultures it results in a nearly perfect density-dependence control (Meijden 1989). Finally, the importance of the genetic variation in the control of species populations cannot, as it has been in many publications, be ignored (Harper 1977; Meijden 1989). Increase in mortality of individuals less resistant to the effects of either biotic factors (disease, herbivores) or abiotic factors (drought, frost, wind) may result in the persistence of species populations at a higher or lower density than previously. Meijden quotes the examples of the evening primrose (*Oenothera biennis*) and ragwort (*Senecio jacbea*). In the former, individuals with abundant but light seeds flowered when densities were very high, those with fewer larger and heavier seeds did not. In the latter, individuals with high levels of specific alkaloids were avoided by herbivores.

BIOMASS STRUCTURE

The relative abundance and distribution of plant life forms determine the structure of the biomass. Structure can be defined as the spatial distribution (both vertical and horizontal) of life forms in a given biomass. Stratification or the vertical distribution of the component life forms (Fig. 5.6) is the most obvious structural characteristic in both terrestrial and aquatic plant biomass. In the former it is a

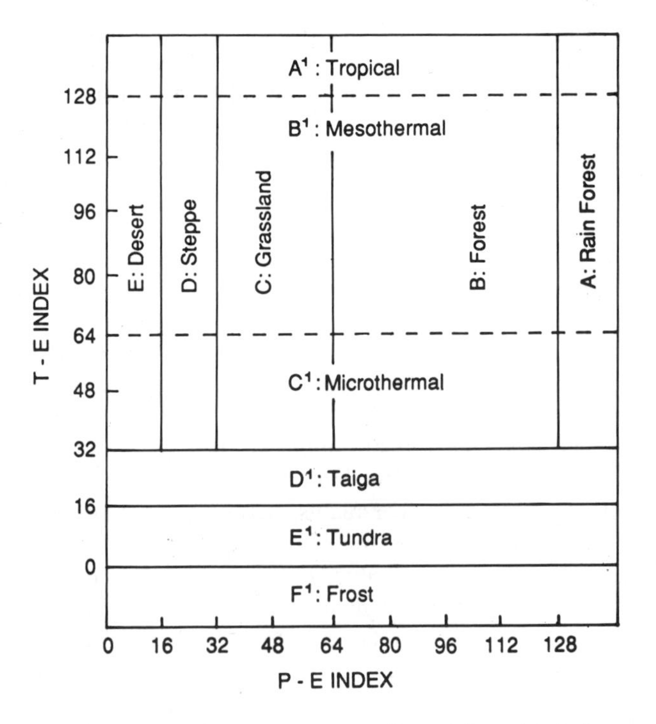

Fig. 5.6 Temperature and humidity provinces according to Thornthwaite's system of classification (from Trewartha 1954)

function primarily of the competition for light by plants of varying growth height and *leaf area index* (i.e. ratio of the total leaf surface area to ground surface area) and the resulting vertical decrease in light intensity from the top of the phytomass to the ground surface. The number of strata (i.e. of trees, shrubs, tall herbs, short herbs, mosses) is dependent on the density and seasonal variations of shade cast by the highest strata. Stratification in above-ground biomass is frequently paralleled by that below ground where root systems of varying vertical and horizontal extent draw nutrients from different soil horizons. Stratification is of ecological importance because it increases the number of plant niches as well as the number of habitats for epiphitic plants and for animals. Stratification is also a characteristic of the aquatic phytobiomass. In this case, it is more a function of vertical variations in temperature, salinity (and hence density) and in light intensity in sea-water which determine the vertical variations in the composition of, particularly, plant plankton of varying size and shape. In addition, many species of zooplankton undergo vertical migrations related to varying light intensity, to availability of food resources and to avoid predation.

DOMINANCE

Those plant species and/or life forms which, because of their abundance and/or size, comprise the largest percentage of the total phytobiomass are the dominants. *Dominance* is a long-established ecological concept. It can and has been expressed as absolute abundance (density), relative (percentage) abundance, productivity of a species or life form. Because of the time and scale problems involved in measuring biomass, *cover abundance* is most frequently used as a surrogate. The latter is the percentage of a sample area, when viewed vertically, that is covered by a particular plant species or life form and normally ranked on a 5- or 10-point percentage scale (see Table 5.1). The cover abundance of either the highest strata or each distinct vertical stratum in the plant biomass can be assessed. Because of their superior biomass the dominants can exert the greatest influence on the physical habitat and on the other plant forms that can grow in company with them. By virtue of their size and bulk they contribute most to, and take most from, the soil. They modify the light, temperature, humidity and air movement of the space occupied and thereby create a particular microclimate above and below ground surface.

Dominance in plant biomass is mainly a function of competition. The dominants are those species or forms which at a given time and place can compete most successfully for the available environmental resources. Grime (1979) defines dominance with reference to the 'stresses' produced by larger on smaller neighbouring plants. He also distinguishes between *competitive dominants* and *stress-tolerant dominants*. The latter replace the former as an increasingly greater proportion of nutrients become locked up in the biomass and the competitive dominants decline in vigour. Although *animal dominance* is thought to be controlled by mechanisms which resemble those in plants, its effect is less obvious than that of plants on the total biomass. Competitive dominance among animals is related to social dominance in the 'scramble' for limited resources which is

Table 5.1 Climatic and vegetational characteristics of Köppen's five major climatic regions

	Climatic regions	*Vegetation*
A	Tropical climates; no cool season; temperature; of coldest month > 18 °C	Megathermal plants; tropical rain forest and savanna
B	Dry climates; excess of evaporation over rainfall, semi-arid and arid	Xerophyllous plants; steppe and desert
C	Warm temperate rainy climates with mild winters; average temperature of warmest month < 18 °C; average temperature of coldest month > 10 °C	Mesothermal plants; temperate forests
D	Cold climates; snowy; severe winter; average temperature of warmest month > 10 °C; coldest month < −3 °C	Microthermal plants; Boreal forest
E	Polar climates; no warm season; average temperature of warmest month < 10 °C	Hekistothermal plants; tundra vegetation; permanent ice-caps

dependent on successful attack, fight, chase and supplanting action (Ricklefs 1973). Animal dominance can be indirect – as exemplified in the case of the control of smaller by larger herbivores as a result of the depletion of food resources and high animal density.

In aquatic and terrestrial environments the concept of *keystone animal species* can be fundamental to biomass structure. These are single species which play such a critical role that their disappearance or removal allows the establishment of another ecological dominant which can compete more successfully for space and hence displace other existing species. Examples quoted by Krebs (1985) are the starfish (*Pisaster ochraceous*) on rocky intertidal zones, the near-shore sea otter (*Enhydra latris*) of western North America, and the lobster off the east coast of Canada. An example of a terrestrial keystone species is the African elephant (*Loxodonta africana*).

THE COMMUNITY CONCEPT

The term 'community' is used in two different senses. In one it refers to any assemblage of two or more species populations living together. In the other it means an interrelated interdependent group of species populations with a functional identity. In either sense a community is characterised by a number of attributes (emergent properties) not possessed by the species populations. Plant communities have traditionally been identified on the basis of those biomass properties previously analysed, i.e. species diversity, relative abundance and dominance, physical structure (stratification) and habitat. Animal communities identified by species composition are functional (trophic), structure and food (vegetation) and habitat. The functional community is still a major ecological paradigm. However, given the 'physical' dominance of the phytobiomass in terrestrial habitats the emphasis will be given to the formal features of the plant community in the following sections.

COMMUNITY COMPOSITION AND STRUCTURE

Variations in community composition and structure can be related very broadly to three groups of factors: (1) the nature of the physical environment; (2) the environmental history; and (3) past and current human activities. Where the environmental resources are limited by climate and/or soil conditions communities tend to be simpler in composition and structure, as for example in arid and very cold habitats where a one-layered, often discontinuous or 'open' vegetation cover may be composed of a few highly specialised types of plants adapted to these extreme conditions. It is also reflected in the difference between the Boreal forest of high latitudes and the tropical rain forest of equatorial regions. The former is dominated, over extensive areas, by a relatively small number of hardy coniferous species, with a correspondingly simple substratum. The latter is characterised by a tremendous diversity of trees – as many as 50–100 different species per hectare have been noted – and a complex stratification. Generally, the more favourable the physical conditions the greater the diversity of composition and structural complexity of the community.

However, within comparable physical habitats the existing nature of the plant communities may vary because of its environmental history. Communities of northern Europe, and particularly Britain, are less rich in species and often less complex in structure than those in ecologically similar areas in America or Asia. In the former, as has already been noted, the effect of the Pleistocene glaciations and barriers to subsequent plant migrations have left a legacy of relative floristic poverty. Also, the time that has elapsed since a particular habitat became available for plant colonisation (as on newly created mud-flats or recently formed volcanic islands) may not have been sufficient for the most diverse and complex plant communities possible to develop. Human activity has been, and continues to be, one of the most powerful factors in the simplification of plant communities. As a result of the intensification of burning and grazing the number of species capable of maintaining themselves has been drastically reduced in many areas, and a formerly more complex structure has been disrupted by removing and inhibiting tree and shrub growth. Among the simplest types of plant communities are those composed of a particular type of agricultural crop or planted tree.

The boundary between one community and another may be gradual or abrupt. In the former case one assemblage of species will tend to grade into another across a *zone of transition (an ecotone)* whose width is dependent on the steepness of the environmental gradient involved. The ecotone is often characterised by greater species diversity than in either of the neighbouring communities. This is considered a function of the *edge effect* resulting from an inter-mixture of species some of which are the most tolerant members of the adjacent communities and others which are edge species peculiar to the ecotone itself. However, abrupt changes from one community to another are not infrequently associated with continuous, steep environmental gradients. This is revealed in the sharply demarcated zonation of communities in mountain, coastal and lacustrine environments. In these cases the rapid replacement of one community by another may be the result of the narrow range of tolerance (or niche width) of the component species to environmental conditions such as temperature, salinity, duration and depth of inundation. In other cases clear-cut biotic boundaries are the result of severe physical disturbance (e.g. wind-blow of trees) or human activity (e.g. fire).

The concept of the community, however, has long been a contentious issue among plant ecologists. The two most extreme and opposed views are those developed by Clements (1916) on the one hand and by Gleason (1926) on the other, particularly in relation to plant communities. The former maintained that the plant community was so highly integrated a system that it could be compared to a superorganism or quasi-organism, subject to predictable development and homeostatic regulation. In contrast to this organismic (holistic), equilibrium concept, Gleason's approach was an individualistic (reductionist) one. Although he recognised the existence of specific communities, he maintained that they were coincidental, continuously varying collections of species resulting from random seed dispersal and continuous environmental variation. He rejected the idea of a predictable development and argued that disturbance was so frequent in time and so variable in space that an equilibrium stage could never be attained or maintained (Crawley 1986).

Today, current community ecology is less preoccupied with the defence of one or other of these views. It is more concerned to understand the complex ecological interrelationships within and between assemblages of plants and animals. As Crawley (1986: 50) notes, 'recognising which communities are stable (equilibrium) assemblages dominated by biotic interactions . . . and which are non-equilibrium assemblages dominated by chance or chronic abiotic disturbances represents a major challenge for the plant ecologist'. The individualistic and holistic views, however, still influence plant ecology and particularly the analysis of spatial variation in the world's biomass.

CLASSIFICATION OF COMMUNITIES

'Plants form the basic biological matrix of all communities' (Krebs 1985: 462) and hence provide the primary means of identifying and classifying terrestrial vegetation. The two principal criteria used are the form or physiognomy and the floristic composition of the plant biomass (or vegetation). The former has long attracted the attention of biogeographers because of the possible relationship between the life forms of plants and the environmental conditions with which they are associated. In addition, the ability of plants to modify their physical habitat is partly dependent on their size and form while their competitiveness is frequently a function of height and leaf area.

PLANT FORMATION

A community distinguished by the dominance of a particular growth form or combination of growth forms is called a *plant formation* – in general terms 'a type of vegetation'. While formations can be readily observed in the changes in the general appearance and colour of the vegetation cover, their description and classification suffer to an even greater extent from the same problems as those posed by the form of the individual plant. Formations may be described in lesser or greater detail by one or more attributes of the dominant growth forms, as for instance forest, broad-leaved forest, or broad-leaved evergreen or deciduous forest. Alternatively, they may be characterised by the predominant growth form in association with a distinctive feature of the physical habitat in which they occur, as in acid grassland or desert scrub, and often regional terms which are indicative of a particular form of vegetation and associated environmental conditions which include such categories as heath, moorland, prairie, steppe, garrigue, llanos and savanna.

Comparable formations associated with similar, though widely separated habitats constitute a *formation type* (i.e. group or class). On the basis of the most obvious contrasts of appearance, the world's vegetation cover has been subdivided into a few major formation types of which the most commonly recognised are forest, grassland, wooded grassland (savanna), scrub and desert, each of which can be subdivided into smaller units according to differences in the form of the dominants. However, the term 'formation' has been reserved by some authors for those major units of vegetation, such as the deciduous forests of north-west Europe, the broad-leaved, evergeen forests of equatorial Africa or the grasslands of central North America which comprise the characteristic and most prevalent type of vegetation over extensive areas. The formation type is then synonymous with what are usually regarded as world types of vegetation. These major formations together with the associated animals are designated *biomes*. A distinction, however, can be made between formations associated with a major climatic region (and hence the largest type of plant community possible) and local formations related to soil conditions and/or the activities of man.

CONVERGENCE OF GROWTH FORMS

The predominance of comparable types of vegetation in similar, though widely separated, climatic regions is a distinctive feature of world vegetation distribution that can hardly be ignored. The dominance of a particular growth form often common to a great number of different species in the major plant formations of the world is usually attributed to convergent evolution under similar climatic conditions. In other words, during the course of evolution the species which have become dominant are those with a form that has best fitted them to compete successfully under the prevailing climate. The degree of convergence of growth forms in any climatic region varies and is by no means complete. It is a feature of the tropical rain forest formations of South America, central Africa and South-east Asia which has been stressed by many students of world vegetation. Each area is composed of a tremendous number of different species, but at the same time is characterised by a marked predominance of evergreen broad-leaved trees often of remarkably similar appearance. It has been suggested by plant geographers that this high degree of convergence is the result of a long uninterrupted period of unchanged climatic conditions (see p. 73). Under constant conditions of temperature and humidity which are optimal for growth, the evolution and survival of other forms have been inhibited. Convergence of growth form is illustrated both in the predominance of needle-leaved conifers in the Boreal forest (taiga) formations of America and Eurasia; and in the broad-leaved, winter- or summer-deciduous forests in temperate and tropical climates with a marked seasonal regime of temperature or rainfall respectively.

Convergence is probably least developed in arid and semi-arid scrub formations. Although drought resistance is a characteristic of the dominant plants, this may be expressed in a great variety of different growth forms such as deciduous or evergeen sclerophyllous shrubs and many varied forms of succulent plants occurring under similar climatic conditions. Other 'climatic' regions are characterised by a mixture of contrasted forms, as in the mixed broad-leaved deciduous and coniferous forests of north-east America or by a combination of trees and grasses in tropical savanna. Nor would it appear that the dominance of one particular growth form is restricted exclusively to one type of climatic region. As Eyre (1968) points out, needle-leaved coniferous formations exist under a wide variety of climatic conditions ranging from cold to warm, from humid to semi-arid. Similar climatic regions are not invariably distinguished by comparable formations. Beadle (1951) early noted that in similar climatic conditions in central California and south-east Australia, grassland is the natural formation in the former, eucalyptus woodland in the latter. He also suggested that the former has presumably remained grassland because of the absence of a suitably adapted woody species. The currently very successful spread and regeneration of eucalyptus species, introduced as shade trees into California and the veld (South Africa), gives point to this observation. In the broadly comparable climatic regions of north-west Europe, British Columbia and southern Chile, the major vegetation formations are broad-leaved coniferous, mixed evergreen, broad-leaved and coniferous forests respectively.

The broad and general correlation which, in spite of certain anomalies, exists between the distribution of formation types and climatic regions has profoundly influenced the development of plant geography, ecology, biogeography and climatology. It has been the basis of the concept that climate is the dominant, if not the controlling, factor determining the major features of the world's vegetation cover.

VEGETATION AND CLIMATE

Students of world vegetation have long been preoccupied with its relationship to climate (Mather and Toshioka 1968). Many attempts have been made to express, in quantitative terms, the characteristics of desert, grassland and forest climates and the climatic limits of these major formation types. Such efforts have been based almost exclusively on temperature and rainfall (or evaporation expressed as a function of temperature) as being the main climatic factors limiting plant growth. The climatic definition of vegetation boundaries has, in the main, been concerned with those between the major formations of tundra, forest, grassland and desert. By inspection of the climatic data available, attempts have been made to assess what quantity (and/or duration) of either temperature and/or rainfall coincides most closely with these particular vegetation boundaries. For instance, the general but by no means exact coincidence of the 10 °C isotherm for the warmest month and the boundary between Boreal forest and tundra in the northern hemisphere was early noted. It has been suggested, however, that a closer fit is obtained with a limit of 3 months with a mean temperature of 6 °C (the generally accepted threshold for plant growth in temperate regions). Similarly, by the inspection of temperature and rainfall values along the boundaries between forest and grassland and between grassland and steppe, many efforts have been made to calculate the amount of effective rainfall (expressed in terms of the relationship between rainfall and temperature or rainfall and evaporation) which coincided with these boundaries.

The assumption that vegetation provided a faithful reflection of climate has also influenced many systems of climatic classification. Initially, it was considered by biologists and geographers alike that, in the words of Miller (1950: 90) 'a satisfactory classification of climate must reflect the climatic control of vegetation: that climatic provinces must coincide as closely as possible with the major vegetational regions of the globe'. Efforts were made to define the boundaries of climatic types in terms of those rainfall and/or temperature parameters which revealed the highest degree of correlation with the limits of major vegetation types. Types of climate are still designated by vegetational characteristics, as for instance tropical rain forest, tropical savanna, steppe, tundra climates, etc. Conversely, the classification of the major types of world vegetation was constructed on the basis of climatic characteristics!

One of the still commonly used systems of climatic classification – that of the biologist, Köppen (1918) – was developed from an intensive study of the climatic limits of the major types of world vegetation. On the basis of mean annual and monthly rainfall and temperature data he delimited five major climatic regions which were intended to correspond to major vegetation regions, as summarised in Table 5.1. While the main humid forest climates were delimited on the basis of temperature, the boundary between these and dry climates was defined by effective rainfall. From a study of rainfall and temperature data available along the forest–steppe boundary, Köppen constructed formulae which expressed the effective rainfall in terms of mean annual temperature, mean annual rainfall and its seasonal distribution. This gave a value which showed a fairly consistent fit with the vegetation boundary: e.g. mean annual rainfall (R) = (0.44 mean annual temperature (T)) − K (3 when rainfall maximum in summer; 14 when in winter). It is worth noting that other climatologists and biologists have proposed alternative $R : T$ ratios which they purport gave a more accurate measure of the same boundary (Bagnouls and Gaussen 1957).

A later classification by the American climatologist, Thornthwaite (1948) also attempted to define climatic boundaries in terms of their coincidence with the limits of major vegetation types. His approach differed from that of Köppen in being based on a combination of three criteria which aimed to provide a more realistic measure of climatic efficiency for plant growth. These are precipitation efficiency (expressed as monthly rainfall divided by monthly evaporation), temperature efficiency, and the seasonal incidence of rainfall. On this basis, Thornthwaite defined humidity and temperature provinces, the limits of the former being those precipitation/evaporation (P/E) values calculated for the areas which lay along the boundaries of the major types of vegetation. In cold regions where temperature is the more significant factor limiting plant growth, a primary division was made on the basis of temperature provinces. In those with a temperature efficiency index of over 32 the basic division is by humidity provinces (see Fig. 5.6).

These are but two of the most familiar and frequently used of the many systems of climatic classification which are really vegetation regions climatically defined. Their shortcomings, and the criticisms to which they have been subjected, stem basically from the problems inevitably inherent in any attempt to correlate types of climate and vegetation and to define one by the characteristics of the other. Even if formation types are assumed to be expressions primarily of climate, the latter is a complex of variable and interrelated factors among which ambient temperature and rainfall are only two which influence plant growth. Also, in many parts of the world reliable information about even these two elements is still relatively scarce, while the characteristics, distribution and exact limits of the major world vegetation types have had to await the recent availability of Landsat imagery. Attempts to define vegetation or climatic boundaries quantitatively must necessarily tend to be unrealistic. Both are zones of transition, of varying width which may, as in the forest–tundra ecotone be as much as 160 km wide. In addition, it has been suggested that vegetation, particularly in high latitudes in the northern hemisphere, may still be in the process of readjustment to post-glacial climatic fluctuations and that the arctic timber-line may not be stable. There is evidence in Alaska that the colonisation northwards of trees is still in progress (see Ch. 10).

Finally, much doubt has been expressed about the validity of defining the limits of the major vegetation regions by climatic data alone. The vegetation of any area is the expression of the sum total of all environmental factors operating through time. While climate is undoubtedly one of the major factor complexes, in many parts of the world historical and biotic factors have been equally if not more important in the determination of existing types of vegetation. The influence of past climatic change and of geographical barriers to plant migration has, as already noted in Chapter 4, resulted in different formations in similar climatic regions.

Also, it has become increasingly obvious that human activity has so modified the vegetation cover that in many parts of the world its form cannot be explained in terms of climate alone. In the first place so much of the cover has been removed or modified that the distribution of the pre-existing 'natural vegetation' cannot be reconstructed with certainty. In many areas long-continued grazing and burning have been instrumental in favouring some types of vegetation at the expense of others. There is now strong evidence to support the theory that much of the tropical savanna grasslands, as well as the more humid temperate prairie and steppe, owe their existence to burning which, rather than climate alone, prevents tree growth. As a result, the existing boundaries between forest and grassland, grassland and desert are 'man-made' rather than climatically, or otherwise naturally determined. As Miller (1950: 90) pertinently remarked, 'the correlation [between climate and vegetation] remains a will o' the wisp whose presence we may sense but whose outlines remain blurred and indefinite'. The realisation of the problems and pitfalls of assuming a direct relationship between climate and plant formations has motivated more recent recommendations and attempts to classify world types of vegetation and climate on the basis of their own characteristics rather than those of each other.

PLANT ASSOCIATIONS

Plant communities can be characterised not only by their form but by their floristic composition. While discrete units or types of vegetation can be distinguished on the basis of the dominant growth forms of which they are composed, they are not necessarily – indeed are rarely – composed of the same species throughout. A particular formation or type of vegetation is usually composed of a variety of communities which may be of similar form but which are distinguished by different assemblages of species. In British and American ecological literature, a fundamental distinction has long been made between plant formations characterised by plants of a particular growth form and associations characterised by particular assemblages of species populations, of which one or more may be dominant or co-dominant. However, even within associations so defined there may be considerable variation as a result of the species which accompany the dominants. Oak-wood associations in Britain vary from those with well-developed, species-rich shrub and field layers to those in which the shrub layer may be absent and the ground layer is composed of a few species of heath plants and/or mosses.

These subordinate species tend to reflect minor variations in habitat more sensitively than do the more tolerant dominants (see Table 5.2). Among the commonest and most widespread types of moorland vegetation in Britain are the *Agrostis–Fescue* dominated grasslands, and the *Calluna*-dominated heaths. In the former, there are communities characterised by the presence of such species as moor mat grass (*Nardus stricta*), heath-bedstraw (*Galium saxatile*), tormentil (*Potentilla ertecta*) on poor acid or heavily grazed habitats; by thyme (*Thymus vulgaris*), clover (*Trifolium repens*) on base-rich soils, and meadow grass (*Poa pratensis*) on sites heavily manured by grazing animals. Similarly heather-dominated communities range from those in drier sites, accompanied by varying

Table 5.2 Variation in an even-aged oak wood (*Quercus robur*) (Arrachymore Point, Loch Lomond)

	1	*2*	*3*	*4*
Dominant tree species	Oak/birch	Oak	Oak	Oak/birch
Average tree height (m)	13	12–13	12–13	13–17
Canopy depth (m)	5	5	7	7
Canopy cover (%)	55–60	50	50–75	
Shrub layer	None	None	Scattered individuals (oak, ash, hazel)	Scattered individuals (rowan)
Bracken/bramble (percentage cover)	20	65	30	45
Field layer (number of herbaceous species)	8	3	15	11

Sites 1 and 2 on soils derived from acid quartzose schists; 3 and 4 from base-rich schist. Differences in number of herbaceous species related to (i) density of bracken/bramble cover and (ii) soil base status. As well as supporting a richer ground flora, sites 3 and 4 were characterised by species, e.g. *Brachypodium sylvaticum* (false wood brome grass), *Mercurialis perennis* (dog's mercury) and *Rubus* spp. (bramble), not found on sites 1 and 2.

proportions of bell heather (*Erica cinerea*), crowberry (*Empetrum nigrum*) and blueberry (*Vaccinium myrtillus*), to those on wet peaty areas where the more common associates are the cross-leaved heath (*Erica tetralix*), bog-cotton (*Eriophorum* spp.) and deer-sedge (*Scirpus/Trichophorum caespitosus*).

The term 'association' has unfortunately been, and still is, used in a number of different senses. On the one hand it can denote an actual assemblage of species, a particular plant community growing in an arbitrarily defined site or a distinctive habitat. In this sense it means a concrete association or, more correctly, a vegetation stand, i.e. any particular piece of vegetation cover distinguished by its relative homogeneity of physiognomy and species composition. On the other hand association, in the strict ecological sense, means an association type, i.e. similar assemblages of species or stands which may recur in similar though widely separated habitats. In this sense the association type is regarded by phytosociologists as the basic, natural vegetation taxon comparable to the species. American and British plant ecologists have tended to classify associations on the basis primarily of dominant species expressed in terms of relative abundance and cover value (Table 5.3). Scandinavian phytosociologists have placed more emphasis on community structure and strata dominance classifying associations (or, as they are termed, 'sociations') on the basis of characteristic combinations of strata dominants.

Table 5.3 Braun-Blanquet cover abundance Scale (I) and Domin abundance–vigour–cover Scale (II) as modified by Dahl (from Braun-Blanquet 1932; Evans and Dahl 1955)

	I	*II*
10	—	Cover $^9/_{10}$ to complete (91–100%) total area
9	—	Cover $^3/_4$–$^9/_{10}$ (76–90%) total area
8	—	Cover $^1/_2$–$^3/_4$ (51–75%) total area
7	—	Cover $^1/_3$–$^1/_2$ (34–50%) total area
6	—	Cover $^1/_4$–$^1/_3$ (26–33%) total area
5	Cover over 75%	Cover about $^1/_5$ (11–25%) total area
4	Any number of individuals covering 50–75%	Cover up to $^1/_{10}$ (4–19%) total area
3	Any number of individuals covering 25–50%	Occurring as numerous individuals but which cover less than 4% total area
2	Very numerous: cover value at least 5%	Occurring as several individuals: no measurable cover
1	Plentiful but small cover value	Occurring as one or two individuals with normal vigour: no measurable cover
+	Sparse or very sparsely present, very small or negligible cover value	Occurring as a single individual with reduced vigour: no measurable cover

Species dominance, however, is a more characteristic feature of plant communities in northern latitudes, where the flora is much poorer, than in other more climatically favourable regions. In northern Europe this relative poverty has been reinforced by the effects of the Pleistocene glaciations and by human activities. Not only are there fewer species adapted to the more severe conditions but competition is less intense and those species with a wide range of tolerance can maintain their dominance over an extensive range of soil and climatic conditions. This is illustrated in the subarctic Boreal forest, and in the range and extent of heather-dominated communities in Britain. In these communities the same plant may be dominant in a variety of different habitats but be accompanied by varying assemblages of subordinate species in each. The dominant plant is usually less sensitive to environmental changes than the subordinate species whose range of tolerance is much less. For this reason the classification of communities on the basis of dominance alone is considered by some workers to be superficial and lacking in ecological significance. On the other hand, the conspicuousness of the dominant species is the main criterion whereby major contrasts in the vegetation cover can most easily be observed.

In other parts of the world, however, where environmental and particularly climatic conditions are more favourable, the diversity of species is greater, dominance over extensive areas of one or a few species is much less common. This is true, as has already been noted, in the floristically richer deciduous and in the mixed broad-leaved evergreen and deciduous forests of eastern North America. It is nowhere more strikingly demonstrated than in the tropical rain forest. In the floristically richer areas of southern Europe advocates of what became known as the Zurich–Montpellier school of phytosociology placed less emphasis on dominance and more on total floristic composition in their attempts to classify types of vegetation. They took into consideration other criteria such as the constancy (presence) and fidelity (exclusiveness) of species in the classification of stands. The former is expressed as frequency, i.e. percentage of stands in which a species occurs which is a measure of the evenness of species distribution. On the basis of 'constancy classes' (Table 5.4) species can be described as rare or accidental, accessory or constant.

Table 5.4 Braun-Blanquet's five-point rating of constancy classes, e.g. percentage of total stands examined in which a given species occurs irrespective of its abundance or relative importance (from Braun-Blanquet 1932)

1	1–20	Rare (or accidental)
2	21–40	
3	41–60	Accessory
4	61–80	
5	81–100	Constant

Associations are classified on the basis of distinctive assemblages of species. The French school of phytosociologists, in particular, have attempted to identify associations by *characteristic or faithful species*. These are either the most abundant or those which attain their greatest vigour only in particular stands. They are exclusive to certain communities and as such are important diagnostic species of an association. To this end Braun-Blanquet (1932) classified species into five fidelity classes:

1. *Strangers*: appearing accidentally;
2. *Companions*: indifferent species without a pronounced affinity for any community;
3. *Preferents*: present in several community types, but predominantly in one;
4. *Selectives*: present particularly in one community type, but occasionally in others;
5. *Exclusives*: found most exclusively in one community type.

Classes 3, 4 and 5 are the so-called 'characteristic' or 'faithful' species. The recognition of association types on this basis has been subjected to considerable criticism because it is theoretically only possible after an intensive investigation of a subjectively identified particular type of vegetation has been completed!

A comparable classificatory method was developed by Poore (1955, 43) and adapted by McVean and Ratcliffe (1962) to the classification of Scottish moorland vegetation. On the basis of standardised sample plots, lists of all species and their relative significance in terms of dominance or abundance were recorded for different types of vegetation. Stands characterised by assemblages of species occurring in much the same proportion (e.g. cover abundance plus constancy) are classed in the same *community type* or *nodum*. The latter is an 'abstract' vegetation unit (or taxon) of any category. It summarises the characteristic features common to a group of similar, though not necessarily absolutely identical stands of vegetation. The nodum, like the association type, is to the concrete plant community or stand what the species is to the individual plant.

CLASSIFICATORY TECHNIQUES

The study of vegetation has been accompanied by the development of increasingly sophisticated techniques for sorting and classifying the various attributes by which it can be described. This has resulted in a trend away from subjective, descriptive and manual (or visual) to more objective, numerical and computerised means of ordering samples of stands into associations (Pears 1968; Shimwell 1972; Frenkel and Harrison 1974; Harrison *et al.* 1986).

The earlier approach based on, or derived from, the Braun-Blanquet school of phytosociology, involves the selection and description (in terms of cover abundance) of samples taken from stands which appeared to be homogeneous. The raw data, presented in the form of species lists, were originally (and still are if the sample is not too large) sorted by visual inspection into hierarchical groupings on the basis of differential species. More systematic, and particularly more rapid, methods of computer-sorting of individual lists are now employed. This technique, however, has been subjected to considerable criticism. First, the use of homogeneous stands as the basis for the selection of samples entails a preliminary visual and hence subjective classification which must influence subsequent sorting. Second, the size of stand sampled can vary considerably. Third, the estimation of cover abundance is subjective, allowing for a wide range of variation within both the 5- and 10-point scales. Finally, not only is the manual sorting of species lists extremely time-consuming but it is not easy to ensure or test the comparability of the hierarchical levels derived by this technique.

The availability of computers in the 1950s and 1960s led to the rapid development of numerical techniques for classifying vegetation, usually on the basis of one (monothetic) or several (polythetic) characters. These techniques can be divisive or agglomerative. Association analysis is divisive with a total population of samples broken down into increasingly smaller groups, on the basis of the presence or absence of divisive species. The latter are those with the highest index of division at a particular level – the index being derived from a selected statistical measure of homogeneity or central tendency. Nearest-neighbour analysis (or single linkage sorting) and centroid sorting are polythetic, agglomerative classifications. The former involves choosing those attributes of the vegetation to be studied and grouping pairs of individual samples on the basis of their degree of similarity with the initially most similar pair. This produces groups or clusters of samples (individuals) which can eventually be built up into one group which includes the entire statistical population. Centroid sorting differs in that groups or clusters of samples are formed by calculating the degree of similarity between each new group and the remaining individuals.

Both the divisive and agglomerative techniques are dendritic, hierarchical types of classification. The raw data are derived from systematically or randomly selected equal-area samples (*relevés*). The main advantages of such numerical techniques is that, at any given level in the hierarchy, the amount of information and the degree of homogeneity for any group or class are known and both can be described and tested by standardised statistical techniques. Fundamental to vegetation classification is the occurrence of discrete stands with common characteristics of composition and/or structure distinctive enough to allow them to be grouped into a community type and to be clearly differentiated from other types. Methods of classification have undoubtedly been influenced by the practical requirement of cartographic representation and the firmly entrenched hierarchical approach to taxonomy. Although classification is essential to a fuller understanding of the nature of vegetation it has, as in so many other disciplines, often tended to become an end in itself directed towards the study of the taxonomy of vegetation *per se*. Classification, however, should be a means to an end. As Frenkel and Harrison (1974) point out in their detailed review, the technique or combination of techniques adopted depend on the purpose and scale of study – be it for description, mapping or the analysis of the relationship between vegetation and one or more environmental parameters. Subjective/descriptive and objective/numerical techniques of classification are complementary in that the former are more applicable to large-scale, the latter to small-scale studies where the amount of detailed data is high.

CONTINUUM CONCEPT OF VEGETATION

Major vegetation types (e.g. formation types) distinguished by common physiognomic characters can, particularly on a large scale, be fairly readily identified. In contrast, the range of variation in species composition within a type of vegetation is often such that change is continuous rather than discontinuous. The recognition of clearly delimited community types is therefore difficult, if not virtually impossible. The American individualistic school of ecologists was influenced by the fact that within the vast area of mixed deciduous forests in eastern North America it was not possible to delimit discrete stands clearly. While recognising the existence of communities they rejected the concept of association types as discontinuous natural units and regarded the association as an arbitrary unit of classification of subjectively selected stands of vegetation. In reality, vegetation is composed of continuously varying assemblages of species populations. Hence while generally similar, though not necessarily identical, stands of vegetation may occur within a range of comparable habitats, variation from one to the other will be so gradual (i.e. continuous) that a particular type of vegetation will be composed of a continuous series or sequences of community types, one overlapping with and merging into the other through gradual changes in species composition. The whole then forms what has been termed a *vegetation continuum* rather than a mosaic of clearly defined association types. A vegetation continuum, in the words of McIntosh (1967: 171) 'Although not formally defined, ... is plainly described as a gradient of communities in which species are distributed in a continuously shifting series of combinations and proportions in a definite sequence pattern.'

The development of the continuum concept owes much to the Wisconsin school of ecologists among whom Bray and Curtis (1957) have been the most outstanding protagonists. They applied two methods to the analysis of variation in the composition of vegetation and its relationship to spatial environmental changes. One is direct gradient analysis of changes in species abundance along established environmental gradients, e.g. altitude, moisture, temperature and soil acidity, as shown in Fig. 5.7. In the analysis of an area of vegetation associated with a

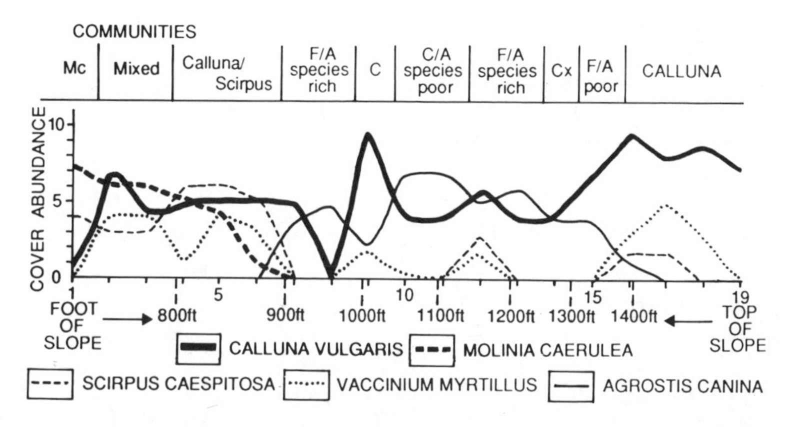

Fig. 5.7 Spatial ordination of m^2 quadrats along a continuous transect of a hillslope in south-west Scotland showing variation in cover abundance of selected species from Mitchell 1973)

complex of environmental gradients of varying steepness and direction, ordination techniques were developed to rank species and samples in relation to one another. Bray and Curtis (1957) calculated an importance value (derived from the sum of the relative density, frequency and dominance) for each species in each sample of a mixed conifer–hardwood forest in northern Wisconsin. The absolute importance value multiplied by the rank number (or climax adaptation number) gave a *continuum index* which could be used as a measure of a complex environment gradient against which the importance value for all species could be plotted. These methods were subject to considerable criticism not least from the protagonists of the association school. The forest stands studied were young and even-aged consequent on recent disturbance. Also the techniques were not regarded as adequate for a rigorous testing of association types (Daubenmire 1968).

Nevertheless, the continuum concept stimulated the application of ordination techniques to the analysis of vegetation, not least in Britain. For instance, in their summary of the grass moorlands of the forest and subalpine zone in Scotland,

King and Nicholson (1964) gave an example of the range and nature of variation in the widespread *Agrostis/Festuca* grassland on the Pentland Hills, south of Edinburgh. When groups of sites are arranged in an 'environmental sequence' from the most leached (low pH) to the most flushed (high pH) (see Table 5.5) the changes in the species composition is shown to be continuous. No one group is clearly differentiated from the next in the sequence. No species is exclusive to any one group. The dominant genera *Agrostis* and *Festuca* are characterised by a high specific frequency in all sites. The frequency of several grasses tends to increase or decline from one end of the series or continuum to the other, while others are most abundant in the intermediate habitats. The gradation of species composition is even more strikingly illustrated in an *Agrostis/Festuca* continuum on the Cheviot Hills. Figure 5.8 illustrates the gradual changes in cover abundance of a number of

Table 5.5 Grass species composition in terms of specific frequency of a series of seven groups of sites arranged in order of the soil sequence from the most leached (group 1) to the most flushed (group 7) (from King and Nicholson 1964)

Site group	*1*	*2*	*3*	*4*	*5*	*6*	*7*
Log soil water (%)	1.35	1.43	1.67	1.61	1.53	1.77	1.69
pH surface soil	5.04	5.50	6.03	6.05	5.73	6.10	6.23
Mean no. species per site	9.30	13.0	18.2	16.70	15.50	14.60	12.70
Agrostis canina ssp. *canina*	—	3.3	—	—	2.0	—	—
A. canina ssp. *montana*	9.0	6.7	—	—	—	—	—
A. stolonifera	—	—	—	—	—	5.5	37.1
A. tenuis	91.0	93.3	84.3	98.3	96.0	92.7	65.7
Anthoxanthum odoratum	65.0	86.7	57.1	48.3	70.0	34.5	8.6
Briza media	—	—	42.9	6.7	4.0	1.8	—
Cynosurus cristatus	8	3.3	18.6	35.0	44.0	29.1	28.6
Deschampsia caespitosa	4.0	—	10.0	8.3	16.0	10.9	8.6
D. flexuosa	5.0	—	—	—	—	—	—
Festuca ovina	95.0	86.7	87.1	56.7	44.0	5.5	—
F. rubra	54.0	76.7	94.3	100.0	100.0	100.0	100.0
Holcus lanatus	—	—	2.8	8.3	6.0	—	—
H. mollis ($2n = 21$)	—	—	—	3.3	—	—	31.4
H. mollis ($2n = 35$)	23.0	—	—	—	20.0	14.5	11.4
Poa annua	—	—	—	—	—	3.6	34.3
P. pratensis	22.0	6.7	14.3	23.3	48.0	29.1	25.7
P. trivialis	2.0	—	5.7	15.0	22.0	49.1	77.1

species common to this type of grassland, in relation to a sequence of soil profiles showing a continuous variation from the more acid peat-podzols and brown-forest soils to the less acid brown-forest soils and gleys.

Variation in vegetation, however, is the product not just of one but of a complex of a great number of interacting environmental variables. Simple, unidimensional ordinations, while none the less valuable, are obviously inadequate to cope with the analysis of multiple correlations.

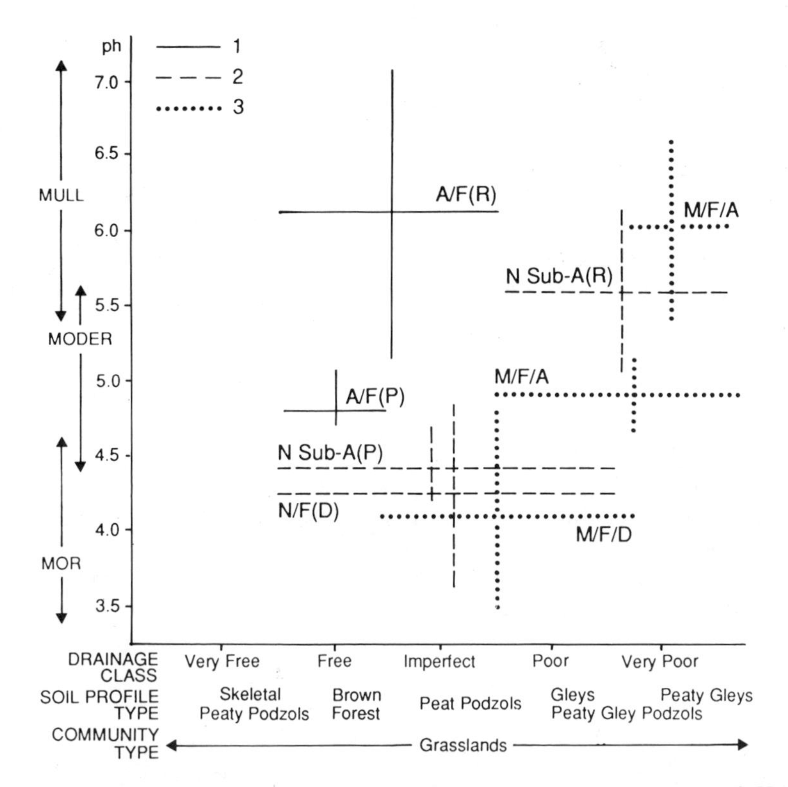

Fig. 5.8 Interrelationships of Scottish grass moorland community types in terms of pH and soil-drainage class. Dominants: (1) *Agrostis/Festuca*; (2) *Nardus stricta*; (3) *Molinia caerulea*; A = Agrostis; F = Festuca; M = Molinia; N = Nardus; D = Deschampsia; R = Species rich; P = Species poor; Sub A = Subalpine (adapted from King and Nicholson 1964)

The combination of two or more environmental gradients in a multidimensional ordination as in Fig. 4.6 provides a partial solution to this problem. This summarises the interrelationships of the grassland community types and their relationship to two edaphic gradients. However, as King and Nicholson (1964) point out, ordination relation to only two environmental axes cannot account for the total floristic variation. Hence, it should not be assumed that adjacent or overlapping types are necessarily as closely related as their proximity in one plain might suggest. Also, because the boundaries of community types may arbitrarily exclude intermediate stands, Fig. 4.6 does not provide any evidence for either continuous floristic gradients, or for discontinuities. Since then the development of electronic computers has made possible the application of more sophisticated statistical techniques for multivariate and factor analysis to the investigation of an even greater number of environmental and phytosociological variables.

Vegetation, because of the multiplicity of its attributes, is among the most difficult of natural phenomena to describe and analyse. Classification and ordination are two means to this end. However, as McIntosh (1967) is at some pains to emphasise, the community and continuum concepts are not mutually exclusive. They represent two different but complementary approaches to the same problems. He maintains that the purpose of both the traditional hierarchical classification systems and of the ordination techniques is 'to arrange or organise a set of observations and to seek meaningful patterns of relationships. Both are explanatory devices designed to elucidate a large mass of observations whether qualitative and subjective or quantitative and objective' (p. 136). It has been pointed out that while community types can be subjected to ordination analysis, the latter can be used to establish either subjectively or objectively the possible divisions that may need to be made for a particular purpose within a vegetation continuum.

Poore (1962: 56) summarises the problem of vegetation analysis very aptly:

> Many methods have been developed for the description and characterisation of communities, which vary from subjective to qualitative to elaborately quantitative. The choice of method is a personal matter influenced partly by the purpose of the investigator, partly by the particular view taken by workers of the nature of vegetation. But it should be remembered that however exact the method, it cannot describe all the attributes of even the simplest community at one moment of time. Every description is an abstraction from available data.

Chapter

6

Biological productivity

PHOTOSYNTHESIS

The amount of biomass (or standing crop) at any one time and its growth (or production) over time is a function primarily of phytosynthesis – the most fundamental of plants' various growth processes (Bjorkman and Berry 1973). It is the process whereby chlorophyll-containing green plants use solar energy to build up highly complex organic material from the simple inorganic elements of carbon dioxide, water and mineral nutrients. Photosynthesis (light or carbon assimilation) can be summarised simply as follows:

$$12H_2O + 6CO_2 + \text{solar energy} \rightarrow (C_6H_{12}O_6) + 6O_2 + 6H_2O$$

(solar energy: absorbed by chloroplast cells in green parts of plant; $(C_6H_{12}O_6)$: carbohydrates)

The carbohydrates provide the basic building-blocks from which more complex organic substances including proteins and fats are formed. About a sixth of the light energy absorbed by the green plant is used in the process of photosynthesis; the remainder is converted into the chemical (or potential food) energy of the plant tissues. The total amount of energy fixed over a given period of time is the *gross photosynthetic production* (GPP). Much of this is expended in the plant's metabolic processes and is dissipated as heat in the reverse process of respiration (oxidation), i.e.

$$(C_6H_{12}O_6) + O_2\,[\text{metabolic}] \rightarrow 6CO_2 + 6H_2O + \text{heat energy}$$

[enzymes]

The remainder, stored for a longer or shorter period of time in the plant tissue (phytobiomass) is the *net photosynthetic production* (NPP).

Photosynthesis 'harnesses' solar energy; respiration releases it as heat.

Photosynthesis, moreover, constitutes the basic difference between the nutrition of green plants and that of animals. The former are *autotrophic*, i.e. wholly dependent on the sun and inorganic materials; the latter are *heterotropic*, dependent on organic food. Until humans can duplicate the process of photosynthesis, efficiently and economically, outside the green plant they are no less dependent on plant food than the lowest type of animal.

PLANT FORM AND PHOTOSYNTHESIS

The necessity for plants to carry out photosynthesis under a wide range of environmental conditions has played a major role in the evolution of the diverse forms of life that exist today. Plants absorb light and carbon dioxide through their surfaces and their efficiency in this respect is dependent on the ratio of their surface area to that of their volume. Fossil records leave little doubt that plant life originated in the oceans. The remarkable similarity of chemical composition between sea-water and that of the living cell has elicited the remark by one biologist that 'land and fresh water organisms are packets of sea water that have found ways of maintaining themselves in different environments' (Bates 1964: 96). The oceans provide a much less extreme and variable medium for photosynthesis than does the land. In the sea carbon dioxide, water and mineral nutrients are contained in one enveloping medium, while on land they must be drawn from the two separate sources of soil and air. In the sea the plant is entirely surrounded by water which is much richer in dissolved carbon dioxide than the atmosphere. It is, therefore, not altogether surprising that the plant life of the oceans is the most prolific form of life on earth. It consists, however, almost entirely of the lower and more primitive (in the evolutionary sense) types of plants. The bulk is composed of microscopic unicellular organisms – the *plant plankton* – which float in the upper light-absorbent layers of the sea. Their minute size ensures a very large absorbing surface in relation to their volume. In the light of the efficiency of this simple form of photosynthesis and the remarkable uniformity of the environment, the absence of a greater variety of plant forms in the sea is perhaps understandable.

On the other hand, the evolution of land plants required successful adaptation not only to a different but to a much more extreme and variable set of environmental conditions. Plants growing on the land are subjected to much greater variations of temperature, from place to place and from season to season, than in the oceans. There is, in addition, a separation of the atmospheric environment, from which the plant obtains light and carbon dioxide, and the soil from which water and mineral nutrients must be drawn. It is necessary that the land plant be fixed in one spot and have the means of extending into and extracting water from the sunless area of the soil. It must also be tall enough to attain sufficient light in competition with other plants, and at the same time have as large a photosynthetic area in relation to volume as possible. The elaboration of the external form by branching and the subdivision of leaves has provided a means of maintaining a maximum photosynthesising surface with increasing size. In the sea, large size would not be of any great advantage. On land, competition for light among immobile plants obviously gives those with greater size, and particularly height, an advantage. Height, however, creates problems associated with the transport of water and mineral nutrients from the soil to the photosynthesising organs and, lacking the buoyancy of marine plants, with the mechanical support of ever longer stems. Also, the necessity for the absorbing cells in the leaves to be protected from desiccation by an impermeable skin (cuticle) while at the same time allowing carbon dioxide entry through minute openings called stomata, inevitably exposes the land plant to the loss of water by evaporation. The problems of attaining maximum photosynthesis on the land have therefore created those of maintaining an adequate water supply in the plant. The variety and complexities of form which have developed in the higher cone-bearing gymnosperm and seed-bearing angiosperm plants have enabled them to cope most successfully with the problems of attaining maximum efficiency of photosynthesis in competition with other plants under a diverse range of physical conditions; they have thus become the dominant plants in the terrestrial vegetation cover today.

RATE OF PHOTOSYNTHESIS

The rate at which a plant photosynthesises depends on its inherent capacity to utilise the available light energy; its ability to intercept light; and the availability of the limiting resources of water and mineral nutrients.

Less than 1 per cent of the solar radiation reaching the earth's surface is available for photosynthesis. Of the light falling on land plants it has been estimated that, on average, 20 per cent is reflected, 20 per cent is radiated as heat and 49 per cent is used in evapotranspiration. Only 10 per cent is transmitted through the leaf, the exact percentage dependent on the colour, thickness and internal structure. Reflection and absorption of light by water are high; even when it is clear only some 50 per cent of that impinging on the surface penetrates to a depth greater than 10–12 m. This rapid decrease in light intensity limits photosynthesis to a maximum depth of about 50 m in aquatic habitats. Of the light falling on the leaf surface only 50 per cent of the visible (white) light (i.e. that within the wave-bands 400 and 700 mu) is photosynthetically active radiation (PAR)) and the rate of carbon assimilation increases with increasing PAR. The minimum light intensity is that below which uptake of carbon dioxide is inhibited by stomatal closure. The *light compensation point* (see Table 3.3) is reached when temperature and light conditions are such that the uptake of carbon dioxide in photosynthesis equals that of carbon dioxide output in respiration. The optimum light intensity or *light saturation level* for most plants falls below the potential maximum light intensity and is that at which the carbon dioxide concentration of the atmosphere becomes the limiting factor for photosynthesis.

Minimum and optimum light intensities for photosynthesis vary among species and even between leaves on the same plant. There are those which attain maximum photosynthesis in low light intensities, others in high intensities. The minimum requirements of unicellular algae – such as plant plankton – are much lower than those for the higher land plants. Some deep-water algae, and those algae and mosses which inhabit caves, can photosynthesise in light intensities no greater than that of moonlight. Plants which require high light intensities for optimum photosynthesis are referred to as 'sun loving' (i.e. *heliophytes*). Most trees, cereal crops and many grasses and herbaceous weeds are heliophytic, while a great number of mosses and ferns, together with herbaceous and shrubby plants of woodland, are *sciaphytes* (shade plants). Many plants are obligative in their response to light intensity while others are facultative. The ubiquitous bracken fern (*Pteridium aquilinum*) is a classic example of a facultative sciaphyte. It grows in deciduous woodland. However, once the trees are felled it grows and spreads with very much greater vigour. Its tall fern canopy and thick litter layer make it an aggressive competitor on abandoned farmland and on heavily grazed and burnt moorland. Also, the light requirements of a particular species (or its tolerance of shade) can vary during its life cycle. While many trees are heliophytic, their seedlings are frequently sciaphytic. Light and shade leaves on the same plant are not uncommon with the sun leaves having higher saturated photosynthetic rates and higher light compensation points than the shade leaves.

Limiting resources

The effects of light and temperature conditions on the rate of photosynthesis are closely related. The light compensation point for photosynthesis is temperature related and the rate of photosynthesis increases as temperature increases to an optimum which varies between species. Mooney (1986), for instance, notes that the optimum for desert perennials exceeds 40 °C while those for Antarctic lichens are close to freezing-point.

Temperature, through its effect on the relative humidity and evaporation rate of the atmosphere, influences the availability of moisture and hence the rate of water loss from the land plant. Water availability affects the rate of photosynthesis both directly and indirectly. The proportion of water taken up by the plant that is used in the biochemical processes involved in photosynthesis is relatively insignificant compared with that which is transported through the tissues and transpired from the stomata or through the cuticle. Water loss which maintains a continuous stream of water through the plant is necessary to maintain the turgidity of the tissues, and particularly that of the guard cells which control the size of the stomatal apertures. Stomata tend to close and leaves to wilt when the water balance in the plant is not maintained. However, while there is an optimum temperature limit to the rate of photosynthesis, water loss from the plant tissues is a function of the rate of evaporation of the atmosphere. As the temperature increases so also does the ratio of water loss to carbon dioxide uptake. This relationship is expressed in the concept of *water-use efficiency*, i.e. the amount of water lost by transpiration per unit of carbon fixed per unit of time. In those environments where water is the most important limiting factor for photosynthesis high water-use efficiency combined with methods of curtailing the rate of water loss are essential strategies for plant growth and survival in arid habitats. In addition, evaporative water loss dissipates heat from the leaf surface. Other environmental factors that influence the rate of photosynthesis are the availability of nutrients, particularly nitrogen, and atmospheric and soil pollutants which may either stimulate or depress the process.

THE PHOTOSYNTHETIC SURFACE

The photosynthetic capacity of land plants depends on their ability to intercept solar radiation. This is related to the total surface area of their foliage, together with the disposition, duration (longevity) and age of the individual leaves (see Figs 6.1 and 6.2). The total leaf area of a plant is frequently greater than that of the ground surface beneath it. Increase in leaf area, however, is inevitably

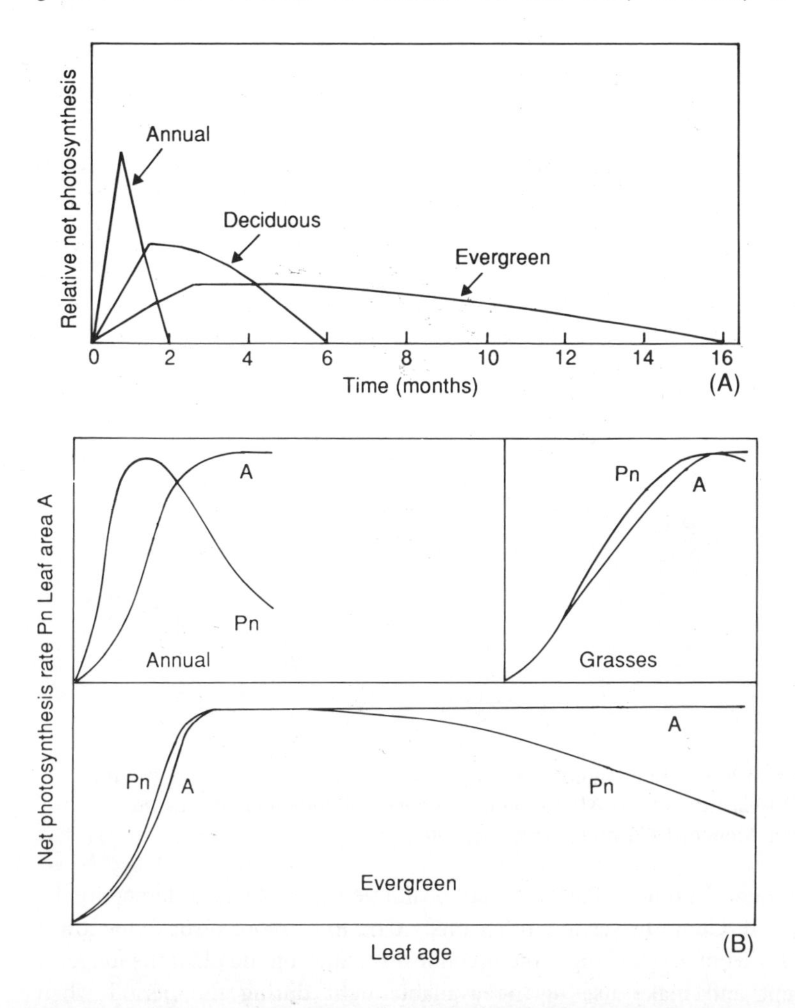

Fig. 6.1 Generalised relationships between (A) leaf area development and photosynthetic capacity (from Mooney and Gulman 1982); (B) leaf duration and photosynthetic capacity (from Mooney and Gulman 1982)

accompanied by mutual shading of the lower by the upper leaves. To a certain extent the higher photosynthetic rate of the upper sun leaves compensates for the lower rate of the shade leaves. *Leaf area indices* (LAI) (i.e. the ratio of leaf area to ground area) of 8 or over are not uncommon for trees with large canopies. The optimum LAI for photosynthesis, however, tends to be lower than the actual maximum because of wastage of light falling on bare ground and/or reduction of photosynthesis because of mutual leaf shading. The amount of light intercepted is also influenced by the disposition of the leaves. Horizontal leaves tend to intercept more than vertical ones; optimal LAI rises as leaf angles become more vertical, thereby allowing more light to reach the lower leaves. Many plants are *solar trackers* (Mooney 1986) with leaves which change their orientation in relation to

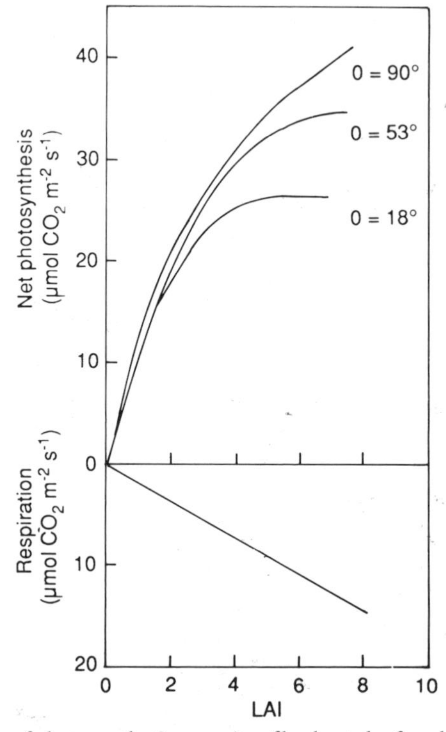

Fig. 6.2 Relationship of photosynthetic capacity of barley to leaf angle (from the ground surface) and to leaf area index (LAI). Respiration increases as total amount of tissue increases (from Mooney 1986 after Pearce *et al.* 1967)

the angle of light. In others, the leaves are positioned to maximise interception by the whole plant. Upper leaves in a tree canopy tend to be more vertical, the lower to be more horizontal. The longer the green leaf remains on the plant the longer it can intercept and make use of the available light during the period when temperatures and/or moisture are favourable for growth. However, there is an inverse correlation between leaf duration and photosynthetic rate. Short-duration leaves of annual plants and deciduous trees have a higher photosynthetic rate than those of long-lived but slow-growing evergreen trees (Fig. 6.1(A)); as a result the amount of carbon fixed by both during the growing season can be similar. Finally, the photosynthetic capacity of the leaf changes with age and reaches its maximum just before the maximum area is attained (see Fig. 6.1(B)).

PHOTOSYNTHETIC PATHWAYS

The rate of photosynthesis also varies as a result of differences in the biochemical process or carbon pathway employed by different plants. In the late 1960s the existence of C_3 and C_4 plants was first noted. Most species belong to the C_3 group in which the initial step in carbon dioxide conversion is the production of a carbon-3 compound (the Calvin pathway). Others, which occur in at least 100 genera and 10 families (Bjorkman and Berry 1973), belong to the C_4 group in which the initial process is the production of carbon-4 molecules. This difference is reflected in leaf anatomy. In the latter the chloroplasts are concentrated around the veins, in the former they are widely dispersed throughout the leaf tissue. Also, C_4 species can use carbon dioxide and water more efficiently and attain maximum growth rates that can be two to three times those of C_3 (Table 6.1). In contrast,

Table 6.1 Photosynthetic pathways and related characteristics of C_3, C_4 and CAM plants (from FAO 1978)

*Photosynthetic pathway**	*C_3*	*C_4*	*CAM*
Temperature response (°C)			
Optimum	15–20	30–35	25–35
Operative range	5–30	15–45	10–45
Radiation intensity at max. photosynthesis (Cal cm^{-2} min^{-1})	0.2–0.6	1.0–1.4	0.6–1.4
Max. net rate CO_2 exchange at light saturation (mg dm^{-2} h^{-1})	20–30	70–100	20–50
Max. crop growth rate (mg dm^2 h^{-1})	20–30	30–60	20–30
Water use efficiency (g g^{-1})	400–800	150–300	50–200

* C_3, carbon-3; C_4, carbon-4; CAM, crassulacean acid metabolic SM.

some succulent desert plants with a crassulacean acid metabolism (CAM) have lower photosynthetic rates; carbon dioxide is taken up mainly at night, when stomata are open, and is photosynthesised during the day when stomata close in order to minimise water loss.

More is known abouit the carbon pathways of cultivated than of wild plants (Tivy 1991). The geographical range of C_3 and C_4 crops tend to overlap. However, some of the former like rice can give high yields in tropical monsoon climates because the highest temperatures and light intensity coincide with the crops grain-filling growth phase at the beginning of the dry season. Some of the latter like maize can give high yields in continental climates with high summer temperatures. In general, however, C_3 cultivars do not yield well or at all in subtropical areas while many C_4 are precluded from temperate areas; C_3 are characteristically long-day and C_4 short-day cultigens. The gradual displacement of the latter by the former has been noted in the passage from the south to the north of the grasslands of the Great Plains of North America. Sisal and pineapple are the only known CAM crops.

PHOTOSYNTHETIC EFFICIENCY

The ratio between the energy value of the organic matter produced and that of either the light available (or that actually absorbed) is a measure of the plant's *photosynthetic efficiency*. The most frequently quoted estimates suggest that the mean annual net photosynthetic efficiency for the earth's surface is somewhere in the order of 0.1–0.2 per cent, depending on whether calculations are based on total light available or that of suitable wavelength. While this would appear to be a very low order of efficiency it has, nevertheless, been suggested that it represents an annual conversion of over 200 billion tonnes of carbon from carbon dioxide into sugar (equivalent to about 100 times the combined weight of the world's chemical, metallurgical and mining industries!).

However, any estimate of the mean annual global efficiency of photosynthesis must obviously obscure wide variations from one part of the earth's surface to another. There are deserts and ice-caps where solar radiation cannot be utilised because either water or temperature conditions are insufficient for plant life; there are other areas where they limit or severely restrict photosynthesis for part of the year. On the other hand, annual net efficiencies of 2–5 per cent have been recorded for such crops as sugar-beet, sugar-cane and maize and of between 0.3 and 2.7 per cent in British forestry plantations. It has further been suggested that a good, well-fertilised, well-watered field crop might be expected to have a photosynthetic efficiency as high as 7–10 per cent at its period of maximum growth. Certain algal cultures grown under laboratory conditions have indicated that at low light intensities, efficiency of energy conversion of as high as 25 per cent can be obtained, but the actual yield of organic matter in terms of quantity produced per unit area per unit time is low.

Expressions of photosynthetic efficiency vary according to the measurements on which the ratio between light and food energy are based. The former may be total incident light, that available to, or that actually absorbed by, plants; the latter may be expressed in terms of either gross or net plant production. Further, expressions of efficiency will depend on the unit of time during which the amount of photosynthesis is calculated, be it annual, growing season, monthly or daily. For these reasons most biologists prefer to employ either GPP or NPP rather than efficiency when studying the conversion of solar radiation to plant material.

BIOLOGICAL PRODUCTIVITY

Biological productivity or the rate at which organic matter accumulates and an organism grows in size, is usually expressed in terms either of the amount of carbon assimilated or of the dry weight or energy equivalent of the biomass produced per unit time per unit area (or volume). Plants or *primary biological productivity* is a function of the net photosynthetic production plus loss of parts of the plant biomass (*B*) by death (*L*) and by grazing (herbivory) (*G*). Figure 6.3

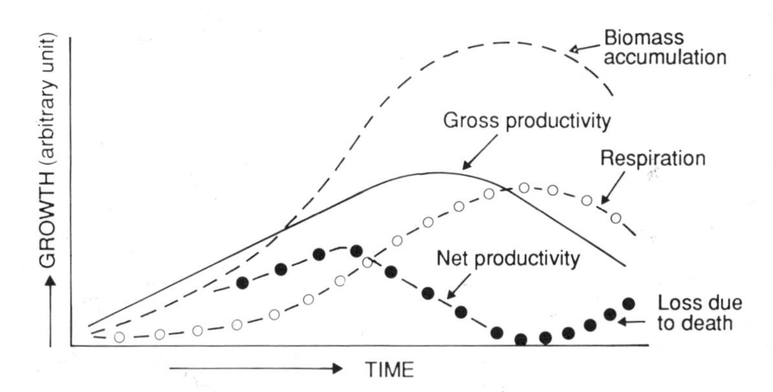

Fig. 6.3 Trends in gross primary productivity and biomass in an herbaceous plant (from Whittaker 1975)

illustrates trends in, and relationships between, GPP and NPP and biomass in a herbaceous plant. The productivity of a given amount or area of plant biomass is expressed as the sum of that of the individual plants, i.e.

$$\Sigma \text{ NPP} \quad \text{or} \quad (Bt^2 - Bt^1) + L + G$$

(i.e. the net community growth or ecosystem production). The latter, however, may not bear a relationship to NPP. In a fast-growing forest 30–60 per cent of the total NPP may accumulate in 1 year; in an old mature forest, NPP may be high but ecosystem production low or zero because of the high loss of tissue and individuals. Plant plankton composed of small short-lived organisms subject to high consumption rates by grazing has a very high NPP but low biomass production. Variation in NPP can be expressed by the biomass accumulation ratio (i.e. biomass : annual NPP) which is less than 1 in many aquatic ecosystems and can range from 50 to zero dependent on the age of individuals in terrestrial vegetation.

MEASUREMENT OF BIOLOGICAL PRODUCTIVITY

Measurement of primary productivity in the field is a difficult, time-consuming and expensive procedure. The most commonly used techniques are harvesting, gas exchange (CO_2 and O_2) and growth measurements. The simplest of these is the harvest method, i.e. the calculation of the difference between the dry weight of plant biomass in a given area at the beginning and end of a selected time period. In some cases the below-ground (roots/stems) biomass may be included; more frequently only the above-ground biomass change is measured. The harvest method is commonly used for herbaceous, seasonal vegetation such as grassland. In the case of shrubs and trees, growth measurements, i.e. increase in girth of bole, height, etc. complemented by gas exchange, are more applicable. Measurement of aquatic primary productivity is complicated by the minute size, short life cycle and mobility of plant plankton. In this case the harvest method is not feasible except on a very local scale and the gas exchange technique can more easily be used on small samples of sea-water enclosed in collecting bottles.

In 1964 the International Biological Programme (IBP) was set up under the auspices of the International Council for Scientific Unions (ICSU). Entitled 'The Biological Basis of Productivity and Human Welfare', its object was to promote the world-wide study of (a) organic production on land, in fresh waters and in the seas, and the potentialities and use of new as well as existing resources; and (b) human adaptability to changing conditions, in order to reconcile the need for increased food production with that for the rational use and conservation of organic resources. However, it is only within the last two decades that the data, on which a realistic assessment of the productivity of the biosphere could be based, have become available.

GLOBAL BIOLOGICAL PRODUCTIVITY

Although actual measurements are accumulating rapidly, the number of satisfactory long-term measurements of primary productivity is still limited. Recourse has, therefore, been made to the construction of models by which primary biological productivity can be predicted from known environmental parameters, such as temperature and precipitation, for which there is a reasonably dense global network of long-term records. World maps of potential primary production on this basis include the Miami model using average annual temperature and precipitation; the Thornwaite Memorial (or Montreal) model using evapotranspiration; and the Hague model based on the length of the growing season and using phenological rather than climatic data (Whittaker 1975).

However, there are now sufficient data available from which to make reasonable generalisations about the comparative productivity of the world's surface. Variation in global productivity and biomass is illustrated in Table 6.2. Although the land area is only a third of that of the total global surface its primary productivity is about 20 per cent higher than that of the oceans. Krebs (1985) quotes an average biomass of 50 g m^{-2} in the sea, 50 kg m^{-2} on the land. These differences have been attributed to the proportionally greater loss of material in respiration by the minute plant plankton than by land plants, together with the fact that a higher proportion of the incident light is absorbed and reflected by water than by the atmosphere. In addition, nutrients are more important determinants of primary production in the oceans than on land, and plankton growth frequently depletes mineral nutrients in the superficial layers of the seas faster than they can

Table 6.2 Primary biological productivity of main ecosystem types (from Whittaker 1975)

Ecosystem type	Area (10^6 km^2)	LAI (m^2 m^{-2})	Net primary productivity per unit area ($g\ m^{-2}\ year^{-1}$)		World net primary production 10^9 t $year^{-1}$	Biomass or standing crop kg m^{-2}		World biomass (10^9 t)
			Normal range	Mean		Normal range	Mean	
Tropical rain forest	17.0	6–16.6	1000–3500	2200	37.4	6–80	45	765
Tropical seasonal forest	7.5	6–10	1000–2500	1600	12.0	6–60	35	260
Temperate evergreen forest	5.0	5–14	600–2500	1300	6.5	6–200	35	175
Temperate deciduous forest	7.0	3–12	600–2500	1200	8.4	6–60	30	210
Boreal forest	12.0	7–15	400–2000	800	9.6	6–40	20	240
Woodland and shrubland	8.5	4.2	250–1200	700	6.0	2–20	6	50
Savanna	15.0	1–5	200–2000	900	13.5	0.2–15	4	64
Temperate grassland	9.0	5–16	200–1500	600	5.4	0.2–5	1.6	14
Tundra and alpine	8.0	0.5–1.3	10–400	140	1.1	0.1–3	0.6	5
Desert and semi-desert scrub	18.0		10–250	90	1.6	0.1–4	0.7	13
Extreme desert rock, sand and ice	24.0		0–10	3	0.07	0–0.2	0.02	0.5
Cultivated land	14.0		100–3500	650	9.1	0.4–12	1	14
Swamp and marsh	2.0	11–23	800–3500	2000	4.0	3–50	15	30
Lake and stream	2.0		100–1500	250	0.5	0–0.1	0.02	0.05
Total continental	149			773	115		12.3	1837
Open ocean	332.0		2–400	125	41.5	0–0.005	0.003	1.0
Upwelling zones	0.4		400–1000	500	0.2	0.005–0.1	0.02	0.008
Continental shelf	26.6		200–600	360	9.6	0.001–0.04	0.01	0.27
Algal beds and reefs	0.6		500–4000	2500	1.6	0.04–4	2	1.2
Estuaries	1.4		200–3500	1500	2.1	0.01–6	1	1.4
Total marine	361			152	55.0		0.01	3.9
Full total	510			333	170		3.6	1841

Units: km^2 = square kilometers; g = grams dry weight; t = tonne.

be replenished and to amounts which limit production. In the oceans nutrient reserves, particularly of nitrogen and phosphorus, are either limited or unavailable; the latter is the more critical in the open ocean, the former in coastal waters (Bunt 1975). The most nutrient-rich water occurs where there is an upwelling of deep water such as occurs off western continental coasts and to a lesser extent at the equator; and where turbulence maintains nutrients in circulation between the sea floor and the surface zone. Productivity is lowest where there is a pronounced and stable thermal stratification of water with depth. This is most marked in the tropics where a sharp transition is maintained between the warm surface water and the cold water below and hence nutrients lost from surface water cannot be recirculated. Primary productivity is higher in temperate and cool than in tropical latitudes, because of rapid plankton blooms in spring and autumn when water depleted of nutrients by the end of the growing season are replenished at the beginning of the next by vertical mixing. In addition, the NPP of the plant plankton is lower than that of land plants because of the proportionally greater energy loss by respiration from the small biomass.

Terrestrial primary productivity is, at the global scale, correlated with climate, being highest in warm humid, lowest in cold and arid areas of the world. The tropical rain forest accounts for over 40 per cent of the global terrestrial net productivity on only 16 per cent of the total land surface; comparable values are recorded only for wetlands, and algal beds and coral reefs (see Table 6.2). Under similar environmental conditions, the productivity of evergreen conifers exceeds that of deciduous trees. In the former, the evergeen habit allows photosynthesis to proceed, albeit very slowly, when deciduous trees are leafless, and to take immediate advantage of the onset of favourable conditions when deciduous species have still to expend considerable energy producing non-photosynthetic organs. Similarly, warm deserts are twice as productive as the cold tundra deserts; in the latter both light intensity (albeit compensated by long day-length) and temperature curtail the growing season in the winter, while the slow-growing perennials or rapidly growing annuals, adapted to the limited water supplies of the warm deserts, can make use of high light intensity at any time of the year. Odum (1983) identifies three orders of magnitude of primary productivity: (a) low, less than 1 $g\ m^{-2}\ day^{-2}$ characteristic of deserts, deep oceans and lakes; (b) medium, 1–10 $g\ m^{-2}\ day^{-1}$ – grasslands, coastal seas, shallow lakes and 'average agriculture'; (c) high, 10–20 $g\ m^{-2}\ day^{-2}$ – shallow water systems, moist forests and 'intensive' agriculture.

SECONDARY BIOLOGICAL PRODUCTION

Secondary (i.e. heterotrophic) production is, like primary production, an expression of growth, i.e. the accumulation of animal biomass over time. It is dependent on the amount of food available, the efficiency with which plant or animal food is consumed, assimilated and stored by the particular type of organism and the relative importance of herbivores (grazing), carnivores (flesh eating) and saprovores (decomposing) within a particular plant biomass. However, while secondary production is initially dependent on primary production neither the amount nor the rate of the former are correlated with that of the latter. Secondary production (*SP*) is dependent on the relationships between food consumption (*C*), waste (*W*) (i.e. faeces material not digested and urine – the by-products of metabolism), and respiration (*R*) i.e.

$$SP = C - (W + R)$$

Hence growth will occur when $C > (W + R)$ i.e. when the food energy assimilated is in excess of that required for animal biomass maintenance. The efficiency of secondary production is dependent on the ratio of food assimilated (*A*) to that consumed, i.e. the *A/C* or *assimilation efficiency* (Table 6.3) and of the biomass

Table 6.3 Assimilation efficiency (*A/C*) expressed as a percentage of the ingested food that is rejected by different herbivores (from Petrusewicz and Grodzinski 1975)

Herbivore species	
Field mouse (*Apodemus agrarius*)	11
European hare (*Lepus europeaus*)	22
Wild boar (*Sus scrofa*)	24
Orthoptera (*Chortippus dorsatus*)	30
Elephant (*Loxodonta africana*)	60
Lepidoptera (*Croesus septentrionalis*)	79

produced to that assimilated, i.e. the *P/A* or *growth efficiency* (Table 6.4). The former is a measure of the efficiency of food use which varies mainly with digestibility and nutritive value of the material consumed and, in the case of plant food, the season of the year, and the type of animal. It is lower in most herbivores (30–40 per cent) than in carnivores (63–80 per cent). The *P/A* varies with the life cycles and habits of the particular animal species and reflects the proportion of assimilated energy used in growth, as compared with that in reproduction, maintenance and work (movement). Both ratios vary widely in different taxonomic and, as shown in Table 6.4, trophic groups (Heal and MacLean 1975). Among invertebrate animals the lowest *P/A* values are characteristic of the very active predatory insects and of those with long periods of little or no growth in the course of their life cycles. High values are associated with short life cycles and, particularly, larval growth. While vertebrate carnivores have high assimilation

Table 6.4 Variation in growth efficiencies (*P/A*) in animal groups (from Pimm 1988)

P/A efficiency (%)	*Animal group*
60	Carnivorous invertebrate
40	Herbivorous invertebrate
10	Long-lived ectotherms
10	Endothermic birds and mammals

efficiencies comparable to those of invertebrate carnivores, both homeothermic herbivore and carnivore animals have low growth (*P/A*) efficiencies which reflect their very high respiration energy costs (Pimm 1988).

Animal populations are often composed of overlapping generations of individuals of varying age and growth rate, from fast-growing young to mature adults in which growth is very slow or negligible. The total secondary population production then will be the sum (in weight or energy) of the growth of the constituent individuals over a given period, i.e.

$$SP = \overset{n}{\Sigma}\,(Bt_2 - Bt_1) \quad \text{or} \quad \Sigma\,(Bt_2 - Bt_1) + (Et_2 - Et_1)$$

(where n is the total population size, B the biomass or body material, t the time and E the loss of biomass by death, and loss of body material by emigration).

In the absence of measurements of assimilation and growth by which the productivity of a heterotropic population can be calculated, the *biomass turnover rate* (*P/B*) or *relative production* is a particularly useful alternative measure of the rate at which the whole animal population functions. The *P/B* varies more with the size and life history of the species individuals than with the environmental variables (Table 6.5).

Table 6.5 Biomass turnover rate or relative production of two animal groups varying in size and life history (from Heal and MacLean 1975)

Low *P/B* large, slow-growing (0.2) long-lived organisms	Caribou with seasonal breeding
High *P/B* small, rapidly growing (6.0) short-lived organisms	Brown lemming with year-round breeding

TROPHIC LEVELS OR GROUPS

Measurements of secondary (heterotrophic) productivity for *trophic levels* or groups compared with those for particular taxonomic (species) groups and domestic livestock are still relatively scarce. Analysis of spatial variation has, therefore, had to be estimated from known species productivity in different ecosystems. In general, the primary biological production sets a limit to that of secondary production. The plant biomass is the 'base' of all food chains, and in the passage of food from plants to herbivores to carnivores there is a considerable loss of energy as heat and waste material. If only a percentage of the energy in one trophic level is transferable to the next the biomass that can be supported is reduced. Lindemann's (1942) early model of energy conversion derived from a study of a freshwater spring assumed a 90 per cent energy loss at each conversion stage of a food chain or trophic level and a retention of only 10 per cent of the energy of food eaten by the consumers. This so-called *Law of the Tenth* is not revealed in reality. However, some would maintain that deviations may be due, to some extent, to variations in methods and precision of measurement.

The limitation of trophic levels to three to four in terrestrial and to four to five in marine ecosystems (see Table 6.6) has long been attributed to these energy-

Table 6.6 Relative exploitation of autotrophic net production by the first-order biophage populations of different ecosystems. Above ground (terrestrial or open-water) does not normally include food chains containing first-order saprophages (from Weigert and Owen 1971)

Type of system	*Number of trophic levels*	*NPP* *Characteristics of autotrophs biophages (%)*	*to first-order*
Mature deciduous forest	3	Trees, large amount of non-photosynthetic structure; long generation time; low biotic potential	1.5–2.5
1–7-year-old South Carolina fields	3	Herbaceous annual plants; medium biotic potential	12
30-year-old Michigan field	3	Perennial forbs and grasses; medium biotic potential	1.1
African grasslands	3	Perennial grasses; small amount of non-photosynthetic structure; rapid growth rate when environmental conditions are favourable	28–60
Managed rangeland (Grass-cattleman)	3	Perennial grasses; small amount of non-photosynthetic structure; rapid growth rate when environmental conditions are favourable	30–45
Ocean waters	4	Phytoplankton; small and numerous; high biotic potential; short generation time	60–99

conversion losses. However, while trophic levels can be identified there is no unanimity of opinion as to the maximum number that can be supported by the available plant biomass. This is because some animal species may feed from more than one trophic level and hence it is not always easy to designate every organism to a particular level. Nevertheless, the numbers appear constant although much less than the potential. Pimm and Lawton (1977) suggest that the number of levels is determined by population dynamics in a community. If food chains were longer the population fluctuations would become so severe that it would be impossible for higher-order predators to exist. Also, within all ecosystems there are two main types of interrelated trophic subsystems (Heal and MacLean 1975):

1. *The grazing system* (or *herbivore*) in which plant biomass is consumed by herbivores which are in turn eaten by carnivores;
2. *The detrital* (or *saprovore*) system in which dead organic material (DOM) is consumed by scavenging or *saprophytic* organisms.

The relative importance of these two subsystems varies from one ecosystem to another.

The grazing or herbivore system

Terrestrial herbivores in general graze a relatively small but variable proportion of the available above-ground biomass in any ecosystem. The amount consumed can range from 5 per cent in forests to 30–35 per cent in grassland (Table 6.7); and of

Table 6.7 Primary production and utilisation by herbivores in terrestrial ecosystems (from Crisp 1975)

Habitat	*NPP (kcal m^{-2})*	*Herbivore ingestion as fraction of production above ground*
S. Carolina grass field	1075	0.12
Michigan grass field	1360	0.01
Spartina salt-marsh	6585	0.08
Savanna		
Tanzania		0.28
Uganda	750	0.60
Managed range maximum exploitation		0.45
Forest		
Coniferous	2150 (litter)	0.19
Warm temperate	5650 (litter)	0.28
Temperate (Canada)	1570 (foliage)	0.05–0.08
Temperate deciduous (*Liriodendron*)	1640 (foliage)	0.056
Vaccinio–Myrtilli–Pinetum	3030 (foliage and litter)	0.096
Pine–oak–alder	5060 (foliage and litter)	0.17
Potato crop	676 g dry wt (?) leaves	0.115

this 95 per cent is eaten by vertebrate herbivores, 5 per cent or less by invertebrates. The amount of herbivore biomass consumed by carnivores is larger and there is a closer correlation between carnivore prey and carnivore predator biomass than between plant and herbivore biomass. In contrast, in aquatic and particularly in marine ecosystems, 80 per cent or more of the plant (i.e. phytoplankton) biomass is grazed by herbivores. The latter, because of the microscopic size of the phytoplankton, are either very large vertebrate filter-feeders (e.g. whales), sedentary filter-feeders or microcrustaceans of which the *copepods* are the most important marine herbivorous phytoplankton. It used to be assumed that very little (5 per cent or less) of the plant biomass in the sea was not grazed. However, it is now recognised that the copepods – the dominant zooplankton in the temperate oceans – consume, during the period of highest primary production, more phytoplankton than can be assimilated. As a result a relatively high proportion of the primary production grazed is evacuated as faeces and finds its way into the decomposing subsystem.

The decomposing or saprovore system

This is the dominant trophic subsystem in the majority of terrestrial ecosystems. However, since DOM is such a complex substance and the micro-organisms can operate simultaneously at a number of levels, the identification of distinct trophic levels within the system is virtually impossible (see Ch. 8). It has been estimated in a mixed prairie in Colorado that almost half of the above-ground NPP passes into the decomposing system in the soil either via surface litter or plant roots to the soil organic matter. Most of the saprovores live below the ground surface and are invertebrate, particularly microscopic organisms such as bacteria and fungi which are responsible for the decomposition of organic material into its original inorganic components. And the soil animal biomass can far exceed that above ground. This subsystem differs from the grazing subsystem in being more conservative of energy; DOM not used at one trophic level or that contributed by dead saprovores or carnivores is not lost from the system and can be recycled. Heal and MacLean (1975) note that this combined with the high growth efficiency (*P/A*) in micro-organisms gives rise to very high production levels in the saprovore subsystem.

PRODUCTIVITY OF MANAGED ECOSYSTEMS

Agricultural and forest plantation ecosystems are managed to produce a harvestable plant or an animal product. Although their primary and secondary biological productivity is influenced by the same environmental variables as are unmanaged systems, they differ in two fundamental respects from the latter. First, their primary biological productivity is expressed as *crop yield* – the amount in weight of plant material recovered or harvested per unit area per year. That of long-duration trees is measured either as amount of wood (m^3) laid down each year (the current annual increment) or as the average since commencement of growth at a given age (the average annual increment).

The *harvest of crop index* or crop yield (i.e. the ratio of the utilisable or commercially valuable to total recoverable plant) depends on the type of cultigen. In root crops, green vegetables and grasses it is high, 85–100 per cent, close to or equivalent to the primary production of the whole plant. In the seed crops (cereals, pulses, cotton and oil seeds, etc.) it is usually less than 50 per cent and may be as little as 25 per cent. Second, these systems are managed with the object of producing the most 'satisfactory' yield in terms of maximum financial return or amount of subsistence food. This is achieved:

(a) by the inputs (subsidies) of non-solar energy in the form of human and animal labour, machinery, fuel and power, and biochemicals, etc. which create as potentially productive an agrohabitat as possible given the particular natural environmental resources and the stage of human technical development;

(b) by crop breeding to increase the yield potentials;

(c) by ensuring that environmental resources will allow the crop to realise its yield potential, i.e. ensuring the 'best fit' between the crop and the resource base.

VARIABLES AFFECTING CROP YIELD

The environmental variables affecting crop yield are the same as those for uncultivated plants. In the first place, it is dependent on the length of the crop-growing period relative to the climatically favourable growing season. Bunting (1975) makes a distinction between crops whose harvestable yield accumulates in the leaves, stem (e.g. forage grass, sugar-cane) or underground storage organs (potatoes, carrots, turnips, etc.), and those whose yield is in the seeds or grains of cereals and other seed crops. In the former, a high leaf area index can be maintained throughout the growing season; in the latter, the growth phase is shorter and terminates when the flower head (inflorescence) – the yield-accumulating stage – commences. In the second, deficiency of either heat or water or nutrients during growth will limit potential yield.

In addition, two other factors which can depress yield and which have a greater impact in agro-ecosystems and forest plantations than in unmanaged ecosystems are the so-called *negative biological factors* – the weeds, pests and pathogens – and the weather. From the agronomic point of view the former are those organisms which can, when they exceed certain population thresholds, reduce the quantity or quality of the crop yield to an unacceptably low economic level. The most successful weeds are those which are well adapted to take advantage of the disturbed agricultural habitat and hence to grow with greater vigour than they would have done elsewhere. Adaptive strategies common to many agricultural weeds are a short annual life cycle, rapid growth, high seed production and viability, and efficient seed-dispersal mechanisms. Many of the more difficult weeds, however, are perennial grasses (e.g. couch (*Agropyron repens*), wild oats (*Avena fatua*)) with underground storage organs (stems, rhizomes) the growth of which is, in fact, stimulated by cutting and fragmentation when the soil is tilled. Indeed, the success of many agricultural weeds is because of the close adaptation of their growth form and growth phases to that of the crop and its associated methods of cultivation, either because they germinate before the crop can shade them out or do so at the end of the harvest period and before the succeeding tillage. Other weeds, particularly of permanent grassland, are, because of their toxicity or unpalatability, avoided by livestock and hence have a competitive advantage over those grazed.

The pests and pathogens are the 'unwanted herbivores', plant-eating insects, mites, soil nematodes, slugs and snails, etc. whose large populations are a function of a concentrated food supply in an optimal physical environment for pest reproduction, and of a paucity or absence of natural predators. Continuous cultivation of a genetically homogeneous crop variety characteristic of modern intensive agro-ecosystems provides ideal conditions for the proliferation of pests and crop-specific diseases. The problem is further exacerbated by the use of herbicides and insecticides to which these organisms rapidly develop a resistance. It has been estimated that losses incurred by weeds, pests and pathogens may account for a reduction by *c.* 50 per cent of the world's total pre-harvest crop yield. Spedding (1975) gives global pre-harvest yield losses of 36 per cent (37 : 64 weeds : pests and diseases) for maize, 24 per cent (40 : 60) for wheat, and 40 per cent for rice (23 : 77). Weeds depress yield by competition with the crop for primary resources; pests and pathogens by consuming the photosynthetic organs (leaves/stems) and/or reducing the quantity and quality of the harvestable organ.

Crop yields are also depressed by unfavourable weather conditions which can vary to a greater or lesser degree from the climatic average during the course of the current growing season and from one season to another. This also has a greater effect on the productivity of the agro-ecosystem than on that of the natural ecosystem because of the genetic uniformity of the crop population and hence lack of adaptation of some individuals to weather variations (i.e. drought, lack or excess of heat, frost, high wind force). Yield loss due to 'unfavourable' weather will depend on the severity and duration of the conditions and particularly on the crop growth phase when they occur. The most critical phase varies; for green crops a temperature or water deficit at any time during growth, if severe enough, may depress the final yield. In cereals it is just before and during the yield-forming flowering stage and in soft fruits at the beginning of the fruit-swelling stage.

CROP YIELD AND MANAGEMENT

In addition to the preceding factors, spatial and temporal variation of crop yields also depend on the type of agro-ecosystem and its level of management. In those climatic regions where there is a long growing season multiple cropping (i.e. two or more crops of the same or different type on the same land area) can give high annual yields per unit area that compensate for the low productivity of the crops grown. Double and triple cropping is characteristic of areas with a monsoon climate. Irrigation too can dramatically increase crop yields (provided there is no nutrient deficiency) in semi-arid and desert areas of the world where, with an unlimited water supply, the year-long thermal growing season can be fully exploited. However, the harvested crop results in a continuous drain of nutrients out of the agro-ecosystem and, unless this is made good by some form of nutrient replacement, the potential productivity (i.e. fertility) of the soil and its associated crop yield will decline.

Several methods of nutrient management are employed. The oldest and that still characteristic of primitive systems in the humid tropics is that of shifting agriculture. After a period of extractive cultivation the land is abandoned for a period long enough to allow the original type of forest or woodland vegetation to rehabilitate the nutrient cycle; the 'new' nutrient store is then released when the area is burned preliminary to another period of cultivation. Rotation cultivation, in fact, is basic to fertility renewal in all save the most modern agro-ecosystems. In older European and many existing tropical systems a period of cropping alternated with, one during which the land is left fallow for 1 or 2 years. From the mid-eighteenth to mid-twentieth century the predominant type of farming was that in which cereals, roots and a sown grass–legume crop were rotated. The latter was the 'break-crop' which received the dung of animals pastured on it and which, over a period of 1–6 years, built up an organic nutrient store in the soil (see Ch. 19) and with addition of manure and other fertilisers (lime and superphosphate) established and maintained a higher soil fertility than previously.

However, since the Second World War there has been a dramatic increase in crop yields consequent on the availability of biocides (weed killers, pesticides) and of inorganic fertilisers (particularly of nitrogen, phosphorus and potassium) on the one one hand, combined with the breeding of high-yielding crop varieties (HYVs) on the other. It has been estimated that world consumption of inorganic fertilisers increased about five times between 1939/40 and 1979/80 and that 35–55 per cent of the yield of barley, wheat, potatoes and sugar beet in Britain today is a consequence of increased use of inorganic fertilisers (Hood 1982). However, it is generally considered that the increase in the harvest index of the short-stemmed large-headed cereal HYVs was the most important contributor to yield increase. And the production from the 1960s of HYVs of rice, wheat and maize adapted to humid (or irrigated) tropical farming initiated what became known as the 'Green Revolution' in the less developed countries.

EFFICIENCY OF CROP PRODUCTION

In terms of biological or photosynthetic efficiency crop plants, as illustrated in Table 6.7, are comparable to that of uncultivated temperate forest. In terms of the economic yield the efficiency is only about 0.3–0.4 per cent and, although harvested land accounts for *c.* 11 per cent of the global land surface, harvest production represents less than 1 per cent of the global primary biological production. This reflects the very considerable loss of energy because of the differences between the actual crop, the recoverable (harvested) and finally the economic crop. In addition, the energy efficiency (i.e. ratio of input energy subsidies to output food energy) of intensively produced HYVs is much lower than that in less intensively produced as conditions indicated in Table 6.8.

The low biological efficiency of agro-ecosystems compared with that of natural photosynethic yield in similar physical environments is the high price that has to be incurred in the production of high crop yield. Further, the relationship between increasing energy input and increasing yield is not linear – it follows the sigmoid 'growth' curve so characteristic of all biological processes. There is an optimum point beyond which the yield : input ratio starts to diminish and, if continued, can result in yield decline because of overcultivation and deterioration of soil structure. Since a lower proportion of photosynthetic material could not produce a higher harvest (food accumulation) index, higher potential yield must then be dependent on a means of increasing photosynthetic efficiency, a problem which still awaits a solution. However, actual record yields only attain near-potential yields under ideal experimental farm conditions. Average actual regional, national or global

Table 6.8 Comparison of photosynthetic efficiency for types of vegetation and selected cultivated crops (from Cooper 1975)

Crop or ecosystem	*Location*	*Growth period (days)*	*Photosynthetic efficiency (%)*
Natural ecosystem			
Tropical rain forest	Ivory Coast	365	0.32
	Denmark	180	
Pine forest	UK	365	1.95
Deciduous forest	UK	180	1.07
Crops			
Sugar-cane (March) (C_4)	Hawaii	365	1.95
Elephant grass (C_4)	Puerto Rico	365	2.66
Maize (two crops) (C_4)	Uganda	136–435	2.35
Maize (one crop) (C_4)	Keyna (uplands)	240	1.37
Soya beans (two crops) (C_3)	Uganda	135–435	0.95
Perennial ryegrass (mean of six C_3 cuts)	UK	365	1.43
Rice (C_3)	Japan	180	1.93
Winter wheat (C_3)	Holland	319	1.30
Spring barley (C_3)	UK	152	1.49

yields still fall well (50–60 per cent) below their potential either because of biophysical limitations or of low levels (intensity) of management.

LIVESTOCK PRODUCTIVITY

The productivity of domestic livestock, as in the case of undomesticated animals, cannot be measured per unit land area, since it is an expression of the annual production per animal or group of animals of the animal products of meat, milk, eggs, etc. Pasture carrying capacity (hectares needed per breeding animals) is sometimes used as a surrogate measure. However, since livestock vary in size and feed requirements comparisons can only be based on *livestock equivalents*, i.e. units of food energy required by different animals or annual maintenance and production. A very general rule would then equate two cows per horse, four sheep per cow and so on.

Livestock productivity is dependent on the rate of animal growth and/or milk/egg production and is dependent on the amount of food digested in excess of that needed to maintain the animal. The efficiency with which plant food is converted into animal tissues (the energy conversion ratio) varies with types and breeds of livestock. However, productivity can be curtailed by low food intake, environmental conditions (excessive heat which depresses appetite and cold which increases loss of energy) and pests (particularly parasites) and disease which, in extreme cases, result in death. And in the tropics the pests and disease hazards are such as to inhibit livestock production.

Under intensive livestock farming systems, particularly in temperate Western countries, livestock productivity has experienced the same recent dramatic increase as have crops. This is the result of breeding varieties with high reproductive and growth rates (so that meat, milk and number of eggs per breeding animal per unit time is increased) but which need a higher proportion of high-energy (protein-rich) feedstuff to achieve their productive potential than the less productive animals of the same type. The energy costs of animal protein food at the expense of primary biological productivity are even higher than those for crop production.

To maintain or attain a particular level of primary or secondary natural or managed biological productivity requires the continuous renewal of environmental resources.

Chapter

7

The biological cycle

Primary biological productivity is dependent on the constantly renewed input of solar radiation and on available sources of those inorganic elements essential for photosynthesis. The flow of energy through the biosphere is one-way and conforms to the first and second laws of thermodynamics in that: (1) it is neither created nor destroyed but merely transformed from one form (i.e. light) to another (e.g. chemical, heat); and (2) the efficiency of energy transformation (or conversion) is never perfect since, in its transformation from a concentrated form (chemical or food) to a dispersed form (heat), there is a loss of energy at each conversion stage. In contrast, the supply of inorganic elements is finite and, in many instances, limited in amount. To maintain primary productivity these elements must be constantly reused; and a very high proportion of the plant's energy store is, in fact, used in the process of the decomposition of organic matter. The alternating uptake and biosynthesis of inorganic elements from the physical environments, release as a result of organic decomposition and eventual reuse constitutes the *biological cycle*. It involves the circulation (or flow) of elements within the ecosystem between the organic *pools* in which they are retained for varying periods of time and between these organic pools and the inorganic *reservoirs* from which they were initially drawn (see Fig. 7.1). The main stages in this circular route are: (1) the production and accumulation of dead organic matter (DOM); (2) decomposition; (3) absorption of minerals by the biomass.

DEAD ORGANIC MATTER

About 95 per cent of the annual terrestrial DOM production is derived from plant material, the remainder coming from animal faeces plus the tissues and skeletons of dead animals. Some DOM is composed of standing dead material (Satchell 1974a) and some of senescing or dead plant material attached to the living plant. Recently dead material which accumulates on the ground surface is the *litter*. It is composed mainly of leaves together with flowers, glumes, seeds, fruits, small

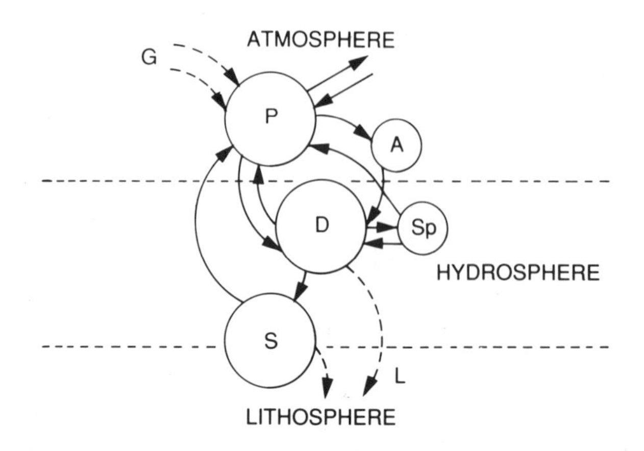

Fig. 7.1 The global biological cycle – nutrient cycling between primary inorganic reservoirs and organic pools (circles): P = plants; A = animals; D = detritus; Sp = soil micro-organisms; S = soil organic matter; G = gains; L = losses

twigs and occasionally larger branches and tree boles (Bell 1974). Leaves contribute from 60 to 80 per cent of the total litter fall of trees (see Table 7.1) and shrubs to over 80 per cent of that of herbaceous plants. Below the ground surface DOM includes dead, particularly fine, roots and root hairs, sloughed root tissues, mucilageous material (*mucigel*) and root exudates (amino acids) and animal tissues (Coleman 1976).

The annual litter production is primarily a function of the productivity of the existing biomass and the longevity of its constituent parts (see Table 7.2). It has

Table 7.1 Annual leaf and litter fall in forest vegetation (after Jensen 1974)

Forest type	*Locality*	*Total leaf litter fall* $(ha^{-1} (year^{-1})$
Cool temperate		
Alder (*Alnus* spp.)	Oregon (USA)	3.64–6.39
Birch (*Betula* spp.)	Finland	0.94–1.27
Beech (*Fagus* spp.)	S. Sweden	3.57
Oak (*Quercus* spp.)	England	2.13
Warm temperate		
Eucalyptus regnans	Australia (Vict.)	3.6–4.2
Nothofagus sp.	New Zealand	2.14–6.91
Q. ilex	S. France	2.45–2.72
Tropical rain forest	Nigeria	3.7
	Ghana	7.0
	Thailand	11.9

Table 7.2 Mean biomass of litter layer in major world types of forest and ratio of biomass litter layer : litter fall (from Rodin and Bazilevich 1967)

Vegetation	*Litter layer mean biomass* ($t\ ha^{-1}$)	*Litter fall as % biomass*	*Biomass ratio litter layer: litter fall*
Shrub tundra	83.5	1.7	92
Taiga pine forest (south)	44.5	—	20
Taiga spruce forest (north)	35.0	4.0	17
Taiga spruce forest (south)	30.0	2.0	10
High-oak forest	15.0	1.5	4
Subtropical forest	2.0	5.8	0.7
Tropical rain forest	—	—	0.1

traditionally been used as an index of net production. Variations in litter production parallels but is generally lower than that of NPP except in the case of grassland communities where virtually the whole of the annual above-ground shoot production may die down at the end of the growing season. At the global scale litter-fall patterns vary with climate and vegetation type, being markedly seasonal in cool temperate and subtropical zones. In the former the main leaf-fall period is at the end of the thermal growing season (Oct./Nov.), in the latter at the end of the dry season (Aug./Sept.). In the equatorial forests leaf fall is continuous except during longer dry periods.

Because of the difficulty of assessing the annual volume and production of root detritus, the contribution of the below-ground biomass to the detrital pool has been either neglected or underestimated in published literature. However, during the past 25 years or so, assessment of root productivity and activity has become an important focus in ecological research. Analysis of existing and of recent data (Rodin and Bazilevich 1967; Coleman 1976) has revealed that roots make a significant, and often major, contribution to the total DOM in many ecosystems (see Table 7.3), particularly in grassland and desert vegetation.

DECOMPOSITION OF ORGANIC MATTER

DECOMPOSERS

Decomposition is the process whereby the complex organic molecules in dead plant and animal tissues are ultimately broken down into their original inorganic constituents of carbon dioxide, water and mineral elements. It is effected by a large variety of *saprovores* that feed directly or indirectly on the dead and decaying organic matter (see Fig. 7.2). The most important group of general-purpose, primary decomposers are the micro-organisms – bacteria and fungi whose role can be compared to that of the grazing herbivores above ground. The bacteria comprise a vast array of species of minute, single-celled organisms which, given favourable environmental conditions, can proliferate very rapidly. Most are

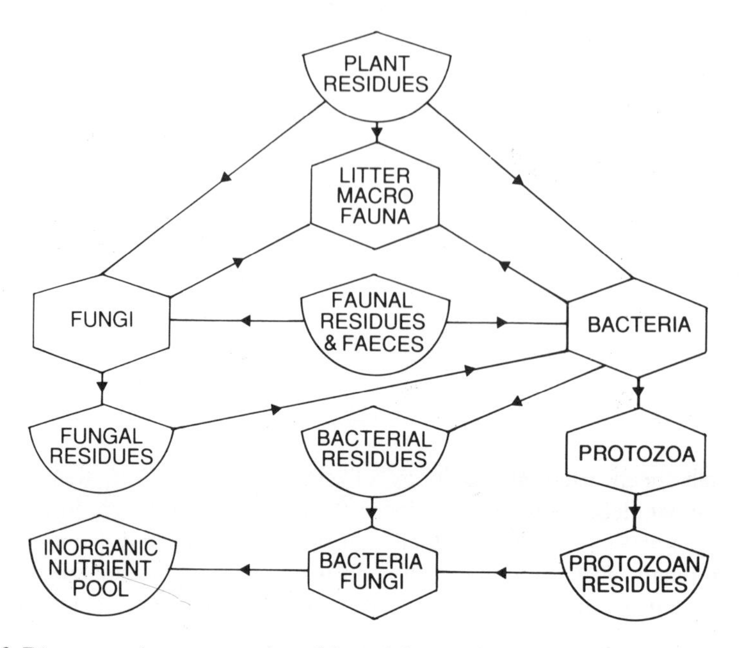

Fig. 7.2 Diagrammatic representation of the soil detritus food web greatly simplified (from Richards 1974)

Table 7.3 Distribution of DOM production in four terrestrial ecosystems (g m^{-2} $year^{-1}$) (from Coleman 1976)

Production site	*Arctic tundra (Alaska)*	*Deciduous forest (Tennessee)*	*Lightly grazed short-grass prairie (Colorado)*	*Cool desert shrub (Utah)*
Above ground	150	700 (300)	118	120
Below ground	100	900	576	270
Ratio below-ground : Total	0.4	0.56 (0.71)	0.83	0.69

aerobic. Some are facultative aerobes in that although they normally require oxygen, they can adapt to oxygen-deficient conditions; others are obligate aerobes that only function in the absence of oxygen. Certain species of bacteria are practically entirely responsible for three vital processes: the oxidation of sulphur compounds (a process by which they obtain energy); the fixation of gaseous nitrogen; and the nitrification or release of inorganic nitrogen from organic compounds. The fungi (or so-called 'moulds') are filamentous forms. Although very abundant in all types of organic substrate they frequently become dominant in acid (i.e. pH < 6.0) soils. Their growth rate, however, is usually slower than that of the bacteria and they are less efficient nitrifiers. They are most abundant in the litter layer on or near the soil surface, a location which is related to their function as more efficient cellulose reducers than bacteria.

In addition to the bacteria and fungi a diverse group of meso- and macrofaunal saprovores infest DOM and thereby aid the process of decomposition. Some live permanently in the detritus, others are only temporary residents using the soil for shelter, feeding or breeding. Some feed on primary detritus, others on faecal matter (*coprophytes*). They are primarily responsible for the maceration (i.e. mechancial comminution) of litter which aids leaching from and the microbial invasion of cells. In addition, they contribute partially decomposed and nitrogen-enriched faeces and urine to the substratum. The meso- and macrofauna also facilitate the downward movement and mixing of DOM with the underlying mineral soil. The amount and diversity of the soil faunal biomass vary with the volume and nutritive quality of the detritus available. It is limited by cold and nutrient-poor, acid, litter in northern latitudes, by lack of moisture in arid and semi-arid regions and by high temperatures in the humid subtropical and equatorial regions. The maximum occurs in nutrient-rich soils in temperate deciduous woodland and grassland (Satchell 1974a).

The most important group of macrofauna are earthworms in humid temperate climatic regions and termites in tropical humid and arid regions (Swift *et al.* 1979). There are numerous species of earthworms (e.g. about twenty-five in Britain). The dominant, however, is *Lumbricus terrestris*, found particularly in detritus with a low carbon : nitrogen (C : N) ratio, a high content of soluble carbohydrates and a low tannin content. It feeds exclusively on DOM, ingesting it with the mineral matter through which it burrows and, migrating between the detrital and mineral layers in the soil, brings about a very efficient mixing of the two materials. In a nutrient-rich, well-aerated soil there may be as many as 25 000–1 000 000 earthworms per hectare; the weight of worms in pasture land may exceed that of the grazing livestock. Worms are the 'natural tillers' of the soil. They progressively bring soil from below and redeposit it on the surface. Evidence from archaeological sites indicate that this may, in some instances, be at the rate of 15–20 cm per century. Earthworms are normally scarce in acid, nutrient-deficient and waterlogged soils where less efficient mesofauna become dominant.

The ecological equivalent of the earthworm and the typical and often dominant members of the soil fauna in tropical, subtropical and to some extent in warm temperate (i.e. Mediterranean) climatic regions is the termite (*Isoptera*). It feeds on a wide range of both living and dead organic material. It can digest cellulose and is responsible for the breakdown of the abundant woody surface litter in tropical forests. In the assimilation and digestion of organic matter termites (particularly wood-feeding species) produce faeces with a very high C : N ratio. In some species an extremely high proportion of this is used in hive or nest-building. Together with salivary excretions it is used to line the internal walls and galleries of their nest systems (Lee and Wood 1971; Wood 1976). The majority of the nests are subterranean; all or parts of others appear as mounds of varying height and shape on the ground surface, dependent on the species involved.

Wood (1976) notes several differences between termites and invertebrate saprovores. First, a very large quantity of plant material is eaten particularly in the tropical savanna before decomposition, and second, recycling of nutrients via the faeces is slow because they can only become available when all or part of the nest is abandoned. Termite nests can persist for less than 10 to over 80 years and many are highly resistant to decomposition after abandonment. In contrast, other meso- and macrofaunal decomposers normally return *c.* 70–90 per cent of their ingested food to the substratum where it is rapidly decomposed by micro-organisms. As a result termite nests can form virtually closed nutrient systems in which unhealthy and dead individuals are consumed in the nest and losses by predation and breeding flights are relatively small.

DECOMPOSITION PROCESSES

Organic decomposition involves three basic processes (Mason 1977):

1. The rapid leaching of soluble products either from the foliage or the decomposing detritus;
2. Mechanical breakdown of DOM by freezing and thawing, wetting and drying, etc.;
3. Biological degradation – the maceration (comminution) and ingestion of DOM by meso- and macrofauna and oxidation by microbial respiration.

The main temporal and spatial stages in this process are illustrated in Fig. 7.3.

Once mechanical and/or biological degradation is initiated, leaching from, and rapid microbial invasion of, exposed cells proceeds rapidly. Enzymatic secretions accelerate the process, particularly the breakdown of carbohydrates. As a result the C : N ratio of the organic matter decreases. This stimulates the proliferation of the microbial decomposers and soil fauna. Decomposition, however, involves not only the breakdown of organic material but the synthesis of new animal and microbial tissues. The latter, in time, constitutes an increasing proportion of the soil organic matter (SOM) and may prove as, or more, resistant to degradation than the original detritus (DOM). This is the result of two processes:

1. *Mineralisation* (or oxidation) of organic compounds during microbial respiration;
2. *Immobilisation or remineralisation*, i.e. assimilation of the inorganic nutrients released into new microbial tissues (see Fig. 7.4).

Mineralisation proceeds more or less rapidly dependent on two interacting variables. The first is the relative decomposability of the detrital material. Young, feebly lignified, nutrient-rich tissues will break down more rapidly than old, nutrient-poor, woody and/or resinous material. The second is the environmental conditions, particularly temperature, humidity and aeration of the substratum. 'Flushes' (i.e. rapid increase in the rate) of decomposition are triggered off when unfavourable conditions are followed rapidly by favourable ones. However, during the initial stages of decomposition there is often severe competition for nutrients, particularly for nitrogen, phosphorus and sulphur, between growing plants and the increasing microbial populations. Plant growth may as a result suffer a temporary check. This is significant in agro-ecosystems where an initial decrease in crop

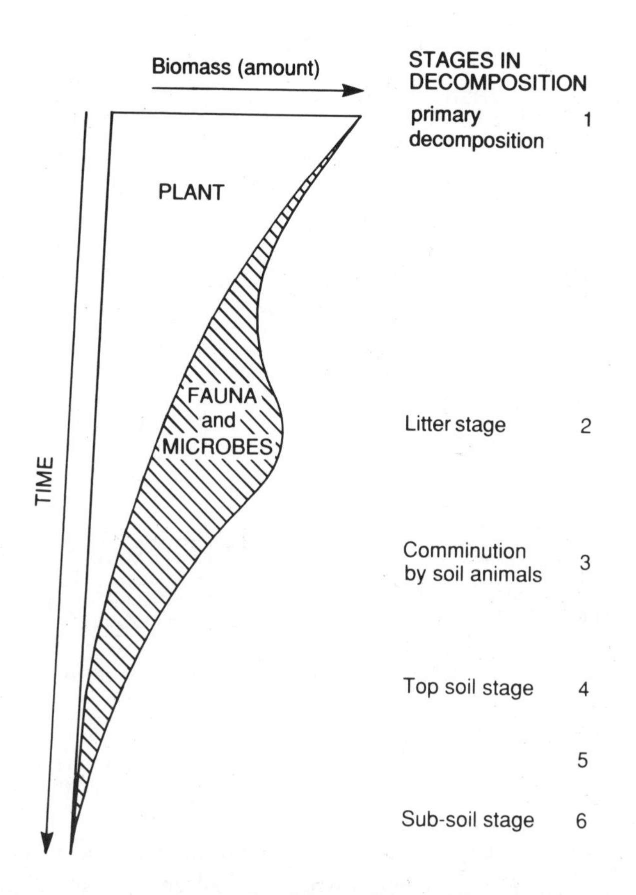

Fig. 7.3 Stages in terrestrial decomposition (adapted from Stout *et al.* 1976)

yield often accompanies an over-generous application of fresh manure to a newly cultivated soil.

Mineralisation is accompanied by immobilisation. Some of the nutrients released but not absorbed by plant roots or absorbed by the soil colloidal complex are recycled by the microbial decomposers. The large populations of diverse species with varying life spans ensure a long period of recycling within the DOM, accompanied by slow mineralisation. Immobilisation may account for the retention of 40–80 per cent of such elements as potassium, nitrogen and phosphorus which would otherwise be easily leached and hence lost from the system (Witkamp and Ausmus 1976). Immobilisation is also essential to the process of *humification*, i.e. the production of humus from the more resistant fractions of the initial DOM and from the new by-products of remineralisation. *Humus*, the end-product in a continuum of decaying organic matter, is very difficult to define or to isolate and is also particularly resistant to further decomposition. It is a dark, completely amorphous, colloidal substance composed of varying proportions of large, complex organic molecules of humic and fluvic acid, and of humin. The acids,

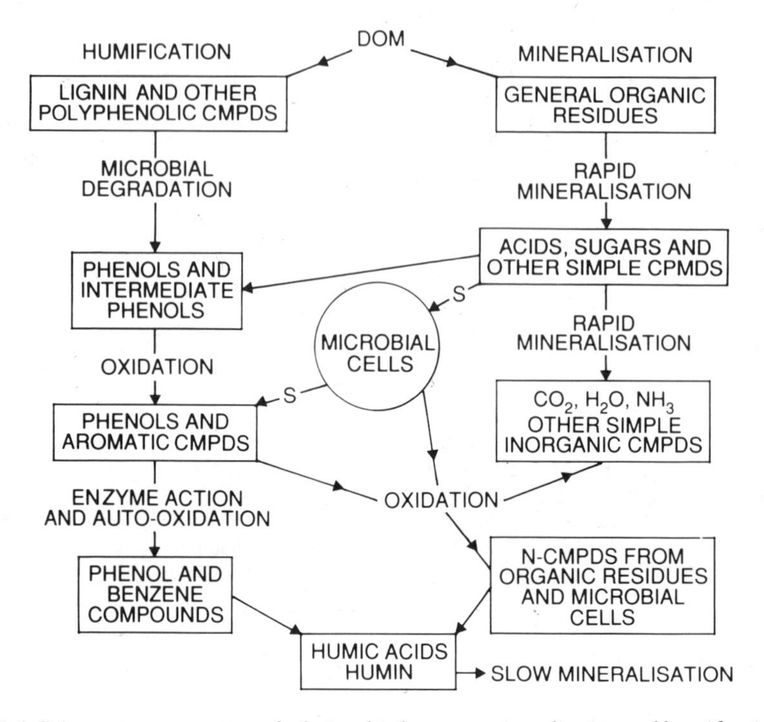

Fig. 7.4 Schematic presentation of relationship between mineralisation and humification in biological degradation; S = remineralisation or microbial immobilisation (simplified from Pitty 1979)

which may comprise 50–80 per cent of the total humus, can be isolated by very alkaline (over pH 10) or acid (less than pH 2) solvents. In some circumstances humus can form organometallic compounds or *chelates* with iron, aluminium and manganese in the soil; in others, a stable association with the clay fraction (particles < 0.001 mm diameter) of the mineral soil.

The type of humus produced is largely dependent on the composition of the original DOM. A well-aerated, nutrient-rich litter with a low C : N ratio will tend to produce a well-decomposed *mull humus*, which becomes well mixed with, and distributed through, the soil mineral matter. It can combine with the clay fraction in the soil to form what is known as the *soil colloidal complex*. As such, it promotes a stable crumb structure which enhances the capacity of the soil to adsorb and retain nutrient cations and hence check mineral loss by leaching. In contrast, the decomposition of resistant nutrient-poor, acid plant tissue is slow and both mineralisation and humification are curtailed. A poor relatively thick brown-black acid *mor humus* forms on the soil surface in which a litter (L), humifying (H) and fermenting (L) layer can be distinguished. With a paucity of soil macrofauna, particularly earthworms, little mixing of the organic with the mineral substrate takes place. Further, the high production of strong organic acids can effect the disintegration of both the clay minerals and the humus with the consequent mobilisation and leaching of the normally less soluble ions of iron,

Table 7.4 Characteristics of mull, moder and mor humus (from Duchafour 1965)

Humus type	*Soil horizon*	*Structure*	*pH*	*C : N*	*Mineralisation*	*Animals*
Mull	A, grey-brown	Granular	5.5	< 20 (12–15)	Rapid	Worms
Moder	A_0 2–3 cm (brown) A, black	Weak cohesion	4.0–5.0	15–25	Moderate	Insects
Mor	A_0 5–20 cm (brown) A, 2–5 cm (black)	Fibrous	3.5–	A_0 30–40 A, 25	Slow	?

aluminium, manganese and humus down the soil profile. Table 7.4 summarises the main morphological characteristics of mull and mor and the intermediate moder types of humus.

RATE OF DECOMPOSITION

The rate of decomposition is normally expressed as the percentage loss of dry weight of litter through time. The absolute daily rate is known as the *relative decay rate*, the average annual rate as the *annual fractional loss rate* (Satchell 1974a). It is dependent on the interaction of a number of variables, including the chemical composition and age of the DOM and the prevailing environmental conditions. The composition of the DOM and, particularly, the relative proportion of easily decomposable carbohydrates (sugars/starches) and protein to the more resistant cellulose, lignin, fats, waxes and tannin exercise a major control on the rate of decomposition. Further, the nitrogen requirements of the microbial decomposers is relatively high and hence nitrogen deficiency is a significant limiting factor. The C : N or carbon : mineral (C : M) ratios have long been used as indices of potential decomposability. However, as Paul (1989) notes, a high C : N ratio may indicate that the actual nitrogen concentration is low. However, it could also mean that although insoluble carbon is present in excess, the amount of nitrogen is nevertheless favourable for decomposition.

Recently, more emphasis has been put on the effect of the lignin content of the DOM on decomposition rates (Metello *et al.* 1984). Meentemeyer (1984) uses climate (expressed as annual evapotranspiration) and lignin concentration (at low 5 per cent, intermediate 15 per cent and high 30 per cent) of organic matter to predict the distribution of decomposition rates in humid North America. Arid areas were excluded because of the effect of termites whose relative control has not been established. His statistical analysis of distributions based on the three lignin concentrations indicates that at the continental scale climate is the most important determinant of rate of decomposition. At more local scales the chemical composition of the litter becomes the controlling factor.

The rate of decomposition is also influenced by the same environmental factors as are all other organic processes. Decomposition will be at a maximum when optimal temperature and moisture are combined with a well aerated and nutritive habitat. There is a close but not completely perfect synchronisation of the rate of decomposition with that of primary biological productivity. Both increase with increasing temperature, provided neither nutrients nor moisture are limiting. Optimum temperatures (30–35 °C) for decomposition, however, exceed those for productivity (20–25 °C), ensuring an adequate supply of nutrients particularly in hot humid climatic conditions where growth rates are very rapid. However, with lower temperatures the rate of decomposition declines more rapidly than that of litter production. Decay rates decreases 25-fold from the tropical rain forest to the Boreal forest; primary production does so only 2.5-fold according to Kononova (1961). As a result DOM matter accumulates. This is most evident where low temperatures are combined with resistant mineral-deficient litter as in the tundra. The ratio of non-humified to humified organic matter also increases as does that of the litter-layer biomass to the litter-fall biomass (see Table 7.2). In the tropical rain forest the nutrient capital available is relatively small but the turnover rate is sufficiently high to meet the requirements of the continuously high primary production. In colder climates the capital accumulation is larger and the turnover rate slower.

Finally, the rate of decomposition declines as the age of the initial detritus increases and, consequently the amount of nutrients available for microbial decomposers diminishes. After the initial, relatively rapid phase of decay, in which mineralisation is dominant, the decomposition of the more resistant, and particularly, humified soil organic material becomes progressively slower. In addition, a proportion of the organic matter can, as already explained, become 'bound' to inorganic elements in the soil in such a way as to protect them from decay. Humus adsorbed on to clay particles may persist for indefinite lengths of time. Carbon dating has revealed that as much as 50 per cent of soil organic matter is over 1000 years old (Paul 1989).

PEAT

The amount of DOM and SOM organic matter is then a function of three interrelated variables: litter production, decomposition of DOM and the fixation or binding of organic matter by the inorganic soil fraction. It is assumed that under relatively stable biophysical conditions an equilibrium is reached at which the amount of organic matter on and in the soil will remain relatively constant. However, when soils are waterlogged (saturated) decomposition is retarded by lack of oxygen, the rate of accumulation exceeds that of decomposition and slowly decomposing DOM accumulates to a depth at which it is designated *peat*. Peat is generally defined as a surface layer over 30 cm thick and with an organic matter content of at least 80 per cent of its dry weight content. It is mainly formed of plants such as the bog mosses and sedges, and woody shrubs – willow and alder adapted to growth in waterlogged habitats. They form a relatively shallow-rooted actively growing surface zone. Predominant among the peat-forming plants are the bog mosses (*Sphagnum* spp.) which can absorb and retain (saturate themselves!)

water from the atmosphere and so maintain upward growth. So absorbent is bog moss that it used to be used as a surgical dressing!

The rate of peat growth is difficult to establish because of compression of lower by upper layers, as well as differences in the rate of decomposition of the constituent material. Peat formation is, however, primarily a low-altitude, temperate to cold region phenomenon. In these climatic areas depth of peat tends to decrease with altitude and latitude parallel with the decrease in plant growth and DOM accumulation.

Peat accumulates under two particular conditions. First, in climatic conditions where evapotranspiration exceeds precipitation and atmospheric humidity is high for most or all of the year and where surface drainage is retarded by the gradient of the ground surface. In Britain Pearsall (1950) suggested that *blanket peat* can form on land surfaces with a gradient of less than 15 degrees in those areas where mean annual rainfall was over 55 inches (140 cm) per annum. Second, in depressions and on valley floors i.e. *valley* or *basin peat* where the water-table is at or above ground level for all or most of the year. Under some conditions the peat can grow to a level above that of the water-table, allowing the surface to become drier and to be colonised by shrubs and trees adapted to drier soil conditions. However, in climates with particularly high atmospheric humidity, the bog mosses (*Sphagnum* spp., *Polytrichum* spp. etc.) can absorb moisture and maintain the upward peat growth. This results in what is called *raised peat or bog* in which blanket peat forms on the original basin or valley peat. In many peat deposits there may be two or more layers in which the recognisable remains of trees and shrubs are concentrated. In some cases these are thought to indicate as in Scotland alternating past dry and wet periods; in others such as the Boreal forest (see Ch. 10) cyclic growth of peat and trees.

Peat can vary in both physical (structural) and chemical composition. The first is dependent on the type of plants of which it is formed. Mosses and sedges in which the C : N ratio is relatively low decompose into a structureless material while the woody bog-plants produce a more fibrous and often laminated structure. The second is dependent on the nutrient status of the water source. Water derived from the atmosphere and/or a nutrient-poor source will be *oligotrophic* (acid), from a nutrient-rich source, *eutrophic* (base rich); the latter is sometimes referred to in Britain, as *fen peat*.

All peat, whatever its origin or composition, has certain distinctive characteristics which distinguish it from a mineral soil. First, its bulk density is very low (*c.* 0.2–0.3) so that when dry is of light weight. It is then porous and easy to cultivate. Second, it has an extremely high water-holding capacity, on average two to four times its own dry weight but up to twenty times in some moss–sedge peats (Brady 1974). In addition, most of the water (*c.* 90 per cent of its dry weight) occurs in a colloidal combination with the organic matter – and it is so tightly held as to be unavailable (so-called 'dead water'). Lateral movement of water through peat is slow and artificial drainage can be difficult. Drainage which needs to be more intensive than on most mineral soils results in shrinkage and lowering of its surface relative to the adjacent non-peat land. Third, by reason of its colloidal characteristics peat has a high nutrient adsorptive capacity. However, although its cation exchange capacity is considerable, it is an extremely well-buffered material which puts a severe constraint on attempts to decrease acidity.

Finally, due to the anaerobic conditions both mineralisation and humification are slow. The former tends to be confined to the better-aerated surface litter layer. The latter involves a very slow fermentation process with the production of complex organic compounds (humic acid) and bog-gases (methane (CH_4) and hydrogen sulphide (H_2S)). The relative rates of fermentation and humification vary, dependent on the C : N ratio and the lignin content of the original DOM. However, because of the slow decomposition rate a high proportion of nutrients remain tightly bound in the large but relatively 'inactive pool' of peat. Work at Moor House Research Station in England has shown that in acid peat found there, only 10 per cent of the nitrogen content is recycled.

Once the peat surface dries out and the protective vegetation cover is either deliberately removed or degenerates, dry peat can become very susceptible to wind erosion – a particular hazard on the well-cultivated granulate structure of the fen peats.

NUTRIENT UPTAKE

The inorganic elements released in the process of decomposition may be:

(a) leached out of the place in which they were produced and deposited further down the soil profile beyond the reach of the deepest plant roots or be removed entirely from the soil system;
(b) absorbed by soil micro-organisms and plant roots;
(c) temporarily retained by adsorption on the soil colloidal complex.

The *labile* or *available pool* of most mineral elements is small compared with the total mineralomass bound up in the living and dead organic matter. Most of the readily available minerals occur in the soil solution as electrically charged positive cations or negatively charged anions. The cations (or exchangeable ions) can be adsorbed on to the negatively charged clay and/or humus particles, or on to the root hairs. The anions are not so readily absorbed as the cations and hence tend to be more susceptible to loss by leaching.

The rate and ease with which nutrients are absorbed by the living plant biomass will depend not only on the rate and stage of plant growth but on the prevailing temperature and moisture conditions, the pH and the nutrient status of the soil. The optimum pH at which most nutrients become readily available is *c.* pH 6.5. As indicated in Fig. 7.5, with increasing or decreasing pH some minerals become more or less available. Availability can also be affected by antagonist reactions between mineral elements. For instance, the uptake of boron is curtailed in calcium-rich soils, zinc and iron by an excess of copper or a surplus or deficiency of molybdenum. Ultimately, however, the amount and balance of nutrients absorbed will be a function of the plant species and the particular soil and climatic environmental conditions to which it has become evolutionarily adapted.

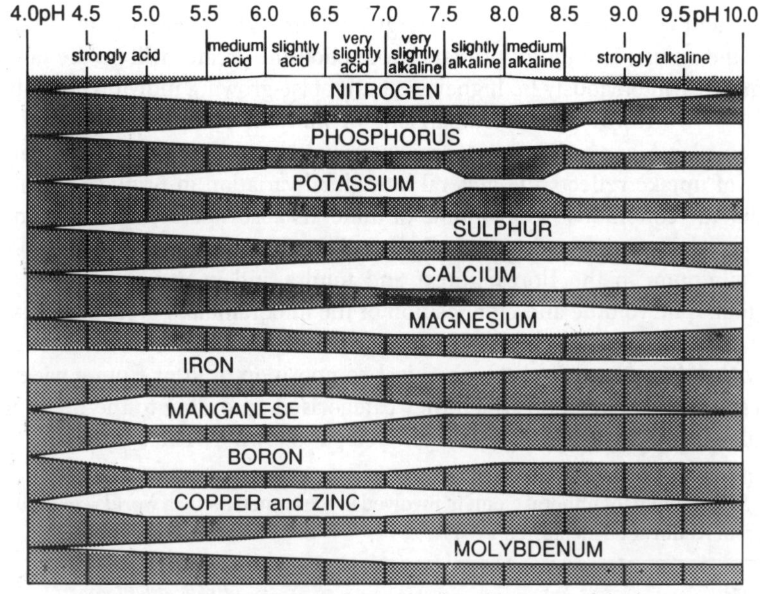

Fig. 7.5 Soil reaction influence on the availability of plant nutrients

Mineral uptake can also be facilitated by *symbiotic associations* between living organisms such as occurs between *mycorrhizal fungi* and the roots of certain trees and shrubs. The mycelia of the former are often particularly active in decomposing litter and provide a direct route for the rapid passage of nutrients into the plant. This association is particularly well developed in the tropical rain forest where rates of decomposition are very high and where a dense tree root-mat with its symbiotic fungi is concentrated on or near the soil surface (Witkamp and Ausmus 1976). The mycorrhizal association may result in 35 per cent increased uptake of phosphorus, 75 per cent of potassium and 85 per cent of nitrogen by trees (Duvigneaud and de Smet 1970). Many coniferous trees (*Pinus* spp.) and shrubs (*Calluna vulgaris*) cannot regenerate successfully on acid soil in the absence of mycorrhizae. Another important symbiotic relationship is that between nitrogen-fixing bacteria and the root nodules of leguminous plants. The former provide the plant with nitrogen, while the plant provides food energy for the micro-organisms.

THE NUTRIENT CYCLE

While the process of nutrient cycling is universal, the volume, composition and intensity of the cycle varies dependent on the species and life forms of the organisms involved on the one hand and on the physical and chemical conditions of their habitat on the other.

VOLUME

The nutrient cycle parallels that of energy flow through the ecosystem. Hence the greater the annual amount of primary production the greater will be the uptake of nutrients. It will obviously be higher in young, fast-growing individual plants than in older slow-growth ones. Comparisons of nutrient cycles must, therefore, be based on plants or vegetation at similar growth stages. At the global scale, the volume of uptake reflects the general latitudinal variation in primary productivity with climatic conditions. The volume of minerals absorbed attains a maximum in the tropical rain forest – where it is three to four times that in temperate forests – and a minimum in the Boreal forest and tundra and in dry steppe and desert vegetation. The volume and composition of the mineralomass cycle also vary from one ecosystem to another. In their analysis of the data available Rodin and Bazilevich (1967) compared the chemical composition of litter from a wide range of types of world vegetation which they categorised according to the predominant mineral element involved in the annual cycle, as shown in Table 7.5.

Table 7.5 Predominant mineral element involved in the annual cycle in world vegetation types (from Rodin and Brazilevich 1967)

Vegetation	*Predominant mineral*
Tundra	Nitrogen (very predominant)
Temperate and subtropical deciduous forest	Nitrogen
Broad-leaved temperate forest and semi-desert shrub	Calcium
Steppe, savanna, semi-desert shrub with grass, tropical rain forest	Silicon
Desert	Chlorine

They also noted that the proportion of organogen minerals (carbon, potassium, phosphorus, sulphur) in litter ash is higher (85–90 per cent) in the mineralomass of forest and shrub vegetation than in steppe, dry savanna or deserts. Nitrogen and calcium are the most abundant elements in forest and shrub, except in the case of tropical forests in which a preponderance of silica (50–60 per cent) together with a significant proportion of aluminium and iron (20–50 per cent) distinguish them from temperate forests. The nutrient cycle of the latter, including both deciduous and coniferous stands, is dominated by calcium (40–50 per cent) which in the tropical rain forest accounts for only 15–20 per cent of the mineralomass. Dry steppe, savanna and desert vegetation, in which grasses are the principal or an important element, are also characterised by a high proportion of silica. But overall, the percentage of biohalogens (sodium + calcium + excess silica) tends to be higher (over 50 per cent and up to 90 per cent) in alkaline deserts than in forest cycles. Despite its ecological significance, particularly for biomass increase, nitrogen only becomes the first ranking nutrient in the Boreal forest and tundra ecosystems.

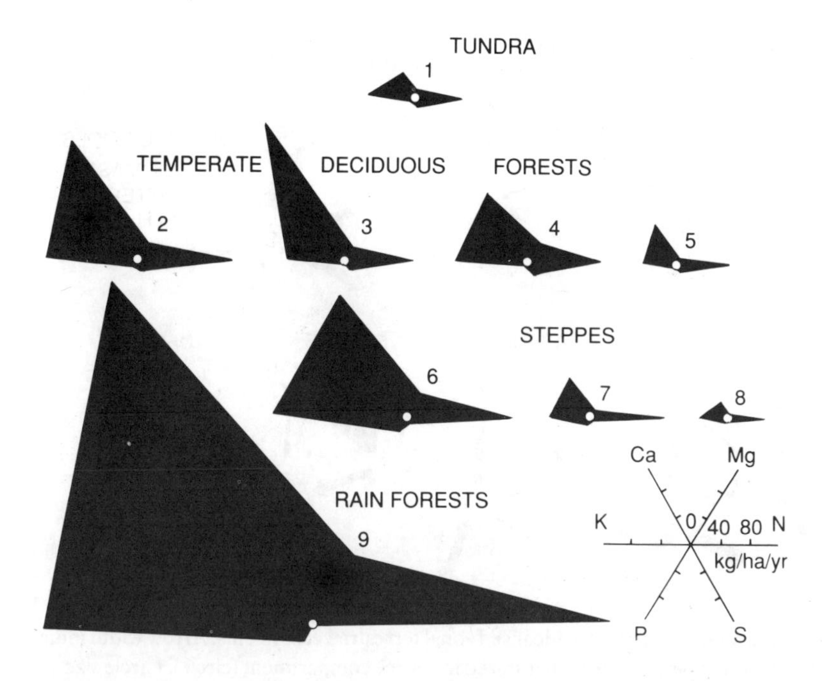

Fig. 7.6 Volume and composition of annual uptake of the six main macronutrients (Ca, Mg, N, P, K, S) in major terrestrial ecosystems (from Duvigneaud and de Smet 1970)

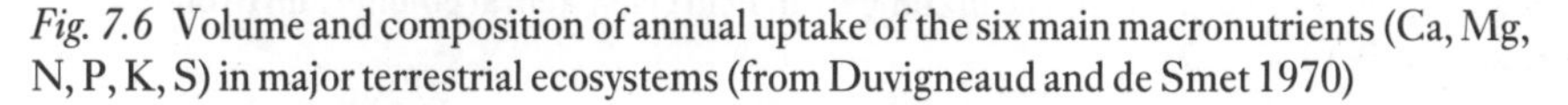

Duvigneaud and de Smet (1970) made a graphical comparison, illustrated in Fig. 7.6 of the volume and composition of the annual uptake of the six main macronutrients (calcium, magnesium, nitrogen, phosphorus, potassium, sulphur) in some of the major terrestrial biomes. The size and form of the graphs reflect the plant/soil relations. At the ecosystem level the tropical rain forest is distinguished by the size of the cycle in which calcium, nitrogen and potassium are abundant and phosphorus and magnesium are scarce. Among the temperate deciduous forests of comparable age, on base-rich soils, calcium uptake is higher than that of nitrogen while the reverse can occur on poorer acid sites. The range of variation in the mineralomass of steppe vegetation is comparable to that in temperate deciduous forests, with large cycles in long grass vegetation and much smaller ones in short grass steppe in which nitrogen is the dominant element. The grassland graphs fail, however, to take account of calcium and chlorine which in these ecosystems are more important than the macronutrients selected.

NUTRIENT FLOW

Nutrient cycles differ not only in total volume but in the size of the nutrient pools (i.e. biomass, litter and soil organic matter respectively) and the volume and rate of *nutrient flow* or *flux rate* between them. Gersmehl's (1976) diagrammatic modelling of the relative size of nutrient pools and flows in the main types of world vegetation shown in Fig. 7.7 highlights the paucity of nutrients in the tundra

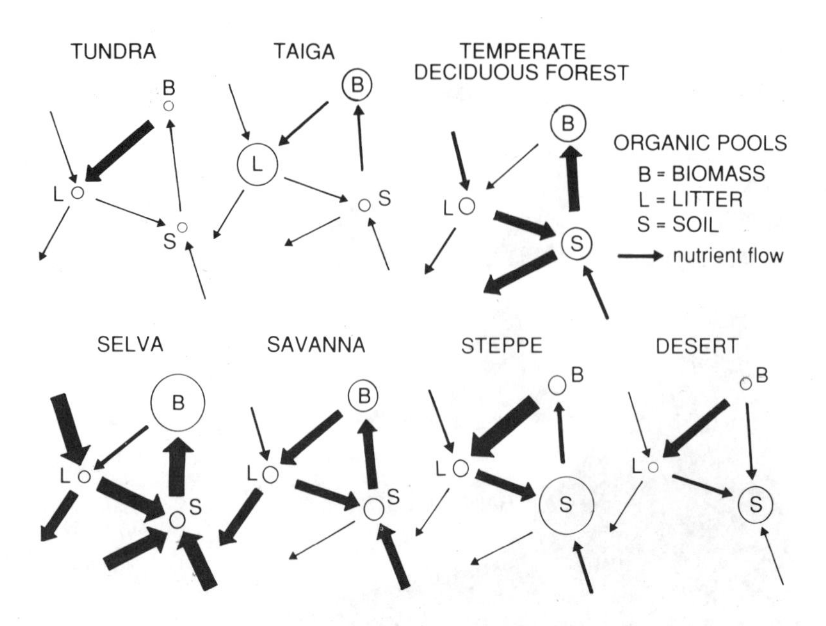

Fig. 7.7 Nutrient circulation in idealised zonal terrestrial ecosystems. Arrow width indicates nutrient flow as a percentage of that stored in source compartment (circle). Circle size denotes amount of nutrients stored in compartment in ecosystem at equilibrium (from Gersmehl 1976)

and the limited flows from the litter and the soil compared to that from the biomass, the relative importance of the soil pool in grassland and desert, and of the biomass in the tropical rain forests and the taiga; and the more even balance of pool and flow within the others. The volume of flow provides a comparative measure of the extent to which nutrients are released in the course of one year. However, as Gersmehl notes, actual nutrient cycling can deviate from these models in at least four ways. Since the relationships are non-linear, a doubling of nutrient input may not give rise to a proportionate increase in volume in nutrient flow; similar changes may cause different responses at different times; nutrient mobility varies with the chemical (e.g. pH) as well as the climatic environment. Biotic effects also vary in that different plants create different conditions which affect both storage and movement.

INTENSITY OF NUTRIENT CYCLING

The intensity or rate at which mineral nutrients are cycled is a function of the rate of nutrient absorption or uptake by the biomass and of organic decomposition and release. Many ecologists have, in lieu of more exact measurements of nutrient flux, followed Rodin and Brazilevich in the use of the ratio

L (annual litter accumulation) : *LF* (annual litter fall)

as an index of the intensity of nutrient cycling. The close correlation of this index with rates of growth and mineralisation of organic matter is hardly surprising.

However, the development of more sophisticated techniques have provided the means of analysing nutrient accumulation and flow between the various soil/vegetation compartments more accurately, particularly in deciduous forest ecosystems.

Intensity of nutrient circulation varies within as well as between the main nutrient pools. And in this respect a distinction can be made between short-run (rapid) circulation in plants with a short life span (annuals) and long-run (slow) circulation of nutrients in perennials with storage organs in which nutrients are held for longer periods than in the green photosynthesising tissues. In the case of trees some 30–50 per cent of the annual nutrient uptake in young, fast-growing trees and 10–15 per cent in old trees is retained in the woody tissue with a potential life span of decades in comparison to the leaves with a life span which varies from less than a year to a few years at most. Long- and short-run nutrient circulation also occurs in the pool of DOM, with recent litter mineralising rapidly and older humified material mineralising only very slowly. Similarly the rate and completeness with which the individual nutrient elements are circulated vary according to their relative availability and mobility and the extent to which they may become bound or fixed in the soil. While short-run cycling is usually expressed as *annual flux rate* in weight per unit time per unit area or biomass, long-term cycling must be expressed in *turnover time* in total years or length of half-life (i.e. years needed to reduce the amount of original material by half). Table 7.6 compares the turnover times for carbon, nitrogen and calcium in a

Table 7.6 Comparison of turnover time of carbon, nitrogen and calcium in temperate deciduous forests in Tennessee (from Reichle *et al.* 1975)

	Turnover time (years)		
Component	*Carbon*	*Nitrogen*	*Calcium*
Soil	107	109	32
Forest biomass*	155	88	8
Litter (01 + 02)	1.12	<5	<5
Total†* 54	1815	445	

* Above-ground biomass only.
† Total calculated as sum of elements in living and dead components of the ecosystem: 01 = top; 02 = second litter deposit.

deciduous forest in Tennessee (USA). Nitrogen is distinguished by a mean residence time longer than that for any other mineral nutrient. This is a consequence of the relatively large pool of nitrogen conserved in the soil organic matter from which release is slow (see Table 7.7).

NUTRIENT MOBILITY AND AVAILABILITY

The relative mobility and availability of the macronutrients varies. Calcium, magnesium and to a lesser extent potassium occur in mineral form, are rapidly

Table 7.7 Turnover time (years) of nitrogen in various parts of the global terrestrial ecosystem and some selected ecosystems (from Rosswall 1976)

Ecosystems	*PP*	*PL*	*SOM*	*MO*	*IN + N*
World	4.9	1.1	177	0.99	0.53
Tundra mire	5.6	1.7	372	0.32	0.30
Oak–hickory forest	4.1	2.9	150	0.15	0.19
Mixed deciduous forest	4.0	5.1	109	0.02	0.23

PP = primary producers; PL = plant litter; SOM = soil organic matter; MO = micro-organisms; IN + N soluble and exchangeable inorganic nitrogen.

released in the mineralisation of organic matter and are readily available as cations, although potassium can be immobilised in chemical combinations with the clay minerals. Calcium and magnesium are more mobile, being readily leached by only slightly acidulated rain-water (i.e. $H_2O + CO_2 = H_2CO_3$). Most of the nitrogen, phosphorus and sulphur, however, is held in organic form. Phosphorus is the least mobile and available. Although it is less susceptible to leaching than the others it can be chemically fixed and form insoluble compounds with iron, aluminium and manganese in acid or alkaline soils. Phosphorus deficiency is a characteristic of soils derived from old highly weathered and eroded acid igneous and metamorphic parent material of the tropical regions of the world. Nitrogen (see Fig. 7.8) differs from the other macronutrients in that the routes by which it enters and leaves the cycle are more numerous and complex. Further, it exists in several important inorganic forms which vary in mobility and availability. 'After the carbon cycle, the nitrogen cycle is arguably the most important to living organisms' (Sprent 1987: 3). For these reasons it will be considered in some detail.

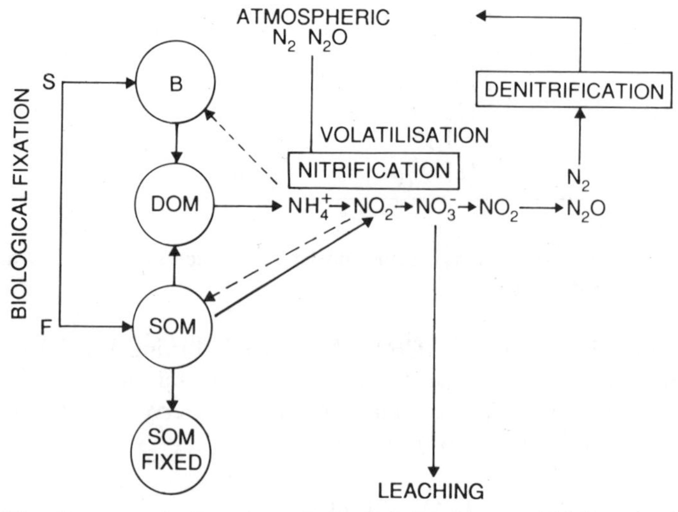

Fig. 7.8 The nitrogen cycle. Organic pools (circles); B = biomass; DOM = dead organic matter; D = decomposers; SOM = soil organic matter. NH_4^+ = ammonium; NO_2 = nitrite; NO_3^- = nitrate; N_2 = gaseous nitrogen; N_2O = nitrous oxide

Although gaseous nitrogen (N_2) comprises *c.* 72 per cent of the existing atmosphere this is a very stable, inert form which is unavailable to higher plants. It can be 'mobilised' in electrical storms and enter the soil in precipitation. This is particularly characteristic of tropical regions with a high frequency of high-energy thunderstorms. Most of the nitrogen, however, reaches growing plants via nitrogen-fixing soil macro-organisms – the *prokaryotes* and *archaebacteria* (or metabacteria). Some are free-living; others, in which most nitrogen fixation takes place, occur in a symbiotic relationship with those leguminous plants on whose roots nodules are formed by *Rhizobium* spp.; and with a few non-legumes nodulated by *Frankia.* These micro-organisms derive energy (carbon) and nutrients from the plant; the latter obtains nitrogen directly from the nodules or indirectly when roots and micro-organisms die and are decomposed.

Mineralisation of organic nitrogen results in the production of ammonium ions (NH_4^+) which may be: (1) volatalised; (2) converted to N_2 or N_2O; (3) absorbed directly from the soil solution; (4) temporarily adsorbed by the soil colloids; or (5) reduced in the process of *nitrification* to nitrate (NO_3^-). The latter is the most common nitrogen compound in agricultural soils, deserts and water (Sprent 1987). Many plants, particularly the *nitrophiles* need nitrate for maximum growth. Indeed, unless it is absorbed fairly rapidly nitrate is particularly susceptible to leaching. Nitrification can be impeded by acid DOM with a high lignin and C : N ratio; indeed the production of ammonia promotes acidity. Under anaerobic conditions denitrification can take place whereby nitrate is reduced and returned to the atmosphere as nitrous oxide (N_20) or nitrogen (N_2) gas. Any of the reactions in the nitrogen cycle can be constrained by one or more of a number of the variables summarised in Table 7.8, thereby controlling the overall rate of the nutrient cycle

Table 7.8 Factors which may affect rate-limiting steps in the nitrogen cycle (from Sprent 1987)

Factor	*Example*
Environmental conditions unsuitable for growth of particular organisms	Excess or deficiency O_2 pH, H_2O or heat may differentially affect N-cycle organisms
Inhibition of one organism by another	Various allelopathic effects, e.g. natural nitrification inhibitors
Lack of substrate	Insufficient free NH_4^+ in some soils for nitrifying bacteria
Lack of essential co-factors	MO deficiency on N_2 fixation and NO_3^- reduction

and the turnover of nitrogen in terrestrial organic pools. Ellenberg's (1971b) classification of world ecosystem types on the basis of the main forms of mineral nitrogen (Table 7.9) reflects the effect of climate, vegetation type and soil drainage conditions and particularly the difference between the more or less nitrogen conservative ecosystems.

Table 7.9 Ecosystem types classified by the main forms of mineral nitrogen (from Ellenberg 1971b)

NH_4^+	NH_4^+–NO_3^-	NO_3^-
Taiga, dwarf shrub tundra	Many temperate deciduous forests on loamy soil	Moist tropical lowland forest
Sub-alpine coniferous forest	Alluvial forest	Temperate deciduous forest on calcereous soil
Coniferous peat forest	Alder fen (*Alnus glutinosa*)	Fertilised meadows where soil is not wet; most gardens
Oak–birch forest	Many grassland types	
Calluna heath	Dry grassland on calcereous soil	Ruderal formations
Many swamps	Tropical savanna	
Raised *Sphagnum* bog	Some tropical forests	

AQUATIC NUTRIENT CYCLING

The nutrient cycle in aquatic ecosystems is dependent on the same reservoirs and follows the same routeways as in terrestrial ecosystems (Steele 1985). However, the former differ from the latter in respect to the relative importance of the organic pools and the nature and rate of nutrient recycling. The relative contribution of animal to plant biomass is much higher in aquatic habitats. A very high proportion of the plant biomass passes via the grazing route to zooplankton and macrofauna. The latter, in particular, constitute a much larger pool of nutrients than do the terrestrial macrofauna, but one in which the residence time is short compared to that in the long-lived biomass pool in terrestrial forest ecosystems.

The process of decomposition is the same as that on land but it takes place through a greater depth of water than the depth of soil on land. Biological productivity is greatest in the upper euphotic zone of the ocean (see p. 276) and, unless very rapidly mineralised, the DOM (composed of faecal pellets, particulate organic matter and dead organisms) tends to sink below the zone of nutrient uptake and to be deposited on or in the bottom sediments. The rate of decomposition decreases with depth because of:

1. Low temperatures (2–4 °C) and the slow growth of even cold-adapted bacteria;
2. High hydrostatic pressures;
3. The persistence at greater depths of the older, more resistant organic compounds.

Also, with increasing depth, oxygen may become insufficient to cope with the demand from anaerobic decomposers resulting in a slower release of nutrients than in a well-aerated soil.

The spatial separation of the decomposing zone from the primary productive zone is probably the most significant contrast between aquatic and terrestrial nutrient cycles. Cycling of nutrients in the former is dependent on the local, regional and, in some cases, global circulation of water to replenish the photic zone with nutrients. The efficiency and rate of circulation depend on the temperature and the mobility together with the depth of water. In warm climatic conditions the establishment of a *thermocline* between the warmer surface and colder waters below reduces conventional overturning and hence impedes nutrient return. In calmer, tropical seas and lakes thermocline can be a permanent feature, while in temperate latitudes it is a seasonal phenomenon. Over the shallower continental shelves water can be turbulent enough to bring nutrients relatively rapidly and continuously from the floor to the surface. At great ocean depths minerals may become 'locked up' in bottom sediments and lost to the cycle indefinitely. In this respect the deep ocean sediments act as an important 'sink' or 'reservoir' in the global nutrient cycle.

Chapter

8

Biotic change and soil development

A fundamental and universal feature of the global biomass is its susceptibility to change in amount and composition through time. Factors promoting change can be extrinsic or intrinsic. The former include climatic, geomorphological and edaphic variations, as well as those, directly or indirectly, effected by anthropogenic processes. The latter involve biotic processes such as invasion, competition, predation, parasitism and disease, and speciation (i.e. evolutionary change). All these factors interact to a lesser or greater degree either to increase the rate and/or amount of biomass change (i.e. by positive feedback) or to reduce change (i.e. by negative feedback). The relative importance of the causal factors vary in space as well as time and each factor can operate at a time–space scale (see Fig. 8.1) which may not necessarily be synchronous with those of the other factors. Understanding of the mechanisms of change, however, is still far from complete and there is an

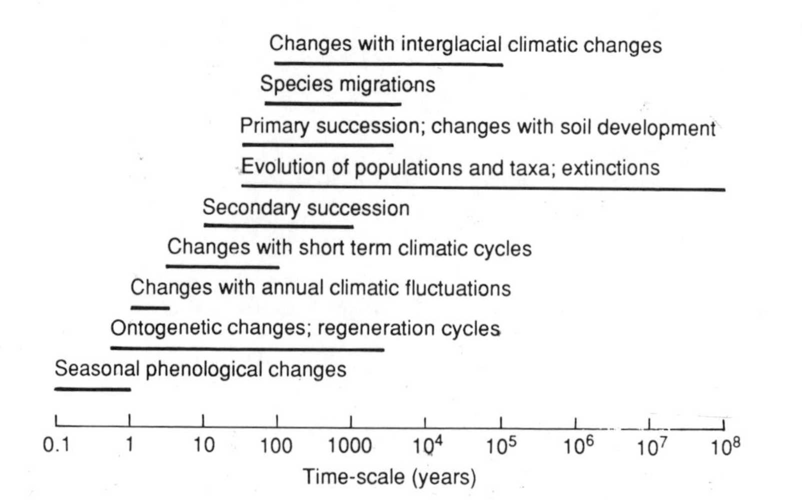

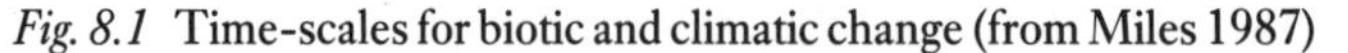

Fig. 8.1 Time-scales for biotic and climatic change (from Miles 1987)

urgent need for a sound theoretical basis on which to manage the use of the world's biomass.

TYPES OF CHANGE

As Miles (1979) points out, the study of vegetation has become bedevilled with a plethora of often confusing terms. He distinguishes three types of change on the bases of the amount and direction of change:

1. Regeneration (or replacement) change is that resulting from the natural processes of germination, growth and death of the component individuals without a change in the amount or composition of the vegetation.
2. Fluctuational (reversible or cyclic) change occurs when the biomass composition varies to a greater or lesser degree from a mean state over shorter or longer periods of time.
3. Successional (directional change, non-reversible) involves a change from one type of biomass to another in time.

The distinction between the three, however, is by no means as clear-cut as the terminology would suggest.

REGENERATION CHANGE

Natural regeneration is normally perceived as a continuous process which maintains the composition and structure of the vegetation and is reflected only in the uneven age distribution of its constituent species individuals. However, as a result, on the one hand, of variations in seed production, germination and plant mortality and, on the other, of variations in the life span of the component species, regeneration may be accompanied by cyclic changes in the composition of the vegetation. The life span of annual and perennial herbaceous plants is short (10 years) compared to that of woody shrubs and trees (10–100+ years). The growth of the latter from seed is much slower than in the case of the former and where a gap is created by the death and fall of a particularly large senile tree, it will tend to be filled by fast-growing herbaceous plants and saplings until such time as these are eventually over-topped by one or more of the original slower-growing but taller dominant trees. In a uneven-aged forest stand variations in the growth stages of youth, maturity and old age may under certain circumstances give rise to what has been called a *regeneration cycle*. This results in a regeneration mosaic of patches of varying composition dependent on the time since the gap was created. Cyclic gap regeneration (or *gap dynamics*) is now accepted as a major process in regenerative vegetation change. It is exemplified in the stages in the regeneration of heather moorland, in woodlands dominated by opaque canopies such as those of beech, evergreen conifers and the tropical rain forest in whose shade seeds of the mature dominants cannot germinate, or can germinate and grow only very slowly.

FLUCTUATIONAL CHANGE

Fluctuational change in vegetation can be a consequence either of seasonal phenological changes in the component species or of short-term environmental variation. Seasonal phenological changes are observable in the sequence of growth phases of plants and animals in habitats with marked seasonal climatic regimes of alternating warm and cold or wet and dry conditions. These are reflected in the varying *aspection* (seasonal appearance) of the biomass. The deciduous habit of perennials results in a seasonal reduction in the amount of plant biomass, while short-lived spring or autumn annual species may disappear at other times of the year. The reduction of plants is accompanied by that of animals many of which hibernate, pupate or migrate during the unfavourable season. Aspection is particularly well developed in temperate deciduous woodland where spring, autumn and winter aspects are reflected in variations in the sequential growth phases of species in the herbaceous, shrub and tree strata respectively.

Of the environmentally induced vegetational change that of climate is undoubtedly the most important. Temporal variation is the rule rather than the exception in many, if not all, climatic regimes. Its effect on the biomass will depend on its intensity and time-scale, i.e. how far and for how long climate deviates from the average condition. Annual variation in temperature and/or precipitation can influence growth and regeneration, particularly of short-lived annual plants and invertebrate animals. The vertebrate and invertebrate animal biomass in cold and temperate regions is frequently determined by the climatic conditions, and hence the amount of plant food produced in the preceding growing season or by the severity of the preceding winter. Short-run cyclic variations of climate over a series of years are a characteristic feature of many climatic regimes and, dependent on the degree of variability, can result in variation in the relative importance of the component species in the biomass. The alternation of sequences of dry with wet years characteristic of the semi-arid grassland areas of the world are accompanied by variation in the relative number and importance of the more with the less drought-resistant or tolerant species.

SUCCESSION

Succession can, very broadly, be defined as sequential change in the form and composition of biomass through time. It is regarded as one of the most universal – albeit contentious – ecological concepts. Although it is an observable site phenonemon, there is no unanimity of definition or of causal explanation for the process. Miles (1987) notes that most recent definitions tend to be broad, and to cover all changes except the seasonal, fluctuational and cyclic. They are, however, mutually consistent in defining succession as a progressive directional change in the structure and species composition of the biomass or vegetation.

Concept of succession

The concept of succession was developed in North America (at the beginning of this century) by Cowles (1901) and Clements (1904, 1916) as a result of studies of sand-dune vegetation along the shores of Lake Michigan and in Nebraska respectively. Clements's descriptive model of the process has had a profound and still persistent influence on ecological thought. He defined succession as a sequence (*sere*) of plant communities (*seral stages*) characterised by increasing complexity of life forms (i.e. mosses, ferns, grass/herbs, shrubs, trees, etc.) the end-stage of which is the *climax vegetation*. Succession, once initiated, was biotically controlled primarily by progressive habitat modification and, later, by interspecific competition. Those species which cannot compete successfully for atmospheric and soil resources with the higher life forms of the succeeding stage will die out.

The final climax stage in succession was regarded as that dominated by the largest and/or tallest plant forms capable of maintaining themselves in the prevailing climatic conditions. Irrespective of the initial substrate, succession would result in the convergence of a type of vegetation dominated by a particular form and composition. The Clementsian model also recognised *sub-climaxes* which, because of local physical conditions, resulted in succession being slower than elsewhere in a climatic region but which would eventually progress to the (climatic) climax. Likewise, extreme human activities could deflect a sere from its natural course and produce a *plagioclimax* or *deflected climax* which, with the removal of the deflecting factor, could be expected to proceed to the climax.

The Clementsian model was not only progressive, deterministic and predictable but it was also organismic in that the development of the climax community was seen as analogous to that of an organism. It has been subjected to considerable criticism ever since its initial formulation. Empirical data collected, particularly over the past 40 years, reveal a lack of conformity in terms of the number and identity of seral stages and of seres associated with a particular habitat. That the model is too simple and lacking an empirical underpinning is hardly surprising, given the paucity of data available about the characteristics and mechanisms of vegetation change at the time. Despite a large and increasing number of site-specific examples of plant succession, understanding of the mechanisms (processes) involved are still far from clear. From the mid-1950s to the end of the 1970s much ecological literature and debate focused on attempts to explain successions (MacIntosh 1980). Four models were proposed by Connell and Slayter (1977):

1. *Facilitation* (habitat or site modification) – the process whereby early-established species modify the habitat making it more suitable for later colonists.
2. *Tolerance* (competition) – the process whereby slower-growing, more tolerant (i.e. competitively superior) species invade and mature in the presence of earlier faster-growing, less tolerant species which are eventually excluded.

3. *Inhibition* – the process whereby the first plant species to become established inhibit further invasion until they eventually die.
4. *Random* – a process of chance survival of different species at the time succession is initiated and subsequent random invasion by new species – all of which grow and mature together at different rates.

The first two reiterate the biotic forces implicit in Clements's model; the third relates to what, in Clementsian terms, was described as a sub-sere or sub-climax; while the fourth restates Gleason's early individualistic concept of vegetation and hence of succession as the progressive growth and death of an unrelated collection of species' populations. A new terminology was coined without any significant advance in successional theory. What, however, has emerged from numerous field observations and recordings is a greater appreciation of the diversity of successional sequences that occur and, allied to this, the increasing difficulty of formulating a universally applicable model. Also, as Miles (1979) notes, the Clementsian model overemphasised the role of biotic site modification (*autogenic*) as compared to abiotic changes (*allogenic*) in imitating succession.

PRIMARY SUCCESSION (OR PRISERE)

The primary succession which comes closest to Clements's initial model, is that initiated on a 'new' biologically unmodified site. Plant colonisation depends on prevailing climatic conditions, seed availability and dispersal and the physico-chemical nature of the habitat. Since, theoretically, it involves the replacement of one vegetation type by another as a result of habitat modification, it has been variously described as an obligatory or replacement succession and as such conforms to the facilitation model. A distinction has long been made between primary *xeroseres* established on dry sites including sand or psammoseres, rock or lithoseres, *hydroseres* on waterlogged sites and *mesoseres* on intermediate conditions. Historical evidence of primary succession following the Pleistocene deglaciation is recorded in the successive changes in the macro- and microvegetation remains preserved in peat-bogs and lake sediments. Although areas where primary succession is currently in progress are relatively limited in extent, a great variety of new habitats are constantly being formed as a result of volcanic activity, river, lake and marine deposition and glacier shrinkage. Recent volcanic debris on Hawaii, Krakatoa and Surtsey Island (Fridriksson 1975) and fresh glacial debris in North America and Europe have provided natural laboratories for the study of plant (and animal) succession. In the Glacier Bay area of Alaska, Cooper (1931, 1939) recorded in some detail at intervals between 1916 and 1935 the vegetation changes which occurred on a number of plots of ground from which the glacier had retreated within a known period of time. He described a succession in which three principal stages could be recognised (see Table 8.1). The progressive increase in density of herbs and of the cover of mat-forming shrubs is illustrated in Table 8.2. Nevertheless, members of all the forms in Table 8.1 were present at the start of the succession. In other similar areas trees have been observed growing on abalation moraine on glaciers, adjacent to the glacier front and on recent mud-flows.

Table 8.1 Principal seral stages in primary succession on a deglaciated site in Glacier Bay area, Alaska, 1916–31 (from data in Cooper 1931)

Stage	*Process*	*Plants*
1. Pioneer (moss–herb)	Colonisation	*Rhacomitrium* mosses Perennial herbs (*Epilobium latifolium* *Equisetum variegatum*)
2. Thicket	Soil stabilisation	Mat-forming shrubs (*Dryas drummondii*) *Salix*
	Nitrogen enrichment	*Alnus*
3. Forest		Trees *Picea sitchensis* *Tsuga* spp.

Primary *allogenic succession* is characteristic of sites subject to continuous environmental changes. It can be observed in tidal estuaries where the gradual accumulation of sediments in the intertidal zone is seen in the spatial stages in the development of salt-marsh vegetation. There is a progressive change in plant communities from the lowest zone where mud-banks or flats are exposed for short periods twice daily, to the highest where silting has built up a surface out of range of all but the highest tides. The lowest outer edge of the salt-marsh, just above low-tide level, is where initial colonisation begins. The pioneer stage is characterised by a few highly specialised plants capable of tolerating an extremely mobile, frequently inundated and highly saline habitat. As they become established, at first sporadically and then in increasing numbers, they tend to check the rate of water flow, stabilise the mobile substratum and accelerate the process of silting. As the surface level increases in height the length of time during which the salt-marsh is subjected to inundation, and consequently the degree of salinity, decreases. The pioneer species of the low marsh are gradually replaced in the high or middle marsh (submerged for a varying period only at spring tides) by an increasing number of species less tolerant of high salinity and a mobile substratum. The highest levels covered only occasionally, during exceptionally high water, can be colonised by non-halophytic rushes and grasses. If the surface level is built up sufficiently the higher and drier surface can be colonised by scrub and woodland vegetation.

A zonation of communities also accompanies the progressive silting of freshwater lakes. As the depth of water decreases, submerged and floating plants which anchor themselves on the lake floor and help to trap and accumulate silt are gradually replaced by reeds and rushes. The continued deposition of silt and, in such waterlogged conditions, an increasing accumulation of organic matter may eventually raise the rooting medium high enough to allow colonisation by willow and alder scrub, and eventually oak-wood.

Table 8.2 Changes in species abundance in eight 1 m^2 permanent quadrats on recently exposed glacial debris at Glacier Bay, Alaska (from Cooper 1939)

	Year			
	1916	*1921*	*1929*	*1935*
Number of individuals				
Arctous alpina	0	1	1	0
Polytrichum sp.	0	2	1	0
Poa alpina	1	16	5	0
Epilobium latifolium	110	10	1	0
Habenaria hyperborea	2	2	2	3
Salix spp.	139	140	71	110
Equisetum variegatum	150	161	175	183
Carex spp.	27	43	51	91
Euphrasia mollis	0	17	8	102
Cladonia sp.	0	0	0	13
Pyrola secunda	0	0	0	19
Marchantia polymorpha	0	0	0	1
Aggregate branch length (m)				
Salix spp.	6.5	15.7	21.1	18.4
Cover (dm^2)				
Dryas drummondi	27	120	249	355
Rhacomitrium spp.	40	52	71	127
Stereocaulon tomentosum	0.5	1.4	22	32

In much the same way the gradual accumulation of wind-blown sand dunes piling up around coasts and lake shores reveals a successive development of communities from those of the youngest mobile to the older fixed and stabilised dunes. Under these circumstances the accumulation and stabilisation of drifting sand are accomplished by plants which can tolerate not only a mobile but a very dry habitat. The pioneers are commonly such drought-resistant grasses as lyme grass (*Elymus arenaria*) and marram (*Ammophila arenaria*). They check wind-speed and trap sand. Their upward growth can keep pace with the continued accumulation of sand. They have, in addition, extensive and rapidly spreading rhizomes and a ramifying network of very fine roots which are constantly being renewed at higher levels. The growth of these 'dune-building' plants not only stabilises the sand but ameliorates its aridity by the addition of water-retentive organic matter. Eventually (and particularly if shelter is provided by the formation seaward of a new line of dunes) plants less tolerant of mobility and aridity can establish themselves; lichens and mosses and a variety of herbaceous plants provide a complete ground-cover and increase further the humus and water content of the dune soils. The formerly mobile dune becomes fixed, the pioneer species gradually die out in face of competition and, where not disturbed by man, shrubs such as hawthorn, gorse, broom and buckthorn can establish themselves on the improved habitat preliminary to the final development of woodlands. In each of these cases the zones of vegetation reflect a gradual and continuous spatial change in habitat conditions. Each zone will tend to be replaced in time by the next stage in the succession as silting continues to extend the fresh or salt-marsh or sand dune progressively further from the shore.

SECONDARY SUCCESSIONS

Secondary successions differ from primary ones in that they occur on sites which have been biologically modified and which may retain some of the previously formed soil organic matter and or vegetation. Initiation of succession is effected by the removal of all or part of the former vegetation/soil cover by deforestation (either by felling or intense wind storms), fire, abandonment of agricultural land or by accelerated soil erosion consequent on severe degradation of the vegetation cover. In the last instance soil erosion may be such that a biologically unmodified habitat results which can only be restored naturally by primary succession. Indeed many of the habitats created by human activities are biologically unmodified as in the case of mining and quarrying spoil or only slightly modified as in the case of abandoned/derelict land and water bodies. Secondary succession is a more widespread process than primary succession. A high proportion of so-called natural vegetation is a product of secondary succession. Much of the world's remaining forest cover is secondary or second growth (indeed in many areas third or fourth growth) which has re-established on formerly cleared land and which, because of time since clearance or continued cutting or burning, has not reattained its original structure and composition.

A not very clear distinction is sometimes made between secondary succession which is initiated and characterised mainly by those plants which were present in the original vegetation and which have persisted in the soil as seeds, rhizomes, etc. accompanied by early invaders, and that in which species not present in the early stages of succession can invade at a later stage because of their competitive ability. Indeed, some ecologists have questioned the validity and utility of distinguishing between primary and secondary succession, firstly because of the long history and ubiquity of vegetation disturbances and secondly because of the similarity in the processes involved.

PROCESSES IN SUCCESSION

Colonisation

Initiation of succession on biologically inert sites is dependent totally on colonisation by invasive species – the so-called *pioneer species* (Fenner 1985; Grubb 1987). These usually have small, wind-dispersed seeds and an ability to grow successfully on open, climatically stressful and often nutrient poor and/or unstable substrates. Grubb (1986, 1987) maintains that the growth form of successful pioneers varies with the nature of the substrate, and makes a basic distinction between the following:

1. Stable mesic sites with adequate water and nutrients with short-lived rapid-growth pioneers characteristic of the conventional succession;
2. Unstable sites in which pioneers are long lived, slow-growing and not necessarily highly dispersible;
3. Resource-poor sites with long-lived, slow-growing pioneers with large seeds and/or rapid vegetative spread.

The first type of pioneers are those designated *r–strategists*, the second *K–strategists* (Grime 1979). In the first case competition for resources is high and rapid growth is necessary to ensure survival. On unstable and resource-poor sites tolerance of unfavourable conditions and avoidance of predators is considered to be more important (Bradshaw *et al.* 1986). On the latter sites shorter-lived species establish themselves once the site is stabilised and/or resource enriched. While there are, obviously, variations and exceptions to this generalisation, it is particularly well illustrated by pioneers such as the mangrove of tropical swamps; perennial grasses (*Pucinella* spp. and *Spartinia* spp.) of temperate tidal flats; perennial grasses (*Ammophilia* and *Elymus* spp.) and sedges (*Carex* spp.) of coastal fore-dunes. Mixtures of longer- and shorter-lived pioneers, however, do occur particularly on sites which are of variable stability and nutrient status.

In contrast, true pioneer species are not characteristic of secondary succession (i.e. on biologically active sites) and a variety of plant forms grow either simultaneously or in close succession from the existing regenerative material. Invasive colonisers originate from seeds and spores which persist in the soil or even on the ground surface between disturbances (Grubb 1987).

Successional trends and stages

Among the successional changes in plant and vegetation attributes most commonly noted are, increase in total plant biomass: rapid initial increase in species numbers and diversity, followed by a decrease in rate of arrival of new species with, eventually, a decrease in numbers as a result of intraspecific and interspecific competition for resources; an increase in the number and diversity of plant life forms and in the structural complexity of the vegetation resulting from the stratification of constituent life forms (Fig. 8.2). In the later stages of both primary and secondary succession to a forest vegetation, a distinction can be made between what have been called secondary (pre-climax) and primary or climax tree species. The former (e.g. birch, poplar, willow, hazel) are characterised by efficient dispersal by light wind-borne seeds, an intolerance of shade, rapid growth in light and a shorter life span than the succeeding primary or climax trees. The latter have larger seeds and a lower dispersal ability. Their growth is slower, but with a longer life span they attain a larger size than the pioneer trees which they eventually succeed. The regeneration of the primary (climax) trees is dependent on a supply of persistent seedlings, rapid growth of which will occur mainly in the gaps created by death of senile secondary trees.

Concepts of succession were originally concerned only with plant or vegetation changes. The changes in the animal components of the biomass were neglected until relatively recently. However, as Brown and Southwood (1987) note, emphasis has been mainly on insects in the early successional stages and on birds

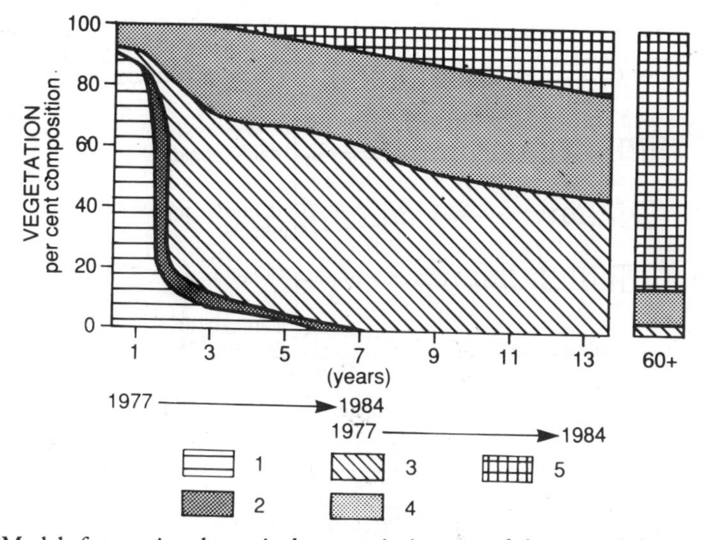

Fig. 8.2 Model of vegetation change in the same site in terms of plant growth form: 1 = annual herbs and grasses; 2 = biennials; 3 = perennial herbs; 4 = perennial grasses; 5 = shrubs and trees (from Brown and Southwood 1987)

and small mammals in the later stages (see Fig. 8.3). They record that of the four major insect guilds, plant suckers are the most abundant throughout succession. The diversity of all phytophagous species, predators and parasites rises rapidly in early successional stages. Although plant diversity falls in later stages, that of animals (particularly of insects) is maintained largely because of the increase in microhabitats provided by the increasing complexity of vegetation structure. According to the same authors, in the early stages of succession the characteristic insects are short-lived, with well-developed flight ability, and many are generalist feeders. Both vertebrates and insects tend to be predominantly plant-feeders and food chains are, as a result, simple. In the course of succession there is an increase in the relative proportion of predatory insects and vertebrates (particularly insectivores to herbivores); taxonomic, morphological and trophic diversity; and in the number of specialist feeders.

While these trends are characteristic of many recorded successions, the occurrence of a clearly defined, orderly sequence of seral communities is doubtful. In addition, there may be several possible species sequences that can develop within a particular habitat from specific starting-points, dependent on the species richness or poverty of the vegetation sources.

RATE OF SUCCESSION

The rate of both primary and secondary succession and the time taken to reach the final stage is extremely variable. Under comparable climatic conditions secondary succession might be expected, particularly in the early stages, to be more rapid than primary succession. The rate of the latter will be dependent on the nature of the substrate – more rapid on nutrient-rich weathered mineral than on nutrient-poor coarse substrate or bare rock – and on the location of habitat relative to the

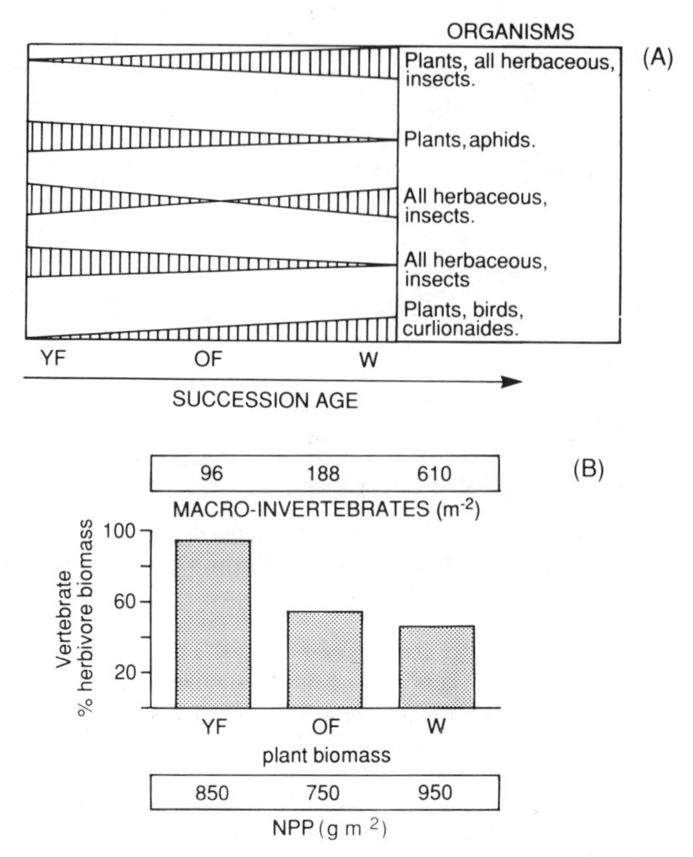

Fig. 8.3(A) Adaptive strategies of plants and animals along a successional gradient; (B) summary of utilisation by vertebrate herbivores and macroinvertebrates when viewed against a constant annual net primary production (NPP). YF = young field; OF = old field; W = wood lane (from Brown and Southwood 1987)

source of potential colonists. The rate of secondary succession will be dependent on the severity of habitat modification by disturbance, the size and internal heterogeneity of the habitat and the time of disturbance (Sousa 1984). Secondary succession can be particularly rapid on formerly cultivated land. For instance, in eastern North America the natural re-establishment of a near-climax deciduous forest on old fields has been estimated to take about 100–250 years. In the humid tropics the process is more protracted. Gourou (1961) notes that after a lapse of 1400 years the forest on the sites of ancient Mayan settlements in Central America are still composed of pre-climax trees. Forest on the site of Angkor Wat (Cambodia) destroyed by fire *c.* 500–600 years ago, while resembling virgin forest, still maintains some differences. In addition, there are situations in which secondary succession may be severely checked or inhibited by the presence or establishment of a type of vegetation which, for one reason or another, is more resistant to the germination of seeds from within or without the area. This can occur as the result of the presence of a thick turf developed under perennial, mat-forming grasses; of the dense shade and a deep litter layer associated with vigorous stands of bracken (*Pteridium aquilinum*); or the presence of allelopathogenic plants.

In contrast, both primary and secondary succession can be rapid in very severe conditions. Cooper (1931) noted in Glacier Bay, Alaska, that the interval between the disappearance of ice and the colonisation of the first trees was only 10–12 years in some areas; in others they appeared on the heels of the retreating ice-front. It has been suggested that in Arctic and hot deserts succession is so limited as to be non-existent (MacMahon 1981). The severity of the climate limits the number plant life forms, the rate of growth and the extent to which the habitat can be modified and succession facilitated. The vegetation, with an incomplete cover and a little weathered or unstable substrate, remains at a pioneer stage. Replacement following disturbance may well be by the same species, a process which has been referred to as *direct succession* (Whittaker and Levin 1977).

RETROGRESSIVE CHANGE

The Clementsian concept of plant succession was unidirectional and progressive. It could be checked or deflected but not, naturally, reversed. Human activity, however, is a greater instrument for change than had been appreciated by the early ecologists. Few areas of uncultivated vegetation have escaped the direct or indirect impact of these activities and those where unmodified primary (virgin) vegetation persists are dwindling rapidly. Among the most important types of human impact are the intensification of livestock grazing; the promotion and use of fire; the direct physical impact of people, domestic livestock and machines (including the automobile); large-scale deforestation; and environmental (soil, water and air) pollution. Anthropogenic factors have not only checked the re-establishment of the original vegetation cover but have brought about changes that would not otherwise have taken place. In many instances this has resulted in a progressive reduction in the species number and diversity, in the amount and the structural complexity of the biomass. In so far as such change is towards increasing simplicity, it can be regarded as ecologically retrogressive or degenerative.

SUCCESSION AND SOIL FORMATION

It is now widely accepted that an important process, at least in the early stages of plant succession, is the progressive modification of the initial physical site. The establishment of a vegetation cover can:

1. Contribute to the mechanical and chemical weathering of mineral matter;
2. Effect the stabilisation of an unconsolidated and potentially mobile substrate;

3. Produce dead organic matter which further contributes to the aggregation and stabilisation, as well as increasing the water-holding ability of the mineral substrate and providing a nutrient pool on which the living plants can draw.

These are, in fact, basic processes in soil development. And the rate and nature of the changes which accompany soil development are influenced by the rate and type of vegetation change.

SOIL FORMATION

Weathering, organic decomposition and the displacement or translocation of organic or inorganic materials in solution or in colloidal suspension result in the formation of the solum, i.e. the pedogenically modified part of the lithosphere (Bridges 1978). This is reflected in the vertical zonation of *soil horizons* which may differ in plant and animal biomass; amount and type of DOM; texture and structure; colour; water content and chemical composition; the combination of horizons forming the *soil profile* (see Fig. 8.4).

Soil formation is characterised by a succession of three main stages (or phases):

1. *The immature stage* of skeletal soil which is dominated by the weathered but pedogenically little-modified parent material. The vegetation cover is normally at an early stage of colonisation with an incomplete, open cover. As a result the contribution of organic matter is limited. There is little or no profile development. Such embryonic soils are usually categorised as *lithosols* (on consolidated rock) or *regosols* (on unconsolidated mineral material).
2. *The youthful stage* is that at which pedogenic processes have begun to operate. However, although a closed vegetation cover may exist, the underlying solum still tends, particularly at the beginning of this stage, to be dominated by the parent material. Horizon differentiation may have been initiated but the profile tends to be relatively simple, with one or more organic-rich horizons in direct juxtaposition with the little modified parent material. Towards the end of this stage horizon differentiation becomes evident. Youthful two-horizon soils are, in some classificatory systems, designated as *rankers*. While some represent a developmental stage, others may be either climatic or slope rankers. The former are characteristic of Arctic and montane habitats where severe climatic conditions and a prolonged period with frozen ground limits vegetation development, organic decomposition, soil-water movement and hence soil development. In the latter case instability and continual renewal of parent material maintain soils in a skeletal or very immature condition.
3. *The mature stage* is, in contrast, characterised by the presence of a sequence of horizons whose physical and chemical composition is thought to reflect the pedogenic processes resulting from the interaction of the more active factors of vegetation and climate, than of the less active geological conditions. This stage, theoretically, marks the culmination of soil evolution and the establishment of a mature soil. It is thought to be stable in the sense that, unless the environmental factors change radically, a dynamic equilibrium can be maintained between

input and output of energy; of nutrient cycling; and between weathering and erosion of the parent material. Mature soils have been, on a global scale, rightly or wrongly correlated with climax vegetation. And a major distinction was early made between *zonal soils* or the *great soil groups* (the equivalent of the climatic climax in vegetational terms) whose profile characteristics reflected, primarily, the interaction of climate and vegetation; *interzonal* soils where the effect of other factors, particularly parent material, slope, etc. modifies zonal development; and the young *azonal* or *skeletal* soils.

SOIL-FORMING PROCESSES

As in vegetation succession, a distinction has been made by some soil scientists between primary and secondary soil evolution. The former is that on formerly unmodified 'new' sites such as recently deglaciated areas, alluvial or estuarine deposits, lava flows, etc. The latter may be associated with one of two conditions. The first is where erosion has truncated either the upper horizons or the entire solum of a pre-existing soil or where a former soil profile has been buried under aggrading material. The second is that initiated by either a change in climate and hence vegetation or in vegetation alone as a result of man's direct or indirect intervention.

The concepts of soil development and of soil maturity are, it must be stressed, difficult to substantiate in reality. They pose the same problems as that of vegetation development. Not least is that of identifying when a soil has reached a stage of maturity and/or equilibrium, since the degree of horizonisation may not necessarily reflect the stage of development with any degree of accuracy. Also, since the soil-forming processes are themselves subject to continuous but varying rates of change, it has been questioned whether the concept of a steady-state soil has any real meaning. It is not, in fact, possible in most cases to establish with any degree of certainty to what extent existing soils reflect present or previous environmental conditions.

SOIL DATING

Dating soils is fraught with a number of complex problems not least of which are: (1) the difference between rates of weathering which produces parent material and of pedogenesis (soil formation) which produces the soil profile and (2) the distinction between time of initiation of formation, and of the production of the existing soil profile. The latter is further complicated by the fact that differing rates of pedogenesis can result in similar profile characteristics (Cruickshank 1972). In most studies zero time is taken as the starting-point of soil development.

Dating can be absolute if the soil contains material by which a datum line can be fixed. The most precise method is by using the *half-life* of *radioactive isotopes* incorporated in organic material in the soil profile. Carbon-14 with a relatively short half-life of 5570 years is useful for accurate dating up to 30 000–40 000 years, well beyond the time required for soil formation. Other methods include the correlation of soil horizons with human artefacts of a known age and/or the date of a major environmental change such as lake draining, tree planting, deglaciation is known. Work on the Mindenhall glacier suggests that it took over 200 years for a young, two-horizon soil to develop; between 1000 and 3000 years for a podzol to form. On the moraines of the Aletsch glacier it has taken 80–90 years for horizonation to reveal itself. In contrast, there is evidence that the rate of secondary development could be faster, as could soil development in the tropics. It has been estimated that podzol formation could take as little as 200 years under pine planted on pre-existing well-drained brown earth soils.

PEDOGENIC OR SOIL-FORMING PROCESSES

Leaching

One of the most important soil-forming processes is that of *leaching* (*eluviation*) or the translocation of material from the upper to the lower parts of the soil profile in percolating water. It is characteristic, to a lesser or greater degree, of all freely draining soils subject to climatic regimes where there is an annual excess of precipitation over evapotranspiration. It involves both the mechanical transport in suspension of colloidal clay, humus and iron particles; and the chemical transport, in solution, of soluble exchangeable ions. Some authors (Cruikshank 1972) make a distinction between eluviation (mechanical movement) and leaching (chemical). Others follow Duchafour (1965) and use the French word *lessivage* for slight leaching (White 1979). However, as Townsend (1972) points out, the two types of transport are not easy to separate as they often tend to proceed concurrently, though under certain conditions one or the other may predominate. Also, the slightly or intensely leached soils are part of a continuum within which distinct class boundaries rarely exist.

Percolating rain-water absorbs carbon dioxide and organic acids, both the produce of decomposition, from the decaying organic matter. As a result, it becomes a more efficient solute than the original precipitation and can effect the chemical weathering of primary minerals, as in the normal weathering process. Mobile materials are thus removed from the upper part of the profile and in the process the modified, eluviated E horizon develops in the lower part.

The degree of leaching is dependent on the mobility of the materials involved and the rate of water percolation through the soil. The intensity of the process is reflected in the *base (cation) status* of the mineral material. The effect of increasing intensity of leaching has been summarised by Townsend (1972) as a series of stages:

Least intense:

1. Loss (if present) of soluble, alkali salts.
2. Solubility, and loss, of free calcium carbonate ($CaCO_3$).
3. Loss of exchangeable bases (calcium, sodium, potassium, magnesium) and their replacement by hydrogen ions with the resultant lowering of the pH in the E horizon.

Most intense:
4. Mobilisation of iron and aluminium.

Cheluviation and podzolisation
With increasing acidity in the E horizon, iron and aluminium are mobilised. Ferric iron is transformed into its soluble ferrous state. In addition to the leaching of cations that of the normally immobile sesquioxides of iron and aluminium is enhanced by the formation of organometallic complexes (e.g. by chelation). The most active compounds in this respect are soluble organic acids produced, particularly, in the freshly fallen litter, or even from the canopies of acidophiles such as heather, pine and spruce. The formation of organocomplexes and their subsequent leaching by these organic acids is called *cheluviation.* In fact podzolisation, which involves the cheluviation of iron and aluminium (normally insoluble compounds in the upper soil zone) is the most intense form of leaching in the cool humid regions. In addition, it is usually accompanied by the final breakdown of the clay minerals, the release of aluminium ions and the dissociation of the humic from the mineral constituent of the colloidal complex.

The result of leaching is first to form an A (or E) horizon characterised by increasing acidity, a decrease in the more and, later, less soluble minerals and a decline in its humic and clay content (see Fig. 8.4). The end-product is the typical bleached (greyish-white), coarse-textured E horizon that has long been the

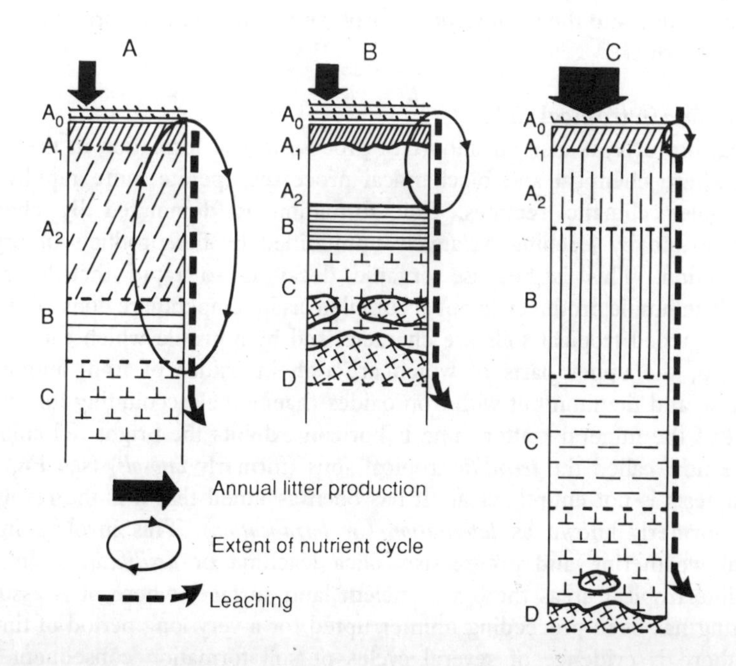

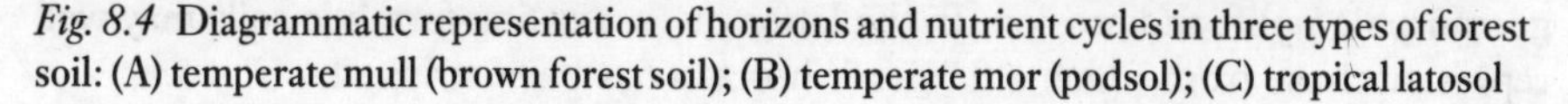
Fig. 8.4 Diagrammatic representation of horizons and nutrient cycles in three types of forest soil: (A) temperate mull (brown forest soil); (B) temperate mor (podsol); (C) tropical latosol

diagnostic character whence the *podzol* ('ash underneath') has derived its traditional Russian name.

Of the materials leached from the A (or E) horizon some of the most soluble bases may be lost from the soil ecosystem in ground-water drainage; some will be absorbed by more deeply penetrating plant roots. However, as the soil environment changes with depth so the mobility of the materials translocated from the upper part of the profile may be checked, or inhibited completely. Deposition or *illuviation* then occurs to form what is known as the *B horizon*. According to the type and intensity of illuviation the deposits may be one or more of the following materials, clay particles, organic matter, bases, iron, aluminium. It is sometimes possible to distinguish within the B horizon several sub-horizons dependent on concentrations of clay, organic matter or iron. The concentration of fine clay particles may be such as to form what is known as a *clay pan* which may check or impede downward drainage. Similarly the deposition of iron and organic matter at a particular level can so indurate or cement the mineral particles as to form a thin *iron pan* which impedes both root penetration and water percolation even more effectively. As leaching becomes more intense so the B horizon is increasingly distinguished by a heavier texture than the E and a reddish yellow-brown colour often accompanied by a distinct organic horizon. Particularly under very acid mor humus, such as that produced by heather (*Calluna vulgaris*) litter, a thin *iron–humus* pan at the base of the B horizon may form, to give what is known as a *podzol with iron pan*. The essential difference between the acid leached and podzolised soils of temperate regions is that in the latter clay is destroyed rather than translocated and the cheluviation of iron and aluminium is more pronounced than in the former.

Desilification in tropical soils

Leaching, however, is also an important process in soil formation in the humid tropics, where chemical and biochemical processes operate more rapidly than under cooler climatic regimes. Soil formation is dominated by chemical weathering and by leaching, relatively unmodified by the products of organic decomposition. This is because organic decay is so rapid that litter and intermediate humic products do not accumulate as in temperate regions. The most intensely leached tropical soils are characterised by a profile which can be over 50 m deep, the upper parts of which are rich in oxides of iron, aluminium, manganese and titanium but with iron oxides (haematite) accounting for over 90 per cent of the mineral matter. The E horizon exhibits the bright red colour of what are now called the *ferralitic* tropical soils (formerly *latosols*) (see Fig. 8.4). Their genesis is not entirely clear. It has been assumed that it is the result of a process formerly known as *laterisation* (or *ferralisation*). This involves intense chemical weathering and progressive silica leaching or *desilification*. In some humid intertropical areas there are ancient land surfaces where, it is assumed, weathering has been proceeding uninterrupted for a very long period of time. In others there is evidence of several cycles of soil formation consequent upon geomorphic uplift and erosion. In the latter cases, existing ferralitic soils may well represent those from which a former E horizon has been stripped.

Also characteristic of many ferralitic soils is the presence of *laterite* in the form of either nodular concretions, extensive but discontinuous lenses or extensive thick, indurated horizons. Laterite can occur at the surface or at varying depths in the soil profile. It is thought to have been formed as a result of an alternation of iron mobilisation and precipitation consequent on extreme fluctuations of soil moisture. Two common modes of occurrence have been noted. One is near the surface as the result of residual iron-oxide enrichment consequent upon a very long period of weathering and leaching. The other is a result of the deposition of iron by lateral drainage along valley sides or at the dry-season level of the ground-water, in a zone naturally subject to the conditions of alternating saturation and desiccation necessary for the mobilisation and precipitation of iron oxides.

ARID AND SEMI-ARID SOILS

Where low annual precipitation is accompanied by high evapotranspiration – as in the arid and semi-arid areas of the world – leaching is very limited in terms of the depth to which the available moisture can percolate and of the mobility of the constituent materials. Only the most soluble cations, of which calcium is the most common, may be leached from the E horizon and deposited to form a calcium-enriched (calcic) B horizon. This is the diagnostic horizon of semi-arid soils of prairie and steppe areas. The depths at which the *calcic horizon* is found are related to the balance between precipitation and evapotranspiration.

In arid areas where evapotranspiration exceeds precipitation for most of the time, leaching is all but absent and soils are saline, characterised by a high concentration of insoluble calcium carbonate (often in nodules or pans) and a high concentration of soluble salts dominated by sodium, chlorine, potassium sulphates, carbonates and bicarbonates and an alkaline status, i.e. $pH > 7$. *Salinisation* occurs naturally where there is a high concentration of soluble salts in the parent material or the ground-water. And it can be particularly intense in large or small depressions (playas) where, after a rainfall event there is a temporary rise in the water-table and the precipitation of salts on or near the surface. The resulting whitish bloom or efflorescence has given rise to the terms *white alkali* (USA) or *solonchak* (former USSR) to describe these saline soils. Should, however, the concentration of sodium exceed 15 per cent of the exchange capacity soil becomes excessively alkaline, with a pH of 8.5–10.0 conditions which few organisms can tolerate.

Secondary salinisation and *alkalinisatiion* of soils is one of the hazards of irrigation farming. Under irrigation the water-table rises, bringing usually salt-enriched water within 2 m of the surface and from which free salts can be continuously precipitated on or near the surface. Irrigation not only tends to increase the total salt content but also results in an imbalance in the proportion of salts in the soil and to result in alkalinisation of formerly less saline soils. A high proportion of sodium to calcium cations causes the soil to deflocculate, lose its structure and any pre-existing organic matter to disintegrate into a black surface scum to which the term *black alkali* has been given. These soils are so alkaline as to be virtually sterile and are extremely resistant to rehabilitation.

HYDROMORPHIC SOILS

In marked contrast to the leached soils are those subject to waterlogging – the *hydromorphic soils* in which the movement of soil water is impeded to a greater or lesser extent. Under these conditions the process of *gleying* (gleisation, gleification) tends to predominate in soil formation. Reduction of iron from its ferric to its ferrous state is primarily a biological process effected by anaerobic bacteria in the presence of the mobile iron-organic compounds produced by chelation. Soil saturation below the organic horizons impedes leaching and, as a result, the translocated iron is precipitated as a mixture of ferric and ferrous hydroxides. This imparts a distinctive, generally uniform, blue-grey to green colour to what is known as the *gley* (G) horizon.

A distinction is usually made between *ground-water gleys*, related to a high water-table, and *surface-water gleys* where impeded drainage and soil saturation are the result of an impermeable horizon consequent upon heavy texture, compaction or induration in the mineral soil horizons. In the former instance, where there is a fluctuating water-table, the alternating saturation and aeration result in a horizon with reddish-yellow mottles in a darker, blue-grey matrix. This mottling is a secondary process resulting from the reoxidation of iron in those spaces (larger pores and root channels) into which air can penetrate most easily. As a result, the *peds* can become coated with iron oxide which increases the tendency towards the formation of a coarse, columnar–prismatic structure. In surface-water gleys, where the alternation of wet and dry conditions is a function of seasonal weather conditions and gleying occurs mainly in the spaces and root channels, the peds are gleyed on the outside but retain their original reddish colour inside.

SECONDARY SOIL CHANGE

Secondary change in soil characteristics can be effected by a change in vegetation resulting from secondary succession or deliberate planting. Slight disturbance may have little immediate or direct effect on the soil; more intense disturbance may, as in the case of fire, modify and/or reduce some or all of the surface organic horizon. In either case, however, the establishment of a vegetation cover different from the original can, as a result of differences in the type of litter produced, effect marked changes in soil formation. The production of a mild, mull humus can counteract the acid effects of a former acid mor humus (Mitchell 1973). The conversion of a brown earth to an acid brown soil or podzol or vice versa can result. Dimbleby (1962a) noted the difference between the mull soil under Neolithic sites in southern England and the podzolic soils which had developed under the surrounding heathland which had replaced a former deciduous forest cover. A change from deciduous to coniferous trees as a result of colonisation or planting can bring about a change from a mull to a mor soil; the invasion of heather moorland, disturbed by burning or grazing, by bracken and/or species-rich *Agrostis* grass can reverse this process.

THE CLIMAX CONCEPT

Plant succession is basic to the Clementsian concept of climax vegetation. This is, theoretically, the stage at which a plant community attains the maximum development possible under the prevailing environmental conditions. Further development or progressive succession ceases and the climax community then maintains itself in a 'state dynamic of equilibrium' so long as environmental conditions remain unchanged. In contrast to seral communities the climax is, therefore, relatively stable. Clements (1916) maintained that the final stage in the development of vegetation was determined by climatic conditions and that the climax community was the most complex that could be attained, and maintained, in a state of equilibrium with the prevailing climate. Within a particular climatic region, given time and freedom from environmental disturbance, plant succession would eventually result in the development of the same type of vegetation irrespective of abiotic variations. The climax community would be dominated by plants of a form best adapted to compete successfully in a given type of climate. This monoclimax concept conceived of only one type of climax, the climatic climax, and of only one type of climax community, the climatic climax formation. Communities which differed from the climax because of particular soil or landform characteristics were regarded as sub-climax. They represented stages in the development of the climax on sites where succession had been arrested or retarded because of more extreme habitat conditions. Clements assumed, however, that these sub-climax communities could develop eventually, though more slowly, to the climatic climax. In time, all the vegetation in a given climatic region would have a similar form, and the climax vegetation would be associated with a mature soil – the zonal soil – which reflected the combined effect of vegetation and climate irrespective of the parent rock type on its formation.

It is, however, doubtful whether climatic or relief conditions have ever remained constant long enough to allow this theoretical stage of complete correlation between climate, vegetation and soil to be attained. Those communities which resulted from the modification of the 'true climax' by biotic factors, particularly the effects of man, Clements designated plagioclimaxes which he considered also had the potential to resume succession to the climax when the disturbance ceased.

While accepting the concept of the climax as a relatively stable plant community in dynamic equilibrium with the prevailing environmental conditions, other ecologists have since rejected, in part or in whole, the Clementsian monoclimax theory and its associated implications of climatic determinism. Advocates of what, in contrast, has been called the polyclimax concept maintain that the climax is not necessarily determined by climate alone but by the interaction of all the factors which influence plant growth and distribution. Within any climatic region there will be a number of climax communities (not necessarily dominated by the same life form) dependent on varying combinations of local climate, relief, parent material and biotic factors. Others defined the climax in terms of the principal factor limiting or determining the maximum development of vegetation, whether climate, soil, relief, biotic or anthropogenic activities. They, however, still regarded the climatic climax as the natural or potential natural vegetation on normal, freely drained, mesic sites. Communities differing from the climatic climax because of extreme soil conditions (excessively dry or wet, extremely acid or alkaline for instance) or because of the continued operation of biotic activities could, if they have reached a stage of equilibrium with the major controlling factor, be regarded as edaphic or biotic (anthropogenic) climaxes respectively. There is no implication that they have the potential to progress eventually towards the climatic climax; in many cases there is no evidence that they can or will. It has been pointed out that communities that would have been regarded as sub-climax by Clements are not only often widespread but are also persistent, particularly when they have been subject to disturbance or modification by man for a long period of time. Indeed, one of the main objections to the climatic climax concept must be the tendency to regard disturbance as unnatural or abnormal and to search for or to reconstruct a theoretical climax which may or may not have existed in the past and which in any case, is unlikely in many areas ever to return.

Later, other workers, including Whittaker (1951, 1975), have suggested that there is no absolute type of climax determined by one or a few controlling factors. Any particular climax community will be composed of an assemblage or combination of species, the composition of which is determined by the sum total of all the interacting habitat factors. There will be, therefore, as many climax communities as there are possible combinations of environmental and biotic conditions. As has been noted, vegetation is subject to continuous change as a result of intrinsic or extrinsic factors. The climax then will be composed of a 'pattern of intergrading communities corresponding to a pattern of environmental gradients; the central or most extensive community type will be the prevailing or climatic climax' (Whittaker 1975: 60). The diagnostic feature of the climax is that it has attained a steady state which will be maintained as long as environmental conditions do not alter drastically. However, the difference between Whittaker's climax concept and that of the polyclimax school would appear to be as much terminological as fundamentally conceptual. A type of climax community will, then, be composed of similar, though not necessarily absolutely identical, populations associated with similar sites. The main criterion distinguishing a climax community, according to Whittaker, is that it should be relatively permanent and self-maintaining so long as the particular combination of environmental conditions does not alter.

The climax concept, however interpreted, implies a degree of stability, a dynamic equilibrium, in the sense that a balance is achieved between not only the plant community and its physical habitat, but all the organisms which are part of the same ecosystem. However, the concept of equilibrium can, in the long term, only be relative since environmental conditions are themselves subject to recurring changes, some of which may be sudden and catastrophic, some of which are gradual and continuous over a long period of time. One of the most variable of environmental factors is that of climate and particularly weather, which fluctuates to a lesser or greater extent daily, seasonally or annually. Short-term fluctuations may not affect mature long-lived perennial plants; they do, however, affect seed production and germination and the number of seedlings and annual plants (as well as the number of animals dependent upon them for food) present in any community from one year to another. Long-term fluctuations involving secular climatic change towards wetter or drier, warmer or colder conditions have been a constant, indeed a normal rather than abnormal, feature of the earth's environmental history, particularly in high latitudes. The climatic changes preceding, during and since the Pleistocene era were such as to initiate quite radical changes in vegetation and to cause the replacement of one type of climax by another.

There must, however, inevitably be a time-lag between climatic fluctuations and the resulting readjustment in the balance of community populations and structure. The effects of a particularly severe winter, or a cool or dry growing season, on seedling mortality in temperate regions will not become manifest until the following season and may then persist for some time after the event. Detailed studies reveal that the recovery of the North American plains grassland vegetation from the effect of the severe and protracted droughts of the 1930s was much slower than the subsequent climatic change back to more humid conditions. In the case of forests with long-lived dominant trees the time-lag must inevitably be ever greater. Climatic change might effect the regeneration of the climax trees fairly rapidly, but it would some time before the mature trees died and new climax dominants replaced them. Because of this time-lag between climatic change and vegetation readjustment, a state of equilibrium may never be attained since the climate may well change before this is effected. Because of this it has been suggested that the climax community is a concept only, never existing in reality either because of the catastrophic initiation of fresh seres or because of an ever-changing environment. Graham (1941) likens the climax to 'a phantom always moving ahead into the future and becoming visible for only relatively brief periods in small areas' and wonders whether the assumption that culmination in an ultimate climax is either the necessary outcome of, or must mark the end of, succession.

PART II ECOSYSTEMS

Chapter

9

Tropical and temperate forest ecosystems

Forest and woodland which presently cover about a third of the global land surface are the largest, most complex and most productive of the earth's ecosystems. Their dominant components are trees, large woody perennials which are arbitrarily distinguished from smaller woody shrubs by height (over 3 m), by the possession of a single main stem (trunk) and by a distinct crown of leaf-bearing branches. Shrubs, in contrast, are normally considered to be less than 3 m, with several basal stems and lacking a well-defined crown. The tree form is shared by a great number of genera and species of different families. However, apart from the tree-ferns (cycads) it is confined to the seed-bearing plants (phanerogams). Of these, all the gymnosperms are woody perennials with the conifers comprising the largest and most prolific group. The majority of conifers are evergreen (with the exception of the larch (*Larix* spp.) and many have small hard narrow needle- or scale-like leaves. The flowering plants (angiosperms), while not exclusively woody, contain the greatest number of tree species, most of which are dicotyledons. Woody monocotyledons include the bamboos and palms; the latter grow in height but not in girth and their simple non-branching trunks terminate in a crown of large leaves.

LIFE FORM OF TREES

SIZE OF TREES

The height and girth of trees are achieved by a very considerable expenditure of energy in the formation of essential strengthening, conducting and storage tissues (see Fig. 9.1). The trunk is formed of three types of tissue:

(a) inner xylem (or wood) composed of cells and fibres reinforced by a mixture of cellulose and lignin. The primary heartwood is inert, the secondary sapwood conducts water and nutrients, and the raywood (bands of living parenchyma cells) store carbohydrates;

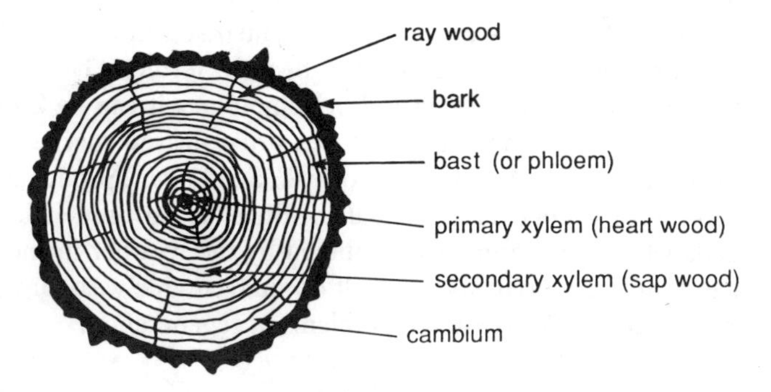

Fig. 9.1 Diagrammatic cross-section of a tree trunk

(b) the outer phloem layer (or bast) contains most of the living cells which store and transport complex organic products;
(c) the cambium layer composed of growth cells which initiate the production of the secondary sapwood and hence the increase in trunk girth.

Seasonal variations in growth are reflected in the width and cell size of the annual growth rings. Finally, an outer layer of epidermal and cortical cells form a protective bark of varying thickness. With increase in girth the outer bark tends to crack, die and, in some cases, to peel off to be replaced by the growth of new epidermal cells. Wind-stress combined with the energy required to raise water from the soil to the crown may determine the maximum height to which trees can grow. While the maximum potential growth height is genetically controlled, the actual height attained is dependent on site conditions. In humid temperate and tropical regions the average maximum height of the dominant trees range from 40 to 60 m. The tallest species, however, are found in humid temperate areas and include the eucalypts of Australia and the conifers of north-west America.

ROOT SYSTEMS

It is hardly surprising that trees have extensive root systems which are needed to anchor the large above-ground biomass and to provide an adequate supply of water and nutrients for growth. Initially all saplings rapidly produce an anchoring tap-root, which may or may not be retained as the tree matures, and subsequently produce a broad mat of ramifying feeding roots. Most of these tend to be concentrated within 18–20 cm of the soil surface. Rooting depth varies with species; spruce (*Picea* spp.) beech (*Fagus* spp.) and poplar (*Populus* spp.) are usually shallow-rooted, while fir (*Abies* spp.), oak (*Quercus* spp.) and many pines (*Pinus* spp.) maintain a deep tap-root. Actual root depth, however, is dependent on soil depth and soil consistency.

As Waring and Schlesinger (1985) point out, the cost of maintaining the supporting and conducting tissues is high. While large storage reserves permit trees to survive unfavourable conditions, the production of reproductive organs may be correspondingly variable. In contrast to their size and longevity, the regeneration of trees is relatively precarious in comparison to that of other plant forms. Most require several years before they become fertile. The age at which flowers are produced varies with species. For some pines it can be 2 or 3 years after germination; for most temperate deciduous broad-leaved species it is rarely less than 10 years and can be 40–50 years for oak. Seed (*mast*) production varies; some species produce a good crop annually, others do so only at longer and variable intervals which are normally determined by the weather in the previous season. Seed and sapling mortality as a result of unfavourable weather conditions and/or of herbivory further increase the problems of regeneration.

LIFE SPAN

The life span of the majority of tree species exceeds that of most living organisms. Although the maximum life expectancy is not known with certainty the age of existing oaks has been estimated at 1500 years, of yews (*Taxus* spp.) and juniper (*Juniperus* spp.) at over 2000 and of some Californian redwoods (*Sequoia* spp.) at over 3000. Tree-ring chronology has established that the bristlecone pines (*Pinus aristata*) of California are the world's longest-living trees; the oldest known is *c.* 4600 years old (Morey 1973). Although no absolute tree age limit has been established, there is an age beyond which the rate of growth declines eventually becoming negligible and the tree becomes increasingly susceptible to disease and wind-throw. However, even after the wood begins to decay, regrowth may be continued by suckers produced by adventitious buds on the fallen trunk and branches.

LIFE FORMS

The three main tree life-form categories are: (1) the evergreen narrow-leaved conifers; (2) the broad-leaved deciduous; and (3) the broad-leaved sclerophyllous trees. The conifers evolved earlier than the others. Of the 650 species (belonging to 8 families and 50 genera) the majority occur in colder areas in the northern hemisphere; in the tropics they are found only on the higher temperate zones of mountains. They are characterised by a growth and wood accumulation rate unsurpassed by the other forms. This is a function primarily of their evergreen habitat; they retain their needles for up to 7 years, but leaf fall is not seasonal. This allows the conifer to commence photosynthesis as soon as environmental conditions become suitable and without using energy to produce new leaves as in the case of deciduous species. As a result they are the world's largest plant forms, attaining tremendous girths and heights of 80–120 m.

The broad-leaved sclerophyllous evergreen and deciduous trees are evolutionarily younger than the conifers over which they became dominant in the hot humid Tertiary era. They are frost sensitive and, hence, now characteristic of humid, warm, tropical to mild temperate regions. The deciduous broad-leaved trees, however, are the main component of the forest ecosystem in humid temperate lowlands with a marked seasonal variation of climate from cool spring–warm/hot summer–mild autumn to cold winter with a distinct alternation of short winter and long summer day-lengths. Their mesophyllous (thin) leaves can photosynthesise rapidly when light and temperature conditions are suitable by reason of a freer and stronger sap flow than in conifers. The deciduous habitat (i.e. annual leaf loss) is an adaptation to cold and drought to which the mesophyllous leaves are intolerant. The loss of leaves, however, is compensated by regular and rapid recycling of their nutrients. Under similar environmental conditions the yield of deciduous trees is rarely half that of conifers and while the growth height may be similar the girth attained is much less – only the poplars compare with conifers in this respect (Edlin 1976).

FACTORS LIMITING TREE GROWTH

Among the most important factors limiting tree growth are an insufficiency of moisture and heat for the production, germination and establishment of seedlings and too short a growing season to allow the production of new wood before the onset of unfavourable conditions. With increasing deficiency of available soil water and/or accessible ground-water, the growth of trees adapted to periodic drought and low precipitation is slow and individuals are small, often deformed with a spreading habit and wide spacing. In addition, insufficient rooting depth, waterlogged soils and high wind-force are inimical to growth. Because of their potential growth height, trees are particularly susceptible to wind exposure; increased evapotranspiration can result in an unfavourable water balance which checks growth. Trees subjected to high winds become stunted with flattened or markedly asymmetrical, wind-sheared crowns. However, the most ubiquitous agent limiting tree growth is direct or indirect human activities such as the promotion of fire and the increase in grazing by domestic livestock and protected (deliberately or accidentally) wild herbivores. Both destroy tree seedlings and saplings and hence inhibit regeneration.

CHARACTERISTICS OF FOREST ECOSYSTEMS

PLANT BIOMASS

By reason of their size and longevity, trees form the earth's most massive and complex ecosystems. Not only is the total amount of plant biomass per hectare high but three-quarters or more of that in a mature forest stand is contained in the trees, with that in the trunks generally exceeding that in the canopy and roots combined. However, the contribution of the component parts of the tree (see Fig. 9.2) to the total biomass varies with the age of the individuals and the developmental stage of the forest ecosystem – the ratio of woody non-photosynthetic to green photosynthetic material increasing with age. A relatively small though variable proportion of the annual photosynthetic production is

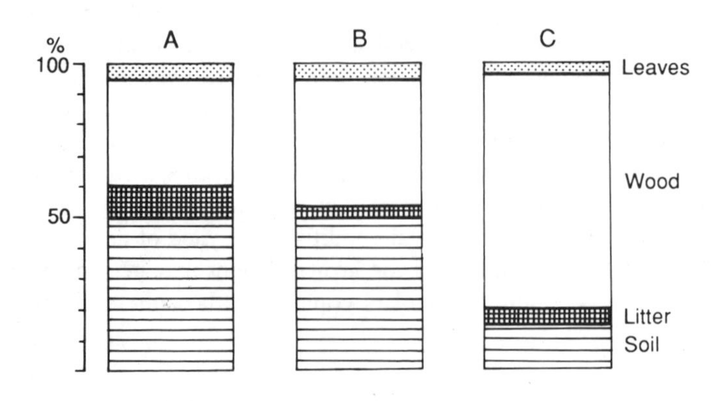

Fig. 9.2 Comparative distribution of organic carbon in coniferous forests (A), temperate deciduous (B), and tropical rain forest (C)

consumed by herbivores, and up to 75 per cent may enter the detrital or decomposer food chain in the form of litter which supports a large diverse population of soil organisms. The volume of this dead organic material (DOM) can be twice that of the fresh leaf tissue.

The structure (or spatial organisation) of the component plant life forms of the forest ecosystem is more complex than that of others. This is most clearly reflected in the vertical stratification of the groups of plants of progressively smaller size and life form (strata or synusiae) each of which is capable of growing and competing successfully for resources at a different level and hence in a different above- and below-ground environment than that above it. The type of stratification and degree of structural development are dependent on the amount of tree biomass and, particularly, on the depth and density of the tree canopy produced in particular environmental conditions. In some circumstances, as in the tropical rain forest, the development of a multi-layered tree canopy may reduce light to an extent which limits the growth of smaller non-arboreal forms beneath it. Structural development tends to be greater where there are spatial and/or seasonal variations in the density of the canopy which allow the formation of shrub, field (herbaceous) and ground (mosses/lichens) strata characteristic of more open, deciduous temperate forests.

ANIMAL BIOMASS

Although the animal biomass of forests is relatively small it is extremely varied, consequent on the diversity of microhabitats provided by the complex structure and diverse food supply of the plant biomass. The most densely populated areas are the canopy and the soil. The canopy and tree trunks combine shelter with a variety of roosting, nesting and feeding niches for a large number of birds, including plant-feeders, insect eaters and predators often brightly coloured and frequently with highly specialised feeding niches. Arboreal animals such as reptiles and, more particularly squirrels, monkeys, marmosets, pine martens and sloths possess morphological adaptations such as prehensile toes and tails. The number of surface-living vertebrates is smaller and is dependent on the openness of the forest and the associated feeding opportunities. In temperate areas the ground dwellers include the fox, wild boar, wildcat, polecat and the red and fallow deer – and larger ground-nesting birds such as grouse, blackcock, capercaillie and turkey (in America). Burrowing vertebrate and invertebrate animals and microscopic decomposers become dominant on the upper soil horizons. The most prolific above-ground animals are the herbivorous, predatory and saprophytic insects whose populations fluctuate from year to year and frequently attain damaging pest levels.

FOREST PRODUCTIVITY

The characteristically high productivity of forest ecosystems is also related to their structural development which allows the use of a large resource volume. The photosynthetic tissues are concentrated in the canopy which, in some instances, can occupy a volume exceeding that of the marine euphotic zone (see p. 276). The weight of leaves can vary from 1 to 10 tonnes ha^{-1} dry weight and the leaf area index (LAI) at maximum crown development can attain a value of 8+. The seasonal duration of active leaves is protracted compared to that of many herbaceous plants. Further, the vertical above-ground development of the forest ecosystem is supplemented by the extent and volume of the root system. This can tap the considerable volume of water and nutrients necessary to maintain the high level of productivity.

However, as has already been implied, a high proportion of the energy trapped and of nutrients absorbed in primary production is stored in the non-assimilating part of the ecosystem. A characteristic feature of all forest ecosystems is the long-term immobilisation of a large percentage of the nutrient capital in the woody tissues. The forest nutrient cycle is, in fact, a two-phase one. Short-run cycling of nutrients takes place via the leaves which have a life span of 1 to about 7 years. The cycling of nutrients stored in stem wood has a long run of hundreds to thousands of years. The annual nutrient uptake by tree leaves and small twigs together with the ground flora is greater than that by the trunk and the main branches (Duvigneaud and de Smet 1970). However, the percentage of the NPP stored annually in the wood varies with age, from a maximum of 30–50 per cent during rapid growth in youth to 10–15 per cent in old age. The cumulative effect is such that in a mature forest 70–90 per cent of the nutrient content of the plant biomass may be locked up in the wood. This has far-reaching implications for the exploitation of forest resources; and the removal of forest biomass at a rate greater than it can be renewed has two main effects. The first is nutrient depletion; because of the cumulative uptake and the high plant : soil nutrient ratio this can be as great a drain on soil fertility (nutrient status) as arable crop production in the same period of time. The second is the modification of microclimatic and associated habitat conditions following the removal of a massive biomass such that regeneration may be drastically retarded or even inhibited (see Ch. 17).

ENVIRONMENTAL IMPACT OF FOREST ECOSYSTEMS

FOREST MICROCLIMATE

As a consequence of its massive biomass, the impact of the forest on the atmosphere and soil it occupies is greater than that of any other ecosystem. The effect of the canopy is to produce a distinctive microclimate within itself, in the space beneath it and in the upper soil layers. Much of the incident insolation is either reflected from, or absorbed by, the canopy. The amount of light reaching the floor may, in a dense forest, be less than 1 per cent of that received at the top of the canopy. Absorption of insolation during the day results in cooler air temperatures inside than outside the forest while the canopy, and the reduction of air movement below it, tend to check outgoing radiation at night. The result is a smaller diurnal and annual temperature range within than outside the forest. The temperature of the ground surface determines that of the soil to a depth of at least 2 m. In a forest the mean daily maxima of the ground surface can be reduced by as much as 10 °C below that on adjacent bare soil. Increases in mean minima are, however, less pronounced. Soils under forest cover freeze later and less deeply than those lacking a cover. Also the amplitude of diurnal and seasonal ranges of temperature decreases exponentially with soil depth; the former become slight at 60 cm, the latter just detectable at 1 m. The time-lag in the occurrence of soil maxima and minima temperatures as compared with the surface increases with depth. Modification of temperature and air movement is accompanied by that of atmospheric humidity; this can be as much as 11 per cent higher in temperate forests, 15 per cent in tropical forests than in the free atmosphere. Dependent on size, structure and density a forest can act as a wind-break around and over which surface air currents are deflected. Wind velocities within the forest may be reduced to 10–30 per cent of those in the open. The degree of microclimatic modification will, however, be dependent on the nature of the forest as well on that of the local and regional climatic characteristics.

CLIMATIC IMPACT

As well as producing a distinctive internal microclimate, a forest can exert a sensible effect on that of its immediate surroundings. *Forest-edge* effects dependent on aspect include shading, increased reflection and radiation, and shelter. The effect of large forested areas on climate, particularly on precipitation, has been a subject of interest and increasing concern. Rapid up-draughts to 500 m and over of warm air above forested areas can occur and, in humid areas, increase the probability of convectional rainfall. In addition, very tall stands of trees may contribute to the orographic lifting of, and condensation from, unstable air masses; or may they promote condensation of water droplets from low-lying cloud or mist – as in the 'cloud forests' of humid mountain areas. Also, forests return water vapour to the atmosphere by evapotranspiration. Indeed most of the water vapour evaporated from the global land area is supplied by those forested regions which receive the highest mean precipitation. It is estimated that *c.* 50 per cent of the precipitation received in the Amazon Basin is produced by evapotranspiration from the tropical rain forest (Waring and Schlesinger 1985).

More recently the significance of the world's forests in maintaining the carbon dioxide concentration of the atmosphere has recently attracted much attention. Increasing use of wood and fossil fuels has been accompanied by an increase in the concentration of atmospheric carbon dioxide, while the rapid reduction of the world's forests removes one means whereby carbon dioxide can be taken up in the process of photosynthesis and stored in plant tissue. Because of tree longevity, variations in wood density and in the isotopic composition ($^{18}O/^{16}O$, ^{14}C, $^{13}C/^{12}C$) of wood cellulose provide a sensitive record of past changes in global climate. It has further been suggested that satellite monitoring of polar and mountain tree-lines and of desert–woodland ecotones may greatly help the assessment of current global climate change by supplementing information derived from individual field sites (Waring and Schlesinger 1985).

IMPACT ON THE HYDROLOGICAL CYCLE

The impact of a forest cover on the hydrological cycle, however, is even more significant and pervasive (Table 9.1). In the first instance some of the precipitation

Table 9.1 Yearly moisture regimes of oak forest aged 220 years on different sites (per cent) (data from Molchanov 1963)

	Plateau	*NW slope*	*SE slope*	*S slope*
Crown interception	14	15	12	11
Transpiration from tree stand	43	35	24	20
Evaporation plus transpiration by ground-cover vegetation	14	13	12	10
Runoff from ground-cover vegetation	4	13	18	32
Soil runoff	6	7	10	12
Ground runoff	19	17	24	15

is intercepted by the canopy and evaporated, the amount being dependent on the type and intensity of precipitation, the density of the canopy and its leaf form. That not intercepted reaches the forest floor by a combination of stem flow and through-fall, the relative proportions being determined by the morphology of the constituent trees. Some of the water may be lost by direct evaporation and/or runoff from the ground surface. Some infiltrates into the ground dependent on the permeability of the soil and the slope of the land. Water retained in the soil is returned to the atmosphere by evapotranspiration. Because of a deep, particularly dense and extensive root system, forests can tap not only a greater volume of soil water than other ecosystems but can also draw on deeper and less variable ground-water sources. However, the net effect of a forest cover on the soil moisture varies with the macroclimate, the relief, the physical characteristics of the soil as well as with the type of forest. On level well-drained sites under comparable climatic conditions soils of similar texture might be expected to be drier than those without a forest cover. On steep slopes, however, runoff would be checked and infiltration facilitated by vegetation and litter; soils would then be moister with than without a forest cover.

IMPACT ON SOIL

The forest biomass also exerts a considerable influence, through its large input of DOM, on the physical and chemical properties of the soil. Litter (mainly leaf litter) can account for 50–90 per cent of the DOM according to composition, structure and age of the particular stand. The chemical composition of the litter can intensify or counteract leaching. A soft-leaved nutrient-rich litter will break down easily to produce a rich well-decomposed mull humus (see Ch. 8). An abundance of soil organisms, (particularly worms, termites, etc.) aids its rapid and thorough mixture with the underlying mineral material and the formation of a colloidal complex with the clay fraction which enhances the cation-exchange capacity of the soil. In contrast, hard-leaved nutrient-poor litter breaks down more slowly and less efficiently to give an acid, moder or mor humus. Soil animals, particularly burrowers, are less prolific and the mixing of organic and mineral matter is as a result limited while organic acids produced during decomposition intensify the effect of leaching.

TYPES OF FOREST

Existing forests vary in biomass, species composition, structure and productivity dependent on their evolutionary adaptation to past and present physical environments, on their stage of development and on the extent to which they have been subject to direct or indirect human modification or change. Types of forest can be and indeed are frequently distinguished and categorised on the basis of one or more of the formal or functional attributes previously discussed, dependent on the scale and aims of the study. For instance, the forester will be primarily concerned with the type of wood and the productivity of the component species. The entomologist may focus his attention on the diversity and distribution of microhabitats and on insect predator–prey relations. Both approaches, however, require a knowledge of the specific composition and structure of the forest community and an understanding of how the particular forest ecosystem functions. At the global scale forests have long been classified and spatially delimited on the basis of one or a combination of three attributes of the mature stand of the actual or theoretical climatic climax:

1. *Environmental conditions*, particularly climate, were the initial basis on which tropical, temperate and cold forest zones were delimited;
2. *The form of the dominant trees* as determined by the size, shape, durability and duration of leaves, e.g. broad, narrow or needle; large or small, divided or undivided; hardness of cuticle (sclerophylly or mesophylly); evergreen or deciduous, or mixed;
3. *The composition* on the basis of dominance or co-dominance of plant taxonomic groups; coniferous (gymnosperm) or flowering (angiosperm); family and/or genus and species, i.e. beech, oak, spruce, pine, etc.

The principal types of forest recognised and whose potential climate zones are still delineated at the global scale include:

(t) Tropical rain forest (selva)
(tm) Tropical deciduous (or monsoon) forest
(tm) Mediterranean evergreen sclerophyllous forest
(t) Temperate broad-leaved evergreen (sclerophyllous) forest
(t) Temperate broad-leaved deciduous (mesophyllous) or nemoral forest
(t) Boreal (taiga) needle-leaved evergreen

Their latitudinal limits are determined either by low temperature (t) and/or seasonal water deficit (m). The main features of the tropical rain and temperate deciduous forests (including Mediterranean) will be analysed in this chapter and the Boreal (taiga) forest together with the tundra in Chapter 10.

THE TROPICAL RAIN FOREST

The tropical rain forest is the most massive, diverse and productive of the earth's ecosystems. Despite its recent rapidly increasing exploitation it still covers considerable areas in Amazonia, the Congo Basin and to a lesser extent the lowlands of Malaysia. It attains maximum development on well-drained (i.e. mesic) lowlands with an evenly distributed mean annual rainfall of at least 1800 mm with little if any seasonal variation and no distinct dry season. Mean annual temperatures remain constantly high (over 18 °C) with the seasonal less than the diurnal range. A distinction is frequently made, on the basis of rainfall, between the humid and the evergreen seasonal forests, which grade with increasing length of dry season into the seasonal deciduous (monsoon) forest and savanna woodland.

Composition

The humid rain forest is above all characterised by the richness and diversity of its flora and fauna. It contains the largest known assemblage of plants and animals in the world. Some 40 to over 50 per cent of the plant species are woody, and 25 per cent are trees – a dominance unsurpassed in any other type of forest. From 60 to 200 different tree species may be found on any one hectare; and there may be three times as many types of vascular plant and four times as many animal as there are plant species. Such diversity reflects the very large number of small, often widely distributed species populations. The floral composition is mixed. Species dominance is extremely low (less than 15 per cent) and, where it does occur, is usually a result of either impeded drainage, or natural or recent human disturbance. Two types of species' mixtures have been recognised (Longman and Jenk 1987; Whitmore 1988): one is a wide random spacing of species individuals, the other small groups or consociations of two or three species which give rise to a multivaried mosaic composition.

The reasons for such high diversity has long been debated. Among the evolutionary factors which may be involved is the higher level of UV-B which can modify the DNA in low latitudes and, as a result, act as an important trigger for genetic mutation (Caldwell 1981). The importance of edaphic factors in an area of relatively uniform climate has also been stressed. Finally, much emphasis has been put on the importance of biotic factors in an environment capable of producing a large plant biomass and hence providing a large number of potential niches for a multiplicity of species. Speciation has then been dependent on the evolution of biotic barriers (e.g. plant growth phases and life-history strategies) which partition time, space and resources in such a way as to avoid excessive competition. This results in a large number of narrow ecological niches with a high degree of specificity of animal–plant relationships. It has also been suggested that opportunities for speciation increased during the climatic changes consequent on glacial advance and retreat in the Pleistocene period. During the maximum dry interglacial period there is evidence that the rain forest in South America was reduced to a number of small refugia (see p. 73) which, with the eventual return of a humid climate, became centres of speciation and dispersion.

While high diversity and plant–animal specificity reduce the probability of explosive parasitic and pathogenic outbursts, small widely distributed populations increase the difficulty of regeneration and the maintenance of a viable population. As Jacobs (1990) notes, the minimum population with sufficient genetic diversity to allow long-term survival of a species requires a large spatial area.

Morphology and structure

The mature rain forest trees exhibit a convergence of form and possess many morphological features exclusive to the tropics. These include smooth, uninterrupted trunks with buttressed bases (see Fig. 9.3(a)); a large horizontal root-mat, 40–50 per cent of which is concentrated on or in the upper 30 cm of the soil (see Fig. 9.3(b)); broad, smooth-edged leaves (10–20 cm long and 5–10 cm broad) terminating in a pronounced drip-tip; and cauliflorous and dioecious blossoms. Wind pollination is rare, the main agents being birds, bats and insects. There is no seasonal rhythm of growth; and flowering and leaf fall proceed at different times throughout the year in different species.

Tree dominance results in a dense canopy in which there is a vertical gradation in tree height in relation to light availability (see Fig. 9.4). However, the existence of a clearly distinguishable vertical structure of three or four strata advocated by Richards (1979) is no longer generally accepted. Indeed, as Longman and Jenk (1987) note, it is very difficult to make a clear distinction between the *rhizosphere* (root sphere) and the *phyllosphere* (leaf sphere) because of the profusion of lianas and epiphytes which may be found anywhere between the base and the crown of the trees. 'Tree-gardens' grow in all available micropockets of water and DOM from the base of the tree to the top of the crown. These authors propose a delineation of the tree canopy according to the amount of illuminance received. Approximately 50 per cent of the rain forest animals live in the canopy, of which 45 per cent are non-flying or gliding forms. The most numerous animals are the ubiquitous ants and termites which are responsible for the comminution and transport of DOM.

Regeneration

In contrast to the relatively simple deterministic models of regeneration and secondary succession advocated in temperate forests, this process is less predictable and more probabilistic in the rain forest. Aubreville (1938) early

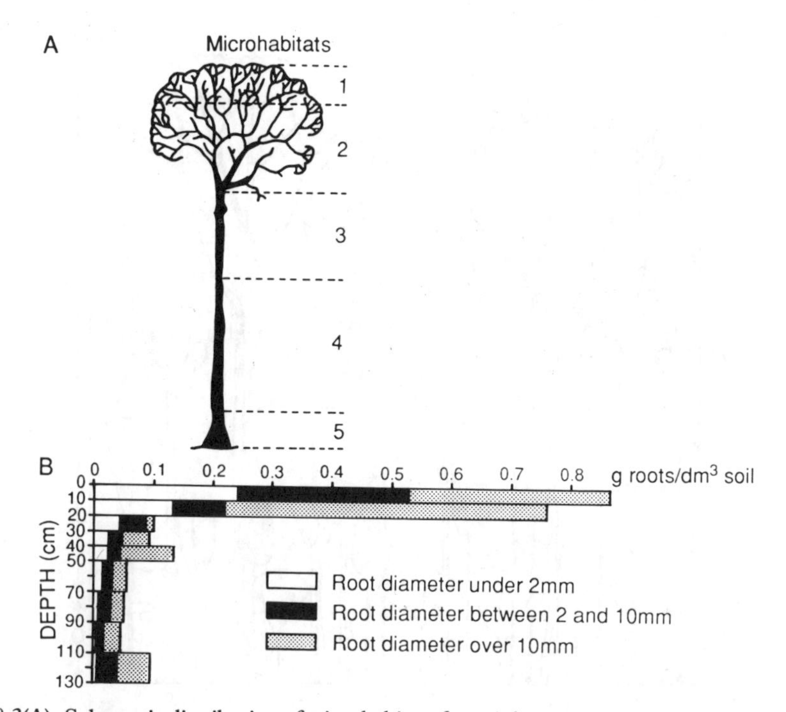

Fig. 9.3(A) Schematic distribution of microhabitats for epiphytes on an emergent tree in the tropical forest. 1 = fully exposed apical portion of crown with microepiphytes; 2 = main zone of epiphytes covering large limbs; 3 = upper drier part of bole with crustaceous lichens; 4 = moister lower part of trunk with lichens and frequent bryophytes; 5 = base of tree covered with bryophytes, particularly between butresses and spurs (from Longman and Jenk 1987); (B) distribution in depth of roots in sorted size classes in the tropical rain forest, Ivory Coast (redrawn from Huttel 1978)

noticed a paucity, even absence, of seedlings and saplings of the uppermost light-demanding trees either in the canopy of the mature forest or on the floor beneath. He attributed this to a cyclic or mosaic process of regeneration in small, natural gaps which varied continuously in location and composition. Large, old, dead or dying trees eventually collapse, creating a gap in the canopy (see Fig. 9.5), disrupting the soil and vegetation and producing what is now termed a 'chablis' (Oldeman 1978) in which a cycle of regeneration (or succession) can be initiated (see Fig. 9.6). At any one time it has been estimated that chablis may occupy 3–10 per cent of the forest area. Changes in the microclimatic conditions (i.e. increase in insolation, in range of floor and soil temperatures and in precipitation) promote a rapid flush of vegetative growth. This involves the germination and growth of heliotropic seeds and seedlings of pioneer tree species. Some come from the existing soil bank, others are immigrants often from other forest areas. A few are seeds of primary (climax) trees whose survival is dependent on rapid germination in an available gap. Growth in height of seedlings is rapid – up to 4 m per annum for individual species – and a total of over 10 m woody growth in 5 years has been recorded (Longman and Jenk 1987). After 5 years the number and density of

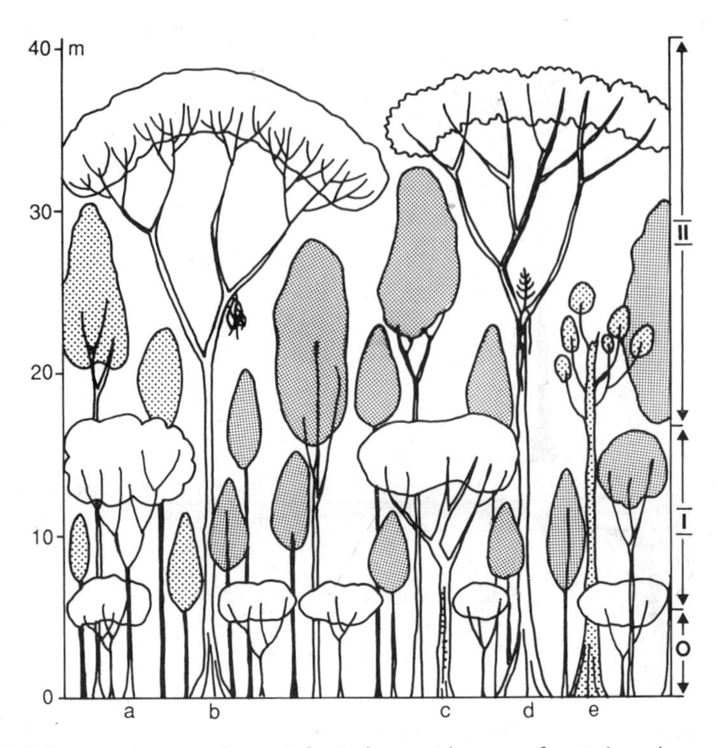

Fig. 9.4 Schematic diagram of a tropical rain forest with mature forest giants (ensemble II) b and d of the present, e of the past; and mature trees of middle height (ensemble I), a and c of the present; and ensemble 0 with trees (unshaded) which remain small when mature. Shaded trees are future trees from I and II (shaded) (from Hallé *et al.* 1978)

secondary species start to decline while those of primary trees, which require shade in which to germinate and become established, begin to increase slowly. The rapid-growth, early-flowering and fruiting pioneer (secondary) species are eventually over-topped by the slower-growing, later-flowering and fruiting primary (climax) trees. The latter mature slowly, needing 50–100 years before the first flowering, and have very long life spans (though the actual length is difficult to assess in the absence of annual growth rings). In the final mature, theoretical climatic climax stage a balanced mixture of shade- and light-tolerant emergent trees is attained. However, because of the species richness the final mix in any one chablis is never exactly the same as that in any other.

Exogenous disturbances such as shifting cultivation, deforestation and elephant browsing also create large gaps in which succession and the regeneration of the primary forest become difficult and in which retrogression and degradation of the forest vegetation may occur. The larger the gap and the further the cleared area is from the forest edge, the less accessible it will be to colonisation by the heavier fleshy-coated seeds of primary trees. Also the impact of a large gap on the microclimate will be more severe than that of a small gap. The opportunities for colonisation by more easily dispersed, secondary, light-demanding species will be greatly enhanced. As a result, larger areas of less species-rich secondary forest can

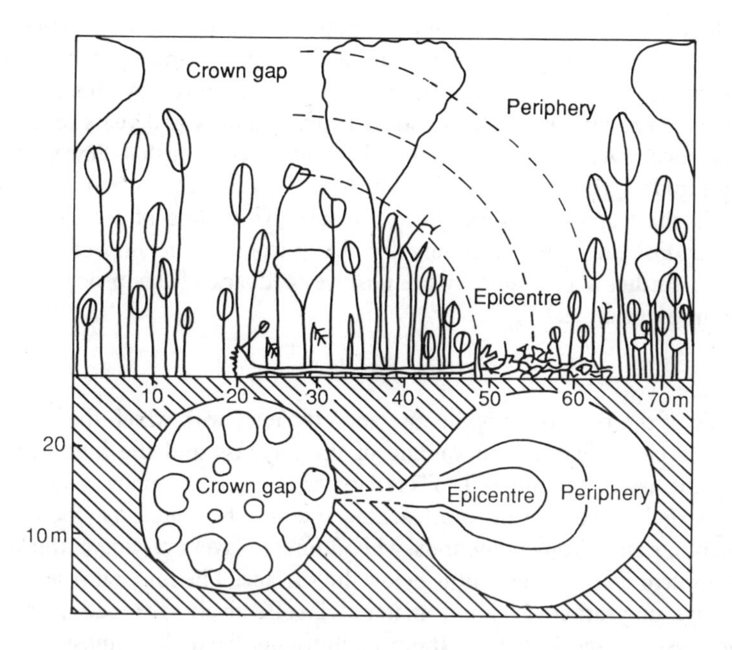

Fig. 9.5 Vertical and horizontal projections of a chablis (from Longman and Jenk 1987 after Oldeman 1978)

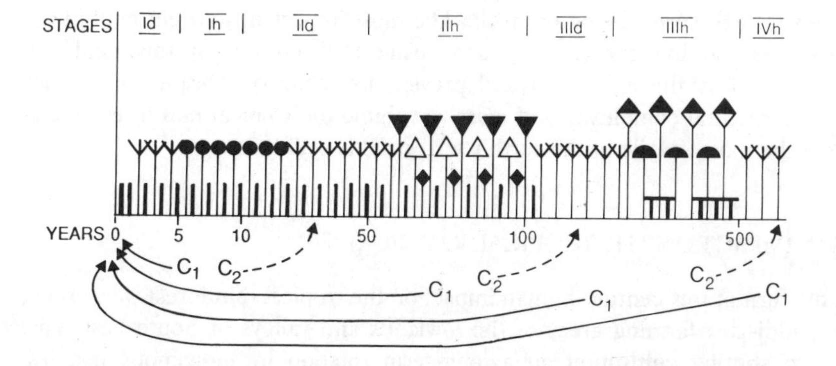

Fig. 9.6 Succession in a tropical rain forest from bare soil to primary forest. Stages: (I) pioneer; (II) young secondary; (III) old secondary; (IV) primary. Duration of stages III and IV estimated. A chablis may form in any stage and regression result in the former stage in development; given sufficient seed accessibility a later stage may be anticipated; d = period of dynamic growth; h = period of homestasis; C = tropical rain forest (adapted and simplified from Hallé *et al.* 1978)

become established. With increased human population the traditional pattern of shifting cultivation, which combines the partial clearing of a small patch of forest land with mixed perennial crop cultivation for 3 years followed by abandonment for 30 years or more, has given way to shorter forest fallows of 3 years or less. In the latter instance regeneration does not go beyond a weedy bush-fallow stage. In addition, a rapidly growing demand for firewood from both urban and rural settlements in the tropics has resulted in the selective cutting of secondary and primary forest instead of wood culling for this purpose. The impact of forest degradation is further accelerated by the positive feedback of this process on the soil and the nutrient cycle.

Nutrient cycle

Characteristic of the rain forest is the rate and efficiency with which the available nutrients are cycled. Despite the volume of the biomass and the high primary productivity, the associated mineral soils are in general nutrient-poor as a result of a very long period of weathering and leaching *in situ*. The solum may be tens of metres in depth, rich in iron compounds, with indurated layers varying in thickness and extent occurring at varying depths below the surface. The mineral reservoir in the underlying unweathered parent material is effectively divorced from the forest's root system (see Fig. 8.4). Input of nutrients from the atmosphere are negligible in relation to demand and leachates from the canopy are variable – less than 25 per cent of the total for nitrogen, phosphorus and calcium to over 50 per cent for potassium and sodium (Unesco 1978). Hence the rain forest is dependent on rapid litter decomposition and absorption of the nutrients released. The efficiency of uptake is testified by the low nutrient content of rivers which drain from undisturbed forests (Stark and Jordan 1978). Uptake is facilitated by the lateral concentration of the root-mat (with which there is associated a dense network of mycorrhizal fungi) in the upper 30 cm of the soil. The rain forest is, in Gourou's (1961) words, living on itself. The most important nutrient pool in the ecosystem is the living biomass. As a result the removal of this pool can effectively destroy the ecosystem and prevent its recovery. Despite its species diversity, structural complexity and massive volume the tropical rain forest is one of, if not the most, fragile of the earth's terrestrial ecosystems.

HUMAN IMPACT ON THE TROPICAL RAIN FOREST

Until the turn of this century human impact on the tropical rain forest was, except in the paddi-rice farming areas of the lowlands and valleys of South-east Asia, limited to shifting cultivation on a long-term rotation by indigenous peoples. Patches of forest felled and burned preparatory to crop planting were small – not much larger than the gaps created by the death fall of primary trees – and widely scattered. A cover of a diverse mixture of perennial crops of varying height replicated the structure of the forest in miniature and protected the soil from accelerated erosion (Ruthenberg 1976). After 3 years' cropping yields declined and the cultivated patch was abandoned and would not be recultivated for at least

30 years, by which time a secondary forest would have re-established itself and its associated soil fertility. Rapid increase in native populations in Africa and South-east Asia has resulted in the shortening of the tree fallow period to as little as 3 years, insufficient time for other than a poor grass shrub vegetation to regenerate, which can easily be further degraded by overgrazing of domestic (sheep/goats) livestock and wild herbivores. Removal of the original forest cover over increasingly larger areas exacerbates the erosivity of the torrential rainfall while increased surface evaporation causes drying and hardening of the exposed mineral soil. In many areas of Africa 'savanna-isation' of the tropical rain forest has occurred with a concomitant decline in soil fertility, an increase in soil erosion, and exposure of hard indurated sterile crusts (*bowal*) near or at the surface.

The decimation of the Amazonian rain forest started later than that in the Old World. Indigenous populations are small and external impact has been greater consequent upon the rapid growth in demand for land for colonists and for beef production and for timber. Wholesale extensive clearance by felling and burning for these purposes has been facilitated by the development of settlement and infrastructure. In this instance the forest has little hope of re-establishment other than by the planting of fast-growing plantations of commercial species. However, by whatever process or at whatever rate the rain forest is degraded, its destruction inevitably leads to a decline in soil fertility that is difficult, and economically unfeasible, to reverse on any but a very small scale.

TEMPERATE FORESTS

In the absence of extensive land areas in the southern hemisphere the temperate (and Mediterranean) forest zone attains its greatest extent in the mid-latitudes of the northern hemisphere in North America, Europe and Asia. To the south it is replaced by tropical and subtropical ecosystems; to the north by the Boreal zone where its limit is determined by too short a thermal growing season for its constituent species; and in the continental interiors by the aridity of steppe/prairie grasslands.

The temperate forests vary in composition in relation to two principal climatic gradients. One is that from the broad-leaved, evergreen and frost-sensitive forest of the warm to the deciduous broad-leaved forest of the cool temperate zone with mixed (coniferous and deciduous) forest transitional to the Boreal and montane zones. The other is that from the harsher more continental to the milder oceanic facies of both the cool and warm temperate zone. The distinctive evergreen sclerophyllous forest and shrub vegetation in the cool wet winter, hot dry summer zone is transitional to the west coast deserts of south-west USA, southern Chile, south-west Australia and Africa and is most extensive in the Mediterranean climatic region of Europe.

The temperate forest ecosystems are less productive with lower species diversity and a less complex structure than those of the humid tropics. Trees make a smaller contribution to the total biomass, woody lianas are scarce or absent; a higher proportion of the nutrients are stored on or in the soil; and the tree strata are composed of more clearly distinguishable, spatially extensive associations of one or a few species, which may be deciduous or evergreen broad-leaved or coniferous.

Origin of the temperate forest

The evolution of the temperate forest ecosystems post-date those of the humid tropics which, until the mid-Tertiary period had, consequent on uniform climatic conditions, a universal distribution. Spatial differentiation of climate following the Alpine orogeny resulted in evolutionary organic adaptation to cooler and drier conditions in the extra-tropical regions of the world. However, the marked but variable decline in number of tree genera and species from warm to cool temperate latitudes is a consequence not only of the decreasing environmental potential poleward but of the time-lag in the northward migration of species following the final retreat of the Pleistocene ice-sheets and the subsequent post-glacial climatic amelioration (Davis 1981). However, the north-west European temperate forest flora is much poorer than that of comparable climatic areas in North America or Asia. This has already been ascribed (see pp. 74–5) to the existence of the transverse barrier of the Alpine ranges which impeded the southward migration of the pre-existing Tertiary tropical flora and fauna in the face of the advancing ice-sheets and associated climatic deterioration. During the post-glacial recovery the Alps and the Sahara cut Europe off from the species reservoirs further south. In contrast, in North America and eastern Asia, many pre-glacial species found refuge further south-east from where a northward migration in the post-glacial period was able to take place. As a result, genera common to existing cool temperate forests such as beech (*Fagus*), lime (*Tilia*), elm (*Ulmus*) oak, ash (*Fraxinus*) and sycamore (*Acer*) are represented by only one or two distinct species in Europe but by three or four times as many in North America; while genera of tropical origin such as *Carya* (hemlock), *Liquidambar*, *Liriodendron* (tulip tree or yellow poplar) and *Magnolia* are native to eastern North America and south-eastern China but not to Europe. Also the dominance of coniferous species in temperate north-west America is considered to be the result of the longitudinal barrier of the north-west Cordillera which together with the aridity of the Great Plains isolated the north-west from the species-rich reservoir in the south-east of the continent. Hence post-glacial colonisers were drawn from the southern Rockies to where the northern conifers had been able to retreat and survive during the Pleistocene period. Broad-leaved evergreen species intolerant of frost are characteristic of the warmer temperate forests in the northern hemisphere; and of the limited southern hemisphere forests of Chile, south-east Australia and New Zealand which are composed of genera (*Nothofagus*, *Podocarpus*, *Eucalyptus*, *Araucaria*) not found in the northern hemisphere.

BROAD-LEAVED DECIDUOUS FOREST

The temperate broad-leaved forest, as the name indicates, is characterised by the dominance of deciduous trees, particularly on mesic soils in what is often referred to as the cool *nemoral* climatic zone. The characteristic tree is the oak. In Europe two species, the sessile (*Quercus petraea/sessilflora*) and the pedunculate (*Q.*

robur), occupy the most favourable sites on the central zone, with beech (*Fagus sylvatica*) and ash (*Fraxinus excelsior*) frequently dominant on shallower chalk and limestone soils or mixed with sweet chestnut (*Castana sativa*) or fir (*Abies* spp.) at higher altitudes. Birch (*Betula* spp.) becomes dominant at the latitudinal and altitudinal limits. Other common species which occur in association with the dominants or alone include alder (*Alnus* spp.), poplar (*Populus* spp.), sycamore (*Acer pseudoplatanus*) and the elm. The broad-leaved deciduous forests of northern China and Japan are composed, with a few exceptions, of ecotypes of the same genera as in Europe. By far the most diverse forest flora is that of eastern North America which contains all the Eurasian genera and some important ones (previously noted) not found in either Eurasia or Europe. In comparison to both these continents it is particularly rich in oaks (about twelve species) and maples (*Acer* spp.).

Physiological and morphological characteristics

The nemoral climate with a marked but not too long a mild cold season (3–4 months) and a long thermal growing season (over 120 days with a mean daily temperature over 10 °C) occurs only in the northern hemisphere (Walter 1973). As has already been noted, the photosynthetically efficient thin mesophyllous leaves are particularly susceptible to both frost damage and to drought. Winter leaf-shedding or deciduousness is an obligate adaptation of trees to the winter cold period. It is accompanied by winter hardening, a physiological process which is initiated by the onset of the first cold nights in autumn. As well as a breakdown of chlorophyll and a change in leaf colour, leaf dehiscence occurs. The sap becomes more concentrated and hence less prone to freezing; and new over-wintering buds develop a protective covering. The average date of leaf emergence varies with species and with seasonal weather conditions. However, deciduous species of the nemoral zone delay leaf-flush until mean daily temperatures are maintained above a critical threshold. This would appear to be an adaptation to a distinct, but particularly variable, spring season during which there is a risk of frost. Tree species adapted to the more continental temperate climatic regime, in which there is no marked spring period and temperatures increase rapidly flush immediately a critical daily mean temperature occurs.

Seasonality of climate is also reflected in the characteristic structure and aspection of the temperate deciduous forests of the nemoral zone. Light and temperature conditions in spring can be exploited by spring flowering herbaceous species which can complete their life cycles rapidly before the trees come into leaf and when the litter layer and upper soil horizon is beginning to warm up. The majority are *geophytes*, with underground storage organs, or *hemicryptophytes* with perennating buds at the base of their shoots. Some can grow after the tree canopy develops in the less shaded areas; others are facultative heliophytes which can grow in shade but tend to be poorly developed and frequently sterile in a shaded habitat, but once the forest is cleared can grow and flower luxuriantly.

Seasonality is reflected not only in the aspection of the deciduous forest but in its well-developed and complex structure. Both, however, vary with the specific composition of the forest canopy. Stratification tends to be most highly developed in mature, little-modified forests, growing on nutrient-rich soil, with an open leaf and canopy density which allows maximum light penetration. Under these conditions distinct shrub, herbaceous, field and ground strata of variable density and composition develop. There may be one or two field layers which tend to be composed of one or a few species populations dependent on competition under varying soil conditions. Where the canopy is dense (as in the case of beech forest) and/or the soil is nutrient-poor, stratification may be limited to no more than a field and/or ground layer.

Fauna

Research in the Hubbard Brook Forest (New Hampshire) has shown that only about 25 per cent of the NPP of this 73-year-old temperate deciduous forest enters the grazing food chain. The majority of herbivores in this mixed deciduous stand are leaf-eating insects (primarily caterpillars); other animals, including chipmunks (intake 39 kcal m^{-1} $year^{-1}$), birds (7.4), shrews (7.1), mice (6.1), deer (4.5), salamanders (1.1) and rabbits (less than 1.0), together consume less than 1 per cent in the majority of years (Gosz *et al.* 1978). Hawks and owls along with skunks, foxes, snakes and otters comprise the main predators. There are relatively few arboreal mammals other than squirrels and chipmunks; surface browsers and small burrowers are more prevalent. Species diversity, however, varies with degree of stratification and habitat diversity. Populations are subject to seasonal fluctuations in response to the reduction in food supply and shelter during the winter months. Some animals, particularly birds, migrate to more favourable climatic regions; others, such as small mammals and reptiles, hibernate, while insects pupate. Those that remain active are susceptible to high mortality.

Productivity of temperate deciduous forest

While the annual productivity of the temperate deciduous forest is lower than that of the humid tropical forest (see Table 6.2), and of temperate conifer or broad-leaved stands, the cessation of winter growth is compensated by the high summer photosynthetic rate of its large thin-leaved canopy and a comparably rapid nutrient turnover consequent on the efficient decomposition of its nutrient-rich litter. The latter supports a large population of soil saprophytes and decomposers. Mineralisation and humification are relatively rapid and the production of a mull humus increases the nutrient retentiveness of the soil and counteracts nutrient loss by leaching. The chemical composition of the leaf litter, however, varies with the tree species – some such as beech (*Fagus sylvatica*) and the sessile oak (*Quercus sessilflora*) and temperate conifers produce an acid nutrient-poor litter whose decomposition is slow and often imperfect leading to the formation of a mor humus and the production of strong organic acids which intensify soil leaching.

Natural regeneration and recovery from disturbance in the temperate forests are generally a simpler and less protracted process than in the tropical rain forest. The spatial dominance of one or a few species ensures their replacement *in situ* either directly or preceded by a rapid-growth pioneer, pre-climax tree. Generally, temperate forests can, as a result of the greater nutrient-storage capacity of their DOM and SOM combined with a less drastic change in microclimate, withstand

and recover more easily from natural or human disturbance than the fragile tropical rain forest. While this may be a reasonable hypothesis when applied to those areas with a mild winter, a long growing season and richer more stable mull brown forest soils, it is less so in the poorer climatic/soil environment of the wetter and cooler and the warmer and drier areas of the temperate zone.

HUMAN IMPACT ON TEMPERATE FORESTS

Deforestation

Although ancient woodlands which may have occupied the same site for hundreds of years (Peterken 1981) still persist, it is doubtful whether any completely untouched, 'virgin' forest or woodland remains in the temperate regions of the northern hemisphere; while those of the temperate rain forest in Australia, New Zealand and southern Chile are fast disappearing. With the growth and spread of population from prehistoric times temperate forests were progressively cleared from lowlands and lower mountain slopes for agriculture and constructional (particularly ships) timber in Asia and Europe. Little now remains in China and Japan. In Europe the existing forest and woodland are highly fragmented, considerably simplified by a long period of use and management, or have been replaced by plantations of indigenous or exotic coniferous species. The persistence of many of these forest/woodland remnants is largely a legacy of social and economic rural structures (Edlin 1976) which served to preserve broad-leaved hardwoods in state (royal), local government (commune) or private forests (estate) as local sources of wood (combined with those for grazing, game, food) up until the beginning of the eighteenth century. Their management on a sustained yield basis by coppicing and pollarding (see Ch. 18) is still reflected in the present form of *c.* 40 per cent of existing broad-leaved woodlands in Europe.

In contrast, large-scale human impact on the temperate forests of eastern North America (and of the southern hemisphere) has only occurred during the last 250 years or so, although there is evidence that indigenous peoples used fire to hunt and to promote grazing. In the early nineteenth century land was rapidly cleared by colonists who at the same time retained and managed large areas of farm wood lots on their holdings as sources of constructional timber and for domestic fuel. Later areas of state forest were established. With the increasing demand for timber in the latter part of the nineteenth century the remaining virgin forests were rapidly cut over and abandoned. As a result, although the extent of existing broad-leaved forest in the eastern USA is considerable much of it is still composed of open-simplified, even-aged second or third growth stands on formerly cut-over forest and/or abandoned agricultural land.

Land degradation

In parts of north-west Europe and throughout much of Mediterranean Europe deforestation has been accompanied by soil degradation and the development of often relatively stable and persistent *heath and dwarf shrub* ecosystems. The former are represented in the lowland heaths associated with nutrient-poor Tertiary sands of south-east England and the north European plain in southern Belgium, the

Netherlands and north Germany and southern Denmark. Archaeological evidence suggests that these may once have carried open deciduous oak–birch wood with associated acid (moder) brown forest soils. Early deforestation by grazing and clearing, coinciding with a period when climatic conditions were wetter than before, resulted in intensified leaching and soil podzolisation which was maintained and indeed exacerbated by the rapid colonisation of such areas by pine and birch. Where regeneration was inhibited by grazing and fire, forest was replaced by heaths (*Calluna* and *Erica* spp.) and associated sclerophyllous shrubs such as broom (*Genista*), gorse (*Ulex*), etc.

Destruction of the former climax forest vegetation of Mediterranean Europe and its replacement by a low drought-resistant xeromorphyllous shrub vegetation (*maquis* or *garrigue*) on soils and/or slopes too poor for cultivation or too steep to terrace, was initiated 10 000 to 8000 years ago. The climax evergreen oak (*Quercus ilex*) and pines were replaced by particularly resistant evergreen shrubs such as the kermes oak (*Q. coccifera*), the mastic or lentic shrub (*Pistacia lentiscus*), rock-rose (*Cistus* spp.), gorse (*Ulex europea*) and broom and, dependent on density of cover, associated spring and autumn flowering geophytes, herbaceous annuals and grasses (Di Castri *et al.* 1981). The shrubs are not only drought resistant, but can recover rapidly from burning and coppicing while their mature shoots tend to be unpalatable to all but hardy browsers such as the goat. The long maintained coppicing for small wood and traditional burning of the shrubland for sheep grazing maintained a stable species-rich secondary (i.e. deflected) climax which because of the absence of seed trees and/or the density of the shrub layer, inhibited regeneration of the former climax forest.

The recent breakdown of the traditional agropastoral systems in the Mediterranean area has already started seriously to affect this characteristic Mediterranean ecosystem. Overgrazing, particularly by sheep, has in some areas been accompanied by a decline in the productivity of the maquis which is reflected in a degeneration of the dense, closed maquis to low open garrigue and eventually to a poor weedy grass cover. With the weakening of the vegetation cover the vulnerability of humus-deficient soils to accelerated erosion by the characteristically torrential autumn/winter precipitation has increased. In other agriculturally abandoned areas, natural colonisation by Aleppo pine (*Pinus halipensis*) or the planting of commercial coniferous or eucalyptus have been widespread.

More serious, fast-escalating impacts on the maquis, are those of the burgeoning population growth. In the period 1950–76 the population of the Mediterranean Basin more than doubled, with tourism increasing at a rate of over 7 per cent per annum and an annual influx of over 100 million visitors each year to usually the most scenically attractive and biologically sensitive areas (Naveh and Liebermann 1984). The tourist pressures on the Mediterranean are the most intense in the world today (Tangi 1977) and the directly and indirectly associated developments have caused a rapid loss of semi-natural habitats. The intensity and frequency of summer wildfires have also been increased by these pressures. Hunting, a popular mass sport for both residents in and visitors to the area, is a very lucrative alternative form of land use on former grazing land. Considerable areas of maquis and forest land are now hunting reserves. There has been a drastic decimation of bird and small mammal species, particularly in Italy (Cassola 1979) and many rare predatory birds (hawks) have either been wiped out or have, as with the Sardinian mouflon and the Abruzzo chamois, become seriously endangered.

In addition, urbanisation and the development of high-intensity irrigated agriculture are already depleting water resources on the one hand and polluting fresh and coastal waters alike – indeed degrading or destroying the natural resources on which these recent developments have been based. According to Naveh and Liebermann (1984) the Mediterranean region of Europe is, after the tropical rain forest, the most critical environment and contains the most endangered ecosystems in the world.

Chapter

10

Boreal ecosystems

Four unifying ecological factors which distinguish the northern circumpolar Boreal ecosystem (see Fig. 10.1) from those of the tropical and temperate forest biomes are as follows:

1. Extreme cold;
2. A non-intermittent winter snow-cover;
3. Permafrost;
4. Relative youthfulness.

COLD

The extreme cold of the Boreal zone is a function of the very large and rapid poleward decrease of insolation combined with a marked seasonal periodicity of winter energy loss and summer energy gain, as compared to that in temperate and tropical zones (see Fig. 10.2). The negative winter energy balance is the result of low income consequent on low light intensity and short days (the period without daylight increases from a day at the Arctic Circle to over 9 months at the poles), and high expenditure as outgoing radiation. The growing season is short, from 3.5 months in the south to less than a month at the polar limits. This is compensated by warm summer temperatures and long day-length; at latitude 75° N polar days are continuous for 3.5 months. Mean daily temperatures remain constantly below 0 °C during the winter, the duration and severity of which increase polewards. Mean ambient temperatures of the coldest month varying from −10 °C in the south to −35 °C in the north while absolute temperatures can drop to −65 °C. Annual precipitation is low. Winter snowfall accumulates and is retained on the surface.

The continuous winter snow-cover is one of the most important, if not the dominant, environmental factor in Boreal ecology. It provides an insulating blanket which protects the ground surface and solum from the full impact of low winter ambient air temperatures, thereby creating a micro-environment of vital importance for the survival of many plants and animals. The duration of the snow-

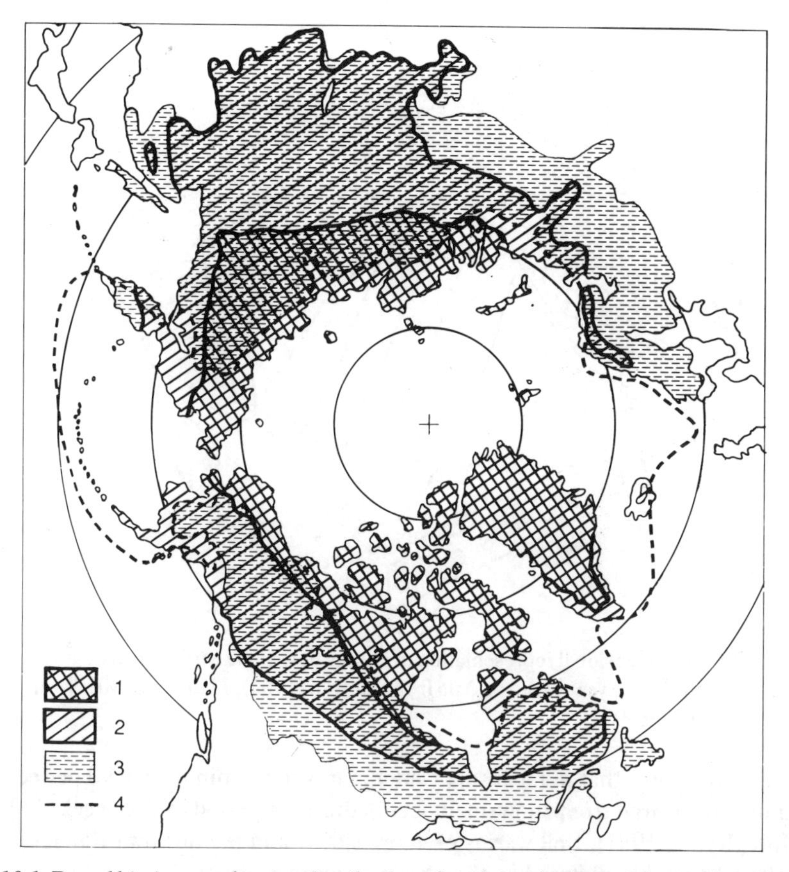

Fig. 10.1 Boreal biotic zone showing distribution of tundra and taiga and the southern limit of continuous and discontinuous permafrost. 1 = continuous permafrost; 2 = discontinuous permafrost; 3 = taiga; 4 = boundary-taiga/tundra (from Kimmins and Wein 1986)

cover increases with latitude. In the south the annual snowfall and hence depth of cover are variable. Further north it is less variable; the depth, however, tends to vary spatially dependent on relief and exposure to high wind-force. The depth at which the insulating snow-cover starts to take effect varies between *c.* 15 and 20 cm dependent on its density (which determines the rate at which heat can be transmitted), as shown in Figs 10.3(A) and (B). Up to this point the decline in soil temperatures, following the autumn turnover, follows that of the atmosphere. At the critical depth – sometimes called the *hiernal* or *winter threshold* (Pruitt 1978) – soil temperatures stabilise at a level dependent on the rate of heat loss during the critical fall period.

Soil temperature remains higher and fluctuates less than in the atmosphere (Fig. 10.4). The base of the snow is warmer than the surface and, together with the development of a *putak structure* (i.e. lattice of ice crystals, 3–10 mm long), provides an important winter microhabitat for many Boreal animals. Snow-melt in the summer is rapid in response to a rapid increase in ambient temperatures.

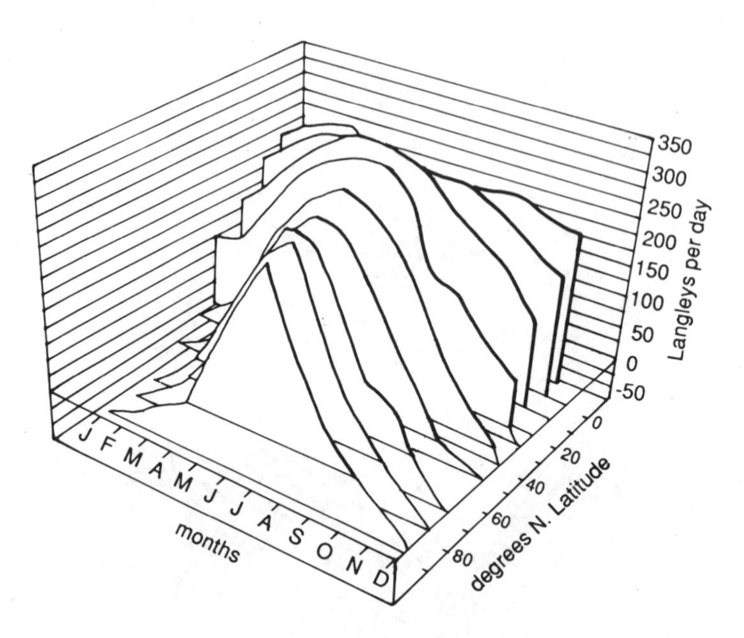

Fig. 10.2 Three-dimensional representation of energy flow at the earth's surface at various latitudes throughout the year based on data from Budyko (1963); 1 langley = 701 W m^{-1} (redrawn from Pruitt 1978)

Studies have shown that in most small catchments in the tundra 80–90 per cent of the annual discharge from snow-melt occurs during a period of 2–3 weeks in June and July (Rydén 1981). Soil warms up more slowly and the downward percolation of meltwater can be inhibited by the presence of permafrost.

PERMAFROST

Permafrost or permanently (i.e. for 2 or more consecutive years) frozen ground is an ecological factor unique to the Boreal zone which affects *c.* 20 per cent of the area. Annual temperature and snow-cover are the major factors determining its distribution. The southern limits (see Fig. 10.1) coincide approximately with a mean annual air temperature of −1 °C and a critical snow depth of 40 cm. Local variations, however, which are particularly important near the southern climatic limits, are influenced by the slope and aspect of the ground and by the continuity and depth of the vegetation cover and/or of the surface organic matter. Permafrost can occur in either an organic or inorganic, consolidated or unconsolidated, substratum. Its distribution may be continuous, discontinuous or sporadic. Beyond the present climatic limits, particularly in central Siberia, moribund or *fossil permafrost* occurs which is a relic of the formerly more severe climatic conditions towards the end of, and immediately following the end of, the last glaciation in the northern hemisphere.

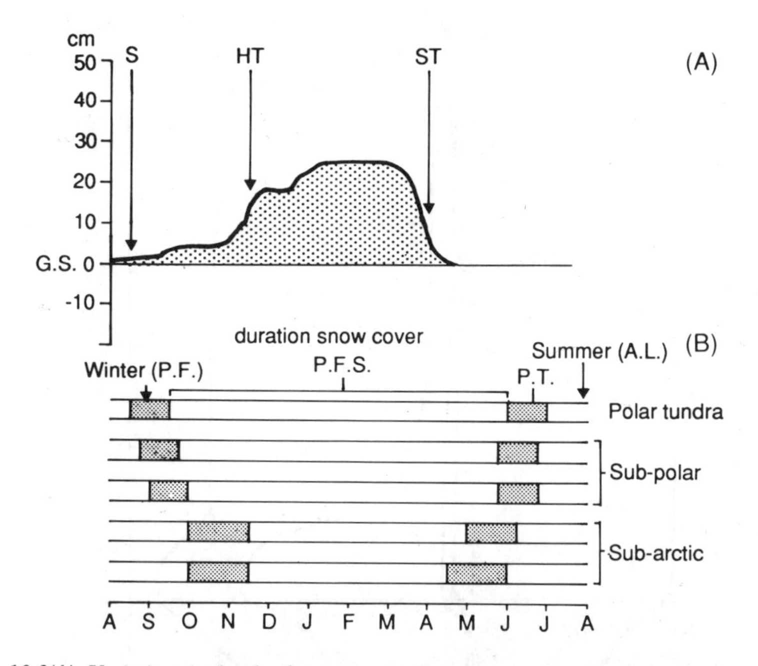

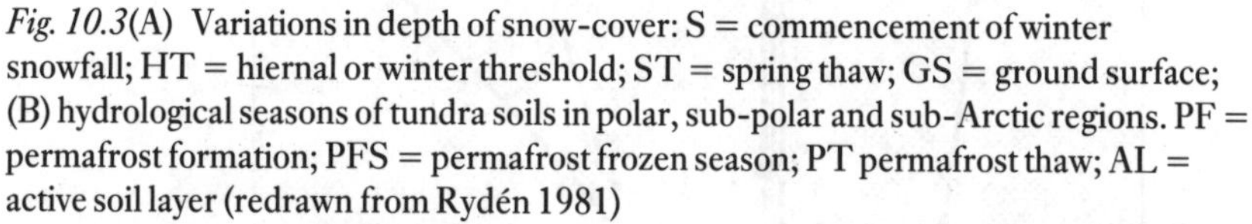

Fig. 10.3(A) Variations in depth of snow-cover: S = commencement of winter snowfall; HT = hiernal or winter threshold; ST = spring thaw; GS = ground surface; (B) hydrological seasons of tundra soils in polar, sub-polar and sub-Arctic regions. PF = permafrost formation; PFS = permafrost frozen season; PT permafrost thaw; AL = active soil layer (redrawn from Rydén 1981)

The depth at which permafrost occurs (i.e. the *permafrost table*) varies from *c.* 20 cm in the north to 1.5 m to over 3 m in the south where, in addition, it becomes discontinuous. It forms an impermeable ice-indurated layer which limits soil processes to an upper *active soil layer* which thaws out to a variable depth and for a variable length of time during the short summer period. Permafrost impedes the downward drainage of meltwater, accelerates runoff and can result in saturated or completely waterlogged soils above. This seasonal freeze/thaw cycle is accompanied by an alternating increase and decrease in the volume of the active layer. Periodic freezing and thawing are also accompanied by soil *cryoturbation.* The soil not only thaws from the surface downwards in summer but freezes in the same direction in the winter. Consequently, the expansion of the surface exerts considerable pressure on the still unfrozen material below resulting in differential movement (shearing/contortion, etc.) in unconsolidated substrata. In some cases variation in texture and hence water content of the substratum gives rise to variations in surface microrelief. Where the temperature of the substratum drops below −4 °C, ice contraction causes soil cracking and fissuring. On the one hand, permafrost, being a heat reservoir, serves to dampen fluctuations in soil temperature. On the other hand, it reduces the rate of soil warming. Although it impedes drainage it is an important source of soil water. To a greater or lesser extent, however, it creates a shallow and particularly difficult, stressful biotic

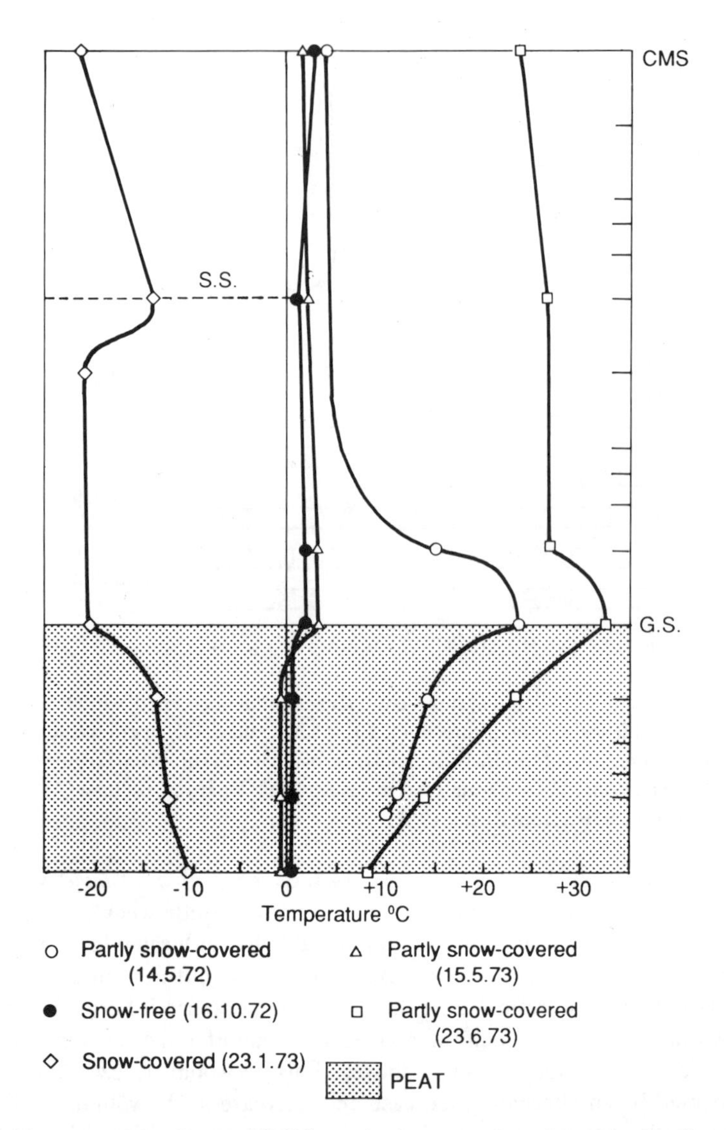

Fig. 10.4 Temperature gradient at 10.00 hours in air, snow, vegetation and soil at Stordalen on snow-face ground; partly snow-covered + snow-covering. SS = snow surface; GS = ground surface (redrawn from Rydén 1981)

environment particularly for root development and growth, and for soil micro-organisms.

YOUTHFULNESS

The Boreal biota are relatively young compared with many of those of other parts of the biosphere. The existing biota remigrated northwards and was re-established as the climate warmed as the last of the Pleistocene ice-sheets retreated *c.* 6000–10 000 years ago. Interglacial tree species had retreated south during the period of maximum glaciation. The Arctic tundra flora, however, persisted in the periglacial zone at the ice margin and at high altitudes further south, and in unglaciated refugia within the ice limits, in central Siberia and Alaska. An analysis of available pollen and radiocarbon data (see Table 10.1) reveals that vegetation

Table 10.1 Rate of northward post-glacial migration of tree species and date at which they each became established in their modern range (from Davis 1981)

Species	*Rate of migration (miles year^{-1})*	*Date established in modern range (carbon-14 age, 000 years)*
Jack/red pine (*P. banksia resionosa*)	400	
White pine (*P. strobus*)	300–350	8/10
Oak (*Quercus* spp.)	350	9/10
Spruce (*Picea* spp.)	250	(12 East) (11 West)
Larch (*Larix lavirinia*)	250	14
Elm (*Ulmus* spp.)	250	10
Hemlock (*Tsuga canadensis*)	200–250	6–10
Hickory (*Carya* spp.)	200–250	6/7
Balsam fir (*Abies balsamea*)	200	11
Maple (*Acer* spp.)	200	6–7
Beech (*Fagus grandiflora*)	200	7/8
Chestnut (*Castanea dentata*)	100	4/5

recovery in North America was relatively rapid (Davis 1981), with spruce the earliest and most aggressive pioneer tree to become established within its present range. The northern extension of the Boreal forest has, however, varied with climatic fluctuations during the Holocene era. The furthest northward expansion, *c.* 150 km of the present-day limit at Lake Ennadai (northern Canada), coincided with the so-called Hypsithermal (or Altithermal) period of maximum temperature *c.* 5000–8000 years ago (Larsen 1980). However, there is evidence that the Boreal forest is not yet stable and there has been a southward retreat as the climate cooled since 5000 BP.

THE BOREAL BIOTA

Palaeobiotic evidence from Eurasia and interior Alaska suggests that in the late Pleistocene a steppe-like environment prevailed in the periglacial zone. This supported a much richer biota than exists today. In central Alaska there were thirty-one mammal species, eleven of which are now extinct and six are no longer native to this region. Some workers attribute the extinction of the latter to a combination of hunting by man and of climatic change 3000–10 000 years ago; others to an abrupt overkill by man. The environmental poverty of the Boreal zone and the rapid latitudinal decline in solar radiation, soil temperature and nutrient availability are reflected in a decline in species diversity on the one hand and in increasing species dominance and species ranges on the other. The relative abundance of mosses and lichens increases polewards. Primary biological productivity similarly decreases to values found elsewhere only in the arid deserts of the world. As the growing season contracts and the length and severity of the winter increase so the problems of survival and regeneration for plants and animals intensify. Both must be able to evade or tolerate low temperatures either by a metabolic rate lower than that for related species in more temperate regions (*cold adaptation*) and/or by being capable of tolerating freezing without injury (*cold hardiness*). Both must be capable of taking advantage of improved conditions and of completing their life cycles rapidly during the short growing season.

Cold tolerance in plants varies with species and growth phase. The seasonal process of winter hardening entails changes in the cell contents which allow survival during periods of varying length when tissues are frozen solid. In some species cellular ice does not form until temperatures drop below −30 °C. Tundra lichens never freeze and hence have a cold resistance greater than any other plant. They can also withstand and adjust to rapid and extreme temperature changes. In addition, they can photosythesise at sub-zero temperatures; Birot (1965) notes that at −5 °C the rate is still half the normal in *Cladonia*, while slight activity has been detected even at −24 °C. Winter hardening also imparts the drought tolerance necessary in face of a cold and frozen soil when evapotranspiration is accelerated by low atmospheric humidity and high wind-speed. Low pulvinate, prostrate or cushion forms can maximise the more favourable environmental conditions near the ground surface and avoid exposure to wind.

An evergreen habit, large vegetative (perennating) buds and rapid opportunistic seed germination allow plants to take immediate advantage of the onset of the growing season. Seed development is variable and vegetative regeneration or viviparous seeds become more common. Many evergreen plants have small leaves with tough, thick or pubescent cuticles which together with a high cellular content of glycerol provide protection against exessive herbivory.

Animals have also evolved a wide variety of strategies which ensure survival. Those dependent on external heat (*ectotherms*) must avoid the winter cold. Invertebrates (dehydrated or with a high glycerol content) have a cold-resistant stage and can over-winter below ground. The warm-blooded *endotherms* include mammals and birds. The *microtine rodents* are the most common small mammals which can survive by foraging under the snow-cover. While their survival rates are low, their extremely high fecundity allows rapid recovery and increase of population in the summer (i.e. r-species). The large ungulate mammals have an insulating coat and/or a mobile habit by which low food quality and quantity in winter can be compensated by a wide foraging range or seasonal migration between winter and summer feeding and breeding grounds as in the case of the caribou (reindeer). Many birds migrate south during the winter. Those that over-winter tend to suffer high mortality rates.

BOREAL ECOSYSTEMS

TAIGA (BOREAL FOREST)

The latitudinal environmental gradient in the Boreal zone is most clearly expressed in the transition from the forested taiga in the south to the treeless tundra in the north. The taiga (from the Russian for swamp-forest) is the most continous and extensive forest in the world today. It is characterised by low species diversity and a marked dominance of a few conifers. Hundreds of square kilometres may be dominated by two spruce species with aspen and birch; and in northern Siberia 90 per cent of the tree biomass over thousands of square kilometres is dominated by larch (*Larix* spp.). Tree species have wide ecological ranges. Forest structure is relatively simple, limited by an often dense evergreen canopy, to an understorey of low to dwarf ericaceous shrubs and a thick ground layer of mosses and lichens. Composition and structure vary with latitude, substratum and with the frequency and intensity of wildfires. Northward, growth is slower and trees become stunted and more widely spaced. Increased soil moisture results in the build-up of peat on the forest floor. Permafrost and associated freeze/thaw cycles in the active layer cause surface instability. Despite well-developed lateral root systems cryoturbation can dislodge and fell trees, producing what is called a 'drunken forest' (Pruitt 1978). Species dominance is dependent on the productivity of the physical site (see Table 10.2). In wide valley plains subject to periodic flooding and nutrient

Table 10.2 Mature forest productivity in terms of above-ground annual tree increment on three sites in the Alaskan taiga (tabulated from data in van Cleve *et al.* 1986)

Site	*Dominant tree species*	*Above-ground tree annual increment* ($g\ m^{-2}$)	*Annual litter fall* ($g\ m^{-}$)
Flood plain	Balsam poplar	264–952	268–653
	White spruce	240–540	127
Upland	Aspen, birch	343–370	27–265
Upland	Black spruce	102–148	29–53

replenishment poplar (*Populus*) and white spruce (*Picea mariana*) with quick-growing alder (*Alnus* spp.) and willow (*Salix* spp.) are most abundant. Well-drained south-facing upland sites without permafrost are characterised by white spruce, aspen (*Populus tremuloides*) and paper birch (*Betula papyrifera*), while nutritionally poorer north-facing, poorly drained sites with permafrost by black spruce (*Picea glauca*) (van Cleve *et al.* 1986).

Wildfire

Finally, within this physical framework the type of species mix is dependent on the stage of development (successional stage) after disturbance. In this respect the most important, inherent and widespread ecological variable is recurring *wildfire*. A large annual litter increment with a high resin content combined with slow decomposition results in the accumulation of a large amount of potentially flammable fuel. Lightning-set fires at short to medium intervals (50–200 years) burn extensive forest areas; over 10 000 ha is not uncommon in North America, 25 000–50 000 ha, and in extreme cases, fires over 400 000 ha in extent in western Alaska. The fire regime depends on climatic and site conditions and/or latitude. Cycles are longer (200+ years) where fuel is wet than where it is dry (50–100 years). They become increasingly longer towards the tree-line. Most species show some degree of adaptation to fire. Conifers have *serotinous* or *semi-serotinus* cones and a long seed-dispersal period. The hardwoods are prolific seed producers and can also regenerate rapidly by sprouts and suckers from stems and root bases. The shrubs and herbs also grow from underground storage organs (stems and rhizomes). The black spruce is less resistant than other tree species because of a thin unprotected bark, a canopy close to the ground and abundant trunk lichens which can carry fire rapidly from the ground to the canopy. The time and routes by which the mature forest is re-established vary, dependent on the site conditions and the original species composition. The most abundant early established plants are perennial forbs such as willow herb (*Epilobium angustifolium*); grasses (*Calamagrostis* spp.), club mosses (*Equisetum* spp.), and the feather mosses (*Polytrichum* spp.), followed by fast-growing pioneer trees such as birch (*Betula* spp.), aspen (*Populus* spp.) and willows (*Salix* spp.). The taiga then is a 'a mosaic of patches of even-aged forest, each patch dating from the last fire but varying in species composition with site factors, propagule sources, and fire variables' (Heinselman 1981: 405).

It is now widely accepted by forest ecologists that fire is an inherent factor – an essential agent of renewal – rather than a disturbance (or perturbation) in the ecological sense (Dryness *et al.* 1986). Charcoal from *c.* 10 000 BP has been found in lake sediments in southern Canada (Larsen 1989) – a period long enough to allow evolutionary adaptation of plants. In much of the taiga fire is necessary both for nutrient cycling and for forest regeneration. On nutrient-poor, humid sites once a closed canopy has formed the initial establishment of a thick ground layer of moss is maintained by the relatively rapid growth of *Sphagnum* spp. which can compete more efficiently for nutrients than trees can. As a result an increasing proportion of the ecosystem nutrient pool becomes locked up in dead organic matter. This accumulates at a rate much greater than it decomposes; it has been estimated that less than 20 per cent of the DOM is decomposed each year. The formation of this insulating layer results in a progressive cooling of the soil with a rise in the level of the summer permafrost table and a decrease in the depth of the active layer. The rate of nutrient cycling slows down and nitrogen fixation becomes dependent on the *cyanobacteria* (better able to tolerate temperature extremes than the normal nitrogen-fixing bacteria) and which can form symbiotic relationships with the tree roots. Tree growth also slows down and if the nutrient cycle were not reactivated by fire, the forest could become moribund in face of the extensive development of peat-bogs.

FOREST LIMITS

The definition of the poleward limits of the taiga vary according to the parameters used; those recognised include:

1. The *timber-line* beyond which felling is uneconomic and detrimental for forest regeneration;
2. The *forest-line* or that beyond which forest (or physiognomic) regeneration is infrequent and variable being confined to good mast years;
3. The *tree-line*, the absolute poleward limit of tree growth – a tree in this instance being defined as a woody perennial at least 2 m in height and with a single stem;
4. The *species line*, the limit of a particular tree species irrespective of its height or form.

While the latitudinal limits of these lines vary from one part of the tundra to another, they are dependent on environmental and particularly climatic conditions. The climatic limits of the taiga have traditionally been correlated with the mean July isotherm of 10 °C and the presence of a permafrost-free active layer of 1 m for at least 2 months (Larsen 1989).

THE FOREST–TUNDRA (LYESO-TUNDRA) TRANSITION

The taiga–tundra boundary is gradual rather than abrupt, spanning a relatively broad transition zone or *ecotone* of variable width which is regarded as a distinctive biome by some workers. It has been the focus of ecological studies in both North America and Eurasia because of the significance of physical and biological processes within it for an understanding of those in the Boreal zone in general; and because of the economic significance of the area for pastoralism particularly in Eurasia. The ecotone has been defined generally as those areas (undisturbed by fire or other events) where an unbroken forest cover occupying at least 75 per cent on uplands or less than 75 per cent on lowlands is unforested, can be identified on an air photo at a scale 1 : 5000 (Hustich 1979). However, the ease with which this criterion can be applied varies according to vegetation patterns. In some areas it is clear particularly where there is a relatively sharp interdigitation of tundra on lowland with forest on higher interfluves; in others where trees are more evenly dispersed and decline gradually in density, it is more difficult. In some regions outliers of forest exist beyond and within the ecotone which are generally considered to be relicts of a previously more extensive cover.

The taiga–tundra ecotone is composed of a mosaic of forested and unforested land. The latter are bogs (dominated by *Sphagnum* spp., *Eriophorum* spp. or *Carex* spp.) which have developed either in thaw ponds above the permafrost or as a result of the *paludification* (swamping by increased soil saturation) of open forest stands or excessively dry and/or steep south-facing slopes with a cover of low shrubs, grasses, forbs and lichens.

TUNDRA

Tundra is the Finnish name for the generally flat, treeless zone between the Boreal forest and the polar ice-caps where a very short growing season and a shallow, unstable active soil layer inhibits tree growth. Ambient temperatures remain below zero for at least 7 months of the year and the mean of the coldest month can vary from −10 °C in the south to −35 °C in the north. Permafrost is continuous and the summer permafrost table is usually less than a metre (*c.* 75 to 30 cm) below ground-level. A high proportion of the biota live within the restricted area at the land/air interface. Here maximum advantage can be taken, particularly at the beginning of the short (less than 3 months) long-day growing season, of the high input of radiant energy and of shelter from high desiccating wind-speeds.

The extreme biological poverty of the tundra is reflected in low and decreasing primary biological productivity, biomass and species diversity. Vegetation is composed of low-growing, woody, herbaceous perennials (with few or no annuals) with longer life spans than in related species in Boreal and temperate biomes and a high and increasing ratio of mosses and lichens to vascular plants. The proportion of green photosynthesising tissues is high and many tundra plants have a lower light compensation point than temperate species. Both attributes allow a rapid metabolic reaction to increased light and temperature at the start of the growing season. The ratio of living underground storage organs (roots, stems, rhizomes, etc.) to above-ground organs increases with latitude as does the ratio of dead to living organic matter.

TYPES OF TUNDRA VEGETATION

Within the latitudinal climatic continuum of the tundra three subzones – the southern low Arctic, the northern high Arctic (including the polar desert) – are usually recognised on the basis of decreasing species diversity (see Tables 10.3 and 10.4), height and continuity of vegetation cover; and depth of peat development. However, within this broad zonation, the tundra vegetation is characterised by a diversity of communities dependent on local variations in the physical habitat and particularly in the depth and duration of snow-cover. In the *low Arctic* variations in relief and substratum are reflected in a mosaic of wetland (grass/sedge marsh); zeric (dry) sites (with dwarf birch (*Betula glandulosa*), ericaceous shrubs, bog-cotton (*Eriophorum*), lichens and mosses; and *fellfields* on rocky exposed ridge-tops (with an incomplete cover of lichens and *Dryas*). In the *high Arctic* snow-cover and the physical nature of the substratum become increasingly important factors in the composition and distribution of the vegetation

Table 10.3 Species diversity of tundra flora as number of species in, and as a percentage (in brackets) of the major taxonomic groups in various communities (from Wiegolaski *et al.* 1981)

Taxonomic plant group	*Polar desert*	*Low Arctic*	*Shrub tundra*	*Forest tundra*
Vascular parts	9(31)	32(47)	25(50)	13(31)
Bryophytes	7(24)	18(26)	15(34)	10(24)
Lichens	13(45)	19(27)	7(16)	19(45)

cover. *Snow-patch catenas* composed of a sequence of species related to snow thickness and hence length of the summer thaw period available for completion of life cycles are characteristic.

Table 10.4 Comparison of environment and biota within latitudinal zones of the North American tundra (compiled from data in Bliss 1981 and personal communication from author)

	Low Arctic	*High Arctic*
Length of growing season (months)	3–4	1.5–2.5
Day-degrees over 0 °C	300–900	150–600
Species diversity (numbers)		
Flora: vascular plants	600	300+
cryptogams	Lichens and bryophytes	Lichens and bryophytes
Fauna: land mammals	10–15	8
nesting birds	30–60	10–20
Vegetation cover (%)	80–100	< 50–80
Peat depth	Several metres	Seldom over 10–20 m

With increasingly severe conditions, the substratum becomes the most important ecological variable. Cryoturbation, causing differential movement and sorting of frost-shattered rock debris, results in distinct soil and related vegetation patterns. Stone polygons (1–4 m diameter) with cores of fine-textured material bordered by stony debris occur on gentle slopes. On slopes exceeding 5 °C solifluction becomes active, producing crescentic downslope lobes or banks of coarse debris. In both instances the vegetation tends to be concentrated on the well-drained coarse permeable material in which the thaw penetrates more rapidly and deeply than on the finer, less permeable material, which thaws later, is colder and is subject to more intense cryoturbation. Plants adapted to such habitats usually have long underground stems whose growth can keep pace with the intense vertical and horizontal movement of the substratum.

NUTRIENT CYCLING

The small amount of above-ground plant biomass and the low productivity of the tundra, illustrated in Table 10.5, is a function not only of the severe climatic conditions, but of an even greater deficiency of available nutrients than in the

Table 10.5 Plant production for the various regions of the tundra biome (from Wiegolaski *et al.* 1981)

Region	*Net production* ($g\ m^{-2}\ year^{-1}$)	*% Production* *Vascular plants*	*Cryptogams*	*Net production, shoot : root*	*Growing season (days)*
High Arctic					
Polar desert	1–10	30–90	10–70	1 : 0.2–1 : 0.5	30–45
Polar semi-desert	10–50	20–95	5–80	1 : 0.2–1 : 1	45–60
Low Arctic–low alpine					
Herbaceous	100–300	60–90	10–40	1 : 1–1 : 3.5	45–100
Dwarf shrub species	150–700	65–85	15–35	1 : 1–1 : 5	50–150
Low shrub	500–1200	70–85	15–30	1 : 0.7–1 : 1	50–150
Sub-Arctic–sub-alpine	150–180	50–90	5–50	1 : 0.3–1 : 1.3	110–365

taiga. Not only is a high proportion of the NPP stored below ground (root : shoot ratios for vascular communities range from 3 : 1 to 10 : 1) but a very high proportion (twenty times that in temperate grassland) of the organic carbon and nutrients (particularly nitrogen) in the tundra ecosystem is stored in below-ground DOM. Bunnell (1981) notes that while 50 per cent of the organic carbon in forests, and 10 per cent in temperate grassland is contained in the living plants, only 2 per cent of the organic carbon in the tundra is 'alive'.

Nutrient input to the tundra by precipitation, weathering of parent material and nitrogen fixation is low compared with temperate ecosystems and decreases poleward (Dowding *et al.* 1981). Nutrient storage (particularly nitrogen and phosphorus) in the SOM is high and also increases with latitude and, with a parallel decrease in the rate of decomposition, the tundra nutrient cycle becomes slower. Organic matter turnover rates are also much lower than in temperate ecosystems; that for carbon is only 1 per cent per annum, i.e. 3 per cent of the rate in the tropics. Since uptake by plant roots in the spring is dependent on nutrient availability rather than on the temperature when growth commences, the former is an important factor limiting primary production. Plants become particularly conservative in nutrient use, a high proportion being translocated to underground storage tissues *before* the cessation of photosynthesis. Litter production is slow and the relatively high proportion of standing dead tissues on living plants testifies to a strategy whereby the decomposition route is bypassed. The annual nutrient flux is dependent on the tundra's seasonal hydrological regime. Spring snow-melt is rapid, lasting 2–3 weeks to as little as 10 days. Downward percolation of water is impeded by a high permafrost table and since much discharge takes place by surface runoff, the nutrient content can be rapidly absorbed by mosses whose decomposition rates are slower than those of the other components.

The tundra nutrient cycle is even more tightly closed than that of the tropical rain forest. Maintenance of the former ecosystem is, however, dependent on a very conservative cycle in which there is a slow turnover of a small nutrient capital. The tundra, then, is a detrital system *par excellence* which stores nutrients (particularly nitrogen and phosphorus) and energy in forms that are difficult for plants and animals to use (Bliss 1975).

TUNDRA HERBIVORES

The most important tundra herbivores are the mammalian grazers (ungulates and rodents) and, in some case, geese (*Chen*) and ptarmigan (*Lagopus*). The mammals are widely distributed, attain high local densities and have a significant impact on the vegetation and the nutrient cycle.

Ungulates

The three ungulate genera common to the tundra are *Rangifera* (caribou and reindeer), *Ovibos* (musk-ox) and *Ovis* spp. (horned sheep) (see Table 10.6). The *Rangifera* have the widest range with seasonal migrations from summer tundra to winter taiga grazings. They appear to be unique among the deer family (cervids) in having population fluctuations with a periodicity of *c.* 100 years (White *et al.* 1981), the reasons for which are still imperfectly understood. The musk-ox and horned sheep have narrower niches. The former are confined to suitable habitats in the high Arctic zone where a thin snow-cover makes year-round grazing possible. The latter are found on the driest, often steep and rocky terrain which once provided protection from predatory wolves. Densities of these large mammals, however, are low; dependent on the amount and quality of the forage, reindeer require 4–8 km^2 per animal.

Rodents

The most important herbivores, however, are the small mammals which account for *c.* 50 per cent of all native mammals in the tundra. Of these, eighteen are widespread and twelve belong to one group – the *microtine rodents* (lemmings and voles). The lemmings (*Lemmus* spp. and *Microtus* spp.) are the most abundant. Widely but unevenly distributed, they can attain higher densities (200+ per hectare) than any other tundra mammal. However, populations fluctuate markedly within a 3–6 year cycle on a given site and between peaks densities drop to less than one per hectare. However, those areas heavily grazed by lemmings are often the most productive. Grazing serves to increase the availability of nutrients by speeding up decomposition as a result of the removal of the standing dead material on grasses and sedges, disrupting the moss ground-cover and depositing faeces. These cycles have been attributed to one or a combination of factors, including variations in winter temperature and snow-cover, nutrition, predation and lemming endocrine levels. Preferred habitats are patchy; populations tend to build up under

Table 10.6 Sub-Arctic ungulate niches (from White *et al.* 1981)

		Habitat types		*Dietetic types*	
Species	*Area*	*Summer*	*Winter*	*Summer*	*Winter*
Sub-Arctic mountain reindeer (*Rangifer tarandus tarandus* L.)	South Norway	Low–High alpine snowbed meadows	Dwarf shrub–lichen heaths	Grazer–browser	Epigeic lichens
Arctic barren-ground caribou caribou, tundra reindeer (*R.t. arcticus*, *R.t. granti*)	North American and Siberian mainland	Dwarf shrub–sedge marshes	Dwarf shrub–lichen heath, conifer–lichen heaths	Grazer–browser	Epigeic lichens
Woodland caribou, forest reindeer (*R.t. caribou*, *R.t. fennicus*)	Newfoundland, Finland, Karelia	River banks, marshes	Climax sub-Arctic Boreal forest	Grazer–browser	Arboreal lichens
Montane reindeer/caribou (*R.t. granti*, *R.t. valentinae*)	West Canada, Altai,	Low–high alpine snowbed meadows	Sub-alpine climax Boreal forest	Grazer–browser	Arboreal lichens
High Arctic island reindeer (*R.t. pearyi*, *R.t. platyrhynchos* L., *R.t. groenlandicus*)	Canadian Arctic, Greenland	Lowland sedge–moss–dwarf shrub tundra	Dwarf shrub heaths	Grazer–browser	Browser–grazer
Muskox (*Ovibos moschatus* L.)	Greenland, Canada, Alaska	Lowland sedge–moss–dwarf shrub tundra	Dwarf shrub heaths	Grazer–browser	Browser–grazer
Dall sheep (*Ovis dalli dalli*)	Alaska	Low–high alpine snowbed meadows	Windswept ridges, dwarf shrub heaths	Grazer–browser	Grazer–browser
Snow sheep (*Ovis canadensis nivicoli*, 16 subspecies	Siberia	Rock escarpments	Rock escarpments		Epigeic lichen

the snow-cover during winter as a result of high natural increase and immigration until, with increasing pressure, mortality increases and/or movement out into less favourable habitats occurs.

Moderately heavy grazing can stimulate the growth of grasses and sedges at the expense of lichens and mosses. However, densities of over fifty per hectare result in overgrazing, a decline in productivity and, in extreme cases, disturbance of the surface vegetation cover as the lemmings grub for roots and rhizomes. The microtine rodents are also critical components in the tundra food web. When abundant they provide the major source of food for carnivores and predatory birds. Although the latter have relatively flexible and broad feeding niches, their dependence on a limited range of prey species results in parallel predator–prey population fluctuations. As a consequence of low species diversity, the tundra food web is simple and easily disrupted and, together with minimal environmental resources and a very slow nutrient cycle, it contributes to the inherent fragility of the ecosystem.

HUMAN IMPACT ON THE TUNDRA

Although humans have occupied the Arctic tundra since at least Neolithic times, their impact on the ecosystem was relatively limited until the pre-Second World War period. Previous to this date self-sustaining cultures had evolved in the Eskimo fish and hunting and in the Fenno-Scandinavian and Russian reindeer herding economies. In the latter semi-domestication and protection from predation resulted in an increase in herd size and, in many areas, to overgrazing. Natural sedge–lichen–grass vegetation has been replaced by grass-dominated communities. In northern Europe and Russia it has been estimated that at the beginning of the twentieth century *Cladonia* lichen pasture had been reduced to 1 per cent of its original area. Early exploration of the Russian tundra in the seventeenth century and later of North America was followed by a rapid increase in whaling and fur-trapping and more recently, pleasure hunting resulting in the reduction or extinction of fish, bird and large mannal populations.

During the last 40 years exploration and exploitation of oil, gas and mineral ores have necessitated the construction of routeways (pipelines, roads, railways and airfields), industrial and domestic buildings and their associated infrastructures and the increasing use of off-road vehicles (ORVs). In many instances animal migrations have been impeded or deflected. Disruption of the insulating cover of soil and vegetation has resulted in localised lowering of the summer permafrost table with the production of an uneven, pitted *thermokarst* surface. On sloping terrain accelerated solifluction has caused massive scree flows and gullying. Increasing instability of the substratum has exacerbated the problems of organic regeneration in a climate marginal for life.

Chapter

11

Grassland ecosystems

Grassland ecosystems are those in which the vegetation is dominated by herbaceous plants of which the most abundant are grasses. The grass family (Graminaceae) contains a large number of genera and species and its range is wider and more truly cosmopolitan than that of any other group of flowering plants. Further, it is distinguished by a greater homogeneity of growth form than in any other family of flowering plants of comparable size except the closely related grass-like sedges (Cyperaceae).

GRASSES

With the exception of the bamboos, all grass species are herbaceous. Their leaf structure is unique (Fig. 11.1). Single individual leaves are produced alternately from the nodes of a cylindrical or elliptical stem. The lower part of the leaf forms a sheath around the stem, the upper part is a narrow, elongated, near-vertical or sub-erect blade of varying length and width. As a result the cover of the grass plant is relatively low compared to plants with horizontally disposed leaves. The disposition of the blade, however, reduces mutual leaf-shading thereby maximising the plant's photosynthetic surface. Grass leaves are structurally tough as a result of a high, though varying degree of suberisation, lignification and, particularly, silification of the cell walls. Many of the coarser xerophytic grasses have prominently ridged surfaces and sharp cutting edges. Further, in contrast to the majority of species, the *meristematic* (or growth tissues) are located at the base of the leaf-sheath and the leaf-blade. Grasses can, therefore, tolerate defoliation better than many other non-graminoid plants and, indeed, growth can be stimulated by cutting, grazing and burning.

The high reproductive and dispersal capacity of the grass family has undoubtedly contributed to its wide range. The grass panicle or inflorescence is generally composed of a large number of small flowers. Most grasses flower every year though some (e.g. salt-marsh and tropical species) do so less frequently. Many species are photoperiodic, with the change from the vegetative to the flowering stage triggered by day-length. Seed production is normally abundant and the characteristically small, light propagules are easily dispersed by wind. Many also possess spines, barbs, etc. which adapt them to transport by animals and humans. Furthermore, many grass seeds can remain dormant in the soil for several years before germination. Tropical grasses, however, seed less freely and have a lower seed viability than those of other regions.

The majority of grass species are perennials with either a tussock or prostrate growth habit (Fig. 11.1). The tussock or bunch grasses are those in which new

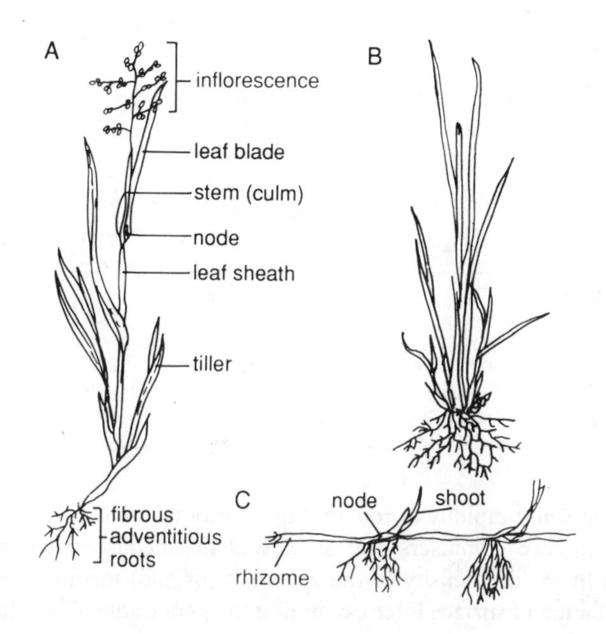

Fig. 11.1 Characteristic grass growth forms: (A) typical annual grass; (B) tussock form; (C) creeping, mat-forming underground stem with shoots and roots developed at nodes

shoots (*tillers*) arise from the basal node of the main stem, near or just below ground surface. With the continued production of new shoots and the accumulation of dead and decaying litter, tussocks can attain considerable proportions. In contrast, prostrate, mat or turf-forming grasses produce new shoots at the widely spaced nodes of creeping underground (rhizomes) or surface stems (stolons).

The root system of most uncultivated grasses is large and extensive in proportion to the size of the shoot (Table 11.1). Adventitious roots are produced either from the base of the stem in annual and tussock forms, or from the nodes of rhizomes and stolons in mat-forming species. Roots tend to be thin, fibrous, freely branching and of similar size, forming a dense ramifying network which is concentrated in the upper 10–50 cm of the soil profile (Table 11.2), though

Table 11.1 Root material as a percentage of mean standing crop of live shoots plus below-ground plant material in tropical and temperate grasslands (from Sims *et al.* 1978)

Tropical		*Temperate*	
Semi-arid	68	Short grass	95
Subhumid	70	Mixed grass	95
Humid	73	Tall grass	91

individual roots can extend to very considerable depth. The efficiency with which this root system can absorb water and nutrients from a large volume of soil gives the grass species a competitive advantage over plants of similar rooting depth.

Table 11.2 Mean contribution (g m^{-2}) of each category of underground plant material found in each layer of a northern mixed-prairie site (from Coupland 1979)

Soil depth (cm)	*Roots*	*Rhizomes*	*Shoot bases*	*Litter*	*Total*	*% Total underground plant matter*
0–10	842	46	141	104	1133	33.2
10–30	528	5	—	19	552	22.5
30–50	331	—	—	8	339	24.2
50–100	478	—	—	12	490	—
100–150	224	—	—	6	230	—

Also, the density and rapidity of root and underground stem growth make grasses particularly effective colonisers and stabilisers of unconsolidated mobile sediments. In addition, the density of the root-mat (or sod) together with the often thick accumulation of surface litter can inhibit the penetration of seedlings of other species.

THE GRASS ECOSYSTEM

The grass ecosystem is characterised by a number of distinctive features which contrast with those of forest or woodland. First, the dominance of grasses is dependent more on their abundance and total volume than on their cover value. The larger brightly coloured flowers of associated forbs, in comparison to the less conspicuous grass flowers, often give a misleading impression of their relative abundance in the vegetation. Studies of near natural prairie in east Nebraska (USA) have shown that although 26 per cent of the total flora (237 species) were grasses (38) and sedges (18), they made up at least 90 per cent of the plant biomass (Sims *et al.* 1978). Second, the structure of the vegetation is simpler than in the forest ecosystem. Stratification is less developed and less apparent, and the dominant field or herbaceous stratum rarely exceeds 4–5 m during the period of maximum growth even under the most favourable growing conditions, as in the humid tropics. In some instances, there may be an open, discontinuous stratum of emergent trees and/or shrubs and ground layer of small forbs, mosses and lichens. However, although the main herbaceous stratum is frequently composed of grasses and forbs of varying growth height, the resultant stratification tends to be obscured by seasonal variation in growth of the component species. A third distinctive characteristic of the grassland ecosystem is the high proportion of the total biomass that exists below ground surface (Table 11.3). At the period of maximum growth the above-ground plant biomass may be only 20 per cent (or even less) of the total and the average for the growing season may be less than 10 per cent. The average ratio of below to above-ground biomass varies from 13 : 1 in the more humid to less than 2 : 1 in arid areas (Coupland 1979). The root system contributes a large amount of organic matter in the soil which supports a large population of soil animals. The ratio of the weight of soil animals, particularly of decomposers, to the total animal biomass is greater than in forests. The energy value of decomposers alone can equal that of the surface vegetation.

Finally, between a quarter and two-thirds of the above-ground NPP enters the grazing route. The proportion of the above-ground plant biomass which is consumed by grazing herbivores can be ten to twenty times or more than in a deciduous forest ecosystem (Weigert and Owen 1971). Many of the grassland herbivores appear to have adapted to the open habitat conditions by either a cursorial or burrowing habit. The former is characteristic of the ungulates, the latter of the rodents which breed and live below ground. In addition, the strong well-developed incisor and broad-ridged molars and continuously growing teeth of grassland rodents and some ungulates are thought to be an adaptation to the tough grass forage.

TYPES OF GRASSLAND

Various criteria have been used to differentiate types of grassland vegetation. The most commonly employed are the absence or presence of trees and/or shrubs; the luxuriance and growth form of the dominant grass species; and the nature of the physical habitat. In the first case a basic distinction is made between prairie and steppe, in which woody growth is absent or negligible and savanna, wooded grassland or parkland characterised by a continuous tree/shrub stratum but in which the herbaceous field layer is dominant. These terms are used by plant ecologists in a strictly morphological sense. Their origins, however, are geographical and they have for long been descriptive of temperate and tropical grasslands respectively. The French bequeathed the term *prairie* (or meadow) to the formerly more extensive grassy plains of central North America; the Eurasian *steppe* derives from the name of one of the commonest genera (*Stipa*) in the grasslands which extend from the Black Sea to eastern Mongolia across central Asia. Unfortunately the two terms are not completely synonymous. In North

Table 11.3 Mean plant material (g m^{-2}) in ungrazed natural temperate grassland sites in USA and Canada (from Coupland 1979)

Type of grassland	*Green shoots*	*Dead shoots*	*Litter*	*Total above ground*	*Total below ground*	*Ratio*
Short-grass prairie	70	129	331	530	1033	1 : 2
Short-grass prairie	70	65	251	386	1716	1 : 4.5
Mixed prairie	79	117	904	1100	1319	1 : 1
Mixed prairie	93	141	496	730	1963	1 : 2
Mixed prairie	178	369	457	994	1715	1 : 1
Mixed prairie	74	411	238	724	2167	1 : 3

America the so-called short-grass formation of the High Plains is not, strictly speaking, covered by the term prairie; while in Russia, steppe denotes all non-forest vegetation including semi-desert grasslands. Within the temperate grasslands a further distinction is frequently made on the basis of height and luxuriance of the vegetation between the *tall-grass* and the *short-grass prairie.* In tropical regions the term *savanna* includes a wide range of vegetation types from the wooded to the treeless grassland.

In terms of origin several types of grassland have been distinguished (French 1979):

1. *Semi-arid grasslands* which occur in areas between the major forest and desert biomes in areas where precipitation is markedly seasonal and highly variable and where available soil water is thought to be insufficient to maintain tree growth;
2. *Mountain (alpine) grassland* at altitudes above which the thermal growing season is insufficient to allow tree growth to develop;
3. *Successional grasslands* which have replaced a former forest or woodland vegetation and are maintained by grazing and/or burning;
4. *Agricultural grasslands* orginally established by planting or improved by drainage and fertilisation and which may be rotational or permanent;
5. *Annual grassland* occurs in areas with a Mediterranean climate where abandoned agricultural land has been colonised by annual grasses (mostly exotic) and perennial forbs.

The most extensive areas of former and existing 'natural' grassland (Fig. 11.2) have three features in common. Firstly, they occur in subhumid to semi-arid climatic regions characterised by a low, variable and markedly seasonal annual rainfall. Secondly, they attain their maximum extent in continental interiors between the humid forest and arid desert zones. Thirdly, they are associated with extensive land surfaces of little relief. However, the factors which determine the boundary between the forest and the grassland formations and the extent to which the latter represents a climatic climax are still debatable.

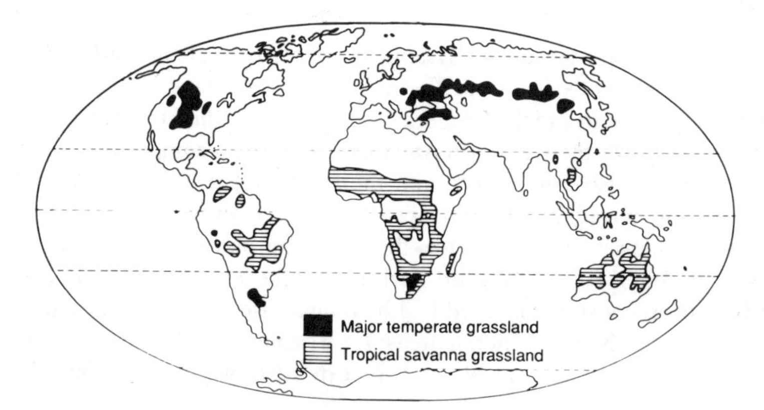

Fig. 11.2 Distribution of existing and former temperate grasslands and tropical savannas

SEMI-ARID GRASSLANDS

TEMPERATE GRASSLANDS

The most extensive areas of existing and former subhumid grasslands occur in central North America and across central Eurasia. The former cover a wider range of precipitation and temperature. The east–west humidity gradient is reflected in the transition from the tall-grass (the prairie) to short-grass communities. The tall-grass prairie is characterised by the dominance of the genera *Andropogon*, together with *Agropyron* and *Stipa* spp.; the short-grass prairie by the gamma grasses, *Buchloe dactyloides* in the north and *Bouteliena gracilis* (blue gamma) in the south. The latter has a C_4 photosynthetic route in contrast to the C_3 of the former. It is also well adapted to prolonged drought when the soil-water potential in the root zone may be −50 bars (French 1979). In comparison to the extensive root system, however, the leaf surface area is small which together with its fine, folded leaves minimise transpiration. The ecotone between these two types of prairie is the mixed grass prairie composed of components from both the long- and short-grass communities and in which the balance between long- and short-grass species can vary with the alternating periods of relative humid and aridity characteristic of this climatic zone.

In these semi-arid grasslands, where low annual precipitation is accompanied by high evapotranspiration, available soil water is limited and the amount and depth to which it penetrates decreases with increasing aridity. Leaching is also limited and only the most soluble cations, of which calcium is the most common, are leached from the upper part of the soil profile and precipitated at a depth which is dependent on the balance between precipitation and evapotranspiration. The resulting calcium-enriched (*calcic*) horizon is an important diagnostic feature of the transition from the humid to the semi-arid prairies (see p. 167). It is accompanied by an upper humus-rich horizon (a high proportion of which is derived from the deep, dense root system), characteristic of the dark *chernozem* (black earth) and *chestnut* soils of the North American grassland biome.

The humidity gradient and the temperature range across the Eurasian grasslands are somewhat narrower than in North America. Variation in composition and form of the vegetation is also less marked and the steppe grasslands of this area are more akin to the mixed grass/short-grass prairies. However, both the North American and Eurasian grasslands together with the more limited areas of the *pampas* (South America), *veld* (South Africa) and the *tussock grasslands* of New Zealand have been considerably modified by human use – for cereal, particularly wheat, cultivation in the more humid areas; for irrigated farming where sufficient ground or surface water is available and for extensive livestock ranching in the more arid parts.

THE ORIGIN OF TEMPERATE GRASSLANDS

It has been suggested that particularly in North America the subhumid temperate grasslands originated as an adaptation to the climatic differentiation during the Tertiary era and the increasing aridity of the continental interior in the late Oligocene and Miocene periods. The evidence, as Dix (1964) notes, is mainly circumstantial and palaeontological. The evolution of large herbivorous mammals (characteristic of existing grassland areas) during the Tertiary era was accompanied by anatomical changes thought to be indicative of increasing adaptation to specialised grazing and rapid movement. The fossil series of horses and other cursorial ungulates has been linked with the development of open grassland habitats. However, both Sauer (1950) and Wells (1965) seriously doubted the existence of a widespread grassland formation in Tertiary North America. In the first place it was a period of considerable geological instability, during which detrital deposits derived from the erosion of the Rockies were accumulating across the Great Plains. In the second, such palaeobotanical evidence as exists is indicative of open woodland rather than of treeless grassy plains. It is more than probable that the grasslands are of relatively recent origin, certainly post-dating the Pleistocene glaciation of North America (and probably also of Eurasia). Advancing ice-sheets would have obliterated the pre-existing vegetation, while beyond their limits there is ample evidence of formerly humid conditions in areas now arid or semi-arid. Furthermore, present grassland soils have developed on transported parent material much of which was deposited in late glacial or more recent times. Pollen analyses of sediments from Illinois and Iowa to as far south as the Llano Estacado (Dix 1964) suggest that the emergence of a grass-dominated vegetation was a post-glacial phenomenon. For instance, in the Sand Hills area of Nebraska investigation of lake sediments has revealed that as recently as 5000 years ago the arboreal (chiefly pine) contribution to the vegetation was as important as that of the grasses.

The extent to which climatic factors account for the establishment and maintenance of existing grassland is difficult to assess. On the one hand there are grassland areas where climatic conditions are currently sufficiently humid for tree growth. On the other, trees are capable of surviving not only in the existing grasslands but under even more arid semi-desert conditions. The former include the tall-grass prairies of the Midwest of the USA from eastern Iowa to western Michigan. Here it is thought that the original grassland and associated soils may well have developed in response to the drier climatic conditions of the post-glacial xerothermic period (about 5000–6000 BP). With a later increase in rainfall, re-establishment of a forest cover was impeded partly by vegetation and soil conditions, partly by man. It has been suggested that given the fine-textured, water-retentive loessal soils of the area, once a grass cover had developed it would be difficult for tree seedlings to penetrate the sod and compete successfully for water. More important, however, for the maintenance of a grass cover was the early regular use of fire by the North American Indian both for game drives and forage renewal. The presence of man in North America during the Pleistocene has now been established and his emergence as a powerful ecological factor in the Midwest at a time when the climate was somewhat drier than at present would undoubtedly have facilitated the replacement of forest by grassland. With the spread of white settlement cessation of prairie fires has been accompanied by an extension of woodland (formerly confined to fire-protected valley-side slopes) up on to the interfluvial surfaces.

The extension of the poplar (*Populus* spp.) on to the black-soil areas of the Canadian prairie region is also considered indicative of a continuing readjustment by vegetation to existing climatic conditions. The *degraded chernozem soils* which occur along the forest–steppe boundary in the south-east of the former USSR are thought to have resulted from the transgression of forest into a grassland which extended further north during the post-glacial xerothermic era. That the vegetation along the northern margin of the steppe was not in equilibrium with existing climatic conditions may explain the success of reafforestation policies, initiated by Peter the Great and resumed at intervals in more recent centuries. The less extensive former grassland areas of the southern hemisphere, particularly the pampas of South America and the Canterbury Plains of New Zealand, defy a climatic explanation. The relative effects of climatic change, soil moisture regimes and man must all be taken into consideration in these cases.

In his analysis of the nature and distribution of woodland within the Great Plains region of the USA, Wells (1965) draws attention to the widespread, though local, distribution of non-riparian woodlands on escarpments and well-marked breaks of slope. All these wooded escarpments are characterised by their height, steepness and length, by thin residual soils derived from a wide variety of parent material, and by a considerable variety of drought-resistant trees, of which the juniper has the widest range. The woodlands break the continuity of extensive grassland areas on gently sloping or flat relief with deep, transported, often fine-grained, loessal soils. The late development of grassland vegetation as indicated by pollen analyses and the survival of drought-resistant native and introduced trees established by planting from North Dakota to Texas led him to conclude that the scarp-woods are relicts of a formerly more extensive woodland, probably of an open, grass, 'savanna' form. The survival of woodland on rocky residual soils is only partly due to the fact that such sites provide less favourable habitats for dense grass growth than the finer and deeper transported soils, and a more favourable one for rapid water percolation and deep penetration of tree roots. The correlation of woodland with the most pronounced escarpments (which tend to be absent from lower, gentler and less extensive breaks) is related to their efficacy as fire-breaks, while the combination of dry, highly combustible grass litter, lack of relief and high wind speeds have concentrated the effect of fire on flat level surfaces. Wells also notes that, as a result of the absence or relative infrequency of fire today, species of the non-riparian scarp woodlands are beginning to spread on to the deeper soils of the adjacent plains.

In the course of earlier investigations in southern California, Wells (1965) had illustrated the way in which fire accentuated the effect of differences in the substratum by destroying a forest canopy and opening up the way for other and more diverse forms of vegetation. In the first instance the increase in herbaceous and shrubby species in a semi-arid climate provides a rapid annual accumulation of inflammable material which increases the fire risk and suppresses tree regeneration. Once trees are eliminated from an extensive area the more difficult re-establishment becomes. This process favours grassland particularly on deeper soils of a fine texture where, because of surface-water retention, grass/herb regeneration is rapid and recurrent fires frequent. In contrast, on coarse, permeable parent material, where water percolation is rapid, herbaceous growth is less well developed and after burning does not provide sufficient fuel to promote fires of such intensity or duration as to prevent the regeneration of woody growth.

His conclusion in the Great Plains that

> a combination of a fire-conducting ground-cover of seasonally dry grasses with extreme flatness and continuity of topography has long been a hazardous environment for woody plants in a region of drought-prone climate and strong winds where the incidence of fire has undoubtedly been increased for at least the last 11 000 years by the presence of man (Wells 1965: 249)

re-echoes that of Sauer. Finally, the influence of the extensive herds of buffalo (the mainstay of the Plains Indians' economy) whose range coincided with that of the grasslands must not be overlooked as an important ecological factor that may well have been co-dominant with fire. The short-grass plains, formerly attributed solely to increasing aridity westwards in the USA are thought, by Larson (1940), to have been developed and maintained by grazing pressures effected in the first instance by the large herds of indigenous wild herbivores and later by the introduction of cattle and sheep. The short grasses are better able to withstand grazing pressure. Also, there is evidence that when this is eliminated or during cycles of wetter years the taller, particularly medium grasses, reappear and become dominant.

TROPICAL SAVANNAS

The origin and status of tropical savannas are even more problematic than those of temperate grasslands. Also, it is of more than academic interest. Extensive areas of the semi-arid/subhumid intertropical regions still support some type of savanna vegetation. It constitutes a major resource in those areas where the rate of population growth is fast outstripping that of food production and famine is recurrent. A greater understanding of the savanna ecosystem is essential in order to ensure its efficient use with the conservation of the highest levels of productivity possible.

Tropical savanna is characterised by a much greater diversity of composition, form and habitat than is found in temperate grasslands. It is generally composed of a dominant herbaceous stratum in which more or less xerophytic perennial grasses and sedges with a pronounced tussock habit, are the principal and sometimes the only components. In addition, savanna may include varying proportions of drought-resistant woody plants ranging from low shrubs to quite tall trees. While the cover of the latter may be as high as 50 per cent, it is essentially an open and discontinuous formation. Huntley (1982: 101) defines savanna as 'all ecosystems in which C_4 grasses potentially dominate the herbaceous stratum and where woody plants, usually fire-tolerant, vary in density from widely scattered individuals to closed woodland broken now and again by drainage line grasslands'. Five major categories of savanna vegetation have been identified: open woodland, parkland, grassland, low tree shrub and scrub. On a continental scale these form a continuum along a humidity gradient from 1500 mm at the rain-forest margin to a minimum of 50 mm in the sparse grassland with scattered thorn-bush. At the regional scale, however, there is often a complex intermixture of wooded, grassy and scrub savanna and every type of savanna can be found in close juxtaposition with humid forest. Huntley (1982) further distinguishes between moist and arid savanna biomes. The former (*sour veld*) is characterised by nutrient-poor herbage with high bulk but low nutritive value which supports a low density of ungulates. The latter (*sweet veld*) is nutrient-rich and large herbivores including elephant and buffalo as well as many ungulate species, attain higher densities. Also, the loss of grazing value during the dry season is greater in the sour than the sweet veld.

While savanna vegetation generally occurs in those intertropical areas characterised by a marked seasonal drought there is, however, little correlation between the vegetation limits and either the annual precipitation or the duration of the rainy season. The absence of a clearly defined zonation of vegetation categories, the persistence of woody shrubs in the driest areas, and the occurrence of treeless savanna within the humid forests have rendered the concept of a climatic climax vegetation untenable. On the other hand, the distinctiveness and stability of the flora of the wooded Sudanese savanna are considered to be indicative of a climax formation. Whatever its determinants, it is now generally accepted that the tropical savanna is a distinct biome. However, with increasing appreciation of the structural diversity of savanna vegetation has come a greater awareness of the number of factors involved in its composition, structure and distribution.

Water

The soil-water balance is undoubtedly the most important factor that not only determines the nature and distribution of the tropical savanna vegetation but distinguishes it from many of the world's wetter biomes. The available soil water is a function of climatic conditions, particularly the annual precipitation amount and the length of the dry season; of soil texture and depth; and of landform. Variation in the structure of the savanna, as in temperate grasslands, is in many cases related to that of soil texture. Fine clay/silt soils can retain sufficient water in the upper few centimetres of the soil to support grass and forb growth but not the more deeply rooted woody plants. The latter are associated with coarse, permeable soils through which precipitation drains rapidly to the water table. In addition, the presence of impermeable lateritic crusts at or just below the ground surface results, under the markedly seasonal rainfall regime, in a corresponding alternation of soil saturation (or even flooding) and desiccation. Under such extreme conditions trees and shrubs are smaller, more stunted and less frequent or are completely absent.

Landforms

There is a close association of savanna with a sequence of plateau surfaces of varying altitude and precipitation. The highest and most humid in central Brazil and Zimbabwe are deeply dissected, capped by weathered ferralitic soils and associated lateritic horizons which form steep scarp cliffs. With decreasing altitude lower plateaux and levels or rolling plains covered with blown sand extend into drier areas (Cole 1986). The coincidence of extensive areas of grass or grass–shrub savanna with the high discontinuous plateau surfaces have been interpreted as relicts of a drier climatic period than at present which have persisted as a result of edaphic conditions unfavourable for the establishment of a denser forest cover. Eden (1974) interpreted the savanna islands which occur in sparse forested areas in southern Venezuela as remnants of a formerly more extensive biome which originated under drier climatic conditions than at present. The similarity of their floristic composition he considered indicative of their former continuity. They have been maintained by fire in the succeeding more humid period. The plateau surfaces are the uplifted, undissected remnants of old land surfaces (erosion surfaces) produced over a very long period of weathering and leaching. Soils are, in consequence, of low fertility, deficient particularly in phosphorus and often containing high and potentially toxic amounts of aluminium and magnesium. Preservation of the plateaux has been aided by ancient lateritic horizons. The composition of the vegetation varies with the age of the plateau surface and soil-drainage conditions (Fig. 11.3). In the core areas of the high laterised surfaces subject to alternating flood and drought conditions trees become smaller, less

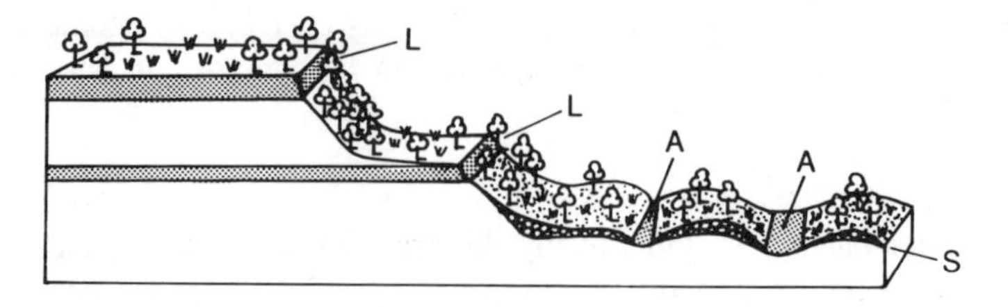

Fig. 11.3 Diagrammatic representation of relationship between landforms and types of savanna vegetation; L = laterite crust; A = alluvium; S = sandy soil

frequent and eventually disappear. On the lower rolling terrain freely drained soils support savanna parkland; on clay soils in depressions or valley floors grass becomes dominant. Plateau uplift and dissection resulted in the breakdown of lateritic horizons and the exposure of deeper, less leached soil material on steep escarpments and in the valleys. The latter supports a denser deciduous woodland/forest which contrasts sharply with the open plateau savanna. The savanna vegetation can thus be interpreted solely as a degraded edaphic climax (Cole 1986).

Grazing

Much of the savanna has been subjected to a long period of grazing, burning and shifting agriculture. The savanna flora can be traced back at least to the beginning of the Tertiary era and its evolution has been accompanied by that of a multi-species fauna of large herbivores which attained its maximum richness in the later Tertiary and Pleistocene eras. Africa retains the largest number, *c.* forty-four species (Table 11.4). In contrast, South America has only three species of deer and one peccary and the present mammal fauna is a small remnant of that which existed when the savanna was more extensive in the Miocene era. The ecological equivalents of the large mammal herbivores in the Australian savanna are the megapod *marsupials*. Most of the African herbivores, with the exception of the elephant and ostrich, are *ungulates* with flocking and migratory habits. The component species vary in body size, population, diet and eating habits (i.e. cropping, browsing, rooting, debarking) which allow a large number of niches to be exploited and a high proportion (40 per cent) of the net primary above-ground production to be cropped annually (Table 11.5).

This is exemplified by the feeding successions of mixed herbivore communities in the Serengeti Park (Tanzania). A large proportion of the herbivores are browsers and bulk-roughage feeders with the capacity to effect cyclic or long-term vegetation changes. It is thought that the elephant may have been an important factor in the degradation of savanna woodland to bush and grassland (Cumming 1982). Capable of felling and uprooting trees to obtain browse (at a rate of 1500 trees per male animal per year) they open up forest clearings which then become subject to more intense grazing and to burning. Lacking any effective predator other than man, the increase of protected elephant populations in game reserves has caused serious overgrazing and depletion of forage for all the wild game involved. Hippo cropping can also effect cyclic vegetation changes such that tree regeneration and fire are prevented until grazing causes hippo and associated buffalo and wart-hog numbers to decrease and allows tree regeneration to take place. As the savanna woodland regenerates increase in grass growth is accompanied and promoted by a higher fire incidence. Over most of the unprotected and tsetse-free African savanna today the mixed indigenous herbivore populations have all but been replaced by high and increasing densities of domestic livestock – cattle, goats and sheep. Much of the northern African savanna have been used by pastoralists for at least 7000 years and for 1000–2000 years in Central and East Africa (Cumming 1982). The number of cattle, determinants of wealth and social status, has increased with that of the human population.

Table 11.4 Large herbivores occurring in African savanna ecosystems (from Owen-Smith 1982)

		Savanna type favoured		
Species	*Trophic category*	*Arid/ eutrophic*	*Mesic/ dystrophic*	*Associated vegetation*
African elephant (*Loxodonta africana*)	Browser and grazer	x	x	
White rhinoceros (*Ceratotherium simum*)	Grazer	x		
Black rhinoceros (*Diceros bicornis*)	Browser	x		
Burchell's zebra (*Equus burchelli*)	Grazer		x	
Grevy's zebra (*E. grevyi*)	Grazer	x		
Hippopotamus (*Hippopotamus amphibius*)	Grazer	x	x	
Warthog (*Phacochoerus aethiopicus*)	Grazer	x		
Bushpig (*Potamochorus porcus*)	Omnivore			x
Giraffe (*Giraffa camelopardalis*)	Browser	x		
Grey duiker (*Silvicapra grimmia*)	Browser		x	
Kirk's dikdik (*Rhynchotragus kirkii*)	Browser	x		
Guenther's dikdik (*Thynchotragus guentheri*)	Browser	x		
Salt's dikdik (*Madoqua saltiana*)	Browser	x		
Steenbok (Raphicerus campestris)	Browser and grazer	x		
Sharpe's grysbok (*R. sharpei*)	Browser	x		
Klipspringer (*Oreotragus oreotragus*)	Browser			x
Oribi (*Ourebia ourebi*)	Grazer			x
Soemmering's gazelle (*Gazella soemmeringi*)	Browser and grazer	(x)		
Grant's gazelle (*G. granti*)	Browser and grazer	x		
Thomson's gazelle (*G. thomsoni*)	Grazer and browser	x		

Gerenuk (*Litocranius walleri*)	Browser	x		
Dibatag (*Ammodorcas clarkei*)	Browser	(x)		
Springbok (*Antidorcas marsupialis*)	Grazer and browser	x		
Impala (*Aepyceros melampus*)	Grazer and browser	x		
Bohor reedbuck (*Redunca redunca*)	Grazer			x
Southern reedbuck (*R. arundinum*)	Grazer			x
Mountain reedbuck (*R. fulvorufula*)	Grazer			x
Kob (*Kobus kob*)	Grazer			x
Puku (*K. vardoni*)	Grazer			x
Waterbuck (*K. elypsiprimnus*)	Grazer		x	
Oryx (*Oryx gazella*)	Grazer	x		
Sable antelope (*Hippotragus niger*)	Grazer		x	
Roan antelope (*H. equinus*)	Grazer		x	
Hartebeest (*Alcelaphus buselaphus*)	Grazer	x		
Tsessebe/topi/tiang (*Damaliscus lunatus*)	Grazer	x	x	
Hunter's antelope (*D. hunteri*)	Grazer		x	
Blue wildebeest (*Connochaetes taurinus*)	Grazer	x		
Bushbuck (*Tragelaphus scriptus*)	Browser			x
Nyala (*T. angasi*)	Browser and grazer			x
Lesser kudu (*T. imberbis*)	Browser	x		
Greater kudu (*T. strepsiceros*)	Browser	x		
Cape eland (*Taurotragus oryx*)	Browser and grazer	x	x	
Derby eland (*T. derbianus*)	Broswer and grazer	x		
African buffalo (*Syncerus caffer*)	Grazer		x	
Total		27	11	10

Table 11.5 Estimated percentage annual NPP consumed and wasted by large native mammals and domesticated cattle in tropical and temperate grasslands

Large native mammals		*Domesticated cattle*	
Uganda (savanna)	3–40	Nigeria (savanna)	45
Tanzania (savanna)	60	South Africa (savanna)	19
Serengeti (savanna)	15–39	Texas (short grass)	29–40
S. Dakota (mixed prairie)	1–10	Western USA (short grass)	40–60

Increased grazing accompanied by a decline in browsing has, in many areas, resulted in the development of dense, often closed scrub vegetation at the expense of grassland and a consequent fall in the livestock-carrying capacity of the land. In the drier savanna of the North African Sahel overgrazing and injudicious cultivation have been accompanied by widespread *desertification*.

Fire

Fire, like grazing, has long been an important ecological factor in the composition and structure of savanna vegetation. Natural, lightning-set fires are endemic in semi-arid regions of the world. The use of fire by humans either to flush game or renew grass forage growth is a long-established practice (Harris 1980). Most savannas are burned frequently – in some areas every year or 18 months – to clear grass litter and stimulate fresh growth. Carbon-14 dating indicates that fire was frequent and widespread in South America at least 1200 years before the Portuguese arrived, while the Australian Aborigines were using fire as early as 40 000 BP (Lacey *et al.* 1982). The high proportion of pyrophytic characteristics of woody plants such as the thick bark, the ligneous sprouting roots (subtrees or xylopodium) and the adventitious buds, together with the protection of perennating buds inside grass-sheaths, are thought to be indicative of a long period of adaptation to burning. On the other hand, fires set by lightning or humans, after exceptionally rainy seasons and a more luxurious growth of grass than usual, can be so intense as to seriously reduce or eliminate growth.

Low-intensity, annual grass fires are common in the moister savanna areas where growth is more prolific and hence grass fuel is more abundant than in drier areas. In the latter, fires are less frequent and there are fewer xerophytes. Where livestock grazing pressures have increased (as in Africa) the fire frequency has decreased because of the reduction in available fuel. Where burning has been reduced as in the eucalypt savannas of northern Australia, the fire intensity has increased because of the accumulation of fuel (Gill *et al.* 1981). The evidence regarding the effect of fire on the balance between grass- and scrub-dominated savanna is, however, conflicting (Huntley and Walker 1982). In some instances fire would appear to favour the maintenance of grassland by destroying saplings. In others, particularly in dry areas, fire is thought to maintain shrubs and trees in a condition and at a height available for browsing animals.

SECONDARY OR DERIVED SAVANNA

Considerable areas of savanna vegetation, particularly on the savanna/humid forest margin or within the humid forest area, have replaced the pre-existing rain forest as a result of shifting agriculture (see p. 186). Progressive shortening of the bush or forest fallow period and/or continued grazing and burning of cleared areas inhibit tree and shrub regeneration. Soils dry out during the hot season and may become laterised and mitigate against re-establishment of a rain-forest cover. A herbaceous savanna grassland (often dominated in Africa by the intractable weed, spear grass (*Imperata cylindrica*), becomes established.

While both temperate and tropical grassland have been interpreted either as a climatic or more commonly a fire climax, it is now clear that the nature and distribution of these grassland ecosystems are the result of a variety of interacting factors whose relative importance has varied in time and place. As Hills (1960) noted in his early review of savanna research, confusion has arisen in the interpretation of the origin and status of tropical grasslands 'because of the failure to distinguish clearly between pre-disposing, causal, resulting and maintaining factors' – a statement no less true of the studies of temperate grasslands. While it is not possible to explain the distribution of natural grasslands in terms of climatic parameters alone, their extensive development in areas characterised by marked seasonal drought and subjected over a long period of time to alternating periods of greater or lesser precipitation cannot be dismissed as fortuitous. Climatic variations and geomorphological evolution have resulted in a soil-water regime and/or soil-nutrient status inimical to the development of a closed forest ecosystem. Within what has been a constantly expanding and contracting ecotone between humid and arid zones, the open woodland and wooded grassland formation and its characteristic fauna of large grazing herbivores would have co-evolved. The relative ecological dominance of the herbaceous and tree strata would have been, and still must be, dependent on a number of ecologically differentiating factors susceptible to variation through time in a constantly fluctuating physical environment. Climate, relief and the original character of the vegetation provide ideal conditions for the propagation of widespread fire. Early set by man, it would appear that in the more arid areas or in periods of more arid climatic conditions, fire tipped the balance in favour of the *pyrophytic plants* of which perennial grasses are the most prolific and, together with the increase in wild and domesticated herbivores, has served to maintain grassland ecosystems. Once established, however, grasslands have revealed a high degree of stability even in the face of changing environmental conditions. Whether it be increase of rainfall or cessation of pastoral activities in both temperate and tropical grasslands, the invasion of trees is often retarded, if not inhibited, by soil conditions unfavourable to the establishment of tree seedlings.

Chapter

12

Deserts: arid and semi-arid

Deserts are those areas where climatic aridity (see Fig. 12.1(A)) imposes severe limitations on biological processes and where crop-based agriculture is not possible without irrigation. Aridity is, however, a relative condition since the amount of water available for organisms is a function of a number of interacting variables, such as precipitation, temperature, slope and aspect of the terrain, and soil texture.

CLIMATE

While a low mean annual precipitation is characteristic of deserts, this can vary from almost none to over 600 mm, the latter being well beyond the commonly accepted limit of 300 mm. As distinctive a feature is the extreme irregularity in amount and in the spatial and temporal occurrence of rainfall. Variability, whether expressed as a coefficient *M/M* (average annual maximum : average annual minimum precipitation) or as the percentage deviation from the mean annual precipitation, is high and increases with decreasing rainfall: e.g. arid conditions *M/M* = 6–20, per cent deviation = 35–70; extremely arid *M/M* = 100+, per cent deviation = 150. Irregularity of occurrence is such that precipitation is described as episodic and unpredictable. It occurs as extremely high-intensity rain pulses; the number of rain days (over 0.1 mm day^{-1}) is low and the time-lapse between pulses can vary from days to years. In addition, rainfall is frequently very localised, i.e. spotty, while 'phantom' rain-storms when precipitation evaporates before reaching the ground surface have been recorded (Evanari 1985a).

The only reliable sources of water in deserts are those supplied by coastal fog, dew and ground-water at a depth that can be tapped by organisms. Condensation from coastal fog can be of the order of 50–300 mm $year^{-1}$ along the coasts of southern California, Peru, Chile, Namibia and Madagascar (see Fig. 12.1(B)). One of the few long-term records of dew-fall are those for the Negev Desert where over a 20-year period the average number of dew-nights was 195 and the average amount of dew deposited was 33 mm (Noy-Meir 1973).

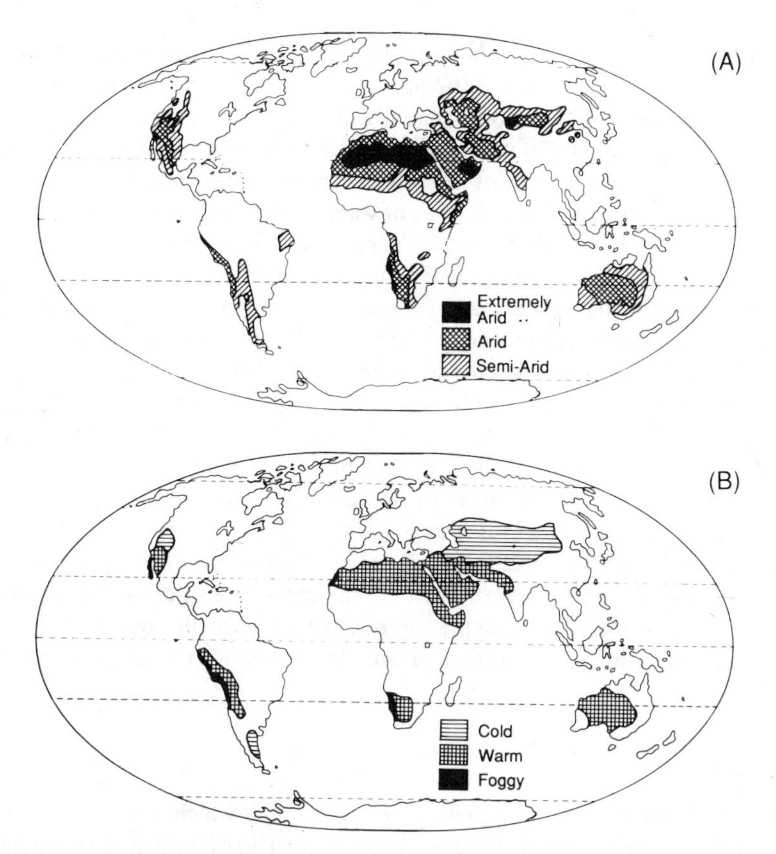

Fig. 12.1(A) Delimitation of arid zones using an adaptation of Thornthwaite's index of humidity (1*h*):

$$1h = \frac{100a - 100d}{ET}$$

Et = potential evaporation; *e* = water surplus given soil retentive capacity 100 mm; *d* = water deficit. Indices: Semi-arid 20–40; arid, less than −40; extremely arid less than −57, no seasonal rhythm of precipitation and 12 consecutive months without rain have been recorded (after Meigs 1953); (B) types of desert regions (from Shmida 1985)

The amount of effective rainfall, i.e. that which becomes available for plant growth, is a function on the one hand of the rate of evapotranspiration as determined by ambient temperature, relative atmospheric humidity and wind-speed, and of the infiltration rate of the soil on the other. The low-latitude, intertropical deserts with low cloud cover experience the highest global intensity of annual solar radiation, except in the fog-deserts where relative humidity (40–50 per cent) is comparatively high and mean annual temperature (16–19 °C) is low. Tropical deserts are characterised by high maximum temperatures (absolute 45–70 °C) and extreme diurnal ranges (50–60 °C). Temperate latitude deserts have, in addition, a wide annual range and experience winter temperatures well below 0 °C. Evaporation rates of 2500–3000 mm year^{-1} are common and in extreme conditions can reach what are probably global peak values of over 4000 mm (Death Valley, Nevada, 4262 mm).

As a result of the high intensity of rainfall pulses and a sparse vegetation cover, surface evaporation and runoff are accelerated particularly on sloping terrain and on impermeable substrata. Runoff tends to accentuate the variable distribution between areas of water receipt and areas of water collection. Infiltration rates are slow except on coarse sandy or rocky debris. There is insufficient water to maintain permanent river flow, and surface runoff, like precipitation, is intermittent, terminating in local depressions or intermontane basins. However, exotic (or allochthonous) rivers whose sources are located beyond the desert can maintain their flow, albeit diminished in volume, across arid areas.

Many, particularly fine- and medium-textured desert soils are highly saline and are characterised by an accumulation of soluble salts (sodium chloride, sulphates, carbonates, etc.) below the surface at a depth dependent on the downward percolation of water and/or the level of the ground-water table from which the capillary fringe can translocate salts near to or on the ground surface. In the former case leaching is incomplete and soluble salts, washed out of the upper soil, are deposited at 10–100 cm below the surface. In the latter case where saline ground-water is near the surface (as in depressions) fluctuation of the water-table with precipitation results in the deposition of salts from below on to or near the soil surface.

SOIL CLIMATE

Effective rainfall has been more precisely defined as that which wets the soil to a depth and for a period which ensures specific seed germination and seedling growth and survival (Monod 1973). The depth to which a given amount of precipitation can wet the soil and the length of time it is retained in the soil is a function of soil texture and the rate of evaporation. In very coarse material, most of the water which infiltrates may very quickly drain down beyond the potential root zone to the ground-water. In finer material, more water can be retained in the upper part of the soil by capillary and hyroscopic forces. The depth to which moisture infiltrates (i.e. *the wetted front*) is deeper in sandy than in loamy or clayey soils. In the latter, water is held more tightly and is less readily available than in loamy soils. As the soil begins to dry out, evaporation slows down; the upper 5–10 cm may dry out in 5–25 days; below this level the soil may remain humid for 10–30 weeks and below 30 cm for months (see Fig. 12.2).

As a result of the low heat conductivity of dry soil, temperatures fall rapidly with depth below the surface. The rate of decrease, however, is influenced by the physical nature of the surface and by the amount and type of surface cover. On bare mineral surfaces maximum daily temperatures can exceed 70 °C in the Sahara. Figure 12.3 compares the diurnal variation in air and soil temperature on a stone-covered site (B) with that on an exposed site (A). The diurnal temperature range from the surface down to 5 cm is much greater than that of the atmosphere and than that below the covered surface. With increasing soil depth, both the temperature and the amplitude of the temperature range decrease more rapidly

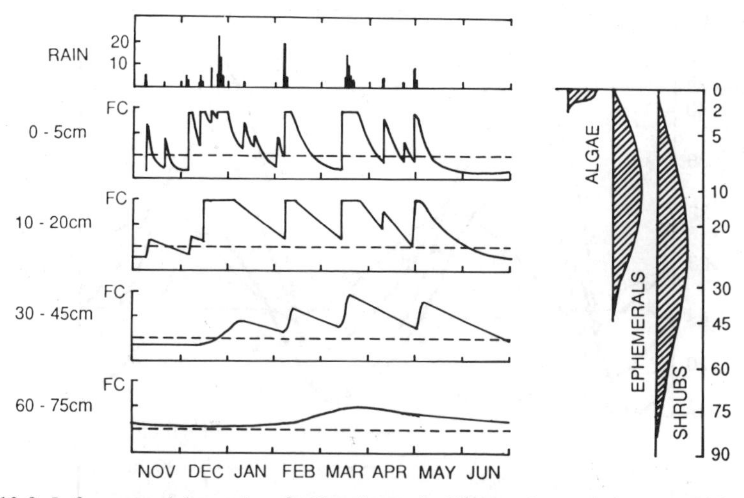

Fig. 12.2 Left: seasonal dynamics of soil moisture in different layers (schematised from data from desert shrubland on loessal plain: dashed line = apparent 'wilting point' and FC = 'field capacity' moistures. *Right:* vertical distribution of activity of plant types (scale distorted to fit left part) (from Noy-Meir 1973)

below the covered than below the uncovered site. At 20 cm both were continuously below that of the air; at 50 cm temperature fluctuations were negligible though the covered site was slightly warmer than the exposed one. At a depth of 100 cm annual temperatures remained virtually constant at *c.* 20 °C (Evenari 1985a). Other factors which influence the variation in soil temperature with depth include slope and aspect (in so far as they affect heat input) and the moisture content because of the higher specific heat of water than mineral matter.

INDICES OF ARIDITY

The definition and delimitation of desert regions in terms of water deficiency are difficult, not only because of the problems of quantifying biologically effective precipitation but also because deficiency and effectiveness are relative concepts. As a result, many biologists have, for pragmatic reasons, favoured the use of mean annual rainfall to define desert limits, as shown in Table 12.1. The range within the class boundaries allows for increase of evaporation with temperature. The boundary between the extremely arid and arid desert is considered to be that between an open, sparse vegetation and a more or less continuous cover; that between arid and semi-arid to coincide with the limit at which dry farming is impossible.

Others, including climatologists, biogeographers and agriculturalists, have attempted to define *water deficiency* or *drought* by means of a combination of related climatic parameters. It has long been assumed that the limit between humid and arid conditions is when evaporation (E) equals precipitation (P). There are, however, still relatively few reliable long-term measurements of evaporation.

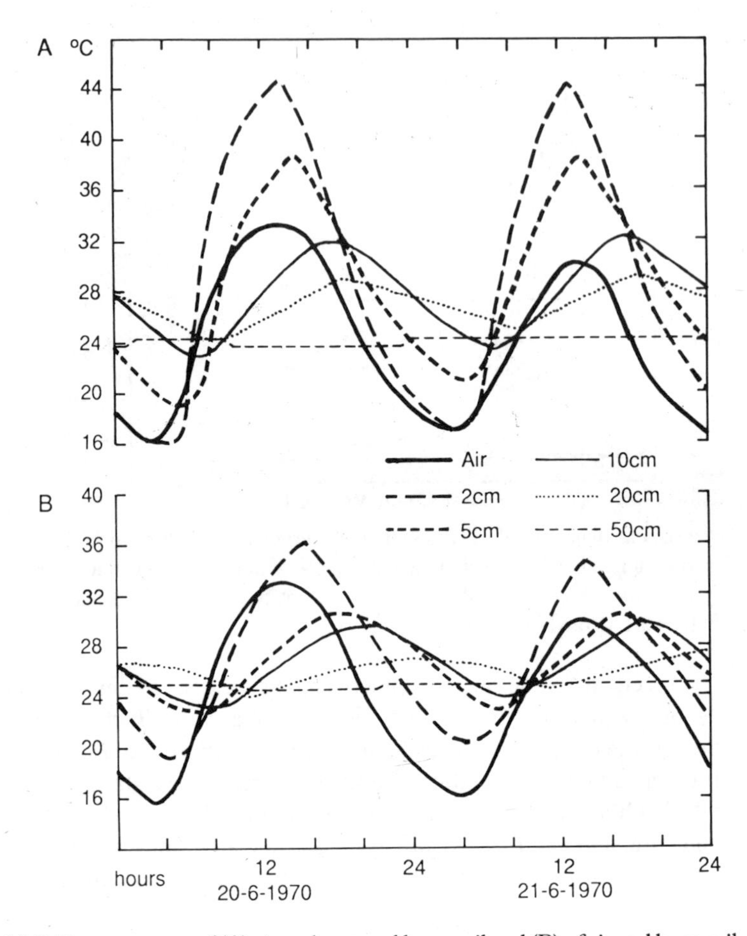

Fig. 12.3 Temperatures of (A) air and exposed loess soil and (B) of air and loess soil covered by a large stone measured at various depths in the course of 2 days at Ardat (Israel) (from Evanari *et al.* 1985a)

Table 12.1 Primary production in arid and hyperarid zones in the Sahara (Le Houérou 1979)

Ecological zone		*Average annual rainfall (mm)*	*Production (kg DM ha^{-1} year^{-1})*
Arid	Upper	300–400	800–1000
	Middle	200–300	450–500
	Lower	100–200	200–400
Hyperarid	Upper	50–100	100–200
	Middle	20–50	20–100
	Lower (eremitic)	0–20	0.0

Consequently, precipitation : temperature ratios have frequently been used to express aridity. Bagnouls and Gaussen (1957) early proposed a classification of biological climates on the basis of mean monthly temperature and precipitation. The threshold between humid and arid periods was that when the mean monthly precipitation (mm) was equal to or less than twice the mean monthly temperature (°C). The obvious limitations of such ratios is that evaporation is not a function of temperature alone. Thornthwaite's (1948) concept of potential evapotranspiration (i.e. the amount of moisture which would be evaporated from the soil and vegetation if there was a constant supply of water) provided a more satisfactory means of estimating soil-water deficiency. Potential evapotranspiration was calculated from known temperature and light conditions. Drought could then be expressed in terms of the soil-water deficit, i.e. the amount by which potential evapotranspiration exceeded the water available in the soil, assuming that the soil has the capacity to store approximately 10 cm of the precipitation received (see Fig. 12.1). Later Penman (1963) developed the concept in his calculation of potential transpiration which allowed for variation in soil-water storage capacity as determined by soil texture in the rooting zone.

ORGANIC ADAPTATIONS TO DESERT CONDITIONS

Organisms which inhabit the hot deserts of the world have to be able to cope with excessive heat and with drought to ensure that neither internal temperatures nor tissue dehydration attain lethal levels. Desert plants and animals have acquired similar morphological, physiological and behavioural strategies which, although 'not unique to desert organisms are often more highly developed and effectively utilised than their mesic counterparts' (Hadley 1972: 338).

The two principal strategies are avoidance and tolerance of heat and water stress (see Table 12.2) The so-called *evaders* comprise the majority of the flora of most deserts (Noy-Meir 1973). They can survive periods of stress in an inactive mature or premature stage or by living permanently or temporarily in cooler and/or moister microhabitats, e.g. below shrubs or stone, in rock fissures or below ground. Of desert animals, 75 per cent are subterranean (*geozonts*) with surface nocturnal crepuscular, or wet surface activity. The tolerant organisms can survive extreme environmental conditions because they possess the means of controlling their temperatures and their water loss.

TEMPERATURE ADAPTATIONS

Desert plants and animals have a more extended metabolic temperature range and are able to function at higher temperatures than their mesic congeners. Herbaceous plants have a maximum tolerance of 50–55 °C in dry air. Some cacti such as the prickly pear (*Opuntia* spp.) can survive up to 65 °C, while many crustose lichens can tolerate 70 °C or more. The upper lethal levels for animals is lower, i.e. 40–50 °C, though arthropods, particularly beetles and scorpions, can tolerate 50 °C

Table 12.2 Adaptations of plants and animals to hot desert environments (simplified from Evenari 1985b)

	Plants	*Animals*
Stress-evading strategies		
	Inactivity of whole plants (geophytes, hemicryptophytes)	Inactivity in time (diurnal and seasonal) and space (geozonts in burrows, other animals in stress-protected above-ground shelters)
	Cryptobiosis of whole plants (poikilohydrous plants)	Cryptobiosis of mature animals (aestivation of snails, hibernation)
	Dormancy of seeds (growing in stress-protected microhabitats)	Cryptobiosis of eggs, shelled embryos, larvae; permanent habitation or temporary use of stress-protected microhabitats
Structural and ecophysiological stress-controlling strategies		
Strategies reducing water expenditure	Small surface : volume ratio	Small surface : volume ratio
	Regulation of water loss by stomatal movements and change in residual resistance	Regulation and restriction of water loss by concentrated urine, dry faeces, spiracular control, reduction of glomerular filtration rate and urine flow rate
	Xeromorphic features	Structures reducing water loss: integuments highly impermeable to water cover by fleece and feathers
	Postural adjustment	Postural adjustment

Strategies to prevent death by overheating	Transpiration cooling	Evaporative cooling
	High heat tolerance	High heat tolerance
	Mechanisms decreasing and/or dissipating heat load	Mechanisms decreasing and/or dissipating heat load
Strategies optimising water uptake	Direct uptake of dew, condensed fog and water vapour (e.g. poikilohydrous plants)	Direct and indirect uptake of dew, condensed fog and water vapour (e.g. arthropods, water enrichment of stored food)
	Fast formation of water roots after first rain	Fast drinking of large quantities of water (large mammals), uptake of water from wet soil (e.g. snails)
	Halophytes: uptake of highly saline water, high salt tolerance, salt-excreting glands	Halozonts: uptake of highly saline water, high salt tolerance, salt-excreting glands
Strategies to control reproduction in relation to environmental conditions	'Water clocks' of seed dispersal and germination	Sexual maturity, mating and birth synchronised with favourable conditions
	Suppression of flowering and sprouting in extreme years	Sterility in extreme drought years

or over. Both plants and animals are capable of modifying the excessive heat-load characteristic of the desert environment by such strategies as the following:

1. Changing the orientation of the whole body or a specific organ in order to minimise the areas and/or the time to which they are exposed to maximum heat input;
2. Light coloration and surface texture to minimise absorption and maximise reflection of light;
3. Surface growth (i.e. spines, hairs, etc.) which can (a) absorb and/or reflect much of the incident radiation, thereby keeping the underlying surface cooler; and (b) create an effective boundary layer of air which insulates the underlying surface;
4. Body size is particularly important in controlling the rate of heat flux between an organism and its environment and that by radiation, convection, evaporation and metabolism are proportionate to the surface area of the plant or animal.

The smaller the organism the larger the surface area to volume ratio and the greater the heat flux. Below a certain size small organisms must find a milder microhabitat either below ground surface or, in the case of birds, at greater altitude in the atmosphere. Large desert animals, such as the camel, oryx and goat, can control overheating by means of evaporative cooling. Cooling by transpiration is also thought to be most effective in the cacti and small-leaved desert shrubs because of their relatively large surface area : volume ratio.

DROUGHT

In face of the extreme aridity desert plants and animals need to be conservative water-users (see Table 12.3) and to reduce water loss by one means or another. Again a small surface : volume ratio is an advantage. Water loss from plants can be controlled by regulation of diurnal stomatal movements and by xeromorphic adaptations such as thick cuticles, sunken stomata and surface hairs. The most drought-resistant higher plants are the succulents (cacti and Euphorbiaceae) which possess well-developed storage tissues, small surface to volume ratios, low cuticular transpiration and rapid stomatal closure. In particular, the stomata of heat-resistant cacti remain closed during the day and open at night when carbon dioxide for photosynthesis is produced by metabolic processess (see p. 113). Similarly, many animals have surfaces which reduce the rate of loss from perspiration and, in addition, their excreta tends to be dry and they produce little and very concentrated urine.

Some plants and many arthropods are drought resistant (or persistent); they can survive and recover from a high degree of tissue desiccation. These are the true *xerophytes* which can withstand both desiccation and near-lethal temperatures. Noy-Meir (1973) distinguishes between the fluctuating and stationary xerophytes. The former have deciduous leaves and stems or produce smaller leaves and stems during drought, e.g. lichens, acacia, chenopods. The latter maintain a constant biomass. The cresote bush (*Larrea tridentia*) is probably the most outstanding example. It can survive up to a year without rain and can tolerate dehydration such that the osmotic pressure of its cell sap exceeds 55 atmospheres. Rapid uptake of

Table 12.3 Evaporative water loss from some desert organisms compared with that from humans (Hadley 1972)

	Temperature (°C)	*Water loss (mg cm^{-2} hour^{-1})*
Mammals		
Human	35	22.32
African oryx (*Oryx beisa*)	32	3.24
Cactus mouse (*Peroniyscus eremicus*)	32	0.66
Reptiles		
Gecko (*Gehydra variegata*)	30	0.22
Lizard (*Uta stansburiana*)	30	0.10
Snake (*Pitnophis catenifer*)	25	0.23
Arthropods		
Scorpion (*Hadrus arizonesis*)	33	0.02
Locust (*Locusta migratoria*)	30	0.70
Plants		
Pineapple (*Ananas cosmosus*)	30	0.75
Century plant (*Agave americana*)	30	1.64
Prickly pear (*Opuntia polycantha*)	26 (RH 90%)	0.90

available water is also a characteristic of *poikilhydrous* desert organisms (particularly lichens and algae) which can absorb water instantly from rain, dew or the atmosphere when the relative humidity is above 70 per cent. Animals can rapidly imbibe large quantities of water, and salt-tolerant plants and animals have high cell osmotic pressures which allow efficient uptake of alkaline water. The roots of many desert plants can exert a greater suction pressure (over 100 bars) than others and can hence extract more water from fine water-retentive soils than can mesophytes. Some plants can produce roots very rapidly with the onset of rain. Others such as the creosote and many cacti have a large volume of shallow, lateral roots that can exploit brief rain-storms efficiently while some desert shrubs such as the mesquite (*Prosposis* spp.) may penetrate 10–30 m into the substratum to tap permanent sources of ground-water.

REPRODUCTION

Survival in the desert environment is also dependent on an organism's ability to reproduce itself and there are a variety of strategies adapted to this end. High seed production and efficient dispersal are even more essential than in humid environments. *Hygrochasy* or the impedance of long-distance dispersal is exclusive to plants in arid regions; it ensures that the seed remains within the same habitat as the parent plant. In some cases dissemination is delayed until after the first few rain days and germination of the seeds from any one species may be spread over a period that may last for several years. This appears to ensure that, despite high mortality rates, there are always some seeds in reserve. In other species seeds are released, a few at a time, after the moistened fruit opens. In addition, seeds possess means of regulating the time and location of germination; many are so-called 'water clocks' with built-in rain gauges which will not germinate until a certain critical amount of water becomes available. Some arid-zone shrubs, such as the paloverde (*Cercidium* spp.) the iron-wood (*Casuarina* spp.) and the smoke-wood (*Cotinus* spp.) of south-west America have seeds with coats so tough that germination can only take place after severe mechanical abrasion during torrential flash-floods. Reproduction in animals also tends to be synchronised with favourable conditions. In both plants and animals reproduction is suppressed in periods of extreme drought.

The diversity of strategies adopted by desert organisms is considerable. No one species, however, exhibits all the features and, as Hadley (1972) points out, the resulting specialisation of survival strategies allows as thorough a use of the available microhabitats and niches as possible. As illustrated in Fig. 12.4, plants with different life forms can utilise different water stores in the soil. In the upper 2 cm, water is so transient that it can only be used by surface algae and lichens which become active on wetting.

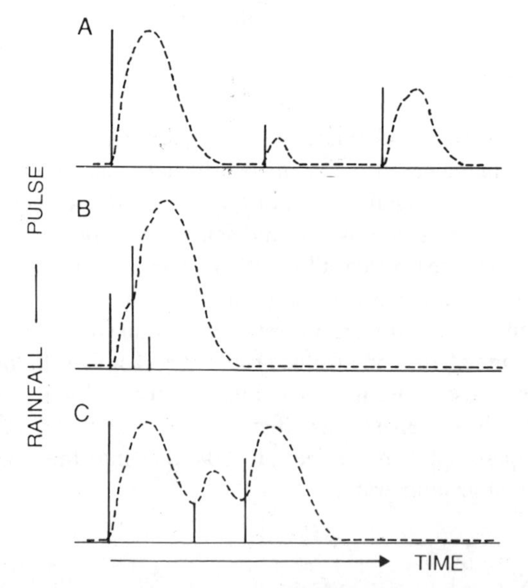

Fig. 12.4 System response to precipitation input pluses; (A) widely separated; (B) clustered; (C) intermediate spacing (from Noy-Meir 1973)

THE DESERT BIOTA

Despite the diversity of life forms, the desert flora and fauna are relatively species-poor. At the continental scale species diversity of lizards and rodents has been correlated with increasing precipitation (Louw and Seely 1982). Few plant taxa are eremic (i.e. have a mainly desertic distribution). Endemism is low and almost all existing families have their centres of origin in contiguous semi-arid regions. Goodall and Perry (1979) note that there are only 28 families found in more than one desert area and of these 15 have only one shared genus, 9 have 2 and 3 have 4. Of those with a large number of shared genera the Chenopodiaceae and Poaceae are the most abundant in species number and percentage cover. However, only a very few species, all of which are perennials, have a cosmopolitan distribution.

There is still considerable debate as to the age of existing deserts. Some workers maintain that deserts existed throughout geological time, certainly predating the evolution of the angiosperms. Arguments for desert antiquity are based on the evidence of continental drift, the early development of arid conditions consequent on orogenic earth movement and on the persistence of sand-dune refugia in coastal areas. Advocates of desert youth point to the evidence for the formerly wide distribution of a moister arboreal flora in existing desert areas in the Cretaceous and early Tertiary periods and to geomorphological evidence of pluvial periods during the Pleistocene. Whatever the date of origin, the paucity of shared taxa and the differences in dominant life forms in existing desert regions, point strongly to the flora of discrete desert areas having evolved largely in isolation one from the other.

DESERT VEGETATION

Desert vegetation is simple in that its structure is poorly developed and its cover becomes increasingly open and discontinuous with increasing aridity. Hills (1960) recognised four categories of arid vegetation which could be correlated with decreasing water availability:

1. Frutescent perennial scrubs. These include the distinctive cactus communities of the south-western USA and low woody scrub.
2. Suffrutescent perennial vegetation composed of a few species of dwarf (30–120 cm) succulents, low shrubs and perennial grasses.
3. Ephemeral or seasonal herbaceous vegetation composed of both annuals and perennials and dominated by grass.
4. Accidental vegetation which occurs as occasional carpets of ephemeral annuals in areas of low and very episodic rainfall on those soils which retain sufficient moisture.

Goodall and Perry (1979) distinguish between steppe (chamaephytic semi-desert) in which dwarf suffrutescent shrubs comprise 10–30 per cent of the vegetation cover, with perennial grasses and cacti; true desert, perennial vegetation with less than 10 per cent composed of dwarf shrubs, or a few tall shrubs accompanied by a flush of herbaceous annuals in the rainy season and with the total cover never exceeding 50 per cent even in the most favourable years; and contracted deserts where vegetation only occurs in wadis (washes) where the ground-water is accessible to deeper rooted plants including palms. Under the most rigorous conditions of aridity, intensity of evaporation and/or extreme soil mobility or salinity may exclude vegetation completely. In the absolute desert areas biologically inhabitable sites are extremely localised and limited in extent. This is the case in the most arid and sterile parts of the Sahara where permanent vegetation is confined to isolated wadis. Particularly harsh conditions also occur along the littoral areas of northern Chile and southern Peru, and on the high plateaux of central Asia. In the former practically the only source of moisture is that supplied by sea mists during winter. In addition, the high salt content of this hygroscopic water can be sufficient to inhibit the growth of non-halophytic plants and to form indurated surface crusts. Best adapted are *halophytes* and, more particularly, succulent epiphytes which can absorb moisture directly from the atmosphere. Local patches of more permanent shrubby evergreen vegetation or seasonal herbaceous annuals may occur where, as a result of aspect and relief, patches of particularly dense fog coincide with water-retentive soils.

The extent to which vegetation can modify or ameliorate the physical habitat in arid environments is limited. Variations in the hydrological regime of the soil as a result of relief, slope, texture and exposure are important factors determining the density and distribution of vegetation in any particular area. The vegetation is often composed of a mosaic of communities related to a variety of physical gradients as in the tundra and alpine biomes. Among the commonest are those associated with the mobility of the parent material as in the case of shifting sand dunes; with playas and salt pans of enclosed depressions where there is a zonation or gradation of vegetation from less to more alkaline conditions; and with soil texture. In the latter instance a striking correlation between the dominance of shallow-rooting grasses and the finer water-retentive soil on the one hand, and of deep-rooting shrubs and coarse debris on the other, has been noted by many authors.

The importance of allogenic factors in determining the composition and density of vegetation early led to the assumption that competition in desert habitats was rare. A contrary opinion, however, maintains that there is intense competition for that most scarce of limiting resources – water. This is thought to be reflected in the even spacing of shrubs in many areas which is attributed to intense root competition. It has, in addition, been noted that root systems occupy most of the underground area even where the above-ground cover may be only 3–5 per cent. The inhibition of shrub seedling growth up to a diameter five times the canopy radius of the mature parent plant, some of which are known to produce allelopathetic substances, has frequently been observed (Noy-Meir 1973).

DESERT ECOSYSTEMS

Noy-Meir (1973) has described deserts as water-controlled ecosystems in which total peak biomass and NPP are the lowest of all terrestrial ecosystems. Biomass and productivity can generally be positively correlated with water availability (see Table 12.4). He suggests that the $P:R$ or $dP:dR$ (management production : additional unit of rainfall) might be a more satisfactory parameter than

Table 12.4 Some major characteristics of six of the world's desert areas (from Louw and Seely 1982)

		Rainfall				
Desert areas	*Location*	*Seasonality*	*Annual minimum (mm)*	*Fog*	*Sand dune*	*Succulents*
Namib	Coastal	Summer to winter	10	3	3	2
Peru	Coastal	Summer to winter	10	3	1	1
Sahara	Continental	Summer to winter	10	1	3	1
Central Asia	Continental	Summer to winter	10	0	3	0
Australia	Continental	Summer to winter	100	0	2	0
North America	Continental	Summer to winter	100	1	1	0

The numbers in the three right-hand columns refer to the importance of the characteristic in each desert: 0 = absent; 1 = present; 2 = important; 3 = very important

the mean range of primary productivity by which to characterise and compare ecosystems of varying aridity. The $P:R$ ratio depends on the percentage precipitation transpired (T) and the water-use efficiency ($P:T$) of the vegetation. The equation

$$P = b\,(R - R_t)$$

where P is the productivity, b the production added per unit rainfall, R the rainfall and $R_t = -a$ (productivity at 0)/b at which $P = 0$, defines the threshold at or below which rainfall is insufficient for growth (or, in saline soil, for leaching of the upper soil layers). Normally $P:R$ is lower for summer than winter rainfall, but may be modified by strategies which increase water-use efficiency in summer, or by variations in surface runoff. The productivity : precipitation ($P:R$) ratio varies with life form and the photosynthetic pathway of the particular species. In general $P:R$ is higher in annual than in perennial vegetation except in years or in habitats with extremely spotty rainfall. Some, albeit a minority of, hot desert species are more efficient C_4 water-users, while the cacti which store water belong to the CAM group.

ENERGY FLOW

Energy flow in the desert ecosystem is controlled by water input which comes in short, discontinuous pulses. There may be only 10–50 rain days per year, occurring in 3–15 rainy periods of which no more than 5–6 are biologically significant (Noy-Meir 1973). The biotic response to a sequence of *rain pulses* depends on the time interval between them (see Fig. 12.4). The NPP is then dependent on the amount and duration of the pulse and, particularly in desert annuals, relative productivity (NPP : biomass) can fall well within the range of temperate and tropical ecosystems.

In some deserts nutrient deficiency (particularly nitrogen and/or phosphorus) may become critical either because the mineral component is (as in Australia) nutrient-poor or because the organic litter is unevenly distributed over the ground surface. In addition, the rapid growth of annuals after a rain event rapidly depletes this store of available nutrients while their return in decomposition is relatively slow. There are many similarities between the patterns of primary and secondary production of desert plants and animals. Animal populations also fluctuate annually. Arthropods with high fertility rates and short life cycles are subject to alternating explosive population growth and rapid decline. The fluctuation of mammals, reptiles and birds is less extreme. In some instances variation of the animal population may be directly dependent on rainfall, in others independent of it. In all, however, the factors such as the quantity and quality of food available, competition, predation and inherent demographic characteristics are involved to a greater or lesser extent. For the majority of animals their food is a major source of nutrients and water and the changing availability of vegetation is reflected in feeding habitats.

The impact of herbivores in deserts is comparable to that in other ecosystems, with *c.* 2–10 per cent of the primary production being directly consumed. However, several studies indicated that 90 per cent or more of seed production may be eaten by *granivores* such as ants and rodents. Nevertheless, it is widely considered that, particularly in the more arid deserts, prey–predator relationships are only weakly developed. Because of the dominance of environmental controls, there is insufficient time for biotic interactions to evolve between populations. The water content and availability of moist prey are as, if not more, important than population density in limiting predators (Louw and Seely 1982). In addition, in response to a variable and unpredictable environment and low primary production, both plants and animals are characterised by opportunistic and flexible behaviour which allows them to take advantage of resources when they become available. As a result, most desert animals tend to be generalist feeders.

NUTRIENT CYCLING

Decomposition, like growth, is dependent on soil-water availability. Microbial decomposers are limited. However, apart from dead wood and tough stems most DOM disappears fairly rapidly and a relatively large percentage of the nutrients are mineralised. Two important processes are involved in nutrient cycling: (1) the typical fragmentation, erosion and transport of DOM by wind and runoff, with the finely comminuted material tending to accumulate around the base of shrubs and in gullies or surface depressions, and (2) consumption of DOM by invertebrate *detrivores* (macrodecomposers) such as termites, ants, mites and isopods which are relatively abundant in the desert. In the absence of leaching labile nutrients may accumulate on and in the upper part of the soil during dry periods. If a second rain spell follows shortly after the preceding one little growth is possible. However, as West (1979) points out, most of the critical elements are held either in young plant tissues or in the fertile islands around the large plants where, as a result of slightly lower temperatures and higher humidity, decomposition is lower and DOM accumulates. Overall rates of nutrient cycling in deserts is comparable to that in

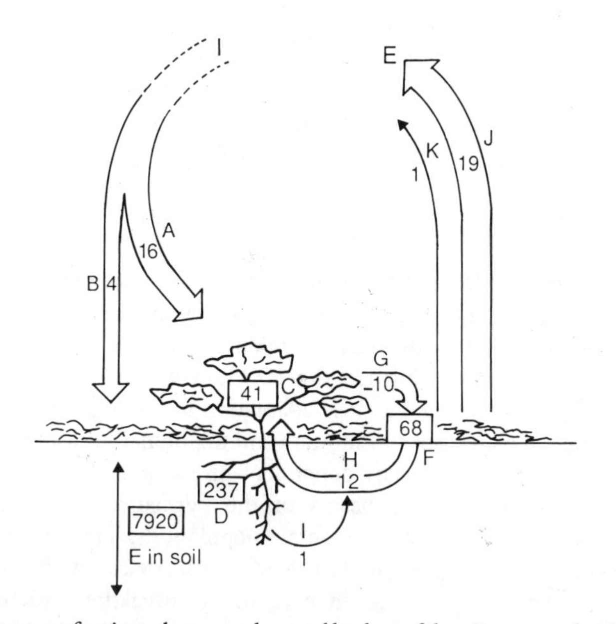

Fig. 12.5 Summary of main pathways and annual budget of the nitrogen cycle in cool desert ecosystems in the western USA under assumed equilibrium conditions and no net erosion. Arrows = annual fluxes; boxes = standing crop (biomass); 1 = import); A = biological fixation mainly by surface cryptograms; B = precipitation and dust; C = above-ground higher plant biomass; D = below-ground higher plant biomass; E = soil up to 90 cm (rooting depth); G = higher plant litter production; F = higher plant litter; H = plant uptake; I = below ground litter; J = export; L = denitrification; M = volatilisation; amounts of nitrogen expressed in kg ha^{-1} (from West and Skujins 1977)

mesic grasslands. As indicated in Fig. 12.5, precipitation accounts for 20 per cent of the annual nitrogen increment. Nitrogen fixation mostly by soil algae and lichens is small, *c.* 5–15 kg ha^{-1}. However, losses by volatalisation are high. In fact nutrient loss can be very high because of uneven distribution. Nutrient pools occur in small tightly structured surface compartments whose recovery rate after use is slow. As a result nutrient cycles are less buffered and are more susceptible to extreme environmental variations than in mesic environments (West 1979).

SUCCESSION

Biomass and DOM are temporally and spatially discontinuous. Hence the effect of the biotic components on the abiotic components of the desert ecosystem is weak. Soils are poorly developed and are dominated by physical processes and the mineral composition of parent material. Unconsolidated sediments are easily transported and sorted by wind and water. Sandstorms and mobile dunes occur to a greater or lesser extent (see Table 12.4). In terms of the traditional concept of vegetation succession, deserts remain for ever youthful. However, despite the extreme short-term variability of the environment, the desert ecosystem is considered, in the long term, to be both stable and resilient. This is because the pulse–reserve nature of primary production is particularly flexible, allowing rapid transformation from an active to an inactive and resistant state in response to environmental fluctuations (Noy-Meir 1973).

HUMAN IMPACT

Archaeological evidence leaves little doubt that humans have lived in the presently desertic areas of the world for many thousands of years, and particularly during the Pleistocene pluvial periods these regions must have provided favourable refuges in face of the advancing ice-sheets. Human survival in the now harsh desert environment necessitated a close adaptation to the natural conditions. The Australian aboriginal collectors and hunters and the nomadic pastoralists of Africa and Asia were directly or indirectly dependent on the limited and variable biotic resouces for food. Only in the basins of the permanent exotic rivers could irrigation agriculture support large sedentary populations. The decline of many of the early riverine civilisations of the deserts of North Africa, the Middle East and Asia have been attributed, at least in part, to the breakdown of the irrigation systems as a result of progressive silting of channels and dams and soil salinisation.

The technological developments of the last 150–200 years have resulted in a rapid exploitation of the deserts throughout the world for agriculture, minerals, urban development and tourism, and for military purposes. The large modern irrigation schemes developed at the end of the nineteenth and during the twentieth century have resulted in widespread soil and land degradation within and beyond the desert areas. Inherent in the process of arid land irrigation is the risk of increasing soil salinity and alkalinity to levels which reduce primary productivity and eventually render the soil virtually sterile for all but a few specialised life forms. Increased salinity is most pronounced where irrigation raises the water-table to near the ground surface where, consequent on high evapotranspiration, mineral salts (sulphates and chlorides of sodium, calcium and magnesium) are precipitated from the water. Attempts to reduce salinity by lowering the water table and flushing the soil unfortunately all too often create even worse conditions. If draining and leaching alter the calcium–sodium balance to the extent that the percentage of sodium ions exceeds 15 per cent soils become very alkaline (pH 10+). The consequent dispersion of clay particles and the breakdown of organic matter result in the loss of soil structure and permeability and the development of intractable black alkali soils.

Other knock-on effects of desert irrigation are the increase of insect populations to pest levels consequent on the availability of a constant food supply and a continuous rather than a short discontinuous cycle of reproduction. Despite the increasing use of biological methods of pest control the impact of previously and currently used pesticides on the above- and below-ground desert biota is still largely a matter of conjecture. In addition, extraction of surface and ground water for irrigation at a rate greater than it can be replenished continues to cause, at an ever-increasing rate, a lowering of the level of the water-table to depths beyond the reach of phraeophytes. It has been accompanied by land subsidence, and in coastal areas the incursion of brackish marine water. A particularly spectacular example of the indirect impact of irrigation is afforded by the recent salinisation and desiccation of the inland Aral Sea in the former USSR (Micklin 1988) as a result mainly of the excessive withdrawal of water for irrigation from the lake's main feeders – the Amu Darya and Syr Darya rivers. Between 1960 and 1987 the level of the Aral Sea dropped 13 m, its area declined by 40 per cent and dry salinised bottom sediments were exposed. The reduced water body has become more saline, its biota has been increasingly impoverished and its biological productivity drastically reduced.

Recently, desert landforms, landscapes and their associated vegetation, and archaeological sites have begun to attract an increasing number of visitors to particular locations, erosion and the loss of areas to communications, tourist settlements and associated facilities have followed. There are already indications that some of the more vulnerable exotic desert species are under threat. For instance, the relatively limited population of the giant arborescent saguaro (*Carnegia gigantea*) of Arizona and the Sonora Desert (Mexico), the tallest of all cacti reaching *c.* 15 m in height, is being depleted to enhance the urban gardens of California.

Finally exploitation of oil and gas reserves, and of uranium and other potentially toxic heavy metals, together with the use of deserts throughout the world for the testing and storage of nuclear weapons, continue to create point sources of air, water and soil pollution which can extend well beyond the present desert limits.

Chapter

13

Island ecosystems

The common denominator of all islands is isolation. It is the basis of their fundamental human attraction and of their biogeographical significance. Islands are, in general terms, areas of land isolated from others of similar character and relatively inaccessible by reason of the extent and character of the surrounding area. The biological significance of islands is related to their contribution to an understanding of the processes of organic dispersal and evolution and to species variability. Within the last 25 years evolutionary biogeography and ecology have been stimulated and revolutionised by the theoretical approach to island biogeography initiated by MacArthur and Wilson (1967). Also, with sharply defined boundaries, islands have provided natural laboratories for the study of ecosystem processes.

Islands vary widely in location, size, physical environment, biotic characteristics and degree of isolation. Although a distinction can be made between aquatic and terrestrial islands, the most common perception of an island is that of an area of land surrounded by water. And until recently the focus of biogeographical inquiry has been the marine island. Despite their range of diversity all share a common 'barrier' of salt water, are limited in area in comparison to adjacent land masses (though the upper limit between a marine island and a 'continental' island such as Australia) is vague, and are subject to an oceanic climate which has a lower temperature range, higher humidity and wind-speed than land areas at a similar latitude. An early biogeographically significant distinction, attributed to Wallace, was made between *oceanic* and *continental* islands. The former are remote, very isolated and surrounded by deep water. They are geologically young, formed of either volcanic extrusions and associated coral reefs and atolls or in a few cases large detached fragments of continental rock. Recent studies of continental drift indicate that these volcanic islands formed on the flanks of mid-ocean subterranean ridges and few pre-date the mid-Tertiary period. They have, therefore, never been joined to a large land area and they can only have been colonised by transoceanic dispersion. In contrast, continental islands are detached portions of the adjacent land mass to which they were connected more than once in the recent past. They are generally located on the continental shelf and are surrounded by comparatively shallow water. Their biota are derived from and are usually very similar to that of the land area. Human settlement has been later, particularly on the oceanic islands, than on the mainland. The relatively sudden breakdown of their isolation has had serious disruptive effects on their flora and fauna.

OCEANIC ISLANDS

Although the overall number of species per unit area can be similar or even greater than that of comparably sized continental areas, island flora and fauna are comparatively impoverished in species and genera. This is particularly evident in the absence of mammals (other than bats), amphibians and freshwater fish (Williamson 1981). Few non-marine animal or plant species can survive submergence in salt water for any considerable length of time. The existing indigenous island species are those with an originally high dispersal capability (Table 13.1) by wind, by floating, by flying (or as a result of transport by fliers) or,

Table 13.1 The methods by which vascular plants have reached various islands (from Gorman 1979)

	Per cent immigration by:					
Island group	*Ocean drift*	*Birds: sticky seeds*	*Birds: mud*	*Birds: internal*	*Birds: hooks, etc.*	*Air flotation*
Hawaii	11.8	10.3	12.8	38.9	12.8	1.4
Galapagos	23.1	8.5	13.7	27.7	22.8	4.3
African Alps	—	6.5	51	0	6.5	35

in the case of animals, by swimming; or are autochthonously evolved descendants of the long-distance colonists or have been introduced deliberately or accidentally by human agency (Pielou 1979). They have been able to establish and maintain themselves in the habitats available and in company with each other. In the early stage of colonisation on new islands (e.g. Surtsey Island and Long Island (New Guinea)) large populations of some species of birds have been noted. Those called 'supertramps' (Diamond 1974) are characterised by their high dispersal capacity across open water and their rapid reproductive capacity. They produce a large number of surplus individuals who become potential immigrants and colonists of other empty or underpopulated islands. In time the original population of supertramps will be displaced by those of species with a lower dispersal capacity but a higher competitive ability.

The relative importance of isolation and competition in determining the number

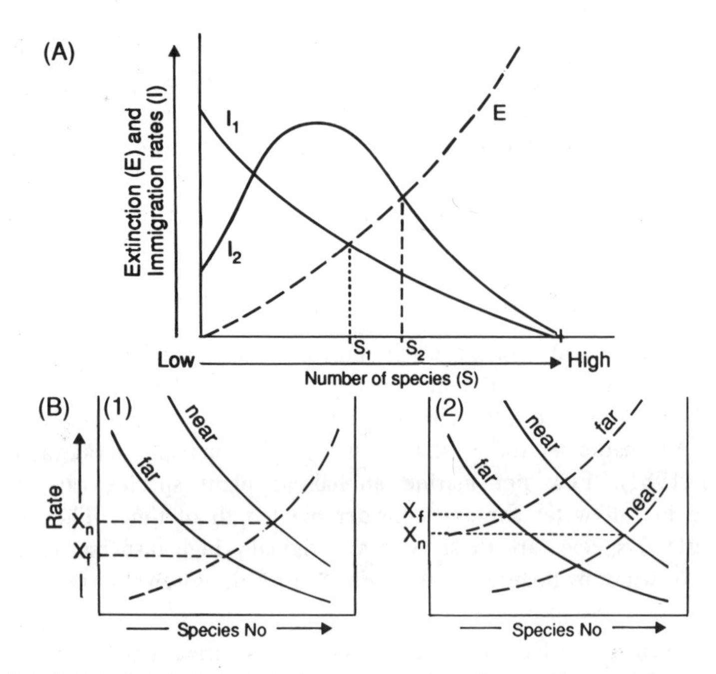

Fig. 13.1 (A) Model of relationships between number of *species* (S) on an island and rate of *extinction* (E) and *immigration* (I). When $I_1 = G, S = S_1$, i.e. the equilibrium value of the number of species in the mainland pool (P); in the case of some organisms (e.g. plants) I_2 is likely to increase when the number of species is low, then $I_2 = I, S = S_2$ (redrawn from Pielou 1979); (B) Modification of model in (A) to show that on islands of similar area G is greater on *far* (f) than on *near* (n) islands because of a 'rescue' effect, and the difference in species turnover rates on near and far islands: (1) simple model with E independent of distance $-X_n > X_f$; (2) modified model with E dependent on distance $-X_n < X_f$ (from Pielou 1979, adapted from Brown and Kodric-Brown 1977)

of island species has long been a subject of ecological debate. MacArthur and Wilson (1967) developed an *equilibrium theory* designed to explain and predict the number of species as a function of a dynamic balance between rates of immigration and extinction (Fig. 13.1(A)). In its simplest expression the theory assumes that the number of species is determined by three factors: (1) the number of species in the mainland source region; (2) the size (area) of the island; and (3) the distance of the island from the mainland. The equilibrium number of species, however, would be expected to increase on a larger island at the same distance from the mainland species pool; and to decrease on an island of a given size with increasing distance from the species pool. As illustrated in Fig. 13.1(B), the larger island would have a lower extinction rate than the smaller one; while the more remote island would have a lower immigration rate. However, although species numbers attain an approximately constant mean level, the composition of the flora and fauna would be expected to change continuously. This is because the loss of some species will be made good by a gain of an equal number of others. The resulting *replacement* or *equilibrium turnover rate* may be large or small dependent on the equilibrium level of the island in question. The average immigration rate (I) of new species per species in the mainland source pool (P) given the number (S) of island species present can be predicted from the equation

$$\frac{I}{P - S} \quad \text{or} \quad I = (P - S)$$

the average extinction rate per species present (E/S); and the rate at which the number of species on the island increases with time:

$$\frac{dS}{dt} = I - E \quad \text{or} \quad \frac{dS}{dt} = (P - S) - S = 0 \text{ at equilibrium}$$

The validity and predictive value of the equilibrium model have been supported by empirical studies (Diamond and May 1976) and by field experiments (Simberloff and Wilson 1969, 1970) involving certain species groups, particularly birds and insects on tropical islands. However, the equilibrium theory has been criticised on several grounds, not least that of its simplicity. It treats all species together with the implicit assumption that they are equal in numbers and constant for a given island (Williamson 1981). Also, it assumes that species number and emigration rate are dependent on island area and are independent of immigration. However, as Pielou (1979) points out, large islands are more likely than small ones to be in the path of migrants. Also, immigration and emigration are not necessarily independent. Particularly on small islands the species may be 'rescued' temporarily from extinction as a result of population enlargement consequent on an influx of immigrants. Also, the theory does not take environmental, biotic or historical factors into account. The relationship between species numbers and island area is less close on temperate islands, where temperature is a more important limiting factor than on tropical islands where the number of vascular species is related to the size of the island and the mean temperature of the coldest month. Also, islands with a large variety of physical habitats have the potential to support a greater number of species than a smaller one.

As a result of empirical studies of animal species on tropical mangrove islands, Simberloff (1969) concluded that a distinction could be made between a temporary equilibrium (when species were non-interactive) followed by a decline in species numbers to a permanent equilibrium as competitive exclusion started to operate and to result in the extinction of those poorly adapted and the immigration of those well adapted to the biophysical environment. This sorting process, it is maintained, would give rise to a 'set of co-adapted' species which was not just a random sample of the original species pool but which was characterised by a constant trophic structure related to the location and ecology of the particular island. Indeed, there are many workers who would maintain that ecological diversity was more important than distance in determining the size of an island's native biota (Wace 1978).

COMPETITION AND NICHE SHIFT

Evidence for the importance of competition in the organisation (structuring) of vertebrate communities is the *niche shift* (expansion or ecological release) characteristic of many island species (Gorman 1979). Island immigrants can spread into habitats not inhabited by their mainland parents and occupy niches which on the mainland would have been occupied by other species (Ricklefs 1973). Niche shift describes the response of an island species (or ecotype) in terms of its feeding, breeding and biotic adaptations to a habitat which differs from that of the area of origin. Lack (1943) demonstrated this phenomenon for island birds in Britain (Table 13.2) and Williamson (1981) quotes the example of the red

Table 13.2 Lack's examples of niche shifts in Orkney birds (from Williamson 1981)

Species	*Normal mode*	*Orkney mode of occurrence*
Fulmarus glacialis (Fulmar)	Nests on cliffs	North Ronaldsay, Sanday Nests on flat ground* and sand dunes
Columba palumbus (Wood pigeon)	Nests in trees	Orkney mainland, Rousay Nests in heather (*Calluna*)
Turdus philomelos (Song thrush)	Nests in bushes and trees	Westray, Papa Westray Nests in walls and ditches*
T. merula (Blackbird)	Habitat: woods plus bushy places	Most islands, rocky and wet moorland
Anthus spinoletta (Rock pipit)	Sea cliffs	Papa Westray Out of sight of sea*
Acanthis cannabina (Linnet)	Bushes and scrub	Sanday, Stronsay, Westray Cultivated land without bushes, reedy marshes

The normal mode is also found in Orkney in all cases.
* Also on other northern islands of the British Isles.

squirrel (*Sciurus* var. *leucenrus*) and the pine marten (*Martes martes*). The red squirrel is a subspecies of the more widespread palaearctic *S. vulgaris*. With the decline of its natural habitat in coniferous forests it had spread to deciduous woods. Its present decline in the latter in face of 'apparent competition' (a hotly disputed point) with the grey squirrel (*S. carolensis*) has been attributed to the fact that it is better adapted to coniferous woods. The pine marten, also characteristic of both coniferous and deciduous woods, expanded its niche on to rocky areas following widespread destruction of its natural habitat in Wales.

ISLAND EVOLUTION

The original equilibrium theory assumed that all island species originated by immigration and did not allow for evolutionary change. However, this is true only for continental islands which were probably oversaturated with species at the time of their separation from the mainland. Fossil evidence supports the view that, on separation, species emigration would exceed immigration and numbers in those islands would fall until an equilibrium level dependent on the size of the island and its distance from the mainland was attained.

The biota of oceanic islands, however, is characterised by a high percentage of *endemic species* (Table 13.3) most of which are the autochthonously evolved

Table 13.3 Percentage endemic flowering plants in selected areas (from Carlquist 1974)

	Number of species	*Per cent endemic*
Continental areas		
California	5529	38.0
West Virginia	2040	0
Continental islands		
British Isles	1666	0
Old continental island		
New Caledonia	2600	(90)?
Oceanic islands		
Canary	826	53.3
Fernando Po	826	12.0
Galapagos	386	40.9
Hawaii	1721	94.4
Juan Fernandez	146	66.7
St Helena	45	88.9
Island-like areas		
Afro-alpine flora	279	?
South-western Australia	3886	(90)?

It should be noted that size of flora and degree of endemism vary with size of area, degree of isolation, physical conditions of the area and adjacent source area: endemism does not correlate with species diversity.

descendants of long-distance migrants. Separated from the parent populations they have evolved independently of them in response to the variety of new habitats available, i.e. by *adaptative radiation*. The number of endemics in a particular taxonomic group and on an island varies. The number of endemic species will tend to be more common in those groups which have, or have acquired, a low dispersal capacity, e.g. wingless insects, flightless birds and seeds without aids for wind dispersal (Carlquist 1974; Gorman 1979). It will also depend on the age, size and degree of isolation of the island. On large islands with a varied range of habitats, the probability that a resident population will be of a size and persistence to undergo adaptive radiation will be higher than on a smaller island (Pielou 1979). The island must be large enough and with sufficient habitat variation to provide effective genetic isolation. The minimum area required will obviously vary between taxonomic groups. It must also be remote enough to result in a sufficiently slow immigration of colonists to ensure that speciation can occur before all the available niches are filled and/or adaptive radiation by one species blocks that of another with similar ecological requirements. Radiation, however, is limited or indeed absent in some taxonomic groups and on some remote islands. Gorman (1979) notes that certain animals at the limit of their dispersal range such as frogs (in New Zealand and Fiji) have not radiated. Even on large islands such as Hawaii, with a high degree of endemism, there are successful colonists with a small number of species.

Adaptive radiation will tend to result in decreasing population size and increasing niche specialisation as exemplified by Darwin's famous finches with over 40 species on Galapagos and the Hawaiian honey-creepers and lobeliad plant group, with 6 endemic genera containing over 150 species (Gorman 1979). With decreasing population size the probability of the species becoming extinct increases as a result of environmental changes or predation. This speculative evolutionary sequence or *taxon cycle*, shown in Fig. 13.2, was illustrated on the

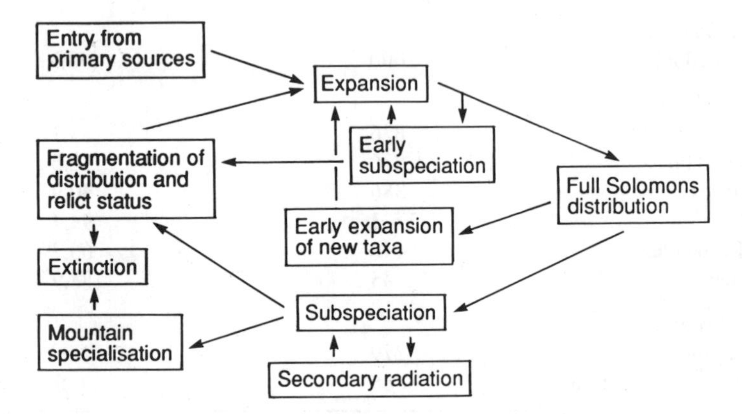

Fig. 13.2 Model of the taxon cycle (from Ricklefs and Cox 1972)

Greater and Lesser Antilles archipelago by Ricklefs and Cox (1972). They classified the bird species in terms of geographic distribution into groups which were interpreted as reflecting the main stages in the taxon cycle (Table 13.4), i.e.

(I) Undifferentiated species spreading rapidly on some or all islands;
(II) Species extinct on a few small islands; clear species population differentiation on larger islands;
(III) Species populations fragmented into separate species or subspecies occupying reduced areas;
(IV) Descendant species populations endemic only on one island; have become relict endemics.

The taxon cycle has been explained in different ways. Ricklefs and Cox (1972) maintain that competition is the main factor 'driving' the cycle in that the initial competitive advantage of a particular species declines as predators (including parasites) and more competitive species evolve. Others (e.g. Diamond and May 1976) argue that the blanketing invasion of an archipelago by mainland bird

Table 13.4 The number of endangered or extinct West Indian species of birds as a function of the taxon cycle (from Ricklefs and Cox 1972)

	Stages of taxon cycle			
	I	*II*	*III*	*IV*
Total number of island populations	428	289	229	57
Number of populations in danger or extinct since 1850	0	8	12	13
Percentage	0	3	5	23

species and subsequent immigrants of the same species could be the result of involuntary long-distance transport by strong winds (Pielou 1979).

ISLAND ECOSYSTEMS

The ecosystems of oceanic islands differ from those in comparable continental physical environments in (1) the impoverishment of their flora and fauna and of their life forms; (2) the relative simplicity of their trophic structure; and (3) the influence of the sea on their energy flows and nutrient cycling. The poverty of the biota (particularly the lack of mammals) of oceanic islands has already been noted. There is also a paucity of the more specialised life forms such as epiphytes, saprophytes and semi-parasites of flowering plants (Carlquist 1974). As a result insular food chains tend to be shorter and simpler than in similar continental habitats. Many niches may be unoccupied because immigrant species have been selected more for dispersability or reproduction than for competitive ability, and in the higher trophic levels consumers tend to be generalist feeders (Wace 1978).

The interaction between oceanic island ecosystems and the surrounding marine ecosystem is of considerable importance because of the high shoreline to area ratio. There is a constant and relatively large transfer of nutrients and energy from the sea to the island and vice versa. Small islands, in particular, can be compared to 'high-energy' coastal environments in terms of exposure to salt spray, high wind and wave force and to tidal variations of sea level. Wind and wave energy effect the import to, and export from, the island of organic energy and of nutrients. Also, the nutrient cycle is considerably influenced by marine life – particularly birds and mammals (seals) and reptiles (turtles). The latter are all carnivores which occupy the top marine trophic level. While they do not feed on the island, it provides their breeding and resting habitat. They effect, via faeces and corpses, an important nutrient input. The amount of dried nitrogen plus excrement – *guano* – produced by birds and reptiles and of rock phosphate deposits, particularly on tropical islands, testify to the importance of the nutrient input. In addition, the coral reefs and atolls which develop in association with oceanic islands in tropical areas represent the accumulation of calcium and other nutrient transfers from the sea to the land.

However, although nutrient input is high, losses from the island to the sea are also large. This is because the size of the land area is small in relation to the energy-exporting agents of wind, wave and surface runoff of precipitation. Also roosting/breeding colonies of the animals previously referred to are often of exceptionally high density and their movement from these to feeding grounds maintain a part or the whole of the island habitat in a disturbed condition. Although the nutrient budget must vary from one island to another dependent on its location and size, it is considered that in general, because the islands are being continuously eroded away by the sea, nutrient transfer from the island to the sea will, in the long term, be greater than those from the sea to the land. And this negative balance has further been exacerbated by the impact of human activities.

ISLAND BIOGEOGRAPHY THEORY AND CONSERVATION

Since its publication in the mid-1960s the island biogeography theory has stimulated two closely interrelated lines of inquiry. One is its relevance for islands other than marine ones, i.e. for the land-bounded *isolates* or *habitat islands*. The other is its potential application to the design and management of nature reserves (Kent 1987). The latter, in particular, has elicited a flood of publications which, though now somewhat diminished, still continues. Most have tended to concentrate on the implications of the theory for nature reserve design without always having due concern for the validity of the theory, the predictions which could be based on it and/or for the management the problems involved.

In his recent and most comprehensive review and critique of island biogeography theory, Shafer (1990) notes that while the theory has provided a conceptual framework for the consideration of a number of related ideas, its practical applications for conservation are, in fact, fairly limited. As has been previously noted, the theory is a simple one, based as it is on the relationship between species numbers and the size and relative isolation of marine islands; and its corner-stone – the equilibrium theory – has not been convincingly substantiated. Of the predictions which have been based on it, i.e. that:

1. The number of species increases with size of area;
2. Species richness decreases with distance from source;
3. The number of species will fluctuate;
4. The turnover rate will depend on the balance between extinction and immigration;
5. Colonisation will be non-random.

Shafer contends that only the first is unarguable.

SIZE OF AREA

All other things being equal, the larger the area delimited of a particular ecosystem the greater will be the number of species contained within it, up to a point. Studies of minimum areas required to sample ecosystems indicate that species number increases up to a point at which it levels off, becoming very slow or ceases. However, given the rate of species and habitat reduction currently taking place, the larger the reserve the better. Since it will be less vulnerable to external impacts. It is more likely to contain viable populations of a greater number of taxa, as well as representatives of the higher trophic levels. Ultimately the optimum and maximum size needed will depend on the particular size of the organisms within a taxon or group of associated taxa, their relative abundance, evenness of distribution and their ecological requirements. The capability of existing reserves to be increased is negligible, while the size of future (as of existing) reserves will inevitably be limited by other than ecological constraints affecting species numbers. Many workers have found a closer correlation with habitat diversity and latitude or a combination of area, habitat diversity and latitude. Habitat diversity appears to be a more important factor than size on the Galapagos Islands. Many relatively small high-altitude volcanic islands have species diversity high in proportion to size because of their diverse range of climatic zones from tropical forest to alpine, as on Tenerife (Canaries).

Size, however, is not the only variable affecting SLOSS – single large or several small. A design problem related to that of reserve area is whether SLOSS reserves of equal area will contain more species. This question has spawned as many diverse opinions as there are ecologists and conservationists combined. They span a continuum from those in favour of one or another to those who consider it of little practical value for conservation (Murphy 1989). Many advocates of the small-sized reserves point to the increased species richness as a result of the longer combined edge than in the single large reserve. However, again, as Helliwell (1976a, b) notes, much depends on the size of the species and the nature of its habitat.

SHAPE OF AREA

Many ecologists have noted the superiority of a circular over a linear shape; the latter certainly gives the maximum area to boundary ratio. It has also been argued that a long thin reserve is likely to suffer from the so-called 'peninsula effect' whereby the number of species decline with distance along an actual peninsula away from the mainland source area. There is, however, no evidence that the latter will contain more species than the former. Nevertheless, as in the case of the SLOSS debate, the advantages of a longer *edge-effect* associated with the linear-shaped reserve is considered an advantage for the management of large game animals (e.g. deer) for whom the ecotone between the complete shelter of a closed wood and the open potential feeding ground is particularly important.

SPACING OF AREAS

Much consideration has also been given not only to the size but to the spacing of reserves and the degree to which they should be aggregated or clustered, as well as to the advantages and disadvantages of connecting corridors between them. As important as the spatial patterns of the reserves are their relationships to the matrix of other unmanaged and managed habitats within which they are set and their local, regional and continental significance.

CRITIQUE OF ISLAND BIOGEOGRAPHY THEORY

During the 1970's island biogeography theory was hailed as an intellectual breakthrough in a discipline in which theories are notoriously difficult not only to formulate but, more particularly, to validate. There was a growing conviction – indeed belief – that island biogeography could facilitate the prediction of the number of species, and the rate of extinction, in reserves of a given size (Shafer 1990). It was seen as the key to conservation management and decision-making problems. Various reserve design guide-lines reputedly derived from island theory have been proposed (Diamond 1975; Wilson and Willis 1975) and adopted without reservation and qualification by the International Union for Conservation of Nature (IUCN) 1980 World Conservation Strategy. Further, it is now embodied as an established principle in many reputable ecological, conservation and planning textbooks!

There is now a growing realisation that the theory related to marine islands is not necessarily applicable to the terrestrial habitat islands (or isolates) of which nature reserves are a particular example. And even its early advocates such as Simberloff and Abele (1976) have more recently seriously questioned the application of a limited and insufficiently validated theory to conservation problems.

FRAGMENTATION AND INSULARISATION

However, even if the theory asks more questions than it can at present answer, it has served to stimulate debate about the relationships between ecological data and conservation needs (which will be discussed more fully in the final chapter) and to highlight the processes of fragmentation and insularisation of the unmanaged habitats and of the landscape as a whole. Since 1950, unmanaged habitats, e.g. forest, woodland, heathland, grassland and wetland, throughout the world are being reduced at an ever-increasing rate either by direct exploitation or gradual encroachment of other land uses on formerly more extensive habitats. Shafer quotes two striking examples – that of the fragmentation of forest in a township in Wisconsin into habitat islands during the period of European settlement. The 55 small forest islands remaining in 1950 had by 1978 been fragmented into 111 forest islands with an average size of 0.009 km^2. A more recent study (Webb and Le Haskins 1980) showed that the area of heathland in the Poole Basin (Dorset) was reduced by 86 per cent; and by 1978 the 10 original blocks containing *c.* 40 000 ha had been reduced to 160 fragments with an average area of 4 ha or more, 608 less than 4 ha, 476 of which were less than 1 ha. As a result not only have the remaining natural areas been reduced to remnant islands in an otherwise managed matrix but, because of the erosion, existing reserves are becoming more insularised and isolated from these islands outwith its boundaries. In these cases conservation must inevitably be more concerned with the management problems raised than reserve design.

HUMAN IMPACT ON ISLANDS

Isolation has protected islands from human impact for a longer time than many other areas. Once broken, however, the depletion of island biota and the breakdown of the island ecosystems have been devastatingly rapid. Indigenous Polynesian populations were early established on the large Pacific islands from Hawaii and Fiji to Samoa. It was, however, not until the circumnavigation of the world and exploration by large sailing ships in the sixteenth and seventeenth centuries that the long isolation of the oceanic islands was broken. They became points from which supplies of water and food were replenished and later, particularly in the nineteenth century, where permanent or semi-permanent whaling stations were established.

As a result of these early contacts exotic (alien) species were accidentally (by escape) or deliberately introduced. Rats (which early attained pest status), cats and dogs are examples of the former. In the latter case pigs, goats, sheep and cattle, deer, rabbits and donkeys were brought in to provide food supplies for transoceanic ships and whalers. Many of the native animals lacked the ability to survive in face of the mammal predators, to compete successfully with large populations of domestic herbivores or to resist the depredation of disease. Some native populations declined drastically in number or became extinct. Among the latter the flightless land birds such as land rails, moas, owls and eagles which filled previously unoccupied mammal niches on islands were particularly vulnerable and suffered a rapid early decimation and extinction. It is estimated that of the ninety original species of land birds in New Zealand 43 per cent are now extinct (Towns and Atkinson 1991). Their 'natural' niches are now occupied by feral ancestors.

The impact on native island floras was no less severe than that on the fauna. Reduction or extinction of plant as of animal species was most marked in those with small populations, restricted habitats, narrow and specialised niches. Many failed to compete with more vigorous immigrants, or to survive the increased incidence and intensity of fire and grazing by domesticated herbivores or the loss of specialised animal pollinators essential for reproduction. On the larger oceanic islands such as Hawaii, Fiji and New Zealand the replacement of native vegetation by exotic forest plantations and large-scale intensive commercial agriculture has been accompanied by an increase in both native and exotic weeds and pests and in the use of herbicides and pesticides to control them.

Today the source and nature of human activities on island ecosystems may have changed but the effect of the initial impact continues. The twentieth-century use of remote oceanic islands for strategic purposes, for research (biological, environmental and defence) and tourism continues to maintain the stresses on, and endanger the survival of, the remaining small populations of rare endemic island species.

Chapter

14

Mountain ecosystems

The island concept can be applied to terrestrial as well as to marine ecosystems. Terrestrial or 'continental habitat' islands (McArthur and Wilson 1967) are those relatively small areas of land or water surrounded by, and isolated from, larger ecologically similar areas by an extensive area of dissimilar habitat. The type and degree of isolation of a terrestrial island, however, is not always so clear-cut as in the case of the marine island, nor can the effectiveness of the terrestrial barrier be assessed as readily as that of the sea. Among the most clearly delimited terrestrial islands are those where wet or aquatic habitats (e.g. bogs, swamps, lakes/ponds, oases) are surrounded by dry or arid conditions, or highlands (mountains) are surrounded by lowlands which are sufficiently different in physical character and extensive enough to make migration across them impossible or extremely difficult for the majority of species. At the global scale the mountain island holds pride of place.

MOUNTAIN CHARACTERISTICS

While mountains owe their identity to their height above sea level there is no universally accepted base altitude, i.e. that above which environmental conditions are inevitably different from those at lower levels. Mountains are more clearly distinguished from their surrounding lowlands by their interzonal character. Rapid change in climatic conditions with altitude is reflected in an altitudinal zonation of the biota which, up to a point, appears to replicate that from the equator to the poles (see Fig. 14.1). The mountain world provides a reduced model or a microcosm of a large part of the total world environment. The ecological importance, if not dominance, of relief is greater in mountains than in any other habitat. Variation in altitude, slope and aspect results in a diversity of sharply contrasting meso- and microhabitats and a biotic variability within mountains that far exceeds that in the surrounding lowlands. At the global scale, mountains combine the roles of a 'pioneer front' and a 'museum' (Rougierr 1962). The

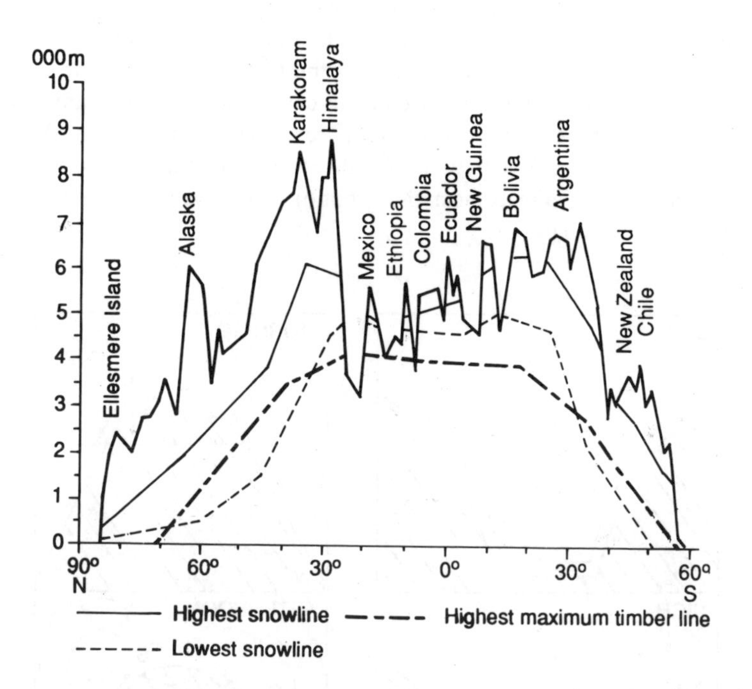

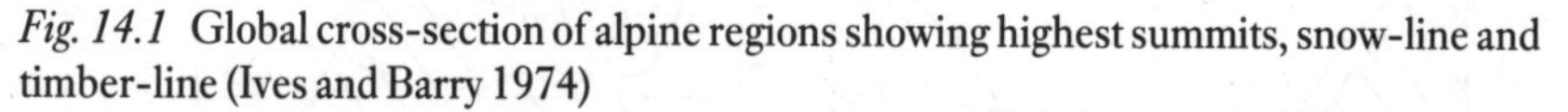

Fig. 14.1 Global cross-section of alpine regions showing highest summits, snow-line and timber-line (Ives and Barry 1974)

former is a function of the severity of environmental conditions which at high altitude eventually become limiting for life. In the latter, mountains and particularly isolated mountain tops are characterised by a high proportion of endemic species many of which are relicts of a formerly more widespread distribution. Finally, although mountains have long provided refugia for people as well as for plants and animals, and despite their recent rapid exploitation, they retain some of the most extensive and still relatively 'natural' wilderness habitats in the world today.

BIOCLIMATIC ZONES

Increase in altitude is accompanied by a general deterioration of environmental conditions. The proportion of direct to diffused solar radiation and of ultraviolet radiation (UVR) increases. Ambient and surface temperatures decrease and the length and warmth of the growing season decrease. In temperate latitudes cloudiness and precipitation increase and an increasingly higher proportion falls as snow while the depth and persistence of the snow-cover increases. Wind-speed also increases with altitude. The resulting environmental gradient is reflected in three distinct and usually clearly delimited *bioclimatic zones* as illustrated in Fig. 14.2: (1) the montane zone; (2) the alpine zone; and (3) the zone of permanent snow and ice. The *tree-line* or upper limit of the montane zone and the permanent *snow-line* or lower limit of permanent snow, glaciers and ice-caps increase in altitude from near sea-level towards the equator. However, at any one place, the actual location of these boundary lines and the biota of the montane and alpine zones are dependent on the rate of climatic change with altitude and relief (i.e. slope (or gradient) aspect and relative exposure of the ground surface). Mountains are characterised by such a tremendous diversity of *topoclimates* (local climates) and *microclimates* (i.e. climate near the ground) as to invalidate the concept of a distinct mountain climate.

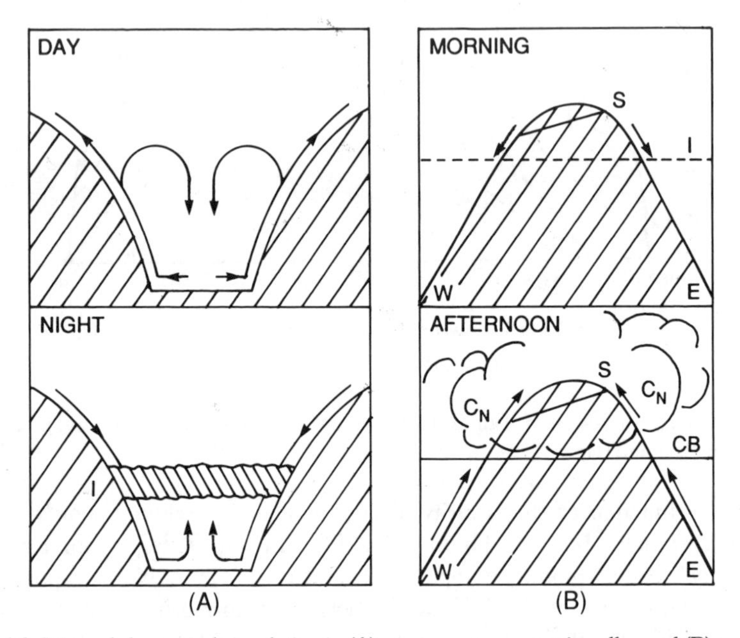

Fig. 14.2 Diurnal slope wind circulation in (A) a temperate mountain valley and (B) on an isolated tropical mountain. S = snow line; I = inversion of temperature; CN = cummulo-nimbus clouds: cloud base (based on Flohn 1969)

Topoclimate

The slope of the ground in relation to the direction and incidence of solar radiation is undoubtedly the major topoclimatic variable. In comparison to level surfaces, the heat and associated water balance on sloping ground at a similar altitude is modified because of the resulting differences in the angle of incidence of solar radiation. The astronomically possible duration and total of daily solar radiation is modified by the screening of the horizon in mountains. Relatively gentle slopes can have a marked effect on the reduction of direct radiation. Also slopes at differing altitudes can initiate secondary (local) patterns of air circulation which further influence the heat and water balance. A measure of the comparative difference between the topoclimate of sloping compared to horizontal surfaces is provided by the concept of *equivalent slopes* (Geiger 1969). The equivalent slope is expressed in terms of the number of degrees of latitude, either north or south, of a given slope where a level surface at the same altitude will have a similar topoclimate. Consequent on aspect, equatorward-facing slopes experience a sunnier, warmer and drier climate than those shady, cooler and more humid poleward-facing slopes (Table 14.1). These climatic differences are accompanied

Table 14.1 Soil moisture as a percentage dry weight of ground at 5–10 cm on the Grosser Stanfenberg (Harz Mts), 17 September 1953 (from Geiger 1969)

	Slope direction			
Height above sea-level (m)	*N*	*ENE*	*S*	*W*
500–540	22	21	14	—
450–499	25	20	14, 15	—
400–449	—	25	14	—
350–399	26	27, 30	21	18

by those in the soil climate and associated biota. In deep, steep-sided alpine valleys with a latitudinal orientation, the tree-line and the transient snow-line attain higher altitudes than on the shaded slopes. Also, slopes with a westerly (in the northern hemisphere, easterly in the southern) aspect have higher ground temperatures than those facing east (or west in southern hemisphere). In the latter a higher proportion of the morning solar energy is expended in evaporation from humid soil, while the latter drier slopes receive intense radiation in the afternoon. Maximum ground temperatures are displaced towards the west and south-west (northern hemisphere).

Periodic variation in the radiation and heat budgets of slopes influences the air circulation and results in characteristic valley climates. As shown in Fig. 14.3, during the day and particularly during the summer, the warmer surface layer of air in the valley bottoms moves upslope. At night, cold air tends to move, albeit very slowly, from the colder summits downslope. The steeper the slope the more rapid the flow which, in mountains, often occurs in periodic 'surges' or 'avalanches' of cold air. The coldest air collects in depressions and valley bottoms where, in the absence of other air movements, temperatures drop below freezing and a sharp *temperature inversion* forms between the cold valley bottom air and a warmer *thermal belt* above (Table 14.2). In the latter the risk of late spring and early autumn frosts is greatly reduced, and the growing season is longer and warmer than on the valley slopes above or below. In tropical and subtropical regions local mountain and valley winds can reach gale-force intensity, particularly when they are funnelled in deep gorges, such as those some 5000 m deep which extend from the Altiplano of Brazil to the Amazon lowlands (Flohn 1969). Above the mountain summits local upslope winds are horizontally integrated into reverse anti-wind systems of varying scales. Over large mountain masses they give rise to large relief-induced and thermally driven mountain systems which at particular times of the year can dominate the weather pattern.

Mountains also exert a passive influence on air movement particularly during

Table 14.2 Observations in the thermal belt on the Grosser Falkenstein Harz Mts (from Geiger 1969)

	Precipitation				*Air temperature*					*Phenology, increase of length of shoots of plants 5–6 years old (mean, 1956–57)*	
		Freshly fallen snow				*Mean temperature May–Oct. 1955*					
Height of station above sea-level (m)	*May–Oct. 1955 (mm)*	*Total (cm)*	*% of melting snow*	*Daily temperature range in may (°C)*	*Absolute min. (°C)*	*Mean min. (°C)*	*Mean (°C)*	*Mean max. (°C)*	*Length of frost-free period (days)*	*Spruce (cm)*	*Beech (cm)*
1307 (peak 1312)	749	(117)	(9)	3.4	−25.2	5.5	8.2	12.1	(118)	—	—
1157	591	131	10	4.5	−20.1	6.9	9.1	12.3	144	15	9
1008	646	127	43	4.7	−19.4	7.6	10.0	13.1	156	12	11
925	651	132	49	5.5	−16.6	8.0	10.5	13.9	157	16	13
850	640	120	46	6.0	−18.5	8.1	11.0	14.6	159	18	15
796	625	119	41	6.3	−19.1	7.9	11.1	15.1	158	22	12
658	619	92	65	9.1	−23.6	6.4	10.8	15.7	109	11	8
622 (valley)	608	96	55	11.7	−28.1	4.3	10.6	16.8	97	6	6

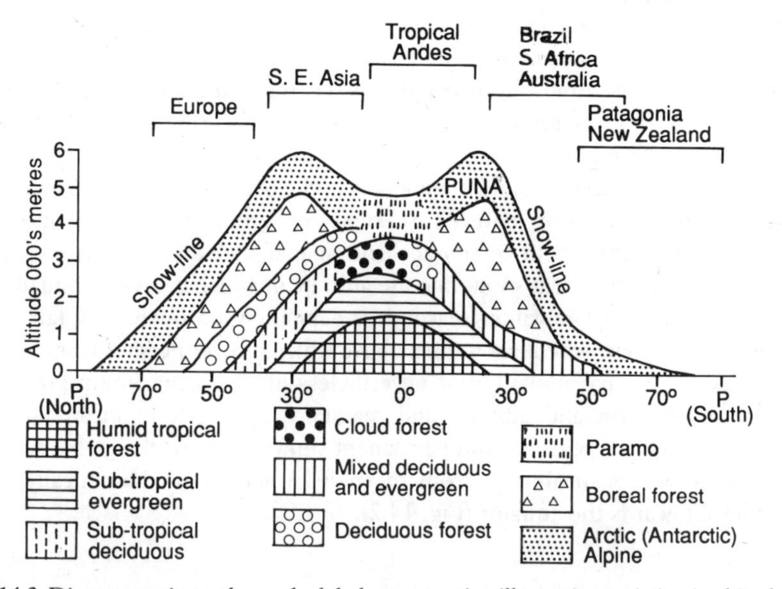

Fig. 14.3 Diagrammatic north-south global cross-section illustrating variation in altitude of main vegetation zones (redrawn from Rougierr 1962)

stormy weather. Wind-speed can be accelerated to storm or gale force particularly where it is funnelled along narrow valleys or over cols. Dependent on wind direction, slopes can be exposed or sheltered from high wind-force. Exposure increases evapotranspiration and exacerbates the depressive effect of altitude and aspect on temperatures. It also exerts a considerable influence on the distribution of precipitation in mountains. The correlation between altitude and precipitation is much less close than that between altitude and temperature. While there is a tendency for precipitation to increase with altitude, in temperate and subtropical climatic regions, this is dependent on the location, altitude, mass and orientation of the mountain in relation to the global air-mass circulation. Mountain ranges can initiate uplift of unstable air masses and result in an increasing amount of orographic precipitation on windward slopes and a rapid decrease in precipitation as a result of the rapid adiabatic warming of descending air on the lee-side or rain shadow. Dependent on orientation, sharp contrasts between humidity and aridity can occur over short distances. In the tropical zone (*c.* 10° N and S of the equator), however, convectional uplift determines the upper limit of the cloud base and results in a precipitation maximum at *c.* 3000 m (Geiger 1969).

Snow

The actual amount of precipitation whether in the form of rain or snow within the mountain area is dependent on the relative degree of exposure to, or shelter from, the prevailing wind. Soil temperatures are determined more by snow depth than altitude. Snow provides a protection against frost. On exposed ridges and summits from which snow is blown the ground can freeze to considerable depth during winter. A high proportion of winter precipitation in mountains falls as snow. At high latitudes winter temperature conditions are such that snowfall is the main form of winter precipitation and the duration of a snow-cover a regular seasonal occurrence. Suitable atmospheric conditions (i.e. high humidity plus temperatures near or below 0 °C at ground level) only occur at a critical altitude which is dependent on the regional climate and the temperature lapse-rate with altitude. Snow-cover can be seasonal and of generally cyclic occurrence or permanent. The boundary between the snow-covered and snow-free ground is the snow-line whose altitudinal limit varies both temporally and spatially. A seasonal snow-cover is characterised by a transient snow-line the altitude of which can vary annually, and during the course of one winter season it is dependent on fluctuating fall, aspect and shelter. The permanent snow-line, above which a cover persists from one year to the next, may vary annually but nevertheless attains a maximum position in relation to landform and altitude and maintains a long-term mean boundary position between the seasonal and permanent snow-cover. At the global scale the snow-line increases in altitude from the poles where it is, theoretically, at its lowest level towards the equator (Fig. 14.2).

VEGETATION ZONATION

In spite of the local environmental diversity the mountain biota is universally characterised by an altitudinal sequence of relatively distinct zones (or stages), each with a homogeneity particularly of vegetation form and composition. Although the number and nomenclature of zones vary, dependent on altitude and zonal climate on the one hand and on the focus and scale of particular studies on the other, five main zones are commonly identified in the European Alps and North American Rockies.

1. *Hill (or basal) zone*: flora and fauna related to that of lowlands; dominance of cultural elements; interregional variation of vegetation communities;
2. *Mountain (or montane) zone*: distinguished by presence of species endemic to a greater or lesser extent to mountains; nearly always forest-dominated, particularly by deciduous broad-leaved trees;
3. *Sub-alpine zone*: comprising one or more 'horizons' dominated by conifers and an upper fringe of shrubs:
   (a) basal horizon – spruce (*Picea*) dominated;
   (b) middle horizon – larch (*Larix* spp.) and pines (*Pinus cembrot*, *P. arole*). The undestorey is, characteristically, composed of heath shrub species, numerous mosses and lichens;
   (c) transitional summit horizon – shrubs (Ericaceae, Rhododendron, etc.) an abundance of mosses and lichens, rosette alpine herbs and some airelles and dwarf trees (*Pinus mugho*), birch (*Betula*), willow (*Salix*) and alder (*Alnus*);
4. *Alpine horizon*: without trees or upright woody plants, the vegetation is sward-

like characterised by a continuous cover dominated by grasses and sedges, small forbs with rosette or cushion growth, short stems, reduced leaves and large brilliant flowers. A further distinction can be made between the well-drained and the marshy waterlogged habitats;
5. *Snow horizon*: between the alpine horizon and the zone of permanent snow; discontinuous vegetation cover with an increased abundance of mosses and with increasing severity of environmental conditions lichens adapted to particular microsites.

However, not all mountains conform to this alpine zonal model. The biogeography of each mountain depends on not only the local and regional environmental conditions but also on the effect of past environmental conditions on the biota. The absolute altitude, number and width of the vegetation zones vary. At the global scale the altitudinal limits increase from the pole to the equator; at equivalent latitudes, from the exterior to the interior of large mountain masses; and from coastal to landward slopes. There is a marked contrast between the zonal flora of the mountains of northern temperate and Boreal climatic regions and that of the intertropical and southern temperate regions. Also, since the mountain ranges of the former were floristically linked during the colder climatic phases of the Quaternary period, the species composition particularly of the montane and alpine vegetation is less varied than in the long-isolated mountains of the southern hemisphere. Among the biotic characteristics common to all altitudinal gradients are:

1. A decline in the number of vascular species with height; but because of high habitat diversity total species richness of mountains is greater than that of any other terrestrial environment;
2. A decrease in the structural complexity of the vegetation cover;
3. A high degree of endemism;
4. The dominance of conifers in the upper montane horizon;
5. The development of alpine meadows (or grassland) which have no latitudinal equivalent.

ALPINE (OR MOUNTAIN) TIMBER-LINE

A distinction is frequently made between the mountain (or altitudinal) timber-line and the latitudinal tree-line (see p. 177). The former is the upper limit of tree and forest growth; the latter the extreme limit of growth of trees or shrubs of 2 m height or more. The tree-line, however, is an ecotone of varying distinctiveness and width between the sub-alpine forest and the alpine zones. As such it is an important focus of biogeographical study by ecologists, foresters and geographers. Wardle (1974: 306) considers it to be 'the sharpest temperature dependent boundary in nature' and hence may provide a sensitive guide to environmental change.

The sub-alpine forest varies in specific composition from one part of the world to another. In temperate and Boreal regions it is dominated by a few hardy conifers. With increasing altitude the growth height of the trees decreases and the forest is eventually superseded by a *krummholz* – a dense thicket of low, contorted, horizontally spreading trees and shrubs – or by an area of sparsely distributed small but upright tress comparable to the tundra *elfin forest*. Coniferous krummholz is particularly widespread in north temperate mountains and is extensive on the Rocky ranges in Colorado and Wyoming. Developmental forms vary from tree islands (where the lower branches of adjacent trees become layered) and low flagged and prostrate krummholz to the isolated cushion krummholz under the most severe conditions. In some species krummholz is environmentally produced, in others it has been proved to be genetically determined. In tropical highlands the sub-alpine forest is composed of arborescent forms of genera which in extra-tropical regions are represented by small herbaceous species. These arborescent megaphytes include species of Euphorbiacae, *Senecio* and bamboo.

On a global scale the altitude of the timber line varies with latitude (Table 14.3) and with degree of continentality. It increases from the poles to the equator, inland from the coast, and from the exterior to the interior of large mountain masses. In the interior of the Rockies and the Himalayas the tree-line attains altitudes comparable to those in the tropics. While the latter has been ascribed to the *massenberg effect* (i.e. decrease in the temperature lapse-rate in larger and higher mountains) the higher summer temperatures of continental compared to oceanic

Table 14.3 Altitude and main species of selected timber-lines at varying latitudes (from Wardle 1974)

Location	*Altitude (m)*	*Main species*
63° N Sweden	1000	Birch (*Betula pubescens*)
60° N Alaska	900	Sitka spruce (*Picea sitchensis*)
50° N British Columbia	1850–1900	*Abies lasciocarpa* (krummholz to 2500)
50° N Rockies (Alberta)	2150–2300	Englemann spruce (*P. engelmannii*)
47° N Switzerland	1900–2000	*Pinus cembra* and *P. abies*
41–44° N Caucasus	Up to 2500	*Betula verrucos*
38° N S. Nevada	3300	*Pinus albicaulis* (krummholz to 3750)
28° N E Himalayan	3800	*Larix griffithii*
19° N Mexico	Mean 3950	
	Max. 4100	*Pinus hartweggi*
9° S Andes	4100	*Polylepi*
6° S New Guinea	3900–4100	*Podocarpus compactus*
19° S N. Chile	4900	*Polylepi tomentella*
36° S Snowy Mts (Aust.)	1850–2000	*Eucalyptus niphophila*
42° S South Island (NZ)	1200–1300	*Nothafagus menziesii*

climatic regimes are considered as, if not more, important. The alpine tree-line has, as in the tundra, been correlated (p. 95), with the 10 °C isotherm for the warmest month of the year by many authors. At the local toposcale the tree-line tends to be depressed on exposed convex slopes and to be higher on sheltered concave slopes. Inverted tree-lines are found in depressions and valley floors where treeless vegetation occurs below the sub-alpine forest zone because of unfavourable conditions created by nocturnal temperature inversions during the growing season. Well-marked temperature inversions in temperate regions may have little effect on vegetation during the growing season because of short nights and temperatures above 0 °C. However, in tropical regions, the less cold-tolerant trees are more liable to damage by long night-frosts which can take place at any time of the year.

The factors and processes limiting tree growth at high altitude have been as contentious an issue as in the case of the tundra tree-line. Table 14.4 provides a

Table 14.4 Comparison of Arctic and alpine environments and vegetation (from Webber 1974)

Component	*Arctic tundra (Point Barrow, Alaska)*	*Alpine tundra (Niwot Ridge, Colorado)*
Latitude	71° N	40° N
Altitude	7 m	3549 m
Solar radiation		
(average July intensity)	0.30 cal cm^{-2} min^{-1}	0.56 cal cm^{-2} min^{-1}
(0.4–0.7 μm for growing period)	10×10^7 cal m^{-2}	9×10^7 cal m^{-2}
Photoperiod (maximum)	84 days	15 hours
Air temperature (July mean)	3.9 °C	8.5 °C
Soil temperature (maximum)	2.5 °C	13.3 °C
Precipitation (annual mean)	107 mm	1021 mm
Wind-speed (annual mean)	19 km $hour^{-1}$	37 km $hour^{-1}$
Water stress	−4 to −5	−6 to −8 bars
Permafrost	Universal	Sporadic
Average length of growing period	55 days	90 days
Number of common vascular plants	40	100
Microhabitat diversity	Small	Large
Average areal vascular production	100 g m^{-2} $year^{-1}$	200 g m^{-2} $year^{-1}$
Average ratio of above- to below-ground biomass	1 : 8	1 : 12
Average area net photosynthetic efficiency	0.5%	0.5%

concise review of the main theories. On the one hand, it was initially considered that tree growth was limited by an insufficiently long period of warmth for the completion of shoot growth so that the requisite degree of winter hardiness (i.e. ability to withstand cold and desiccation) could be initiated by exposure to the first slight winter frosts without risk of damage. Indeed, only four to five species can survive the extreme winter at the high continental tree-line in the Rockies. On the other hand, recent work has put more emphasis on the tree regenerative capacity. Seed production is low and variable and seedlings can only survive if they can withstand or avoid the particularly stressful microclimate at the tree-line. Temperatures near the ground are very much higher during the day and very much lower at night than the ambient temperature above. As a result, plants are subjected to a large diurnal temperature range and to soil frost-heave in the winter. Wardle (1974) notes that microclimate measurements in New Zealand show that seedlings of some species of *Nothofagus* will only become established in shaded sites provided by the more environmentally tolerant mature trees. In this case the tree-line will be composed of erect trees without a krummholz zone. The latter only develops below the regional tree-line where, despite a constant exposure to high wind-force, species seedlings can become established, and survive, but with a slow growth rate and a wind-sheared form.

Among other factors that might limit tree growth, carbon dioxide assimilation balance, soil temperatures and increased UV-B radiation have been suggested but not supported by conclusive evidence. Wardle (1974: 398) summarises his preferred explanation of the alpine timber-line as occurring 'at an altitude where the environmental tolerances of vascular plants and, in particular, their ability to ripen their shoots so as to be able to withstand seasonally adverse conditions are quite abruptly reached'.

The tree-line varies temporally as well as spatially in response to climatic change, to natural disturbance and increasingly to the effect of human impact. Evidence for tree-line fluctuation as a result of climatic change is relatively sparse and scattered. Vegetation change deduced from pollen analysis suggests that the tree-line could have been depressed, dependent on latitude, by 900–1400 m during the coldest phase in the Pleistocene period. Dating fossil tree stumps in the north-eastern Highlands of Scotland have revealed that in the post-glacial Boreal period the tree-line reached 700–800 m, 100 m higher than the present natural tree-line (Pears 1985).

On a shorter time-scale and at the local site-level natural disturbance by regular and severe snow avalanches, intense gales, plagues of phytophagus insects, outbreaks of disease and intense lightning-set fires can all depress tree-lines for longer or shorter periods of times. In addition, the effects of natural catastrophes have been exacerbated by depression and obliteration of tree-lines consequent on deforestation particularly in the long-settled, semi-arid areas of the world from the northern Andes to the western Mediterranean and central Asia, and as a result of the intensification of burning and grazing.

ALPINE ZONE

The Alpine zone is that area above the mountain tree-line which is not permanently snow-covered. It shares with the arid deserts and the polar tundras a severity of environmental conditions that become marginal for life. Similarities in the physical environment, the flora and the vegetation of the tundra and alpine zones (Table 14.4) have received considerable emphasis to the extent that the latter was (and indeed still is in many texts) commonly referred to as the Arctic–alpine zone. There are, however, important and ecologically significant contrasts between the two environments. These include the greater intensity of solar radiation with a particularly high proportion of direct radiation and of ultraviolet light than at high-latitude lowlands. Snowfall and wind-speeds are much higher than those experienced in the tundra while the occurrence of permafrost is sporadic. Consequent largely on varying altitude, aspect and slope the heterogeneity of microhabitats and the ecological importance of the microclimate is even greater in the alpine zone than in the tundra. Both areas, however, are treeless with a low-growing vegetation cover which becomes more open and discontinuous with increasing environmental severity. The flora of both is composed of a high proportion of *cryophytes*, the most abundant of which are herbaceous perennials (particularly grasses and sedges), small or prostrate shrubs, and cryptogams. However, the alpine flora is much the richer of the two biomes. In addition, the alpine vegetation of tropical mountains is both structurally and floristically different from that of temperate mountains. It is composed mainly of grassland with tall, columnar, arborescent or sub-arborescent life forms belonging to genera such as *Lobelia* and *Senecio* in Africa and *Espeletia* in South America. The alpine zone in the tropics, however, is limited in extent and has been studied less than that in temperate latitudes.

The vegetation of the alpine zone in the northern hemisphere is composed of a complex mosaic of plant associations dependent on gradual or steep microclimate gradients such as degree of exposure to, or shelter from, direct solar radiation and wind, the magnitude of the daily range of ground and plant temperature, the annual frequency of the soil freeze–thaw cycles, the length of the growing season, the depth and duration of snow-cover and the availability of snow meltwater on the one hand and on microvariations in the physical condition and the chemical composition of the mineral substratum on the other. The most favourable habitats with the longest growing season are those on gentle but well-drained south-facing slopes above the tree-line. These carry closed, species-rich and relatively productive alpine-meadow (or *alps*). In northern Europe and Eurasia they have long been used for summer grazing by transhumant livestock and as a source of winter fodder (hay) for the valley farms.

The most severe environmental conditions are associated with persistent snow-banks where the thaw commences late and the growing season is very short and with the highest windswept ridges and plateaux. At these limits only cryptogams can survive the long winter cold, the short cool growing season and, on snow-free ground, often intense summer drought. As in the tundra, cryptophytic mosses and lichens can take advantage of the highest daytime temperatures which occur in the interface between the soil and the air and when there is sufficient ice meltwater to permit very slow growth. Capable of withstanding either winter or summer desiccation alpine lichens are even longer lived (up to 1300 years has been estimated for the species *Rhizocarpon geographicum*) than those in the tundra (Billings 1974). The absolute upper limit for plant life is reached between 5000 and 7000 m dependent on latitude and aspect. However, most exposed alpine summits are 'bald'.

ORIGIN OF THE ALPINE BIOTA

Although some alpine plant species have a wide range of distribution such that they can be found in both hemispheres, most are spatially restricted to a few mountain systems. The close genetic relationship between the Arctic flora and that of the alpine zone of the north temperate mountains of Eurasia and North America has long been recognised. The origins of this markedly disjunct distribution of apparently similar species have been a subject of considerable debate among biologists, geologists and geographers. It is now generally accepted that the present alpine floras of the temperate zone are isolated remnants of a formerly more widespread circumpolar crytophytic flora. The latter is thought to have evolved as the climate became colder in the late Tertiary period and to have reached its greatest extent and replaced the pre-existing nemoral flora over much of the land in high latitudes by the Pliocene and early Pleistocene periods (Löve and Löve 1974). The high frequency of occurrence of polyploidy in the small Arctic–alpine flora is thought to be indicative of the relatively recent evolution of the cryophytic flora from pre-adapted species. Inter- and post-glacial fossil flora reveal that as the ice-sheets retreated and the climate ameliorated the formerly continuous cold-adapted flora was disrupted. Remnants were restricted on the one hand to high latitudes, on the other to high altitude habitats. Some species with Arctic–alpine affinities were able to persist in open unstable lowland coastal or riverine habitats, free of competition from the larger more productive post-glacial vegetation. The spatial continuity of the Arctic–alpine flora was, however, maintained and carried far south, particularly on the highest north–south trending mountain ranges such as those of the western Cordillera and eastern mountains of North America, and the Urals of eastern Europe. Isolation of Arctic from alpine and of alpine islands favoured the subsequent evolution of mountain endemics.

The relationships between Arctic and alpine animals are less marked than for plants. Cold adaptation is common to both. While trophic and structural niches are similar in both biomes, alpine vertebrate mammals do not exhibit the cyclic populations common in the Arctic. In addition, they need, at the maximum altitudes, to be adapted to low oxygen levels and to be able to move on steep irregular and frequently unstable slopes. Mammals, in particular, tend to be exclusive to one or the other zone except where they meet in the high latitude mountains. A few species, such as the Arctic or mountain hare (*Lepus timidus*) are regarded as true relicts. It is, however, generally held that the alpine fauna has by and large not evolved from the formerly continuous Arctic–alpine biome. The main centre of evolution of alpine vertebrates is thought to have occurred in the mountains of central Asia. The Tibetan Plateau and adjacent mountains constitute the largest spatially continuous alpine area in the world. Habitat diversity is high. A large part of the area is dry 'alpine desert', and the alpine zone has a wider altitudinal extent than elsewhere. Although quantitative data are lacking the number of bird and mammal species attains a maximum in this area. In addition, it is the only alpine area which escaped complete biotic disruption during the Pleistocene. Local adaptation of montane (sub-alpine) forest and steppe species to alpine conditions also occurred on restricted and isolated mountain tops throughout the world. In general the restricted area and patchy distribution of existing alpine habitats may account for the relative poverty of the vertebrate fauna compared with that of the Arctic tundra.

HUMAN IMPACT

The diversity and isolation of mountain habitats are reflected in that of a wide range of distinctive human societies whose ways of life reflect an early and close adaptation to the particular problems created by a spatially variable and difficult physical environment. Traditional methods of movement and resource use tended to optimise the expenditure of human energy and the conservation of resources (particularly soil and water) on high steep slopes and where level terrain is at a premium. The intricate terracing of slopes for cultivation particularly in many subtropical and tropical mountain areas testifies to the skill with which this was achieved. In many of the long-settled densely populated regions of the Mediterranean Basin and of the Middle and Far East, the mountains were early deforested to supply timber for fuel, ships and construction. Today, little remains of the original forest which has been replaced by shrub or grassland maintained by fire and grazing or by bare slopes ravaged by soil erosion as in North Africa, the Middle East and northern China.

The human impact on the mountain areas of the world has, within the last 200 years, increased in intensity and extent with the symbiotic development of modern methods of communication and resource exploitation particularly for tourism. The latter developed in the nineteenth century with the rise in popularity of summer residences, sanitoria, mountaineering and alpine skiing. Increased mobility and affluence in the post-Second World War years have seen the exponential growth of the ski industry particularly in the alpine zones of mountains with a seasonal or year-long snow-cover of adequate duration, extent and depth. The indirect ecological impact of the construction of roads, buildings and uplift facilities has been greater than the direct effect of 'piste skiing' on the mountain biota. Increased ease of access to the alpine zone in spring and summer, however, has exposed the biota to the direct physical impact of trampling. Alpine developments have resulted in the drastic disturbance of the vegetation in an environment which is particularly vulnerable because of steep slopes, a substratum subjected to frequent freeze–thaw cycles and a biota, as in the tundra, living at the margin and where recovery from habitat disturbance is all the more difficult.

Chapter

15

Aquatic ecosystems

The most fundamental contrast within the biosphere is that between the environmental conditions and associated organisms of terrestrial and aquatic ecosystems. Although the latter cover *c.* 75 per cent of the earth's surface, plant biomass is dominant on land, animal biomass in water (see Table 15.1). However,

Table 15.1 Comparison of size and productivity of terrestrial and aquatic ecosystems

Habitats	*Area* *(10^6 km^2)*	**Approx. volume* *(10^6 km^3)*	*NPP* *(10^9 t year^{-1})*	*SP* *(10^6 t year^{-1})*	*P : C + D*
Terrestrial	145	14.5	110.5	867	
Aquatic	365	1445.0	59.5	3067	1 : 0.001
Ratio	1 : 2.5	1 : 99	1 : 0.54	3 : 5401	1 : 20

* Based on an average depth 4000 m and assuming a terrestrial inhabited zone 100 m deep. NPP = net primary production; SP = secondary (animal) production; C = consumers; D = detrivores; P = primary producers.

while the NPP of aquatic ecosystems is only about half that from the land, aquatic animal production is over 3.5 times that from the land. Nevertheless, despite these contrasts, ecological processes are similar in both habitats and are influenced by the same complex of variables.

THE PHYSICAL HABITAT

Water is in many respects a more favourable and certainly a more equable medium for life than is the land. Desiccation is unknown except in transitional areas between land and water where, because of varying water-levels, organisms must be adapted to exist for varying periods of time in and out of water. Further, organisms living in water are immersed in a solution that contains many of the elements essential for their existence. Oxygen and carbon dioxide are readily soluble and available in water. Relatively easily absorbed from the atmosphere, their replenishment and distribution from the surface downwards is facilitated by water mixing as well as by photosynthesis, respiration and decomposition. Of the two, carbon dioxide is more readily soluble than oxygen. It reacts with water to form carbonic acid (H_2CO_3) and can be easily fixed in carbonate ($-CO_3$) or bicarbonate ($-HCO_3$) form. The concentration of carbon dioxide in water is *c.* fifty times higher than in the atmosphere. That of oxygen is, however, on average much less, 0.001 per cent in fresh water compared to 20 per cent in the atmosphere. Expressed another way, the average dissolved oxygen concentration of water is 10 ppm, i.e. forty times less than the weight of oxygen in an equivalent volume of air. Also, while the oxygen concentration in the atmosphere is uniform, it is extremely variable in water, and is also less evenly distributed in water than carbon dioxide. Its concentration is higher in surface than in deeper water, as a consequence of the concentration of photosynthetic activity in the upper illuminated zone, and in cooler than warmer water. However, except in localised areas in very deep lakes and seas (e.g. fiords, the Black and Mediterranean seas) where bottom water tends to be stagnant and oxygen deficient, water circulation ensures a replenishment of oxygen sufficient to maintain some life even in the deepest oceans. In contrast to lakes and oceans, rivers are relatively shallow and more turbulent. They expose a greater surface area per volume to air and hence, under unpolluted conditions, the amount of oxygen varies little throughout the length and depth of the water body.

SALINITY

Water contains varying amounts of all the minerals found in the earth's crust. Some forty-five mineral elements are known to be present in water; some occur in larger quantities than others. In the sea the most abundant are sodium (30.9 per cent) and chlorine (55.3 per cent) – the main determinants of the salinity of sea-water. Together, with smaller quantities of magnesium, sulphur, calcium, potassium and bromine, they account for over 99 per cent of the total mineral content of sea-water. The remainder, occurring in minute and often variable quantities, include the nutritive elements, such as nitrogen, phosphorus and potassium, essential for plant growth and whose availability determines the nutrient status and hence the potential fertility of the aquatic habitat.

The salinity of sea-water is determined by the amount of salt or sodium chloride (NaCl) measured in grams of salt per 1000 g of water (per cent); there is a continuum in the biosphere from marine through brackish to fresh water with a salinity less than 0.05 per cent. However, given the volume of sea-water, the salt concentration is not excessive and the range of salinity in the open seas is comparatively low. The most saline (*athalassic*) conditions occur in inland lakes (e.g. Great Salt Lake, Dead Sea) and semi-enclosed seas in arid areas subject to constantly high temperatures and evaporation. Among the latter the highest salinities recorded are those for the Red Sea (average 40–41 per cent). Minimum salinity is found in brackish coastal waters where the inflow of large volumes of fresh water from the land or from melting ice causes dilution. In the open seas salinity ranges from only 37.5 per cent in tropical to 33 per cent in polar seas. With similar temperatures, saline water is denser than fresh water and in the absence of movement the former tends to sink below the latter and mixing of the fresh and salt water may be relatively slow.

TEMPERATURE

Water is physically, as well as chemically, a more uniform and less stressful environment for life than the land. The specific heat of water is higher than that of solids; it absorbs heat and loses it more slowly than the land surface. Spatial and seasonal variations of temperature are, however, dependent on the size, depth and mobility of the particular water body. The difference between the surface temperature of the warmest seas (*c.* 32 °C in the Persian Gulf) and the coldest (*c.* 2 °C in Polar areas) is about 30 °C, compared with a temperature range of 87–90 °C on the land. The annual range of oceanic temperature rarely exceeds 10 °C and is normally much less. However, most of the heat derived from solar radiation is absorbed in the surface layers; hence, the seasonal and latitudinal variations of temperature experienced at the surface decrease rapidly with depth. Below about 100 m there is relatively little annual fluctuation and low temperatures persist over extensive areas. On the floors of the deepest oceans, temperatures vary by only a few degrees above or below 0 °C.

The relative uniformity of temperature in the oceans is facilitated by the continuity of the water area, and by the large-scale circulatory movements in the ocean basins. Vertical movement, i.e. *overturning*, is engendered by variations in water density resulting from differences in salinity and temperature. Cold and saline water being denser than warm/and fresh water tends to sink below the latter. Also, under the influence of the rotation of the earth and the friction of winds, surface oceanic waters are kept in constant circulation (see Fig. 15.1). Water heated in tropical latitudes tends to flow away in great anticlockwise or clockwise eddies towards the colder polar north and south latitudes. At the equator, colder bottom water (originating in the polar regions) wells up to take its place. Upwelling also occurs along the west coasts of the southern continental land masses where ocean currents and prevailing offshore winds 'push' warmer surface water away from the continental edge.

Heat radiation is completely absorbed in the top 1–2 m of water. In slow-moving water bodies, the surface water heats up more rapidly than that below. If wind action is insufficient to mix the water, a distinct upper isothermal layer (the *epilimnion*) of varying depth, with a temperature over *c.* 4 °C and hence a lesser density, forms above cooler water below (the *hypolimnion*). The two become separated by a rapid temperature gradient or *thermocline* (+1 °C m^{-1}). The resulting direct stratification may be a relatively short-lived or more persistent phenomenon during the summer in temperate water bodies (oceans and lakes) or

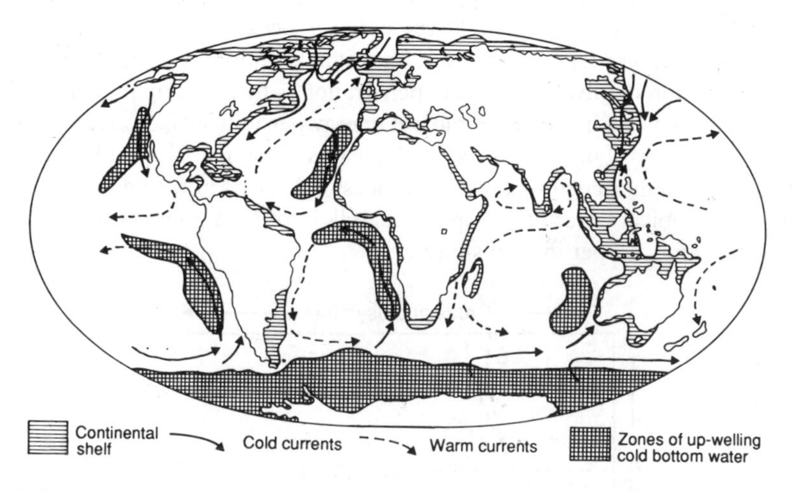

Fig. 15.1 Ocean currents

persist for most of the year in tropical waters. The temperature range beween the epilimnion and the hypolimnion may vary from 8 °C in temperate to 16 °C in continental lakes subject to severe winters and hot summers. Although the range may only be *c.* 4 °C in the tropics, the stratification may be more stable than in water with much wider temperature variations. This is because the change in water density per degree of temperature change is so much greater at high than at low temperatures. The stability of the thermocline is dependent on the depth and mobility as well as on the temperature of the water mass. In shallow (less than 15 m deep) fresh or saline water the ratio of the epilimnion to the hypolimnion will be very much higher than in deep water and stratification may not occur. In tropical climates the thermocline may be permanent or, at least, semi-permanent. Cooling of surface water and the consequent convectional overturning together with wave action disrupt the thermocline in temperate latitudes. In polar areas the development of a marked and seasonally persistent thermocline is inhibited by low temperature. In contrast, during winter freshwater lakes and, under very severe conditions, brackish coastal water may freeze over, trapping slightly warmer (i.e. above freezing) water below.

LIGHT

Although a greater percentage of the total solar radiation reaching the earth's surface falls on water rather than land surfaces, illumination is much feebler than on land. Loss by reflection can be high, particularly at high altitudes where the angle of incidence of the sun's rays is low (90 per cent loss, compared to 2 per cent in calm conditions with an overhead sun according to Boney 1989). Light which passes across the surface is absorbed very rapidly by water, yellow humic substances (*gilvin*), plants and inorganic particulate matter. In addition, scattering by plants and suspended inorganic particles impede downward penetration. Light also changes in spectral composition and irradiance with depth. Red and ultraviolet light are absorbed in the upper layers of clear ocean water; blue–green light penetrates to greatest depths. However, blue light is rapidly absorbed in water with a high proportion of gilvin. Maximum penetration is by green light (Boney 1989). Irradiance or the rate of supply of radiant energy per unit surface area (expressed as $J\,m^{-2}\,s^{-1}$ $(min^{-1}$ or $mol\,m^{-2}\,s^{-1})$ decreases exponentially with depth. In offshore water light penetration can be at least twice that in turbid inshore water. In the clear tropical water of the open ocean light of high intensity can penetrate four or five times deeper than in shore.

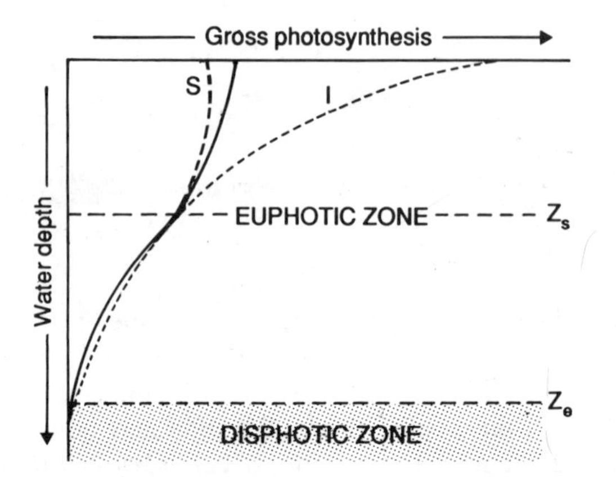

Fig. 15.2 Characteristic relationship between gross photosynthesis and water depth: Z_s = depth at which a white disc disappears from view of observer at the surface; Z_e = limit of euphotic zone, i.e. light compensation point; S = surface inhibition particularly on sunny days due to ultraviolet light; I = irradiance (from Moss 1980)

Two ecologically significant light zones illustrated in Fig. 15.2 can be present in a water body:

1. *The euphotic (or photic) zone* whose lower limit is that at which light intensity is insufficient for photosynthesis to proceed at a rate which compensates for respiration, i.e. the *light compensation point.* The average light compensation point has been generally defined as that at which intensity is 1 per cent of the subsurface light. The depth at which this occurs varies from 10 to 15 m in turbid marginal water, to a maximum in clear, open water of about 100 m, which roughly corresponds to the average depth of the outer edge of the continental shelf. In shallow coastal waters and lakes and in many rivers, the euphotic zone can extend from the surface to the water bottom;
2. *The disphotic (or aphotic) zone* with insufficient or no light for photosynthesis.

While the distribution of species of aquatic plants and animals is determined by their adaptation to salinity, that of primary productivity is influenced more by the nutrient status of the water. Most open oceanic areas are *oligotrophic* (nutrient-poor, particularly in phosphorus and nitrogen). Marine water becomes increasing *eutrophic* (nutrient-rich) from off to in shore because of the input of mineral elements from the adjacent land and because, in certain areas, cold nutrient-rich water wells up at the surface from ocean deeps. In contrast, the nutrient status of fresh water is more directly related to the chemical composition of the substrate, the nature of the adjacent vegetation and, increasingly, the input of phosphates and nitrates as a result of human activities. Oligotrophic lakes and rivers are characteristic of poor, acid soil and/or rock and of particularly deep, rock-bound lakes with permanently or seasonally cold but relatively stagnant bottom water. Eutrophic fresh water is more usually associated with shallow lakes or slow-moving rivers over mineral-rich sediments and in which there is little or no temperature stratification of the water. The most nutrient-rich waters are those in brackish coastal lagoons and estuaries at the interface between the land and the sea.

AQUATIC SUBSYSTEMS

In most aquatic ecosystems three subsystems (Fig. 15.3) dependent on spatial variations in the physical habitat and characterised by distinctive communities can be recognised (Barnes and Mann 1980):

1. The open water habitat (the marine pelagic zone);
2. The substratum or benthic habitat;
3. The fringe or edge habitat of coats, water basins or channels.

The relative importance of these subsystems varies from one type of aquatic ecosystem to another.

PELAGIC HABITAT

The pelagic habitat is dominated by two types of organisms physically capable of living in water. One is the *plankton* ('drifting') organisms with no or limited means of locomotion and whose movements are dependent on those of the water they inhabit. The other is the *neckton*, free-swimming organisms, fish and marine and freshwater mammals. There are, in addition, birds for whom the pelagic fish are an important if not the only food source. The plankton vary in size and function as indicated in Table 15.2. Some are primary producers (algae); some are consumers (bacteria and animal plankton); some are decomposers (bacteria).

Plant plankton

About 90 per cent of primary production in fresh or marine pelagic habitats is undertaken by plant plankton (*phytoplankton*). The most abundant representatives in the sea are minute algae, one-celled diatoms, dinoflagellates and coccoliths; in fresh waters diatoms, dinoflagellates, desmids and blue-green prokaryotes. The latter, either unicellular or filamentous, are nitrogen-fixing. Minute size providing a large area : volume ratio and a slow sinking rate facilitates suspension and absorption of mineral nutrients from the water. The efficiency of this form may

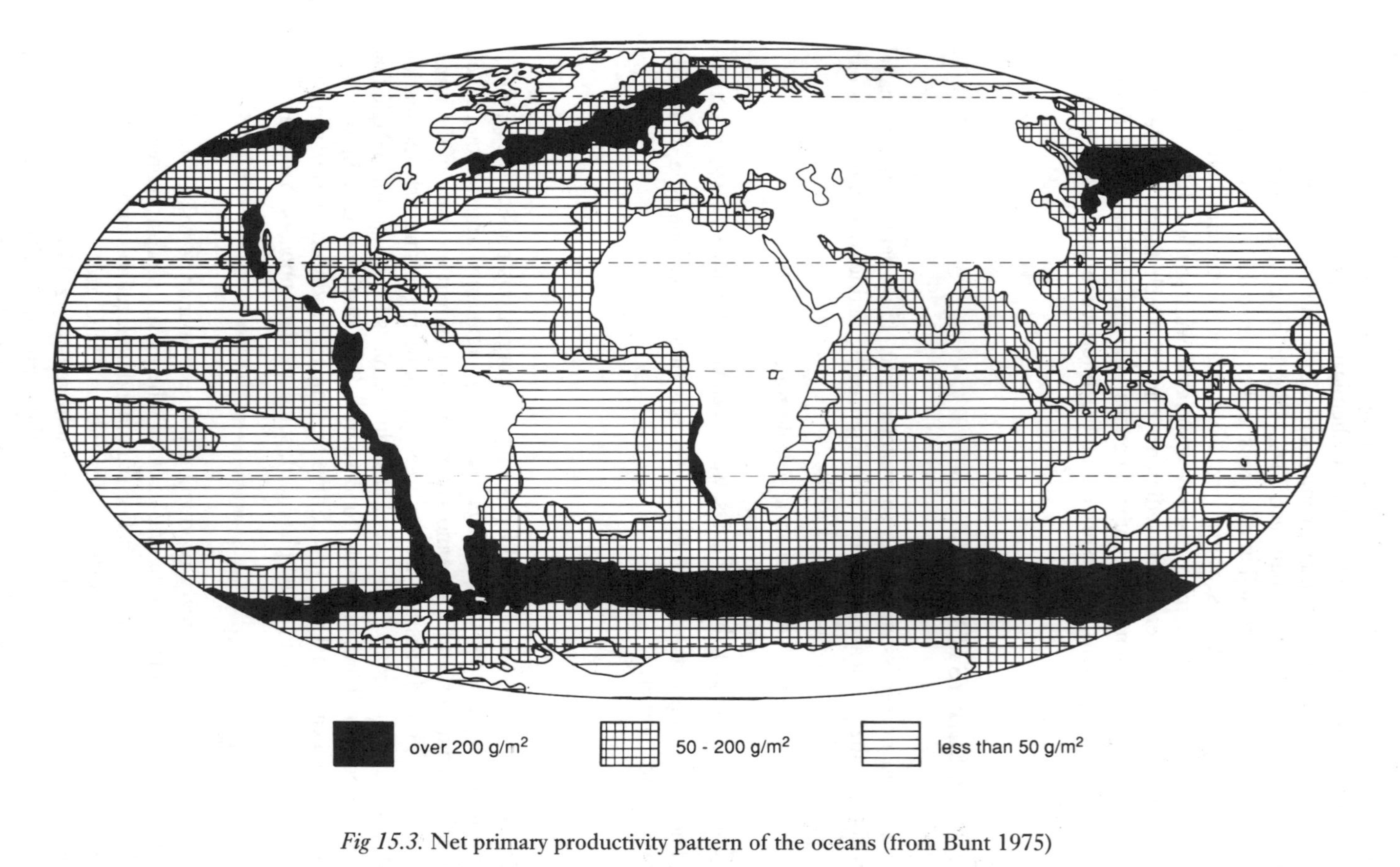

Fig 15.3. Net primary productivity pattern of the oceans (from Bunt 1975)

Table 15.2 Size grading of planktonic organisms (from Boney 1989)

Maximum dimension (μm)	*Plankton 'category'*
Less than 2	Picoplankton (algae, bacteria)
2–20	Manoplankton (animals and algae)
20–200	Microplankton (animals and algae)
200–2000	Macroplankton (animals and algae)
More than 2000	Megaplankton (animals)

well explain the overwhelming dominance of plant life by the phytoplankton in aquatic ecosystems. They can occur in such densities as to form surface concentrations or 'blooms' which colour the water. The 'red-tides' in coastal waters, the appearance of which occasionally hit the headlines, are caused by population explosions of dinoflagellates, usually of toxin-producing species (Boney 1989).

Plant plankton production varies both temporally and spatially – the most important factors limiting productivity are light intensity, the nutrient status and the temperature of the water. Moss (1980) distinguishes between light as the factor limiting the rate (of gross photosynthesis) and nutrients limiting the yield (net photosynthesis). Rate of gross photosynthesis declines with depth from a maximum near the water surface to the compensation point dependent on the initial light intensity and the turbidity of the water mass. In very bright sunlight surface inhibition may occur possibly as a result of the rapid absorption of ultraviolet light (Moss 1980). Rate of growth is also affected by temperature. Some phytoplankton have minimum requirements below 0 °C; some, such as live on mud-flats in tropical regions, can tolerate temperatures of 30 °C. In temperate and cool latitudes seasonal rates of growth are partly temperature-related. The correlation, however, is not perfect. On the one hand, seasonal phytoplankton 'blooms' in cool and cold water exceed the maximum growth in warm tropical water. On the other hand, marked seasonal variation in growth occurs in both Arctic and tropical water where the annual range of temperature in the euphotic zone is relatively low. Explosive growth has been observed in Lake Baikal (former USSR) before the ice breaks (Boney 1989).

The most important factor limiting phytoplankton productivity is the availability of nutrients – particularly of nitrogen, phosphorus and iron (in that order) in sea water and of these plus manganese in freshwater lakes. The nutrient status of water in the euphotic zone is dependent on the rate and efficiency with which the nutrients lost from this zone are replenished. In shore, fresh and salt water tends to be nutrient richer than offshore water because of the proximity to inputs by water flow from the adjacent land surface. Nutrient status is also higher in those pelagic areas where minerals (either from terrestrial input or from organic decomposition) which have become incorporated in bottom sediments are most rapidly recycled, i.e. returned for use by phytoplankton in the euphotic zone. This is most effective where water mixing, as a result of turbulence extends from the surface to the base of the water mass. Such is the case in most rivers, in shallow lakes and in epicontinental seas overlying extensive areas of continental shelf. It is also characteristic of areas where cold nutrient-rich water wells up from ocean deeps. In the oceans this occurs near the equator, where warmer, less dense water diverges (floats) north and southwards, around the margin of the Antarctic continent where surface water drifts north-eastwards under the influence of the strong west-wind drift (i.e. trade winds) and particularly along the western coasts of North and South America, Africa and to a lesser extent Australia (see Fig. 15.1). In these areas the continental shelf is comparatively narrow and deep water lies relatively close in shore and surface water carried off shore by prevailing winds is replaced by upwelling of cold nutrient-rich bottom water. As indicated in Fig. 15.3 and Table 15.3 these nutrient-rich areas have the highest annual primary productivity.

Table 15.3 Net primary production in various marine habitats (from McCluskey 1981)

	Open sea	*Coastal zone*	*Upwelling regions*	*Estuaries (and coral reefs)*
% area	90	9.4	01	0.5
Net primary production kcal m^{-2} $year^{-1}$	50	100	300	1000
1000 g C $year^{-1}$	16.3	3.6	0.1	2.0

In addition to these spatial variations in marine productivity, seasonal variations as a result of thermal stratification are characteristic of both marine and freshwater bodies. This inhibits vertical water mixing and nutrient cycling for longer or shorter periods of the year. In warm tropical waters stratification may be virtually constant throughout the year, resulting in low nutrient status and low primary productivity despite high temperatures and light intensities. Annual productivity is higher, but more seasonal, in cool temperate and cold latitudes (and in high-altitude lakes) (see Fig. 15.4). In the former, increasingly favourable light and temperature conditions for growth in late spring/early summer give rise to an explosive growth (the *spring bloom*) of phytoplankton. This rapidly depletes nutrients in the euphotic zone. Recycling must await surface cooling and wind-induced turbulence in the winter. Disruption of the thermocline may be accompanied (when light and temperature conditions are favourable) by a secondary, smaller *autumn bloom*. In cold Arctic and Antarctic marine water, high turbulence inhibits the development of a marked seasonal stratification and primary productivity is more clearly a function of favourable light/temperature conditions. Constant or seasonal stratification is characteristic of most lakes apart from the very large water bodies subjected to high wave turbulence.

Compared to terrestrial plants, phytoplankton productivity is comparatively low. That in the oceans contributes 35 per cent of the total primary biological productivity of the biosphere. The average productivity (1.0 g cm^{-2} day^{-1}) of the oceans is only slightly greater than that of the world's deserts; that of shallow seas and deep lakes is higher (0.5–3.0 g cm^{-2} day^{-1}). While the productivity of

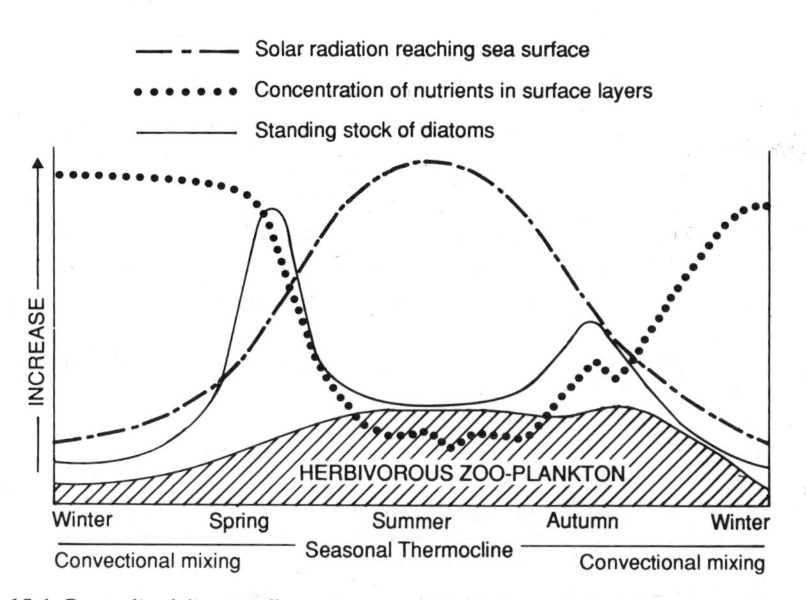

Fig. 15.4 Generalised diagram illustrating seasonal variations of selected parameters in surface water of temperate seas (adapted from Tait 1972)

eutrophic lakes (3.0–10.0 g cm^{-2} day^{-1}) is more comparable to that of terrestrial vegetation (Boney 1989) plants other than phytoplankton make a greater contribution than in the more extensive marine pelagic zone.

The phytoplankton provides all but a small proportion of the primary production in the oceans. In lakes the proportionate contribution depends on the size, depth and the nature of the substratum and tends to be less in shallow sedimentary than in deep rocky lake basins. The nature of the phytoplankton results in plant–food relationships different from those in terrestrial ecosystems. First, because of their microscopic size, direct consumption (i.e. grazing by large herbivores) is limited. The majority of the aquatic herbivores are among the smallest of aquatic animals, of a size capable of capturing and ingesting phytoplankton efficiently. For this reason, food chains tend to be longer in aquatic habitats than on land and the standing crop (or phytobiomass) represents a smaller proportion of the primary plant productivity than in the case of land animals. Second, a higher proportion of the phytoplankton production than of terrestrial plant production is consumed by herbivores; the grazing food chain is more important than the detrital chain and, at any one moment, the animal biomass exceeds that of the plant biomass. This is because consumption tends to keep pace with production. The life span of the phytoplankton whose individuals cells may, given favourable conditions, divide every 10–36 hours is very much shorter than most of the organisms which depend directly or indirectly on it. However, the percentage of the phytoplankton production 'grazed' varies with a number of factors of which the nutrient status of the water is particularly important. In eutrophic water where productivity is relatively high a large amount may be left ungrazed, eventually to be decomposed by the bacterioplankton which inhabit the euphotic and disphotic zone.

Zooplankton

The conversion of the phytoplankton into food particles of a size that can be efficiently used by larger aquatic animals is effected by the *zooplankton* – a group that cannot be compared with any similar group of terrestrial animals. They are defined as animals incapable of maintaining their position against water movement (Barnes and Mann 1980). As such, they comprise an extremely diverse group of organisms varying in size, feeding characteristics, and life cycles (see Table 15.4).

Table 15.4 Salinity magnitude of aquatic productivity in different habitats (from data in Whittaker 1975)

Habitats	*Salinity 8‰ NaCl*	*NPP* ($g\ m^{-2}\ year^{-1}$)	*SP*	*Plant : animal productivity ratio 1 : x*
Aquatic				
Salt water		Mean range	Mean	1:
Open ocean	33–37 (35)	125 (2–400)	8	0.06
Upwelling areas		500 (4–1000)	27	0.05
Continental shelf		360 (200–600)	16	0.04
Algae beds/reefs		2500 (500–4000)	60	0.02
Estuaries*	5–35	1500 (200–3500)	34	0.02
Fresh water				
Lakes and rivers	< 0.05	250 (100–1500)	5	0.02
Swamps and marsh		2000 (800–3500)	16	0.008
Non-aqueous		700 (0–3500)	0–20 (6)	0.003–0.02 (0.008)

* Excluding their fringing marshes.
NPP = net primary production; SP = secondary production.

The smallest are only a few millimetres in length, the largest (jellyfish) several metres. They comprise herbivores, carnivores, omnivores and saprophytes; some are filter-feeders, others are raptors. Some are permanent (holoplankton), others temporary (meroplankton) components of the zooplankton. The former are dominant in open water; the latter are larval stages of invertebrate and vertebrate organisms which spawn in shallow inshore waters. Among the myriad of species which comprise the zooplankton, in both fresh and salt water, the most important are the copepods (one of the main sources of fish food in cooler waters) and the shrimp-like krill, *Euphausia superba* (the principal component of the diet of the Antarctic whalebone whale). In contrast to marine water, fresh water contains fewer phyla of zooplankton and similar species are generally smaller. The horizontal and vertical distribution is uneven. Horizontal distribution is characteristically patchy; shoals of high concentration may extend over small or vast ocean areas separated by those of very low density or absence. Many of the more actively moving species exhibit downward vertical migrations, rising up near to the surface at night and, dependent on size, moving down 10–500 m during the day.

Although the productivity of the zooplankton is dependent initially on that of the phytoplankton, the distribution of the former, particularly in cool to cold water, is not absolutely coincident with the other. This has been attributed to either exclusion or grazing. In the former case, the lack of coincidence of zooplankton shoals with phytoplankton blooms is considered a result of the unpalatability and/or toxicity of large fresh algae growths. There is, however, little evidence to support this hypothesis. In the latter case, grazing is thought to be the operative factor. There is a lag between the spring bloom of phytoplankton and the growth of zooplankton, followed by a rapid reduction of the algae by grazing and/or death resulting in a high concentration of zooplankton. In tropical waters phytoplankton and zooplankton growth is continuous, grazing is constant and the time-lag between fluctuations in algae and herbivores is shorter.

Nekton

Although on a global scale zooplankton and that of plankton-dependent fish (and some whales) are ultimately dependent on the primary phytoplankton production, a direct relationship between these three trophic levels is difficult to establish because of the large numbers of variables involved. The *nekton* include all those adult aquatic animals of a size large enough to propel themselves independently of water movement over relatively long distances. About 19 000 fish species are recognised, of which *c.* 7000 live in fresh water. Tropical waters are characterised by great diversity (70 per cent of the global fish species) in contrast to cool/cold water where diversity is much lower and fish stocks are dominated by large populations of a few species.

All fish travel a greater or lesser distance to and from feeding, spawning and nursery grounds and about 200–300 marine species undertake extensive migration. The model triangular migration route is characteristic of many of the important commercial fish of cool and cold Atlantic and Pacific waters such as the herring, tuna, hake, cod, plaice and also of whales. Different stocks of Atlantic herring (*Culpea harengus*) and the Pacific tuna (*Thunnus alalunga albacore*) spawn at different seasons of the year in restricted, well-defined areas, whose routes are related to regional oceanic circulations (Fig. 15.5). The longest route is that undertaken by whales from their feeding areas in the Antarctic to spawning areas in the Arctic. Some fish, like the Pacific and Atlantic salmon and eels, migrate from salt to fresh water to breed. Migratory distances are generally more extensive in the sea than in fresh waters, where species may move up and down rivers, or between in- and offshore water in lakes. The majority of fish, however, do not migrate extensively. Many polar species spawn in winter, producing a few large eggs which develop rapidly in spring; deep-water fish have buoyant eggs which float to the surface from where the young fish later descend.

The size of fish stocks can fluctuate temporally as a result of variations in environmental conditions, food resources, disease, predation and human exploitation. The aquatic predators which occupy the highest trophic level in the complex

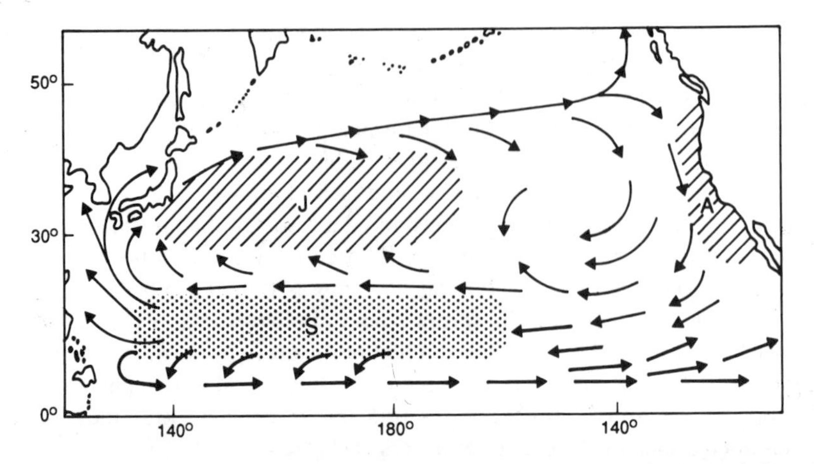

Fig. 15.5 Distribution of albacore (tuna) in North Pacific: S = area in which tuna believed to spawn; A = areas of American fishery on young albacore; J = Japanese fishing in older albacore. Main features of North Pacific subtropical gyre (Kuroshiu and extension, Californian and extension Pacific currents) and the north equatorial counter-current shown (from Barnes and Mann 1980)

aquatic food web are the fish-eating animals such as otter and seal, and predatory birds such as osprey and cormorants. Because of the short life cycles of the planktonic organisms and the efficient grazing, the biomass of the large, long-cycle animals in pelagic habitats is, in contrast to terrestrial ecosystems, much greater than that of the plant biomass.

BENTHIC HABITAT

A varying proportion (estimated at 50–90 per cent) of the phytoplankton is grazed. The remainder enters the detrital route either as exudates from living plant material or as dead or dying individuals, and joins that from dead animals, animal material, faecal pellets, etc. The density of small flake-like particles (*c.* 0.5–25 mm) of organic matter suspended in the water column may be such as to produce a visible fall of what is called *sea-snow*. Decomposition is effected by the activities of a group of micro-organisms (comparable to that in soil) composed of primary decomposers (bacteria or fungi), saprophytes including larger detrivores and mucous-net filter-feeders that consume both bacteria and detritus, and protozoa which graze on detritus and bacteria. Some dead organic material is decomposed partially or completely as it sinks; some, particularly in shallow pelagic water, tends to accumulate on the bottom. In very deep water comparatively little reaches the bottom benthic habitat.

The benthic detritus provides both a habitat and a food resource for a distinctive and extremely diverse community composed of micro- and macroinvertebrate detrital feeders and their predators, and of the detrivores on whom the final release and recirculation of mineral nutrients depends. Some benthic animals (e.g. annelid worms, molluscs, fly larvae) live in the bottom sediment. Others are surface dwellers including the more mobile shellfish (e.g. shrimps, crabs, lobsters, gastropods, starfish and sea urchins) as well as sedentary, attached, forms (e.g. barnacles, sea anemone, sponges). Some live partly in and partly out of the sediments. The constant burrowing of macrobenthic animals results in the incorporation of finely comminuted and partially decomposed material (often as faecal pellets) in the bottom sediments. Where the latter is fine-textured, oxygen-deficient conditions may result because of the high demand by the macro-organisms and the restricted diffusion of oxygen from the water into the fine pores. While anaerobic conditions retard mineralisation of organic matter and nitrification, they ensure the mobility of elements such as ammonia, phosphorus and iron that might otherwise be immobilised in an oxidising environment. The benthic subsystem then plays a role in the pelagic aquatic system comparable to that of soil in the terrestrial ecosystem. It contains the organic pool through which nutrients are recycled to either pelagic or littoral communities.

LITTORAL (OR FRINGING) HABITAT

Littoral (or fringing) habitats occupy the interface between open water and land where the water is sufficiently shallow and calm to allow the growth and, eventually, dominance of macrophytes. These include the large algae (i.e. seaweeds) and the free-floating or rooted herbaceous and woody vascular plants which can tolerate varying degrees and length of submergence in fresh, brackish or salt water. The composition of the biota which occupy this transitional zone is extremely varied as a result of the latitudinal and associated temperature range, frequency and range of water-level, degree of shelter, and the extent and composition of the substratum. Although they occupy a very small area (*c.* 0.6 per cent of the global aquatic habitat) the littoral ecosystems, because of nutrient enrichment consequent upon their ability to trap organic and inorganic sediments, are particularly productive. The estuarine ecosystem is one of the most biologically productive in the world.

Estuarine habitats

Estuaries are long, wide, often 'funnel-shaped' mouths of rivers in which fresh and marine water meet and intermix. Despite extreme variety of scale, geology, geomorphology, climate and biota, the estuarine habitat is distinguished by a number of ecologically important characteristics. First, it is relatively well sheltered from excessive wave action and currents. Second, its water exhibits a salinity gradient from zero to over 32 per cent from the head of the estuary seawards. The steepness of the salinity gradient, however, is dependent on the volume of fresh water entering the estuary, the tidal range and the amount of evaporation. Third, estuarine water normally carries a large load of dissolved and suspended mineral and organic matter brought in from the sea or the land or contributed by organic decomposition in water *in situ*. As a result turbidity is high, particularly in the area of maximum water mixing. The location and magnitude of the turbidity maximum are dependent on the particle size of the suspended load and the water salinity.

Three water patterns, dependent on the interaction of salt and freshwater circulation, are recognised:

1. *Highly stratified or salt-wedge* conditions where the freshwater river inflow is large, and extends well down the estuary, compared to the tidal incursion. In this situation the less dense fresh water overrides the denser wedge of salt water. The vertical junction between the two tends to be sharp, as is the surface change which takes place well down the estuary. Mixing is limited to entrainment mixing along the junction. An extreme form of this condition is found in deep fiords where a pronounced lip at the seaward end of the estuary restricts tidal movements to surface water and results in cold, relatively stagnant, anoxic conditions in the deepest basins in the upper reaches of the fiord.
2. *Partially mixed* conditions occur when salt and freshwater flows are more evenly balanced and undiluted fresh water is found only at the head of the estuary.
3. *Well-mixed or homogeneous* conditions usually occur in very wide estuaries where the Coriolis force results in the horizontal separation and circulation of in- and outflows rather than in vertical stratification.

The circulation and balance of fresh versus salt water in the estuary are also affected by evaporation rates. In this respect, a distinction can be made between *positive estuaries* where (as in temperate latitudes) potential evaporation is less than the volume of outgoing fresh water; and *negative estuaries* (characteristic of tropical latitudes) where potential evaporation is greater than the freshwater flow and the salinity of the water increases up-estuary and restricts the seaward flow and mixing of fresh water; and *neutral estuaries* where the volume of tidal flow, freshwater flow and evaporation are equally balanced to give a rarely observed water regime with static salinity. The salinity at any point within the estuary is a function of the volume of tidal versus fresh water, the tidal range, the form of the estuary and of the climatic conditions. The *flushing time* (*FT*), i.e. the time sequence for fresh water in the estuary to be replaced by river discharge, is expressed as the ratio between the total fresh water accumulating (Q) and the rate of river flow (R), i.e. $FT = Q/R$. The efficiency with which an estuary reoxygenates its water, recycles its large load of organic matter and can deal with pollution is closely related to how well it is naturally flushed.

Estuaries are also characterised by a progressive sedimentation of fine silts and mud consequent on the reduction in water flow and the flocculating influence of salt water on fine particles. This is reflected in a zonation of sediment texture which is closely correlated with water velocity and salinity from the head to the mouth.

Primary production in estuaries is, to a very large extent, dominated by perennial, emergent vascular macrophytes and microalgae. The former are characteristic of salt-marsh vegetation tolerant of varying depths and periods of submergence in brackish or salt water. They also trap and stabilise the relatively mobile and anoxic sediments in which they are rooted. Salt-marsh plant communities are dominated by a small number of species such as mangroves, papyrus and *Spartina* grass of tropical to subtropical latitudes; and grasses (eelgrass, *Glyceria*) and succulents of cool temperate estuaries. Where there is a large tidal range, the fringing vegetation may reveal a marked intertidal zonation of dominant plants from lower to higher areas of the salt-marsh. The microalgae, together with diatoms, are the most important primary producers on and in the upper centimetre or so of otherwise bare mud-flats. The phytoplankton, although present, are limited by the shallowness and, particularly, the turbidity of estuarine water.

Primary productivity is high (see Table 15.4) consequent on a large input of nutrients from both the land and the sea. McLuskey (1981) has described them as 'natural eutrophic systems'. However, although some salt marshes are used for periodic domestic livestock grazing, they support few native grazing animals and a high proportion of the primary production enters the detrital food chain.

The estuarine detritus provides a particularly rich and relatively constant source

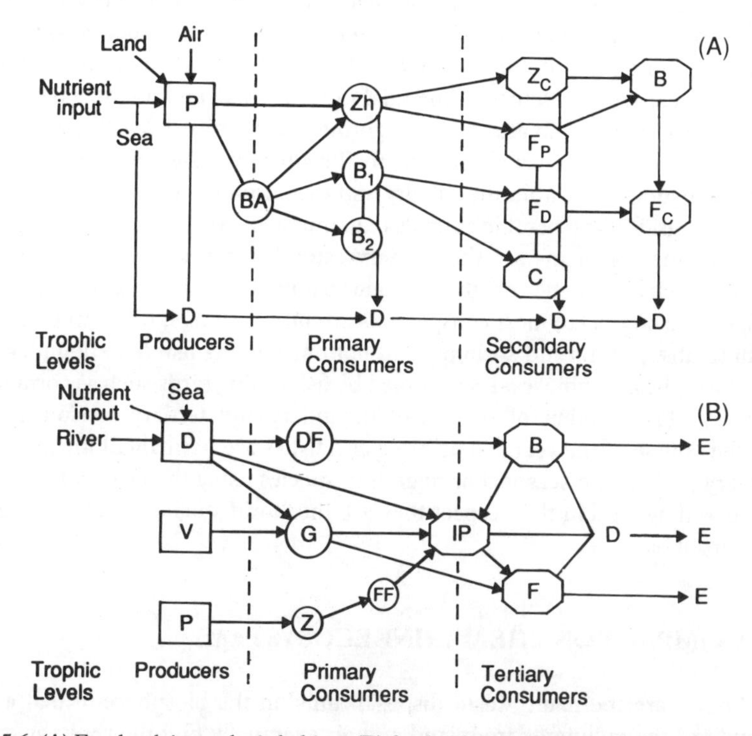

Fig. 15.6 (A) Food web in a pelagic habitat; (B) food web in an estuarine habitat: P = phytoplankton; D = detritus; V = vegetation; BA = bacteria; Zh = herbivore zooplankton; B_1 = macrobenthos; B_2 = meiobenthos; DF = detritus feeders; G = plant grazers; Z = zooplankton; FF = filter-feeders; IP = invertebrate predators; Z_c = carnivore zooplankton; F_p = pelagic fish; F_d = demersal fish; C = other carnivores; B = birds; F_c = large fish; E = export

of food either as dissolved organic carbon from leaves or as fragmented plant material for large populations of detritus-feeding organisms. Indeed, estuaries are *detrital systems par excellence* with a high load of biogenic material derived from salt-marsh plants and benthic animals (such as mussels, oysters and worms) or imported by rivers or tides (Fig. 15.6). McLuskey (1981) identifies two main types of estuary – the European and the American – which represent the extremes of a continuum of variation in terms of primary and detrital energy flows. The European estuary is characteristic of north-west temperate (North Atlantic) conditions. The tidal range is large, a high proportion (90–95 per cent) of the intertidal area is composed of mud-flats, and a considerable amount of the primary production is undertaken by large populations of microalgae living on this substratum. Macrophytic vegetation is normally confined to that area between the highest level of the lowest tide and the highest level of the high tide and often displays a well-marked zonation (or spatial succession) of dominant plants from the lowest most frequently immersed to the highest least frequently immersed area. More of the detrital energy, however, is derived from outside than within the estuary, brought in from the sea and from the land via river and sewage discharge. It is, for instance, estimated that 75 per cent of the detritus in the Wadden Sea estuarine area is imported from the North Sea. While the European estuary acts as a detrital and nutrient trap, the American estuary of the south-eastern coast of the USA is a net exporter of detritus. The tidal range is small and up to 75 per cent of the intertidal zone is dominated by the vigorous marsh grass (*Spartinia* spp.) most of which goes into the detrital food chain. Production, however, tends to be in excess of consumption within the estuary and the surplus is carried out to sea.

The detritus-feeding benthic animals (e.g. mussels, oysters, worms) living on or in the estuarine sediments are the primary estuarine consumers rather than the zooplankton characteristic of the pelagic habitat. Of the many and varied secondary consumers the most conspicuous are birds (waders, wildfowl and gulls), shellfish (crabs, prawns and shrimps) and bottom-feeding fish (e.g. flounders and gobi). The highest trophic level is occupied by fish-eating birds such as cormorants and osprey. The number of species of secondary and tertiary consumers vary during the course of the year. Some birds and fishes are permanent inhabitants of the estuary. Others are seasonal or migratory species using the estuary for feeding and/or breeding, and in this respect they are the world's major fish breeding and nursery grounds.

HUMAN IMPACT ON THE MARINE ECOSYSTEM

Water bodies are the main 'waste-disposal units' in the biosphere in that a high proportion of the pollutants from land and air eventually find their way by one or more routes into coastal fresh or salt-water bodies. Biological impact is greater in shallow lakes, estuarine and coastal waters where water turnover is slow. The most vulnerable are those located in or adjacent to highly urbanised, industrialised and/or intensively farmed land. In this context it is worth remembering that most of the world's large millionaire cities are situated on estuaries.

Water pollution in general will be dealt with in Chapter 17. However, human impact on the ocean ecosystems merits special emphasis. Much less is known about the biological effects of pollution on the major global sink and there has long been an assumption that the volume of inputs are small and of negligible effect in comparison with the water volume. Although there has, to some extent and in some areas, been a reduction in dumping of land waste in deep-sea areas there has been a rapid increase in oil leakage (particularly in ports) and oil spillage in coastal and ocean waters consequent upon the rate of escalation in the past 40 years in the role of petroleum movement by massive tankers.

Oil pollution takes place mainly in marine waters and relatively little is known about long-term cumulative effects of the fractions with long resident times in sea-water. Some may be subjected to biological concentration in food chains; some aromatic hydrocarbons are known to affect the flavour of seafood and induce carcinogenesis under experimental conditions. The effect of oil pollution on the marine biota depends on the amount and type of oil spilt; the rate of evaporation of its component fractions; the season when the spill occurs; the toxic inputs to or outputs from the oil; and the species affected and their habitat.

The ecological hazards of oil pollution are exacerbated by the tendency for oil, once spilt, to form an extensive thin surface layer – an *oil slick*. The effect of the oil spill depends on the volume and scale, the rate of chemical and physical dispersion, and on changes in the properties of the oil as a result of evaporation of lower-density components. This leads to thinning and spreading and to increase in density of the remaining oil droplets. The latter may be biodegradable, sink, or if not dissolved, give rise in turbulent water to oil–water emulsion droplets *c.* 15–20 μm; at concentrations over 40 cc per the oil becomes a very viscous, stable 'chocolate mousse' (Atkins Research and Development n.d.).

The most obvious effects of oil pollution are the physical clogging and blanketing of surfaces and structures which inhibit the movement, respiration and feeding of small animals. Fish gills become coated and aromatic substances irritate delicate gill and gut tissues. Hydrocarbons cover seaweeds while many can penetrate plant tissue and impair metabolism. The most seriously affected by large oil spills are sea birds which perish immediately or die even after cleaning.

The petroleum and associated hydrocarbons that eventually sink become incorporated in or absorbed on bottom sediments. Their residence time will depend on how long they take to be biodegraded. In areas subjected to disturbance of bottom sediments they may be recirculated for some time before this occurs. At present all too little is known about the impact of the heavier oil droplets on the plankton, micro-organisms and benthic mammals of the ocean depths.

PART III BIOTIC RESOURCES: USE AND MISUSE

Chapter

16

Ecosystem stability and disturbance

The idea that an organic community or species assemblage attained, as a result of successional development, a final stage that could be maintained so long as prevailing environmental conditions remained relatively unchanged, was implicit in the early formulation of both the climax and the ecosystem concepts (see Table 16.1); it was initially expressed by the term 'stability'. Stability as applied to an ecosystem is, however, a particularly elusive concept. The definition of the term alone has produced a voluminous literature embodying a thicket of terminology which has, unfortunately, more often tended to confuse rather than clarify the issue (Hill 1975). Table 16.2 summarises the main definitions of stability or, rather the various attributes of stability. Walter (1973) maintains that Holling's (1973) definition of ecological stability as the ability of a system to return to an equilibrium state after disturbance and its corollary of resilience as the system's ability to absorb change in all inputs and still persist are the most useful in terms of the approaches it suggests and the questions it poses. However, the concept of stability raises two basic problems:

1. What biotic attributes contribute to stability?
2. How do biotic communities respond to disturbance?

DIVERSITY AND STABILITY

The assumption that the more diverse and complex an ecosystem is the more stable it would be, early became a widely accepted tenet – indeed an ecological principle – which in the late 1950s and 1960s had a profound influence on the approach to nature conservation, particularly in Britain. It derived from the view then currently developing that a mature ecosystem could be maintained in a condition of *dynamic equilibrium* or *steady state* in face of environmental fluctuations by reason of a multiplicity of negative feedback processes which could, within limits, dampen extreme oscillations in species populations. In other

Table 16.1 A tabular model of ecological succession: some trends to be expected in the development of ecosystems (adapted from Odum 1969)

Ecosystem attributes	*Developmental stages*	*Mature stages*
Community structure		
1. Total organic matter	Small	Large
2. Species richness	Low	High
3. Species diversity	Low	High
4. Stratification and spatial heterogeneity (pattern diversity)	Poorly organised	Well organised
Community energetics		
5. Gross production/respiration (*P/R* ratio)	Greater or less than 1	Approaches 1
6. Gross production/biomass (*P/B* ratio)	High	Low
7. Biomass supported/unit energy flow (*B/E* ratio)	Low	High
8. Net community production	High	Low
Life history		
9. Niche specialisation	Broad	Narrow
10. Size of organism	Small	Large
11. Life cycles	Short, simple	Long, complex
Nutrient cycling		
12. Mineral cycles	Open	Closed
13. Nutrient exchange rate between organisms and environment	Rapid	Slow
Homeostasis		
14. Nutrient conservation	Poor	Good
15. Stability (resistance to external perturbation)	Poor	Good

words, the greater the number of species the greater would be the number of species interactions and, hence, the smaller would be the change in the system as a result of the disturbance in any one biotic or abiotic component.

The assumed correlation between diversity (usually expressed as species richness) and stability was based on theoretical and empirical evidence (Elton 1958) derived from a comparison of more with less complex physical systems. In the first place, it was argued that simple systems were theoretically less stable than complex ones; also the greater the number of routes by which food energy reached a consumer species the less disturbing would be the reduction or elimination of any one of these. In the second place, it had been observed that the probability of a population explosion of pests in species-simple agricultural ecosystems did not occur in the species-rich tropical forest; that marked cyclic population fluctuations were characteristic of species-poor Arctic ecosystems; and that small, remote, isolated and species-poor islands were easily invaded by other species. In addition,

Table 16.2 Definitions of stability (Pimm 1986)

Term	*Definition*
1. Stable (non-dimensional, the system is either stable or not)	A system is deemed stable if and only if the variables all return to the initial equilibrium following perturbation
2. Resilience (units are those of time)	How fast the variables return to equilibrium following perturbation
3. Persistence (units are those of time)	The time a variable lasts before it is changed to a new value
4. Resistance (non-dimensional units)	The degree to which a system is unchanged by a perturbation
5. Variability (coefficient of variation – dimensionless)	The variance of population density or similar measures like standard deviation or coefficient of variation

and based mainly on the comparison of the high species richness of the tropical rain forest with other poorer biota, there was a tendency to explain diversity as a product of environmental stability.

The diversity–stability concept, however, has not stood the test of time in face of more recent and rigorous investigations. As Walker (1989) notes, the palaeontological record of changes in diversity indicate that, at least in the case of animals, stability does not necessarily generate diversity. Also, as has been noted, studies of succession show that maximum species diversity is not necessarily attained at the final stage in ecosystem development. Further, as Pimm (1986) points out, species equitability within trophic levels and/or the presence of keystone species (whose activities may determine community structure, Krebs 1985) may be more important determinants of ecosystem stability than species diversity alone. The reduction or elimination of the population of a top predator may have less serious repercussions than the reduction or elimination of a key animal species or a dominant plant species. Classic examples of keystone species noted by Krebs (1985) include the African elephant – a herbivore dependent on shrubs and small trees supplemented by grass. Over-browsing can reduce the former faster than they can regenerate and, with the increasing dominance of grass, elephant populations decline and other grazing ungulates increase in number. The Canadian east coast lobster and sea otters of the Pacific coast of North America are well-known aquatic keystone species. A reduction of both by exploitation has resulted in an explosive increase in the sea-urchin populations on which they feed and, as a result, the virtual elimination of *Laminaria* (kelp) and *Alaria* seaweeds.

To date, observation and experimental studies of the relationship between diversity and stability have been concerned with population and community variation of a few taxonomic units of micro-organisms, insects and birds. This is a reflection of the difficulties involved in stability studies and the need to select taxonomic units within areas that can be monitored within a reasonable time-scale. Again, as Williamson (1981) points out, even if groups (assemblages) of species within a community were shown to be stable this would not necessarily mean that the community as a whole was stable; and the same would apply to the relationship between communities and ecosystems. In addition, the space-time scale of stability must be taken into consideration. The length of time a constituent taxon will remain constant will be a function of the life cycle of the organisms. Populations of micro-organisms may have a stability time span of a few months, birds of years, large organisms of hundreds of years. Also stability demonstrated on the basis of one small sample area cannot be taken as evidence of the stability of a community or ecosystem occupying a large area.

Most diversity : stability studies relate to animal populations and communities. However, Grime (1979) discusses the use of plant-form and life-cycle strategies as a basis for the prediction of vegetation stability in terms of relative resilience. He makes a basic distinction between resistant vegetation which is lower in ruderal and competitive dominants but higher in stress-tolerant dominants, and resilient vegetation higher in ruderal and competitive dominants and lower in stress tolerant dominants and in which there are persistent seed banks or plant perennating buds in the soil. Also, preoccupation with the evolutionary implications of species diversity and stability has meant that less serious consideration has been given to investigations of the relationships between ecosystem stability and ecosystem characteristics such as biomass, energy flow and nutrient cycling, all of which are amenable to field measurement.

RESILIENCE (OR INERTIA)

While stability is an expression of an ecosystem's ability to dampen down, to regulate variations in its components and processes and to maintain a near-equilibrium or steady state, resilience is an expression of a system's ability to tolerate and recover from internal fluctuations consequent on above-average abiotic, biotic or anthropogenic variations (i.e. disturbances) which tend 'to push' the system away from a mean or steady state. The extent to which a system can fluctuate before being replaced by a different one can be thought of as its range of tolerance to disturbance, the width or amplitude of which will be determined by the system's homeostatic limits as illustrated in Fig. 16.1. However, the identification of the limits within which a system can fluctuate is particularly difficult because each component species or process varies in its tolerance of a particular type and level of disturbances or combination thereof which itself may vary. In addition, no ecosystem is or can be stable in the sense of maintaining a constant state because all exist in, and react to, an environment which varies to a greater or lesser degree both temporally and spatially. The recognition of this has led to a rejection by some ecologists of the concept of an equilibrium-based hypothesis of stability in favour of a non-equilibrium hypothesis. Increasing

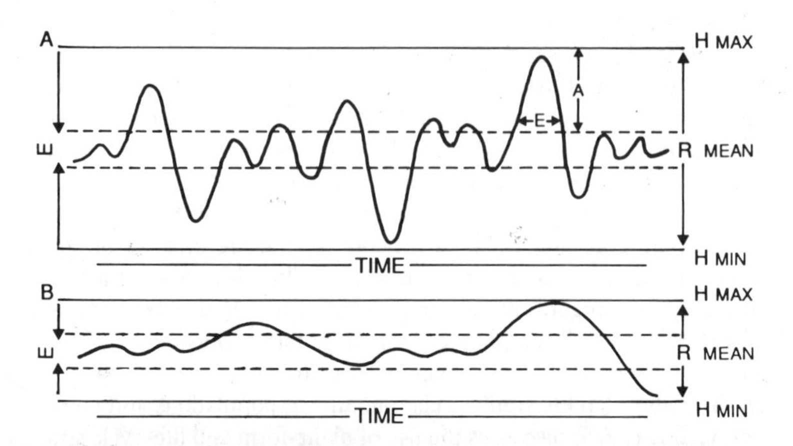

Fig. 16.1 Diagrammatic representation of (A) resilient ecosystem with wide range of tolerance to disturbance and (B) less resilient (fragile) ecosystem with a narrow range of tolerance to disturbance. R = mean range of fluctuation; H = homeostatic limit; E = environment (space); ←E→ elasticity; A = amplitude

emphasis in ecological studies has, as a result, been directed towards the nature of disturbance and the reaction of ecosystems to it.

DISTURBANCE

It is only within the last decade that *disturbance* and its concomitant recovery processes have gained full acceptance as natural, frequent, integrated aspects of all ecosystem behaviour. Reiners (1983) and Sousa (1984) define disturbance somewhat narrowly; the former as the destruction of living biomass or accumulated debris, the latter as a discrete punctuated killing, displacement or damaging of one or more individuals that directly or indirectly creates an opportunity for new individuals or colonies to become established. Bazzaz (1983) combines cause and effect in his definition of a disturbance as a sudden change in the resource base (inputs) of a unit of landscape that is expressed as a readily detectable change in population (ecosystem) response. In the latter sense, disturbance can be reflected in one or more changes in the structural or functional attributes of the ecosystem. Two other terms – *perturbation* and *stress* – are frequently used as synonyms of disturbance; or with perturbation meaning either an experimentally applied or anthropogenic disturbance and with stress applied to factors which produce large negative impacts on organisms. However, as Bazzaz (1983) points out, the distinction between human and natural disturbance is less important than its consequences and the way in which species respond to it over ecological and evolutionary time. Disturbance plays a significant role in the organisation and functioning of all ecosystems in the production of spatial patterns within ecosystems and in the evolution of species strategies adapted to it.

DISTURBANCE REGIME

The reaction of an ecosystem to disturbance depends, on the one hand, on the type and degree of disturbance, i.e. the *disturbance regime*; and, on the other, on the vulnerability of the ecosystem to disturbance. The former is a function of its scale, magnitude, frequency, predictability and the turnover rate or rotation period (i.e. mean time-period) required to disturb an entire area or ecosystem (Sousa 1984). These will be exemplified in relation to fire which is both a natural and a man-initiated process. Also it is one of the most common and most intensively studied types of disturbance.

1. The *scale* or size of a disturbed area varies from the gap created by the fall of one large dominant tree to extensive areas of regional or continental scale and, now, in respect of air and water pollution, to the whole biosphere. The area disturbed by human activity tends to be larger (and is rapidly becoming larger) than that affected by purely natural agencies. All other things being equal, the greater the scale of disturbance the greater will be the resulting changes in microclimate, in the hydrological cycle and consequently in soil conditions. Forest fires can affect areas varying from hundreds to, in extreme cases, hundreds of thousands of acres.

2. The *magnitude* of the disturbance is a function of its intensity and hence the amount of change effected. The intensity of a fire, for example, is a function of the degree of heat generated and the duration of the burn. Both are dependent on the amount and inflammability of the available fuel (i.e. surface and trunk litter). The amount of change effected by a given intensity is related to the type of vegetation, the time interval since the previous fire and, particularly in grasslands, the amount of growth in the previous season. Sclerophyllous leaves, such as those of conifers and eucalypts, decompose only slowly and so build up stocks of ground litter; many of the latter species have a decorticating bark. In addition, both contain oils with exceptionally high flash points. Fire intensity is also affected by the speed and duration of the burn; indeed 'high-energy', rapidly moving crown fires may do less damage than long duration, slow-burning, ground fires during which the underground regenerative plant organs and soil animal life may be very seriously disturbed or destroyed. Other variables which affect fire intensity are the moisture content of the vegetation, litter and soil; wind-speed and direction; and surface relief. Fire intensity is a function of a number of interrelated biotic and environmental variables and, further, it can vary both spatially and temporally in the course of its duration.

3. The *frequency* with which a disturbance occurs in the same place varies from the infrequent 'one-off' event such as a volcanic explosion to a yearly or in some cases almost continuous occurrence (e.g. grazing, air pollution). In general, small natural disturbances tend to be more frequent and regular (or predictable) than the larger less predictable disastrous or catastrophic events. The time and place of occurrence of some forest fires and tropical hurricanes are consequent on modern sophisticated monitoring techniques more amenable to prediction. Fire hazards can be assessed on the basis of weather and ground conditions and/or the amount of

inflammable litter (or fuel) that has accumulated since the previous fire. In this respect it is worth noting that human activity frequently increases the intensity and frequency of a naturally occurring fire disturbance. Initially fire management was directed to controlling all fires with the result that fire frequency decreased but fuel accumulated to an extent that less frequent but larger, more intense and destructive fires were generated.

ECOLOGICAL EFFECTS OF DISTURBANCE

Organisms vary in their tolerance or adaptation to varying disturbance regimes – so that disturbance in the first instance is a selective ecological agent. Survival depends on a resistant life form and/or the ability to produce propagules in the interval between disturbances or immediately after the event. Some species, in fact, require disturbance in order to stimulate seed germination or clone growth.

Effect of fire on biomass

Plants vary in their susceptibility to different intensities of burning. Those which can survive and which, in some cases, are also stimulated by fire are the so-called *pyrophytes* (or fire-resistant plants). Most liable to damage or destruction are annuals and seedlings of perennial plants. Among the most resistant are mature perennial plants with underground food-storage organs from which renewed growth can take place after surface burning. Many of these, such as for example heather, bracken, perennial grasses and trees which produce suckers, have new growth stimulated by the removal of old or dead surface material. For this reason burning has long been recognised and used as a means of promoting the vigorous growth of those plants which are of the maximum use as forage for grazing animals. Above all, fire favours grass and other perennial herbs, at the expense of trees and shrubs, since the former are so much better adapted to withstand its effects. Perennial grasses not only recover quickly, often growing with increased vigour, but they can produce abundant seed in 1 or 2 years after germination. Most trees and shrubs, however, require several years before producing seeds and fire may recur so frequently as to prevent them completing their life cycles and hence regenerating.

Some trees and shrubs are, however, more resistant than others. The woody monocotyledons (the palms and palmettoes of intertropical regions) are particularly well adapted to withstand injury since they do not possess the vulnerable growth tissues (the *cambium*) in their trunks as do other types of trees. Among the latter, a particularly thick or tough bark as in the cork-oak (*Quercus suber*) or the Japanese larch (*Larix leptolepis*) may give greater protection from fire. The foliage and wood of many deciduous and broad-leaved evergreen trees burn less readily than that of the highly resinous conifers and eucalypts. The ability, characteristic of many trees except conifers, to produce new shoots or 'suckers' from the base or lower parts of their trunks and from *lignotubers* on their roots aids recovery from burning. Walker (1982) suggests that the main strategies for tree survival are a resistant crown, root sprouting (from lignotubers) and a heavy investment in seed banks. Some species of pine can produce seed at a very early age, as little as 2 or 3 years after germination. This allows the plant to complete its life cycle before enough debris has collected to initiate the next fire. Others have particularly hard-coated seeds which can withstand very high temperatures. In many of these respects, conifers – and particularly the pines – are outstanding in their resistance to and recovery from fire. A number of pines, such as the North American jack-pine (*Pinus banksiana*), the Mediterranean pine (*P. pinaster*) and the lodgepole (*P. contorta*) pine of western North America, have serotinous fire-cones. The ripe but unopened cones remain attached to the tree for many years, only opening and releasing their seeds when, as the tree gets older, they become desiccated. This process is stimulated by the heat and dryness during and following fires and allows such trees to regenerate rapidly in an area in which other types of plants and their seeds have been destroyed. The animals associated with fire-dependent ecosystems are those better adapted to survive by reason of high mobility, a burrowing habit, or a resistant imago. Many in fact depend on post-fire vegetative growth and it has been noted that some of the highest densities of animals and, more particularly, of higher herbivores plus their predators and scavengers are in fire-affected habitat (Vogl 1980).

While humans have modified and intensified the effects of fire by its deliberate use, they have also, in more recent times, attempted to control and reduce its frequency. In many instances methods of fire protection have reduced the frequency of fires but have, unwittingly, greatly increased the intensity and destructiveness of those that occur. Protection from fire may allow the accumulation of plant debris to such an extent that its severity is very much greater than when it occurred frequently. The effects of fire exclusion have been particularly severe in the Ponderosa pine forests of south-western USA. The trees originally grew in scattered groups in an open, grassy, parkland formation. This pattern had been established and maintained by lightning and fires set by Indians over a long period of time. These low-intensity fires, which probably occurred at regular intervals of 3–10 years acted as a natural thinning agent and kept down the amount of surplus combustible vegetation. The introduction of domestic livestock (which reduces the inflammable grass cover) together with an efficient fire-protection programme has, during this century, eliminated regular fires. The result has been an increase in pine cover and the establishment of often impenetrable pine thickets. The absence of thinning combined with high tree-stocking rates led to stagnation of growth, while the gradual build-up of excess fuel increased the potential fire hazard to dangerous levels.

TYPES OF DISTURBED ECOSYSTEMS

It is now generally accepted that disturbance is a universal phenomenon which affects all ecosystems to a lesser or greater extent. However, the type and role of disturbance in the composition and organisation of ecosystems vary. In this respect three types of ecosystems can be identified:

1. Disturbance controlled;
2. Disturbance dependent;
3. Disturbance maintained.

Disturbance-controlled ecosystems

Disturbance-controlled ecosystems include the idea of periodicity of disturbance, widely fluctuating physical habitats characteristic of intertidal coastal areas and of a highly mobile substratum such as is associated with periglacial freeze–thaw conditions, water movement, steep slopes and all habitats subject, in general, to continuous erosion and/or deposition of weathered mineral material.

The intertidal coastal habitat at the interface between the marine and the terrestrial ecosystems is continuously disturbed by wave motion, the energy of which varies spatially dependent on the relief and orientation of the coast plus offshore marine processes. Not only is it one of the most intensely disturbed habitats but it is one in which variation in intensity is extreme. When the coast is formed of unconsolidated sediments (cobbles, pebbles, sand/or mud) the intensity of disturbance with which living organisms have to contend is further increased by the continous movement of the mineral particles. In this type of habitat the ecosystem is controlled and dominated by the physical environment, and the intensity of disturbance is such as to inhibit the establishment of a continuous biomass and DOM cover which in less disturbed circumstances would stabilise the site and dampen the physical fluctuations. The latter condition is exemplified along sheltered aquatic margins (i.e. estuaries and lakes) where salt or freshwater marsh communities absorb the less intense wave movements and help progressively to trap and stabilise fine sediments outwards from the water's edge.

In the more intensely disturbed habitats, however, the vegetation remains open, discontinuous in time and space at what, in successional terms, would be called a 'youthful' colonising stage. Organisms which live in this environment must also be resistant to high stress from salinity, alternating submergence and desiccation and physical impact. Plant species other than the floating or attached macroalgae (the seaweeds) are few. Vascular species exhibit many of the classic characteristics of ruderals (Grime 1979). Most are annuals or short-lived perennials able to complete their life cycles during the short intervals between disturbances during the climatic growing season. Flowering and seed production are accomplished early in their life cycle, and when physical stresses are particularly high they can be completed, albeit with lowered productivity, more rapidly.

Animal species are also few and specialised. The majority are invertebrates with protective shells and, in many instances, having considerable powers of surface adhesion, e.g. barnacles and limpets. In addition, they have to be capable of non-aquatic respiration. Survival is more difficult for animals than for plants because of the vulnerability of eggs, larvae and juveniles, and in many species the planktonic larval stage is abbreviated or omitted. However, with greater mobility than plants the littoral animals can take advantage of the wide variety of microhabitats on, under and within the sediments for shelter.

Disturbance-dependent ecosystems

In contrast to ecosystems in naturally disturbed physical habitats, there are those which appear to require a certain intensity of external disturbance for their regeneration and hence persistence. This, as previously noted in Chapter 9, is particularly characteristic of unmanaged mature tropical and temperate forests.

These ecosystems not infrequently create a microenvironment and particularly a microclimate in which the regeneration is either retarded or inhibited by competition for light and nutrients, or even the production of toxins, by the dominant trees. Regeneration of the latter and hence of the entire ecosystem in a particular area is dependent on the creation of openings, i.e. gaps in the canopy in which secondary succession can be initiated. The higher light intensity and soil temperatures promote more rapid decomposition of DOM, trigger the germination of dormant seeds which have remained in a viable condition in the soil and stimulate the growth of saplings of the dominant trees.

The creation of such regeneration gaps is a normal process in a forest as trees age and they become more susceptible to parasites, disease and tree-fall as a result of wind-throw or natural collapse. In the humid tropical and temperate forests where trees attain a large biomass and height, tree-fall can create a major local disturbance which may bring down adjacent individuals and damage understorey vegetation. In other ecosystems regeneration is dependent on externally initiated disturbance by high wind-force or fire. In the humid sclerophyllous forests of the Boreal and humid warm temperate biomes, seed dissemination and germination are regulated and the nutrient cycle maintained by recurring fire. In the first case, fire frequency must be within the productive life cycle of the particular species. Too high a frequency may prevent seed production, too low a frequency may allow the growth and dominance of more competitive, longer-lived and more fire-resistant species. The dominance of the eucalypt family in the humid areas of eastern Australia is now attributed to the felling of the former dense evergreen rain forests, in which fire frequency was less than 200 years, and the rapid colonisation of the areas so cleared by the fast-growing and prolific seed-producing gums (Gill *et al.* 1981). The rapid production of highly inflammable eucalypt litter ensures a fire frequency such that the regeneration of the rain forest is inhibited. The dependence of the Boreal forest on fire to release the increasing large nutrient pool that accumulates in the growth of *Sphagnum* mosses on the floor as the forest ages has already been noted (see p. 204).

Recovery from disturbance

The rate and nature of ecosystem recovery from periodic disturbances of equal intensity will depend on the size, shape (form) and location of the disturbed area (or patch). The size and area of the patch influence both the rate and pattern of ecosystem recovery in several ways (Sousa 1984: Pickett and White 1985). Environmental change effected by a particular disturbance will tend to have less impact in a small than in a larger patch. In the former species replacement will tend to be relatively rapid and be almost exclusively effected by the growth of vegetation or seeds from the surviving vegetation. In the latter, where few or no plant propagules may survive, recovery will be slow because of dependence on colonisation from outside the patch and successive establishment of the original vegetation. In areas of intermediate size, recovery will probably be by a mixture of surviving and colonising plants. The balance between the two groups will be dependent on the form and location of the disturbed area as well as its size.

The rate and pattern of post-disturbance colonisation are also a function of the ratio of patch perimeter length to patch size, which is greater for smaller and/or irregularly shaped areas than for larger more regularly shaped ones. In the former case, patch colonisation by the lateral spread of perennial plants and by seed dispersal from the surrounding undisturbed vegetation will be facilitated. In the latter, colonisation will tend to be slower and colonists with a greater seed dispersibility than the species in the undisturbed contiguous vegetation may gain a foothold and become temporarily or permanently dominant. In this case the location of the disturbed area relative to a source of propagules is particularly important where the dispersability of the original dominant plants is low – as in the case of the tropical rain forest (see p. 74). Other factors which will influence recovery will be the time of disturbance in relation to seasonal or annual variations in seed production by the potential colonists; the density of consumer organisms (particularly herbivores) for whom the disturbed patch provides a sheltered food niche; and the degree of environmental variation within the disturbed area itself.

Disturbance is an important ecological process in creating and maintaining local species diversity. It does so because of variation in the time of disturbance and rate of recovery which results in a mosaic of patches of varying species assemblages, and variation in space (i.e. in size, location and microenvironment of disturbed areas) which results in within-patch species diversity.

Disturbance-maintained ecosystems

With respect to the ecological impact of fire Vogl (1980) makes a distinction between *fire-dependent* and *fire-maintained* ecosystems – a distinction which, as already suggested, can be applied, more generally. Disturbance-maintained ecosystems are those where the intensity of disturbance is such that the original ecosystem is replaced by another and recovery is inhibited either by continued disturbance because of a change in the physical habitat such that the previous plant communities cannot regenerate or a lack of available colonists. In the case of the extensive areas of semi-natural temperate and tropical grasslands, of Mediterranean scrub vegetation (*maquis*) of Europe, California (*chaparral*) and South Africa the pre-existing forest and woodland have either been felled or their regeneration prevented by frequent periodic fires and/or continuous grazing by wild and domestic herbivores. The extent to which such ecosystems can be maintained in a relatively 'stable state' is, however, debatable, even if the same intensity of disturbance is maintained. Most are now subjected to an increasing intensity of disturbance consequent upon a continuing increase in domestic herbivores and/or increased frequency of fire. Indeed, many such areas particularly in tropical regions are, as a result of continuing disturbance, being depleted to the extent that biological survival is rapidly becoming endangered. Increasing intensity of human disturbance has engendered growing concern about the ecological consequences and the relative resilience or fragility of both managed and unmanaged ecosystems.

FRAGILITY

Fragility can be defined as the relative ease with which an ecosystem can be disrupted (Tivy and O'Hare 1981). It is a function or counterpart of resilience which Holling (1973) graphically described as a measure of the probability of the extinction of an ecosystem. Both resilience and fragility are complex, relative concepts supportied by relatively little empirical or experimental data at a biome scale and lacking a sound theoretical underpinning. Statements about resilience or fragility are more often than not based on either general observations or on the results derived from small-scale field or experimental data about a particular ecosystem component (species population or community) or function (nutrient cycling). The measurement of the biological response to a given type and intensity of disturbance (or impact) is particularly difficult. Field monitoring needs to be long term, often exceeding the span of the working life of one individual or research group. Experimental manipulation of impacts tend to concentrate on organisms, particularly animals, with very short life cycles. Many field-monitoring studies have been undertaken to assess the impact of wild (rabbits, deer) and domesticated herbivores grazing on vegetation composition on the basis of a comparison between areas subject to known levels of grazing and those from which the herbivore being studied is excluded. A problem basic to all such studies is that the extent to which the vegetation has been previously disturbed by grazing and/or other processes can only rarely be established with any degree of certainty or accuracy. In addition, it is not at all easy to assess and eliminate background noise resulting from the operation of other physical or biotic variables; this is a problem which becomes acute in studies of situations (such as previously noted in the intertidal marine habitat) where natural variability of both physical and biota is extreme in periodicity and in range.

RECREATION ECOLOGY

The rapid growth of informal outdoor recreation, particularly in Western developed countries during the post-Second World War period, engendered concern about its impact on wild or relatively unmanaged ecosystems. Humans, like large ungulate animals, create disturbances by reason of their physical impact on the ground surface. The former, however, differ in that the impact is excerbated by the high local concentration of people within a recreational area and by their associated vehicles (cars, caravans, boats, etc.). The increasingly obvious indications of damage in a range of habitats raised questions about the resilience and *ecological carrying capacity* of such areas for this type of use. During the 1970s attempts were made to correlate changes in the ecosystem components such as species diversity and equitability, plant form, plant vigour, depth of litter and DOM with different intensities of impact (Liddle 1975 a & b; Tivy 1980). The latter was either monitored in terms of the actual number of people traversing per unit area or length (in the case of narrow paths); or simulated by regulated trampling by people or trampling machines compared with untrampled control areas. These studies confirmed and quantified general observations that the ecological changes were dependent on the intensity of impact, the type of vegetation and the physical condition of the site. Low intensity of trampling may effect bruising and flattening from which the vegetation may recover fully the next growing season. It may in fact increase productivity by stimulating tillering of grass species. It is also often reflected in an increased species richness because of the reduction in competition, particularly for light by the taller normally dominant plants. Increasing impact is exacerbated by soil compaction with a reduction in the infiltration of water and, dependent on gradient, increased surface wetness and runoff. The height and total biomass of the vegetation diminish. Plants less tolerant of physical impact are in time eliminated in favour of the more resistant. The latter often include grasses and sedges with tough leaves, underground storage organs or root stocks with perennating buds in or just under the surface litter. Most resistant of all are forbs with flattened rosette leaves and deeply penetrating root systems (such as plantain, daisy and dandelion) (Table 16.3). At the highest intensity all but the most

Table 16.3 Plant species arranged in order of decreasing degree of tolerance to trampling on neutral grassland (from Crawford and Liddle 1977)

**Poa annua*	(annual meadow grass)
Plantago major	(large plantation)
**Poa pratensis*	(meadow grass)
Lolium perenne	(perennial ryegrass)
Bellis perennis	(daisy)
Taraxacum officinale	(dandelion)
Carex flacca	(*flaccid sedge*)
**Cynosurus cristatus*	(crested dog's tail)
Ranunculus acris	(creeping buttercup)
**Deschampsia caespitosa*	(tufted hair-grass)
Prunella vulgaris	(self-heal)
**Festuca rubra*	(red fescue)
**Trifolium pratense*	(clover)
**Hordeum secalinum*	(wild barley)
Phleum bertolionii	(timothy)
**Agrostis stolonifera*	(creeping bent)
**Holcus lanatus*	(Yorkshire fog)

* Grass species.

resistant plants, such as the grasses *Poa annua* and *P. pratensis*, disappear and the proportion of bare ground increases. The latter has been used to identify limits or thresholds of change.

The intensity at which 50 per cent of the trampled area becomes bare ground is considered by many as the critical threshold at which the ecosystem in question changes from a biotic to an abiotic one and at which the rate of degradation of the system by soil erosion increases, i.e. the recovery potential (*elasticity*) falls to zero (Trudgill 1988). The resilience of an ecosystem, however, is dependent on both its biotic and abiotic elements. On physically unstable sites such as mobile sand dunes which are in an early successional stage the critical threshold would depend on the point of balance between erosion and aggradation, and might well be higher than on a peaty moorland habitat where the peat is susceptible to erosion once the relatively shallow actively growing layer is breached (Liddle 1975 a & b).

ECOSYSTEM AND FRAGILITY RESILIENCE

Ecological resilience is an expression of a system's relative amplitude (or flexibility), i.e. how far it can fluctuate in time and space and its relative elasticity (or recovery potential). The higher its resilience the longer it would be expected to persist in face of a given intensity of disturbance; the lower its resilience, the greater will be its fragility, i.e. the ease with which it could be disrupted. Concern about the increasing intensity of human disturbance (to which the biosphere's adaptation or pre-adaptation is low) has served to focus attention on the concept of potentially fragile ecosystems. This concept can be applied also to subsystems; indeed more detailed and quantitative studies have been undertaken on which the 'fragility' of species populations, and more particularly of rare and/or endangered animal species, can be or is assessed.

MINIMUM VIABLE POPULATION

Central to the problem of species or community vulnerability is the question as to how to identify the minimum population necessary for the persistence of a species in a viable state over a relatively long time (Gilpin and Soulé 1986) and below which it becomes increasingly in danger of extinction. However, as these authors point out, the probability of species extinction is not dependent on population size alone. The minimum viable population is determined by a number of sets of interrelated interacting variables which include the demographic characteristics of the population (e.g. size, sex and age ratios, life expectancy, reproductive cycle), the genetic variability of the population and the nature of the environment (both biotic and abiotic).

Populations can be naturally small as in the case of the so-called 'rare' species or have been relatively recently reduced because of anthropogenic disturbance (or stress). The latter include events such as habitat modification, population overkill, reduction disturbance during breeding season and toxic pollution. In addition, random environmental variation may effect a natural reduction (i.e. thinning) of the population. And Gilpin and Soulé (1986) maintain that any environmental change will initiate a chain reaction or downward spiral (vortex) which will eventually culminate in species extinction. They distinguish four 'extinction vortices' (Fig. 16.2):

1. *R*-vortex – in which a decrease in population size (*N*) accompanied by increase in the variability of its growth rate (*Varr*) will increase its vulnerability to further disturbance.
2. *D*-vortex – in which a decrease in *N* and increase in *Varr* can result in the spatial fragmentation of a population and hence an increase in the vulnerability of the reduced 'patchy' populations.
3. *F*-vortex – in which decrease in the genetically effective population (N_e) over several generations can lead to genetic drift as a result of inbreeding and 'loss of heterozygosity'.
4. *A*-vortex – in which a decrease in N_e may reduce the potential adaptability of the population to the existing habitat and/or to habitat change.

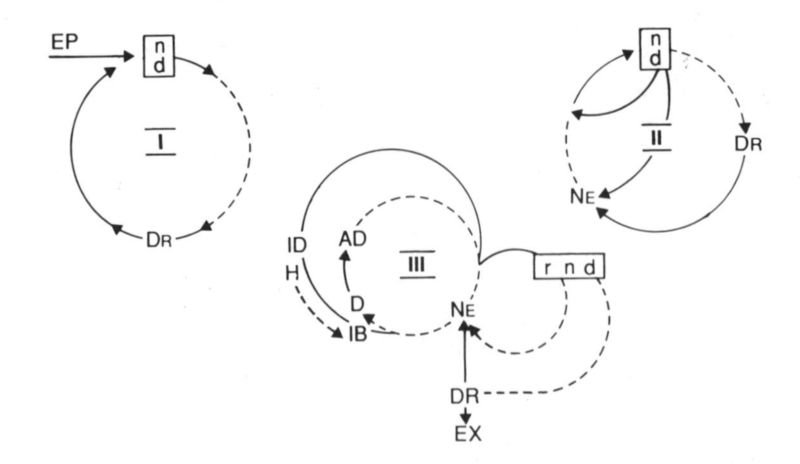

Fig. 16.2 Diagrammatic representation of population extinction vortices: I = R or demographic vortex; II = D or fragmentation vortex; III = F or inbreeding and A or adaptation vortices; EP = environmental push; *n* = numbers; *d* = population; *r* = rate; *ID* = environmental independent; *IB* = inbreeding; *H* = environment dependent; continuous line = increase; dashed line = decrease (Redrawn from Gilpin and Soulé 1986)

While all organisms can be affected by *R*- and *D*-vortices, large animals with naturally low reproductive rates (e.g. whales, elephants) will be more vulnerable to extinction than small ones with short life cycles and high reproductive rates. However, the vulnerability of animals and particularly of large animals will generally be greater than that of many plants such as large trees with a very long life span and high seed production or small harbaceous species with high seed and/ or food reserve production. In contrast, small animals with high reproduction rates such as insects, fish and rodents are, according to Gilpin and Soulé, less susceptible to the *F*- and *A*-vortices, while some large organisms such as trees and large vertebrates which can persist at very low population levels because of longevity may be more susceptible to these vortices. However, irrespective of the taxonomic unit the rate of population decline and local extinction will vary dependent on the intensity of the disturbance and whether it is initiated by a reduction in the size and resource base of the habitat or by a change in environmental variables.

ECOSYSTEM FRAGILITY

Ever-increasing human impact on the biosphere, has, to no uncertain extent, focused attention on the relative fragility of the remaining, but fast-dwindling wild or little managed ecosystems. However, given the organisational complexity of ecosystems it is hardly surprising that statements about the relative fragility or resistance still embody a high level of empirical generalisation based on data drawn from a relatively small number of limited samples. Also ecosystem fragility is all too frequently equated with that of the biotic, and particularly the animal, communities rather than with both the biotic and the abiotic components. It is only relatively recently that the significance of the functional relationships between all the components has been fully assessed in the identification of ecosystem fragility.

The concept of ecosystem fragility raises the problem of its basis, i.e. what characterises fragile as distinct from resilient systems and whether fragility is a function of the combination of all the attributes (i.e. an *emergent property*) or of a 'weak' or limiting attribute (i.e. that with the least resilience) of the particular ecosystem. In the latter case it could be argued that fragility is a function of that attribute most susceptible to disruption or that anthropogenic disturbance in face of which the ecosystem has the least resistance (Tivy and O'Hare 1981).

FACTORS AFFECTING FRAGILITY AND RESILIENCE

Among the factors to which ecosystem fragility (or resilience) has been attributed singly or in combination with one or more others are:

1. Scale (size);
2. Successional stage;
3. Species richness and/or diversity;
4. Reproductive strategies of the biotic components;
5. The volume and distribution of the energy pools;
6. The number of routes by which energy enters and leaves the ecosystem;
7. The variability of the physical (abiotic) environment.

All have been discussed to a greater or lesser extent in Part II in the context of particular types of ecosystems. The aim now is to attempt to assess their general validity and wider applicability.

Scale (size)

Irrespective of their origin, spatially limited ecosystems are potentially more fragile than larger ones. This is a truism when disturbance involves either direct or indirect habitat reduction or removal. However, small ecosystems also have lower critical thresholds than larger ones because species poverty is usually accompanied by small populations of often habitat-specialised endemic species at or very near their minimum viable population threshold. Temporal or spatial isolation further exacerbates the limitations of scale because of a lack of competitive or defence strategies in face of introduced plants and animals. Also, such endemic ecosystems in which the use of abiotic resources is incomplete have only limited powers of recovery and can easily be replaced by a different and often extremely degraded system. However, the corollary that large-scale ecosystems are more resilient to a comparable level of disturbance does not apply except in so far as they may take longer to be disrupted or destroyed.

Species diversity

The lack of correlation between species diversity (richness) and stability holds good also for that between species diversity and fragility for the same reasons of small population sizes, habitat specificity and very narrow niches, all of which have already been illustrated in relation to the tropical rain forest. Indeed, it could be argued that species diversity (richness) alone has relatively little significance for ecosystem fragility. Diversity and more particularly species equitability are more relevant for the relative fragility or resilience of local biotic communities.

Successional stage

It was initially assumed that the early stages of ecosystem development were potentially more fragile than the final mature stage when subjected to a similar intensity of disturbance. This was attributed to either the lack of a complete ground-cover or, more importantly, an unsaturated habitat in which resources had not reached the stage of maximum exploitation and vacant niches were still available. While this may be valid in the case of early seral stages in primary succession in potentially productive habitats, it is less so in secondary successions. Many pre-climax ecosystems are composed of communities dominated by perennial herbs and shrubs (e.g. grasses, bracken, blackberry, hawthorn) which are often particularly resilient and durable even when subjected to high intensities of anthropogenic impact – a resilience related to their nature and the distribution of their energy resources and their high regenerative potential.

Energy pools and routes

The most important universal attribute determining fragility or resilience is the distribution of energy and nutrient pools within, and the energy flux of, the particular ecosystem. Those in which a high proportion of the primary biological energy production is invested in the above-ground labile pool and whose regeneration is precarious, will tend to be more fragile than those in which the below : above-ground biomass ratio is large. For these reasons forest and woodlands are, in fact, inherently more fragile than herbaceous ecosystems. Disturbance may reduce or completely destroy the above-ground biomass and, in addition, reduce or inhibit seed regeneration and sapling growth. The relative fragility of forest ecosystems, however, is also dependent on the energy/nutrient flux through the system.

As has previously been noted (see p. 183), not only is the above-ground biomass of the tropical rain forest the most massive in the biosphere, but it represents a higher proportion of the total biomass than in any other ecosystem. Seed production, dispersability and viability of the primary trees are low. Once the above-ground biomass has been removed the rate of organic decomposition, consequent on microclimatic modificationı, is greatly accelerated and the comparatively small DOM and SOM energy pools are rapidly dissipated. Most of the underlying mineral soils are nutrient deficient. As a result the regenerative capability of the tropical rain forest is weak and resilience is low. Its fragility is such that the ecosystem can easily be pushed beyond its limits to the point of collapse and habitat sterility. In contrast to the tropical rain forest, temperate and Boreal forests are less fragile. With a smaller proportion of biomass above ground, a larger DOM and SOM energy store consequent on a slower rate of organic decomposition and higher seed production and viability they are more resilient than the tropical rain forest. Consequent on the below-surface reserve of energy they are better able to recover from disturbance or removal of the above-ground vegetation.

The most resilient of forest ecosystems are the pioneer, i.e. pre-climax stages. The dominant tree species (e.g. *Pinus*, *Betula*, *Salix*, *Alnus*, *Corylus*, *Eucalyptus*) are characterised by high seed production, dispersability and viability combined with methods of vegetative reproduction (from bole suckers and/or lignotubers) and in which dispersal and germination may be stimulated by a certain level of disturbance by browsing, cutting and by fire. Adapted to absorb and recover from particular types and levels of disturbance, they are resilient to higher intensities of disturbance than other forest ecosystems.

Ecosystems with a high below : above-ground biomass ratio are among the most resilient ecosystems in the biosphere. This is exemplified by the temperate grasslands in which a particularly high proportion of the biomass is below ground. An extensive root and underground stem (rhizomes and stolons) system provides a food reserve from which vegetative growth can be renewed annually (as well as being stimulated by cutting, etc.) and contributes a large amount of DOM and SOM which becomes well mixed and integrated with the mineral soil. The net result is that *c.* 90–95 per cent of the total ecosystem energy/nutrient store is conserved below ground. The very high concentration of grass stems and roots, together with leaf and stem litter, just below the ground surface forms a tough sod of turf which serves to protect the ecosystem from disturbance and from invasion by other plant species. As a result grassland ecosystems are particularly resilient in their ability to absorb and recover from severe environmental impact. These are characteristics common to ecosystems such as heath and moorland, wetlands such as reed and sedge swamp and salt-marsh, and which contribute to the durability of these ecosystems many of which may have replaced more fragile forest or woodland systems.

The abiotic environment

In some instances, the nature of the physical habitat may exacerbate or even be the dominant factor in ecosystem fragility. Two of the most important attributes in this respect are climatic variability and actual or potential soil instability.

Climatic variability. Ecosystems characterised by a marked seasonal or long-term periodic variation in climate are particularly vulnerable to anthropogenic disturbance. The first case is exemplified by Mediterranean and tropical areas with marked seasonal rainfall regimes. Disturbance of the vegetation cover and particularly deforestation expose the surface to the direct impact of characteristically torrential rain. Accelerated soil erosion combined with increased aridity of the exposed ground surface inhibits forest and woodland regeneration. Similarly in the semi-arid temperate and tropical regions subject to alternating sequences of below- and above-average rainfall, disturbance of the vegetation cover can result in a deterioration of soil climate particularly during periods of drought such that ecosystem recovery is endangered and the original grass/shrub land can be displaced by desert conditions as in the Sahel at present.

Soil instability. Soil instability is an important, indeed major, determinant of ecosystem fragility particularly in periglacial areas of the Arctic tundra, mountain alpine zones and on steep slopes elsewhere. In all these situations the abiotic substratum may be at or near a physical threshold that is potentially unstable for biotic regeneration. When, in these habitats, the insulating vegetation cover (which may well have originally been established under less severe or marginal conditions) is disturbed, physical instability may exceed the recovery capability of the vegetation. As has already been noted, disruption of the tundra vegetation exposes the mineral soil to more pronounced frost-heaving and solifluction. The permafrost thaws to greater depths and produces an uneven pitted surface. Conditions become increasingly marginal for the regeneration of naturally slow-growing plants.

Disturbance of the insulating and stabilising vegetation cover on steeply sloping ground increases the susceptibility of the soil to movement by gravity and surface runoff; fragility will be dependent on the intensity of disturbance and the slope gradient. Also, once disturbed, the re-establishment of the former stabilising vegetative root system becomes more difficult and the greater soil instability is maintained for a long, indefinite period of time. Ecosystems in physically marginal environments are particularly susceptible to disruption by human impact. Although adapted to these conditions, the potential habitat instability is such that human disturbance can effect a rapid shift beyond the range of tolerance or homeostatic limits of the natural ecosystem.

Although ecosystem resilience and fragility are relative and still unquantifiable concepts, they are none the less valuable for the focus they bring to bear on the effects of human disturbance. However, too much emphasis in much of the published literature to date has been placed on the problems of definition which have tended to confuse the issue, cast doubt on the validity and usefulness of the concepts and, most seriously, to either detract from efforts to assess the factors which may contribute to the response of ecosystems to human impact. Current understanding of the nature and functioning of ecosystems should make it obvious that a complex of factors is involved in determining fragility or resilience. Although one or another may be dominant neither the dominant nor the relative importance of the factor complex is the same for every type and scale of ecosystem. However, despite that, three general points can be made. First, the rate of energy flux and the size and location of the energy pools within the ecosystem must be as fundamental to the resilience as they are to the persistence of a given ecosystem. Second, the biological approach to the problem has tended to concentrate on the biotic components without giving due or sufficient consideration to the abiotic component and more importantly the interrelation between the two which is basic to the ecosystem concept. Third, no ecosystem is absolutely resilient in terms of recovery when the human impact is continuous, and unless the resulting losses are made good by human management eventual ecosystem collapse is inevitable.

Chapter

17

Human impact: ecosystem exploitation

In comparison to many other forms of life the human population is still relatively small. It, in fact, accounts for only a minute fraction of the earth's biomass. Ecologically, however, humans are the dominant species. Not only are they omnivorous, but their diet is more varied than that of other animals. Also the uses they make of the biosphere are more diverse. As predators humans are without equal; neither lack of food nor the presence of a powerful enough predator has been able to set a limit to their numbers. Because of their superior ability to exploit and modify both the organic and inorganic environment their impact on the biosphere has been greater than that of any other organism. There can be few, if any, parts of the biosphere that have not been affected, directly or indirectly, to a greater or lesser degree by human activities.

The impact of human activities on the biosphere has accelerated rapidly during the last 200 years consequent upon an exponential rate of population growth combined with scientific and technical developments. The latter continue to facilitate the exploitation of the world's organic resources and the modification of its physical environment. The ecological effects of anthropogenic processes will, for convenience, be considered in relation to their impact on the following:

1. Species diversity distribution and evolution;
2. Ecosystems modification and destruction;
3. Environmental pollution.

SPECIES DIVERSITY, DISTRIBUTION AND EVOLUTION

Plant and animal species have been and continue to be drastically reduced in numbers and populations. Many have become extinct and it was estimated in the mid-1970s that over 1000 vertebrate species and subspecies (IUCN 1975) and over

250 000 plant species were in danger of extinction (Lucas and Synge 1978). The main factors involved are ecosystem modification and loss, species over-exploitation, and species invasions.

HABITAT CHANGE

Human activities have effected very considerable changes in the numbers and relative proportion of species populations by either direct or indirect habitat modification. The earliest and most widespread were the use of fire either to flush out enemies or game animals or to open up land for domestic grazing animals or cultivation. As has previously been noted, humans increased the impact of fire and grazing as selective ecological factors, giving a competitive advantage to those species adapted over those not adapted to these processes and thereby causing a reduction or elimination of the latter from the site in question. Ecosystem modification is also effected by extrinsic or intrinsic changes in the physical environment such as drainage or irrigation, land subsidence and environmental pollution. All or part of many pre-existing ecosystems has been replaced by highly managed organic resources (forestry, agriculture, etc.) or man-made 'sterile' surfaces.

SPECIES OVER-EXPLOITATION

In the first instance the reduction and elimination of species have been the result of over-killing – i.e. the direct slaughter of a species population at a rate greater than it can reproduce itself. The over-killing of animals has, to a large degree, been for food; wild game is still the most important though, as yet, not fully developed or rationally exploited source of protein in Africa south of the Sahara. Much indiscriminate slaughter, however, has been motivated by a particular demand from the Western world for limited and highly priced animal products. The late nineteenth-century market for whale oil and whalebone resulted in the virtual extermination of the Arctic white whale. Similar marine animals with low reproductive rates, such as the seal and the turtle, have also suffered serious inroads into their populations. Furs have long been a coveted luxury product and an important item of trade. They provided an early and lucrative incentive to explore and colonise the North American continent. It is hardly surprising to learn that by the end of the nineteenth century, after only 300 years of exploitation, the once prolific beaver had become extremely rare. The demand for more exotic products such as ivory and aphrodisiacs severely reduced the smaller populations of large mammals, such as the elephant and rhinoceros. More recently others, such as the armadillo (the only animal to contract leprosy) and the rhesus monkey, have acquired a high value for medical research.

In other cases animals have been deliberately and systematically slaughtered to protect human life, domestic and desirable game animals, as well as cultivated crops from marauders. Predatory birds and carnivorous mammals have experienced the greatest losses. For instance, the wolf had disappeared from Scotland by the beginning of the eighteenth century and the largest of the predatory birds, the golden eagle, is now a protected rarity. The systematic extermination of the North American bison initially for profit, but more particularly as a means of subjugating the Plains Indians, must represent one of the most rapid and wholesale destructions of a large species population in the world. It has been estimated that in the early eighteenth century, there were at least 60 million bison 'on the hoof'; by 1913 only 21 individuals remained! Until the eighteenth century the killing of animals for pleasure was subordinate to more practical ends. Game animals in many European countries were the preserve of the aristocracy and hunting, although the sport of kings, made an important contribution to the table. Hunting primarily for sport developed during the eighteenth century when 'huntin', shootin' and fishin'' became respectable and fashionable pursuits of the more affluent and leisured classes. The abundance of strange, either dangerous or elusive, wild animals in areas subject to European colonisation saw the development of big-game hunting, particularly in Africa and India. In the latter country it has been estimated that the numbers of the better-known wild animals (e.g. tiger, leopard, deer, blackbuck) have declined to a tenth of what they were 50 years ago. The Indian lion and rhinoceros have been reduced to negligible numbers, while the Indian cheetah became extinct in the late 1940s. Wanton poaching, either to obtain a quick profit or for pleasure, still continues.

The effects of uncontrolled exploitation have been most severe in the case of those species of higher animals such as mammals, birds and fish. This is a result not only of their size but of their smaller populations and slower reproductive rates in comparison with other species. It has been estimated that since the year 1600, 36 species of mammals and 94 of birds have been lost, and 120 mammals and 187 bird species are so rare as to be on the verge of extinction; and that, at present, one species or subspecies of either mammal or bird becomes extinct every year.

SPECIES DISPERSAL AND DISTRIBUTION

Humans have become increasingly powerful agents of species dispersal and distribution. Their effectiveness in this respect has been gathering momentum particularly during the past 400–500 years with exploration, the growth of trade and the development of ever faster means of long-distance transport. The latter has bridged the barriers that formerly restricted the range of many species. While the range of a number of plants and animals has been extended that of others has been drastically reduced or destroyed. Many species of plants and insects have been accidentally transported from one continent to another. Others have travelled as 'stowaways' in grain and feedstuffs; in plant and animal fibres; in timber, ballast and packing materials. Many more have been carried attached to humans or their vehicles of transport. Only a relatively small proportion of those species so transported and dispersed actually become successfully established in an area foreign to them. It is difficult for alien intruders to survive competition in mature, closed ecosystems. However, that some can do so is largely the direct or indirect result of human activities. Establishment of *alien* (or *adventive*) species is and has been most successful in disturbed habitats.

Many alien plants, which in their native homes are only minor or inconspicuous 'citizens', often become much more vigorous, aggressive and abundant in a foreign environment. They may even spread so rapidly and effectively as to compete, not only with native weeds, but with the natural vegetation. The classic example is that of the prickly pear cactus which, deliberately introduced from South America into Australia as a possible fodder plant, overran and seriously reduced the quality of natural pasture land. Somewhat comparable in the USA is kudzu (*Pueraria thunbergiana*), a vine introduced from Japan and used to control gulley erosion in the south-eastern states, which has become a rampant and often smothering weed in woodlands. In Britain the common purple-flowered rhododendron (*Rhododendron ponticum*), introduced from the Mediterranean region in the mid-eighteenth century, has successfully escaped from gardens and invaded sandy and acid soils. Casting a heavy shade and contributing very poor humus, it has become a very troublesome weed in woodlands and newly established Forestry Commission plantations, where it checks tree-seedling growth and is difficult and expensive to eradicate. The aggressiveness and rapid proliferation of alien weeds which do become established, may be due partly to lack of competition, especially in open man-modified habitats and partly to the absence of herbivores and pathogens to which they are normally subjected in their native habitats. Other aliens may cross-fertilise closely related native species and produce hybrids of exceptional vigour. This has been the case with the hybrid rice or cord grass (*Spartinia townsendii*), a fertile polyploid hybrid resulting from the cross between the original British and an accidentally introduced American species of this grass. Within 70 years the hybrid has not only suppressed its parents but has colonised with astonishing rapidity considerable areas of tidal mud-flats around the coasts of southern England and northern France.

The introduction of wild animals from one region to another has often been deliberate, as in the case of the rabbit in Australia and the red deer in New Zealand. Lacking predators or competitors both proliferated alarmingly to the detriment of the natural vegetation cover. Similarly the grey squirrel and muskrat, introduced from North America to Europe, have experienced population explosions and the more aggressive grey squirrel has all but displaced the native red in Britain. Of perhaps even greater significance has been the invasion of alien invertebrates (particularly insects) and pathogenic micro-organisms into areas where natural predators are lacking and where the indigenous organisms do not possess a natural resistance. Where these conditions exist, they become serious economic pests and propagators of disease.

The fluted scale-insect introduced to California from Australia in the late nineteenth century became a serious threat to citrus orchards until its natural enemy was imported. Both the Colorado beetle and the fungus which causes potato blight occur naturally among wild potato species in North America without being particularly harmful. They assumed the proportions of a devastating pest and disastrous disease respectively when introduced to Europe where the cultivated potato crop provided the conditions for their rapid increase and spread. Similarly the aphid, *Phylloxera*, which decimated French vineyards at the end of the nineteenth century, was a natural but inoffensive companion of wild vine species in America which proved fatal when brought into contact with the unresistant European cultivated varieties. Many of the worst insect pests or plant diseases, in fact, are those which have been transported accidentally from one region to another, or native species which have become pests on unresistant 'alien' crops.

The cumulative effect of man (which continues with increasing intensity) has been to favour, either deliberately or accidentally, the evolution and distribution of some types of plants and animals at the expense of others. In Tertiary times, one region or habitat was isolated from another by physical barriers. These barriers allowed an independent evolution of plants and animals within different parts of the world. Now these barriers have been broken down by man. Over great areas of land which were once the habitat of a rich variety of different plants and animals, large numbers of a few specialised species are grown which man deliberately protects or which can tolerate the modified habitat conditions he has created.

Weeds

These open, disturbed habitats, have provided a foothold for those species of plants tolerant of such conditions. Their vigorous growth and survival are favoured by the absence of competition. Of those which occupy such sites, some were originally native only to naturally open and exposed habitats; others were formerly inconspicuous and often minor species repressed by the competition of the other plants with which they originally grew. A great many more have arisen from variants or hybrids to whom these disturbed habitats provided an opportunity for establishment, many are 'camp-followers' – the types of plant designated as weeds. They sometimes are defined as plants growing 'out of place' or 'where they are not wanted'. However, they can also be better described as plants which become established in disturbed habitats where they grow with greater abundance and vigour than in undisturbed ecosystems. Many are herbaceous plants, frequently though not invariably annuals and most, in the absence of competition, are vigorous and aggressive. Prolific seed production combined with a wide range of tolerance of environmental conditions has aided and accelerated their widespread dispersal and distribution by human activity. Among the few species of plant which have attained an almost world-wide range in both tropical and temperate latitudes are some of the notorious weeds of cultivation, such as the dandelion (*Taraxacum officinale*), shepherd's purse (*Capsella bursa-pastoris*), plantain (*Plantago* spp.), chickweed (*Stellaria* spp.) and the braken fern (*Pteridium aquilinum*) which will grow with continued vigour for many years on land once cultivated or heavily manured by animal droppings. The very early association of weeds with humans is revealed by the appearance of pollen derived from such species as the rib-wort plantain (*Plantago lanceolata*) and fat hen (*Chenopodium* spp.) in peat deposits of Neolithic age in northern Europe. In many areas weeds persist as evidence of former human presence. The American Indian called the plantain 'the white man's footprint'. Dense stands of nettles (*Urtica diocica*) frequently remain on or near sites of former habitation where the soil is particularly rich in phosphorus. But deprived of the open, disturbed conditions which favour their rampant growth, weeds can rarely maintain their dominance indefinitely in face of competition from more demanding species.

SPECIES EVOLUTION

In relatively stable habitats evolution is inevitably a slow process. The variant or mutant individuals which develop within any species will only give rise to a new 'race' or 'strain' under conditions which favour their survival and allow them to maintain their identity free of continual cross-fertilisation with other members of the same species. The production of new species by hybridisation is also relatively rare under these conditions. The reasons for this vary; some species cannot interbreed, others may do so but the hybrid progeny is frequently sterile. On the other hand, two species that can be crossed under cultivation do not necessarily produce hybrids when they occur in close proximity in the wild, or if they do the hybrids fail to perpetuate themselves. The failure of hybrids to maintain their identity is, as in the case of variants or mutants, due partly to continual back-crossing or introgressive hybridisation with the parent plants, which tends to cancel out marked differences between species and their hybrid offspring. Also, the optimum requirements and tolerances of the hybrids and variants tend to vary from those of the parent stock. In the absence of a suitable habitat in which they would have a competitive advantage it is difficult, if not impossible, for them to become established and maintain their identity. However, new or changed conditions to which they are better adapted than other plants will tend to favour their survival and allow the emergence of a new species population. Variants and hybrids require, therefore, different or what has been referred to as *hybrid habitats* if they are to maintain their identity successfully.

The simultaneous development occurred not only of plant species which were adapted to disturbed habitats but also of those plants deliberately selected for particular human use and which eventually became fully domesticated, i.e. not capable of reproduction in the wild.

Crop domestication

Disturbed habitats produced by soil tillage were not only necessary for, but also contributed to, the process of the evolution, (i.e. domestication of crop plants) – the *cultigens* and *cultivars*. However, although the range of existing crop plants is large they have been developed from a relatively small number of existing families. The most important food crops – the cereals which supply the bulk of the human carbohydrate requirement – are a few select members of the grass family (Gramineae); while the legumes (Leguminosae) provide the bulk of the plant-protein needs. The continued existence as well as the evolution of cultivated species is dependent on humans. The process of domestication entailed the selection, planting and propagation of particular species, of variants of them or of hybrids between them. Selection favoured individuals best suited to human needs, but which frequently were those least fitted for survival in the face of competition in the wild. The majority of crop plants lack the ability to reproduce or maintain themselves independently. Most (with the exception of such tree crops as coffee, tea, cacao) are heliophytes, intolerant of shade. Apart from those which are propagated vegetatively, the majority are annuals; some were originally annuals, while others have developed this habit under cultivation. A classic example in this respect is cotton, originally a perennial shrub confined to frost-free tropical areas. The selection of forms that fruited early permitted cultivation in areas where summers were warm and long enough for fruiting, but where winters were cold and liable to frost, and this was accompanied by a change from a perennial to an annual habitat in the plants.

Many food crops are dependent on humans for their propagation. Some have been propagated vegetatively for so long that their seed-producing capacity has been drastically reduced. The extreme example is the edible banana. The origin of this highly domesticated cultigen was the selection of a sterile hybrid able to produce attractive fruit but unable to develop seeds necessary for its perpetuation. Characteristic of the cereal crops is the lack of effective or efficient means of seed dispersal – an advantage for harvesting but a decided disadvantage for survival and spread. Many of the wild grasses and primitive species from which the modern grain crops evolved have heads whose central ear-bearing stem is brittle and fragile. When ripe this breaks up (or 'shatters') and allows the seeds to be easily dispersed. In the more evolved cultivated forms, selection has favoured those plants with tough flower stems. Heads remain intact when mature and can hence be easily harvested. Indian corn, or maize, has become so highly domesticated that it is incapable of reproducing itself without human aid. Unlike the other grains whose kernels, or ears, are separate, those of maize occur as a densely packed cluster of seeds enclosed in a tightly wrapped sheath or husk. When this drops to the ground there is no means whereby the seeds can be released, and few survive the excessive competition in the cob or the depredation of animals.

In addition, cultivated plants have in the course of domestication tended to lose those characteristics which would make them unattractive to human taste – for example, hairiness, thorns, toughness or an unpleasant taste – but which in the wild afforded a measure of protection from grazing or browsing animals. Indeed, the increased size and palatability of herbaceous crops make them more liable to predation. Furthermore, the concentration of plants of one kind under conditions favourable for their growth creates optimum conditions for the multiplication and evolution of animal pests and disease organisms. All the major crops and most of the minor ones are ancient in the sense that they have been domesticated since prehistoric times. Forms of cultivated cereals almost identical to those grown today are known to have been in existence at least 7000 years ago. Recent archaeological evidence has revealed the existence of agricultural communities in south-west Asia nearly 9000 years ago (Moore 1967).

ORIGIN AND DISTRIBUTION OF CULTIGENS

The place of origin, the wild or primitive ancestors and the evolutionary development of crop plants are, however, difficult to trace. The evolution of human societies and cultures and their dissemination is inextricably linked with that of plant and animal domestication. The prehistorian, the archaeologist, the anthropologist and the geographer are deeply involved in these problems; they, in turn, must look to the biologist for an understanding of how the 'organic artefacts' of man evolved, where they originated and the directions in which they spread. In the period between the two world wars the Russian plant geographer I. N. Vavilov (1949–50) noted that for a large number of crop plants (and their associated weeds) the greatest number of species and varieties within them, or in closely related wild species, tended to be concentrated in certain centres of diversity, mainly in the Old World (see Fig. 17.1). They are characteristically upland or

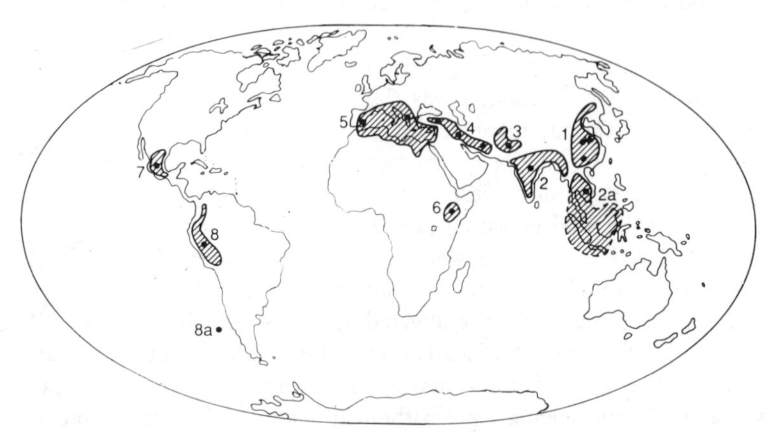

Fig. 17.1 World centres of origin of cultivated plants according to Vavilov: (1) Chinese; (2) Indian; (2a) Indo-Malayan; (3) central Asian; (4) Near Eastern; (5) Mediterranean; (6) Abyssinian; (7) south Mexican and Central American; (8 and 8a) South American. Black dots indicate centres of orgination of form of principal cultivated plants (from Vavilov 1949–50)

mountainous regions – areas of marked physical diversity in tropical or subtropical latitudes. On the assumption that a particular type of plant – a particular genus – would reveal the greater diversity of species and varieties near its centre of origin, he argued that these centres of diversity which he identified were probably the primary or secondary centres of origin of the particular crop plants. The latter would have developed later when already domesticated plants migrated or were carried by man into other areas. Brought into contact with different plants and under different physcial conditions, the original cultigens would then have given rise to a new set of hybrids and variant strains, and a secondary centre of diversity would thus have been created.

The primary crops were those which were originally selected and many of which, as C. D. Darlington (1963) suggests, may have first come to notice as 'habitation weeds'. These included wheat, barley, flax, soya and maize. It is probable that many of the secondary crops, such as rye, oats, mustard, rape and other cruciferous plants, would have made their first appearance as 'weeds or cultivation' (Table 17.1). Under primitive methods of agriculture and rudimentary cultivation, a great variety of other plants must have grown alongside the crops. Fields and plots (as in tropical garden cultivation today where a great variety of different plants are grown together) must have been hotbeds of hybridisation and introgression. The weeds of cultivation – many of them hybrids – would have accompanied people as they moved from one area to another. As in the case of rye and oats, they eventually became the dominant crop under climatic conditions to which they proved better adapted than the primary crop. That the areas which Vavilov identified are centres of marked species diversity for many crops is an established fact. The development of such diversity must undoubtedly be closely associated with the tremendous variety of habitats characteristic of these mountainous areas; it may also, as has been suggested, have been accentuated by the mingling and intermixing of peoples from different areas. This interpretation of these centres of diversity has, however, been disputed as over-simplified. Anderson (1967) for instance, would maintain that the existence of a great diversity of often primitive varieties of particular crops in isolated mountain areas may be due to the persistence of ancient and conservative methods of agriculture – they are regions of survival rather than of origin. Vavilov's survey (on a scale that has never been repeated) was of its nature a pioneer study. It is not, therefore, surprising that many of his preliminary ideas, especially about the centres of origin of particular crops, have been and will no doubt continue to be revised in the light of new archaeological and botanical evidence (Simmonds 1976).

The earliest known archaeological evidence of agriculture in both the New and Old World dates from about 9000 BP. However, as Carl Sauer (1952) notes, this is a relatively late date in the history of the human race. He suggests that humans may well have been vegetative planters before learning to collect and sow seeds

Table 17.1 Origin of secondary crops as weeds of cultivated primary crops (after Darlington 1963)

Primary crops	*Secondary crops*
Common wheat (*Triticum vulgare*)	Rye (*Secale cereale*)
Common barley (*Hordeum vulgare*)	Oats (*Avena sativa*)
Emmer wheat (*T. diococcum*)	
Flax (*Linum usitatissum*)	Spurry (*Spergula linicola*)
	Mustard (*Brassica campestri*) (*Eruca sativa*) (*Camelina sativa*)
Buckwheat (*Fagopyrum esulentum*)	*F. tatarium*
Cereals	Common vetch (*Vicia sativa*)
	Field peas (*Pisum arvense*) (*Coriandrum sativa*) (*Cephalavia syriaca*)

and that the first agriculturalist would have lived where it was climatically easier to exist than in the semi-arid climates of the Middle East. He envisages the former environment as characterised by a warm mild climate with sufficient rain to ensure a reasonably long growing season, by easily cleared and dug soils – light forest and shrub, rather than dense tropical forest or grassland and by a diversity of plants with which man could experiment. For these reasons Sauer has proposed South-east Asia and north-western South America as the areas – the hearths – of initial domestication of plants and animals (see Fig. 17.2). Climatically favourable, these are both highly dissected mountain areas of marked diversity of physical habitat and, as Vavilov and other botanists have confirmed, areas of exceptional plant and animal variety. Both are areas where opportunities exist for settled fishing communities, and both are strategically placed crossroads of human communication. Finally both are areas where planting and vegetative propagation are still significant agricultural techniques. Both too are characterised by the presence of important staple food crops (such as manioc and sweet potatoes in South America and the yam and banana in South-east Asia) which are dependent on vegetative propagation and which were probably originally domesticated in these two respective areas.

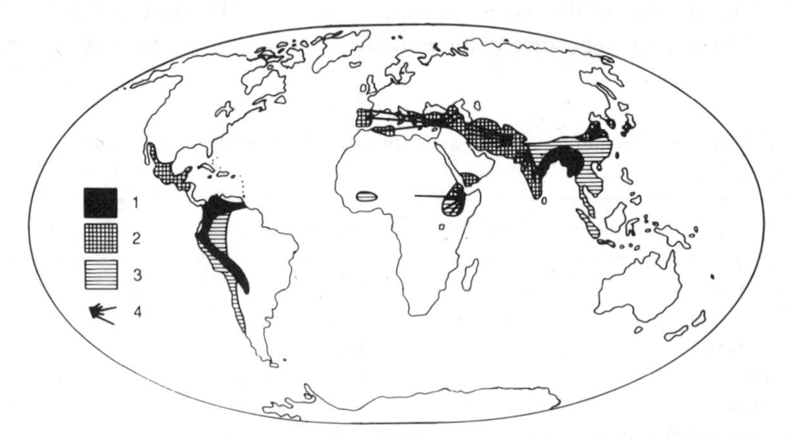

Fig. 17.2 Sauer's hearths of domestication: (1) Vegetative-planting hearths; (2) seed planting; (3) Colonial extensions from (1) and (2); (4) centres of dispersal of seed planting (adapted from Sauer 1952)

As man spread out from these early 'planting hearths' he carried his evolving plants and techniques with him. Moving into new and different areas he no doubt applied his methods of vegetative propagation to other plants that became available. However, as he gradually moved into more rigorous climates which were marginal for the production of the tropical plants he had already domesticated, he began to select species more suited to the new conditions. His attention then probably shifted from perennial tubers, bulbs and roots to annual seed plants – many of which may have first come to his notice as weeds of cultivation. These included the grasses which were to lay the foundation of the major cereal crops, the leguminous species which provided proteins and fats, and a variety of oilseeds and fibre-producing plants. The seed plants, and more particularly the grains, provided a well-balanced highly concentrated form of food. Grain with a very low water content – in comparison to tuberous plants – keeps better, and can more easily be stored and transported. With this at his disposal man could better survive long periods of drought or cold when growth is curtailed.

ORIGINS OF WHEAT AND MAIZE

The origins and development of the two major cereal crops, wheat and maize, have been most intensively studied. Some of the earliest archaeological evidence of agriculture comes from sites in Asia Minor – in the belt of mountains which form the north-western flank of the 'Fertile Crescent' in Iraq and Syria. Here, in the ruins of prehistoric settlements of Jarmo (Iraq) and Tepe Saras (Iran) – dated about 8750 and 8500 BP respectively – remains of primitive barley and of two forms of domesticated wheat have been identified. Of the latter, one is almost identical with an exising wild wheat (*wild emmer*) and the other identical with a still cultivated but very primitive variety (*emmer*). These are the earliest known ancestors of our modern wheats, wild forms of which no longer exist. Modern work in plant genetics has, in this case, confirmed the archaeological evidence that the original centre of wheat domestication was located in Asia Minor. Further, it has thrown much light on how the different wheat species must have evolved. Existing wheats can be classified into three groups distinguished by their chromosome numbers (Table 17.2). The einkhorns, with the lowest chromosome numbers, are considered to be the most primitive. The second group were produced by the hybridisation between einkhorn and a closely related wild grass. The third and most recently evolved, which contain the modern bread wheats (common, shot and club), represent a still later stage of hybridisation. This was not deliberately controlled hybridisation (which had to await the scientific knowledge of the twentieth century) but a result of contacts between crops and weeds, and the survival of the resulting hybrids that humans unconsciously facilitated at a very early stage.

Maize, however, does not exist in a wild form today. The problems of its ancestry and place of origin have been more difficult to solve. Pollen – identified as that of maize – has been found in deposits thought to be some 8000 years old in Mexico. Corn cobs dated from *c.* 7000 BP, which may be those of wild or a primitive cultivated variety of maize, have been discovered in a prehistoric site, also in Mexico. Such findings make it clear that this cultigen originated in the New World and evolved from a wild maize which has long since disappeared. Although the greatest diversity of types of corn found today is in the Andean region of north-western South America, the centre of origin of domesticated maize may, in fact, have lain much further north in Central America.

Crop breeding

The evolution of crop plants has, however, continued with increasing rapidity within the early established groups. Its direction, within historic time, has been greatly influenced by an increasing human population, by the development of more

Table 17.2 Existing species of wheat (*Triticum*); all but wild Einkhorn and Emmer are still cultivated. Only the bread wheats (common, shot and club) and the macaroni wheat are of commercial importance today (adapted from Mangelsdorf 1953)

Latin name	*Common name*	*Chromosome number*	*Distribution*	*Earliest evidence*
T. aegilopoides	Wild Einkhorn	7	Asia Minor, Greece, S. Yugoslavia	Pre-agricultural
T. monococcum	Einkhorn	7	Asia Minor, Greece, Central Europe	4750 BC
T. dioccoides	Wild Emmer	14	Near East	Pre-agricultural
T. diococcum	Emmer	14	India, Central Asia, Europe, Abyssinia	4000 BC
T. durum	Macaroni	14	Central Asia, Asia Minor, Abyssinia, USA	100 BC
T. persicum	Persian	14	Georgia, Armenia, Turkey	None
T. turgidum	Rivet	14	Abyssinia, S. Europe	None
T. polonicum	Polish	14	Abyssinia, Mediterranean	17th century
T. timopheevi	—	14	W. Georgia	20th century
T. aestivum	Common	21	World-wide	Neolithic period
T. sphaerococcum	Shot	21	Central and NW India	1500 BC
T. compactum	Club	21	SW Asia, SE Europe, USA	Neolithic period
T. spelta	Spelt	21	Central Europe	Bronze Age
T. macha	Macha	21	W. Georgia	20th century

specialised farming techniques and by the replacement, particularly in temperate latitudes and modern industrialised societies, of subsistence by commercial agriculture. Of the main types of domesticated crops only relatively few of the available species and varieties have been selected for commercial development. Of the fourteen known species of wheat in existence, the three species of bread wheat (common, shot and club) account for about 90 per cent of the wheat crop in the world today and their main centres of production now lie well beyond their original areas of origin. In addition modern farming methods and commercial agriculture have stimulated, and in their turn have been stimulated by, the revolutionary advances that have occurred in crop breeding within recent years.

The modern phase in the evolution of cultivated plants began with the deliberate breeding of plants for particular purposes. The pioneers in this field were the Dutch bulb breeders who as early as the end of the seventeenth and beginning of the eighteenth centuries were already producing standardised seeds for food crops. The contributions of Charles Darwin and Alfred Wallace to an understanding of the principles of evolution and of Mendel to the laws of inheritance, laid the foundations for the development of new scientifically based techniques of breeding and propagation. By rigidly controlled selection, 'pure-line' varieties of crop plants could be isolated and maintained. By controlled hybridisation new varieties with particular characteristics could be produced. The development of these new techniques has been mainly within this century and their application has been more particularly concerned with the major commercial food crops and ornamental plants of temperate regions. Modern advances in plant breeding were originally most spectacular in countries or regions of recent agricultural colonisation, e.g. the prairie regions of North America and Russia, and in New Zealand and Australia, where crop plants have been introduced from elsewhere; and in areas of intensive commercial agriculture where new crop varieties have either allowed the application of new methods of farming, or were necessitated by them.

The main aim of modern plant breeding is to produce varieties of crops that will give the maximum economic production in terms either of quantity or quality, under the conditions in which they are grown. To this end varieties are produced that are adapted not just to natural environmental conditions but to soils modified by drainage or fertilisation and to mechanical methods of cultivation. They are ecological races adapted to particular agricultural habitats. Early maturing varieties of grain are bred that can be grown in regions where either drought or frost may curtail the growing season; they are bred for hardiness or drought resistance. The overriding objective in plant breeding, however, is to increase yield – to produce varieties of crops that will, if well cultivated and heavily fertilised, build up plant tissues most efficiently. Such highly bred, highly specialised plants of a uniformly standardised genetic composition have, however, little resistance to the conditions to which they are not adapted. A severe drought or prolonged frost may destroy a whole crop, since it will not contain some individuals with a slightly greater tolerance to these conditions, as in the case of more variable natural or less highly domesticated species. For the same reasons they are particularly susceptible to disease. As a result it is necessary to breed disease- or pest-resistant strains, and to continue doing so as ever more virulent pathogens evolve. Cereals with large heads and short stems give a high harvest yield and are more resistant to lodging. Since modern farming is dependent on mechanisation, machines must be designed to handle a particular crop or crops suitable for mechanisation. The successful mechanisation of cotton harvesting in the USA was dependent on the breeding of varieties that would ripen evenly and were of a uniform height. The introduction of drought-resistant grain sorghums into Texas and Oklahoma from Africa necessitated the production of short forms; the African varieties, some 2.5–4 m in height, were too tall for machine harvesting. Today there are few commercial crops which are not fully mechanised.

The evolution of varieties of domesticated crops under modern methods of controlled, directed plant breeding has been particularly rapid in the major cereal crops, and above all in the wheats. In the great wheat-growing regions of Canada and the USA there is hardly a region where the principal varieties of wheat grown today are the same as those used 80 years ago. This has been consequent upon the rapid development of the so-called hybrid wheats, which in their turn contributed greatly to the economic development of the prairie regions of Canada and the spring wheat belt of the USA. The first of these, the Marquis wheat (a hybrid produced by crossing the Hard Red Calcutta (India) and Red Fife (Poland) varieties in the first decade of this century) matured earlier than the spring wheats which were previously planted. They combined high yield with grain that gave flour of a superior baking quality. Marquis wheat has since been used to produce a wide range of hybrids adapted to different regions and for a variety of uses. A comparable, though even more dramatic, revolution occurred 60 years ago with the production of hybrid corn (maize). This came into commercial production in 1933; by 1946 over three-quarters of all the corn grown in the USA was of hybrid origin. Within the same period average yields of corn increased by 50 per cent. In the most fertile parts of the corn belt yields doubled, or even trebled.

High-yielding varieties (HYVs)

One of the most spectacular recent biological developments has been the mass production of true, high-yielding, dwarf hybrids (HYVs) of important cereal crops such as rice and wheat suitable for the tropical areas of the world. The traditional varieties grown in many developing countries are characteristically large, leafy plants with extensive root systems. Their size, however, inhibits close planting. Addition of fertilisers merely tends to stimulate leaf production, and hence mutual shading and results in an elongated stalk which is very susceptible to lodging. New dwarf hybrids have been developed since the mid-1940s. These have small upright leaves, a short, stiff straw capable of carrying a heavy grain head and a shallow root system. These new varieties (HYVs) were the basis of what became known as the *Green Revolution* in the 1950s and 1960s and which, it was hoped, would provide a panacea for many of the Third World's agricultural problems. Outstanding successes have been achieved in respect of wheat and rice yields in Colombia, Mexico, the Philippines and parts of India and Pakistan. They are, however, even more susceptible to disease and retain their competitiveness for only about half the time they would in temperate regions. Since the first rice variety was used in the Philippines in 1966 it has been superseded by nearly thirty more disease- and insect-resistant varieties. The hope for the future now lies with the most recent advances in crop breeding by bioengineering. This aims to produce a crop with selected and, more importantly, stable characteristics by genetic implantation.

It is, however, little use producing 'new' and 'better' crop varieties unless soil moisture and nutrient conditions are suitable for their cultivation. A particular variety of crop may have a potential to give a high yield, but this will not be realised unless its water and nutrient supply is correspondingly high. New plant breeds need new 'environmental' conditions, new methods of cultivation; alone they cannot increase yield. The success of the dwarf hybrids depends on the heavy application of nitrogen fertiliser and the adoption of energy- and capital-intensive types of farming which many areas of the Third World can ill afford. It is, perhaps, significant that the new varieties have had their greatest impact in areas already technically well suited to their cultivation. These are areas of more prosperous, larger, owner-occupied farms, particularly on already irrigated land.

LIVESTOCK DOMESTICATION

In contrast to crop plants, the number of types of fully domesticated (i.e. bred in captivity) livestock is small with cattle, sheep, pig, goat and buffalo comprising *c.* 90 per cent of the world's livestock population and of which the cow supplies two-thirds of the world's animal protein. They undoubtedly owe their dominance to particular physical and/or behavioural characteristics which facilitated their domestication (Hafez 1969). All except the pig are ruminant (cud-chewers), animals capable of digesting plant material with a high cellulose content. The pig (*Sus* spp.), like humans, is an omnivore. It is descended from the wild boar (*S. scrofa*) and together with the dog became an early 'camp-follower' and scavenger of human food. Of the five it is the most efficient producer of animal protein. All, however, convert low-quality plant protein into high-quality meat protein and the older, hardier breeds can subsist on a diet of relatively low nutritive natural vegetation. In this respect they can complement rather than compete with crop plants for land (Tivy 1991). In addition, they are all 'herd' animals – i.e. living in social groups with a low male : female ratio based on a dominance hierarchy.

The origins of livestock domestication are as old as those of crops with whom they shared a common centre in the Middle and Near East where the ranges of the wild ancestors of all five of the main types of livestock overlap (Clutton-Brock 1980). Evidence indicates that the pig (which early became a scavenger of human food remains) – and the cow (a prolific milk producer) were living in close association with humans *c.* 10 000 BP in the uplands of Iraq and that by the time of the early Roman Empire all five had become domesticated in the sense that they had evolved into discrete breeding groups distinct from their wild ancestors. Sheep and goats initially well adapted to poor and the semi-arid mountainous environments were the earliest to be domesticated. While the former are grazers, the latter are predominantly browsers of low trees and shrubs. The goat is the hardiest of all domestic livestock able to subsist on very poor herbage in extremely harsh, climatic conditions. Despite their early origins a number of some less-domesticated taxa as well as the wild ancestors of both sheep (the Asiatic mouflon, *Ovis orientalis*) and goats (the ibex, *Capra ibex*) are still extant albeit in small populations in isolated and often upland areas.

The progenitor of the cow (*Bos* spp.), the wild ox or aurochs (*B. primigenius*) which, as evidenced by Palaeolithic cave-paintings, had a wide range from western Europe to eastern Asia, is now extinct. It gave rise to what are now the two main taxonomic groups – the European (*B. taurus*) and the Asiatic cow (*B. indicus*) (Mason 1984) from which the modern cow has evolved. The latter are morphologically most clearly distinguished by a neck hump (*B. zebu*) or a neck and chest hump (*B. sanga*). They are also better adapted to hot climates than the European breeds.

As with crop plants the evolution of domesticated animals has been accompanied by morphological and physiological changes dependent on the purposes for which livestock were selected and bred. As a result, distinct breed groups each with heritable characteristics have emerged. However, early selective breeding was a slow process dependent on isolation of one breeding population from another either in geographically distinct areas or in enclosed fields. In Britain, as in other parts of Europe, controlled breeding had to await the enclosure of farmland. Since the Second World War livestock breeding has become, with the aid of artificial insemination, more controlled and more specialised. This has been accompanied by a continuing reduction in the number of commercially important breeds derived in the case of dairy cows from a small number of centres or by hybridisation as in the case of pigs. Of the 145 indigenous breeds of cattle in Europe and the Mediterranean region 115 are in danger of extinction (FAO/UNEP 1975). The production of even higher-yielding cattle has necessitated a greater use of concentrated feedstuffs in the form of the cereal food crops of wheat, barley and maize.

DEFORESTATION

One of the earliest and still continuing human impacts on the biosphere is the removal of the original vegetation cover and its replacement by either another or by man-made structures. At the global scale the ecologically most significant impact of this kind is deforestation. It has been estimated that about half of the forest or woodland that originally covered two-thirds of the earth's surface has already been removed and reduction is being maintained at a rate which outstrips replacement.

Deforestation commenced in prehistoric times when the stone axe, fire and domestic livestock (particularly sheep and goats) facilitated the process of clearing land for agriculture, for fuel or to flush out wild animals and enemies. With increasing population and technical development demand for wood increased. Four general though not necessarily completely discrete phases in the advance of the 'deforestation front' can be recognised. The first was in prehistoric times in the Far and Middle East and the Mediterranean areas. While goat browsing has long been quoted as one of the main agents of deforestation in the latter region, the demand for large timbers to build the considerable Greek and Roman fleets may well have been more important; grazing would, however, have aided the process and have effectively inhibited tree regeneration. The second phase across western Europe was initiated in the Neolithic era; it gained momentum in the subsequent Bronze and Iron Ages and later in the medieval period. It continued until nearly the end of the nineteenth century, with the clearing of land for arable and pastoral farming, the demand for constructional timber for buildings and ships and, until the discovery of coal, for charcoal. The third phase came with the discovery and colonisation of the 'New World' when, between the end of the seventeenth and the beginning of the twentieth centuries, forest clearance was stimulated by the opening up of agricultural land and the growing global demand for wood and wood products, not least of which was pulpwood. And the so-called North American 'timber barons' left a legacy of cut-over forest land as they moved across the country from east to west. The final current phase of deforestation is that which has been proceeding at an ever-increasing rate since the 1950s in the humid tropics. This is most acute in the tropical rain forest. It has already been reduced by a third of its former extent in this period and if the rate of deforestation is maintained there could well be only scattered remnants left by the end of this century (Whitmore 1988).

The environmental significance of deforestation is related to the particular attributes of the forest ecosystems of the world, the relative size of the forest biomass, which accounts for 75 per cent of the total global plant biomass and its carbon-storage capacity. The forest biomass is distinguished by the volume of its plant material per unit area. It has, as a result, a very considerable effect on associated microclimatic and hydrological conditions. The atmosphere within a forest is insulated from that outside. Wind-speed and surface evaporation are reduced, atmospheric humidity is higher while annual and seasonal temperature ranges are dampened down. Dependent on the depth and density of the canopy, the amount and intensity of sunlight and precipitation are reduced. Deforestation can then effect a more drastic change of microclimate than the removal of any other types of vegetation. Exposed to full sunlight the ground surface and the soil experience a greater temperature range than before; in consequence, the rate of organic decomposition is speeded up. In the absence of an intercepting canopy the amount and particularly the intensity of surface precipitation is increased. The proportion of that reaching the ground which infiltrates the soil or runs off the surface varies dependent on the type of ground litter, the soil texture and the land gradient. Light sandy soils with a large organic content will absorb more water than bare and particularly heavy soils into which infiltration is slow and which quickly become saturated so that surface runoff is increased. However, in the event, the steeper the surface gradient the faster will be surface runoff following deforestation.

The specific nature and intensity of the microclimatic changes and of their ecological impact in a particular physical habitat will be dependent on a number of interacting variables, such as the type of forest and the depth density and seasonal duration of its canopy, the regional climatic conditions, the remaining amount of surface vegetation and litter. In addition to its impact on microclimatic conditions large-scale deforestation can also affect local and regional climates. Further, deforestation causes a greater disruption in the nutrient cycle than in other ecosystems. This is because, during the long growth cycle of the trees, a varying amount of nutrients become stored in the wood and are eventually exported with the timber harvest. This can cause a progressive nutrient drain from, and reduction in, the potential primary productivity of the soil comparable to that of an agricultural crop. The forest wood is also an important carbon 'sink' or reservoir. About a third of the carbon circulating through the biosphere is retained for a shorter or longer period in terrestrial plants – mostly in trees. The effects of deforestation on this cycle are complex and not yet fully understood. However, it has become increasingly clear that the felling and burning of trees in the tropical rain forest for shifting agriculture, for fuel and for large-scale land clearance in Amazonia has contributed to the increase in the carbon dioxide content of the atmosphere during the last 150 years. It has been estimated that a total burn within the next 50 years of the remaining forest would release twice as much carbon dioxide as that by burning fossil fuels over the same time and assuming no forest replacement (Scurlock and Hall 1992).

ACCELERATED SOIL EROSION

Deforestation and, particularly, land clearance for agriculture initiated the process of accelerated soil erosion which can, within the space of a few years, seriously deplete or destroy a soil which has taken centuries, or longer, to develop. It has been described as one of the most destructive of processes effected by human activities which, according to Butzer (1974) has been a latent, if not chronic environmental problem for some 10 000 years. Its distribution and spread parallel that of deforestation and land clearance already noted. The same author suggests that, before the sixteenth century, soil erosion was relatively limited in extent and was confined to the heavily populated areas of Eurasia. With the rapid, ruthless exploitation of forest and soil resources following the opening of the 'New World', however, it became both intense and widespread. Perception of the process and its global dimensions did not emerge until the 1930s when the earlier extension of cultivated land westwards into the semi-arid grassland of North America during a period of above-average precipitation was subject to a decade of prolonged drought and the bare soils started to 'blow' creating the *Dust Bowl* conditions of the south-western states of the Great Plains. Soil erosion became as well publicised and as emotive a term as pollution and global warming are at present.

The two principal causes of accelerated soil erosion are the degeneration and/or removal of a former closed vegetation cover and a reduction in soil stability as a result of the progressive decrease in the soil organic matter. The first exposes the soil, the second makes it susceptible to accelerated erosion. The severity of the process is a function of the *erosivity* (erosive energy) of the agents involved and of the *erodibility* of the soil material.

The two main agents of soil erosion are wind and water which although not mutually exclusive tend to be predominant in different climatic regions. Arid and semi-arid regions (and soils which dry out rapidly in all cases) are more susceptible to removal of soil particles by wind. The intensity of erosion depends on the soil texture, surface features and the wind-speed and duration. Sandy soils, with over 60 per cent unaggregated particles less than 0.1 mm in diameter are most liable to blowing. Those particles small enough to go into suspension in the air may be carried hundreds of kilometres before being deposited. Larger particles are moved primarily by *saltation*, a process whereby, given sufficient wind strength, finer particles are projected into the air as a result of the force with which they are hit (bombarded) by the larger material.

Water erosion tends to be potentially more severe and prevalent in the humid and subhumid climatic regions. Its effect is a function not so much of the amount and seasonal distribution as of the intensity of precipitation in the first place. Intensity is expressed in the rate of rainfall which is mainly dependent on raindrop size. The erosivity of 100 mm falling in 1 hour will exceed that of the same amount in 12 hours. The kinetic impact energy of the former can break up the exposed soil aggregates into finer particles which seal or 'cap' the soil surface, thereby reducing the rate at which water can infiltrate. The rainsplash impact can then project the soil particles upwards and forwards, removing material evenly downward on sloping ground as sheet wash. This can account for up to 90 per cent of all soil erosion. With increasing amount and duration of precipitation surface runoff becomes channelled by microrelief and can effect downward erosion of temporary rills or of more permanent gullies and ravines.

The rate of soil erosion within any climatic region is a function of the interaction of a number of variables whose relative importance varies from one site to another, including the steepness and length of slopes, the nature and amount of surface vegetation and DOM and the erodibility of the soil. The latter is defined as the resistance of the soil to both detachment and transport (Morgan 1979) and is more specifically dependent on the texture, structure, infiltration capacity and the organic content of the soil and on land use and soil management. The latter factors are particularly important in that they are often the critical factors in creating a stable soil structure (see p. 361).

The effects of soil erosion are numerous and diverse. A decrease in soil depth is accompanied by erosion of soil fertility as a result of the loss of nutrients, particularly of nitrogen, phosphorus and sulphur which tend to be concentrated in the upper part of the soil profile. On naturally shallow soils little-weathered parent material or sterile impermeable pans may be exposed at the surface. Severe gullying can reduce or even destroy the workability of the land. The effects of soil erosion, however, extend beyond the area of removal to those where the eroded material is deposited. Some is deposited windward of areas of origin often burying previously good agricultural land and, in arid areas, creating sand drifts and mobile dunes. Much of that eroded by water finds its way into streams and rivers to be eventually deposited over flood plains, on river beds, in river deltas and on the offshore continental shelf. Silting increases water turbidity reducing light penetration and consequently, primary production. Fish spawning beds may be damaged, river and irrigation channels and reservoir capacities reduced. Periera (1973) has illustrated how relatively small areas of extreme erosion can cause damage over a disproportionately extensive area. In the case of the Paraná River watershed in the High Andes of South America, where soil erosion affected only about 4 per cent of the total area, it was shown that 80 per cent of the sediment carried downstream originated from an overgrazed area at the head of the Bermejo tributary – where the Argentinian cattle industry originated some 400 years ago. The ever-increasing load of sediment has caused serious silting in the port of Buenos Aires while the delta of the Paraná River, which grew at a rate of 46 mm year^{-1} in the period 1873–97, attained 84 mm year^{-1} in 1900–60 without an increase in river flow. At present the increasing rate of the interrelated processes of deforestation, mountain surface-water runoff, accelerated soil erosion, flooding and siltation in the tropics are reflected in the frequently recurring terrestrial mud flows, soil and rock avalanches and valley flooding in tropical areas such as Nepal and Bangladesh, the Amazonian Andes, Philippines and

Madagascar, where the eroded red tropical soils colour offshore waters for several miles out in the Indian Ocean. As has previously been noted, accelerated soil erosion in the humid tropics can, within a relatively short period of time, reach a stage at which the recoverability of the ecosystem is retarded or even inhibited.

Chapter

18

Human impact: environmental pollution

Environmental pollution can be defined as an addition of any substance to part or whole of the physical environment such as to alter its physical and/or chemical composition temporarily or permanently. As such, a distinction can be and often is made between natural and cultural (or anthropogenic) pollution. The former is due to natural processes such as volcanic explosions and the transport and deposition of organic and/or inorganic matter from one part of the biosphere to another. The latter, resulting from human activities, involves not only an increase in the amount and production of naturally occurring substances but also of synthetic materials.

The effect of pollution on organisms depends on the type of pollutant and where and in what form it is emitted. Some are benign, some highly toxic. Some are biodegradable, others are not. Some break down rapidly, others are more persistent. The sensitivity of an organism to a particular pollutant is dependent on its genetic, physiological and morphological characteristics on the one hand, and on the prevailing environmental conditions and the concentration of the particular pollutant on the other. Assessment of the level or threshold (i.e. concentration per unit time) at which a pollutant causes a particular reaction in organic growth, reproduction and survival is, however, extremely difficult because of the problems associated with the following:

1. The identification of anthropogenic pollution from normal but variable background levels of a naturally occurring substance;
2. The accurate, continuous long-term measurement of pollution in air, water or soil;
3. Measurement of the amount and rate of pollutant absorption by organic tissues and its correlation with organic changes.

The two most commonly used methods are the observational and the experimental. The former correlates observed changes in species diversity, density and vigour on an increasing gradient of pollution from an unpolluted to a known observed or quantified pollution source. The latter subjects a particular species to varying levels and intensities of a selected pollutant under varying laboratory-controlled environmental conditions. However, pollutants rarely occur or act independently of other pollutants or non-pollutant substances, and interactions may be additive, synergic or antagonistic.

SOURCES OF ANTHROPOGENIC POLLUTION

Anthropogenic pollution is caused by (1) the production of domestic, agricultural and industrial waste materials, and (2) the use of biocides and aerosols. Both have increased rapidly in amount and distribution during the past century and pollutants continue to be emitted into the atmosphere, fresh- and salt-water bodies and the soil.

AIR POLLUTION

The principal sources of air pollution include the combusion of fuels (wood and fossil fuels) in the production of heat and power, exhaust transmissions from road and air transport and the use of synthetic chemicals. The constituents are either particulate (see Table 18.1) or gaseous, some of which are deposited (*dry*

Table 18.1 Types and definitions of atmospheric particulate pollutants (from Dix 1981)

Grit	Solid particles; suspended; over 500 μm diameter
Dust	Solid particles; suspended; 0.25–500 μm diameter
Smoke	Gas-borne solids; particles usually less than 2 μm
Fumes	Suspended soils; less than 1.0 μm and normally released from chemical or metallurgical processes
Mist	Liquid droplets with less than 2.0 μm diameter
Aerosol	Solid or liquid particles in suspension in air, or some other gas, < 1.0 μm

deposition) on the land surface and some of which are 'washed out' (*wet deposition*) of the atmosphere in precipitation on to land or water. Since the end of the nineteenth century the number of pollutants has increased while their relative contribution to the total air pollution has changed. Clean-air legislation and reduction of domestic coal fires and open industrial furnaces in Western Europe and North America resulted in a drastic decline, particularly since the 1950s, in the notorious black urban smog so characteristic of the heavy industrial areas in the pre-Second World War period. Increase in road transport, however, has been accompanied by the production of other pollutants – such as peroxyl-acetyl-nitrate (PAN) and ozone which give rise to the blue, photochemical smog now so prevalent in cities such as Los Angeles and Sydney.

Of the primary air pollutants more is known about the ecological impact of sulphur dioxide (SO_2) than of any other. Not only is it the main single pollutant, but emissions (with those of particulate matter) have been recorded in and around urban areas at more stations and for a longer period of time than any other pollutant. Also, the variable sensitivity of lichens to air pollution and the occurrence of *lichen deserts* in the most polluted city areas had early been recognised. The negative correlation of sulphur dioxide levels with the number, cover and condition of lichens in particular habitats (corroborated by laboratory experiments) confirmed their status as 'indicator species' (O'Hare 1974). Sulphur dioxide pollution is also thought to cause leaf chlorosis and necrosis (visible symptoms of injury occur above a minimum threshold of 0.3–0.5 ppm per 3 hours' duration), the early leaf fall of some urban deciduous trees, the annual or biennial shedding of conifer needles and growth reduction without evidence of injury in grasses (Treshaw and Anderson 1989). Recently reported (*The Independent* 11.9.91) results of research by National Environmental Research Council (NERC) on the effect of air pollution suggest that air pollution (particularly nitrogen dioxide and sulphur dioxide) can make some conifers more prone to damage by fungi and cause population explosions of aphids (common plant-sucking pests).

During the last two decades attention has become increasingly focused on the nature of pollutant reactions in the atmosphere and the ecological significance of those recently identified global pollution problems of acid rain, aerosols and the greenhouse gases.

Acid rain

Increased rainfall acidity, initially noted in the 1870s in Manchester, was attributed to the formation of sulphuric and nitric acids from the pollutant oxides of sulphur and nitrogen. The initial term *acid rain* is now used to describe precipitation having a pH value below 5.6 (the pH of pure water in equilibrium with atmospheric carbon dioxide) and, in addition, containing elevated concentrations of many chemical components. The latter include: acidifying ions of sulphate SO_4^{2-} (60 per cent) and nitrate NO_3^- (30 per cent) derived mainly from anthropogenic emissions of sulphur and nitrogen oxides respectively; chlorine Cl^- (10 per cent); a wide variety of minor and trace elements including heavy metals (lead, vanadium, zinc, cadmium); and other organic micropollutants. The result is a 'chemical cocktail', a substantial part of which is emitted from tall chimney-stacks and is subject to long-range transport in the upper atmosphere.

The ecological effect of acid rain is dependent on the way in which these substances are removed from the atmosphere – on the method of *acid deposition*. The latter can be dry, i.e. the direct transfer to, and absorption of, gases and particles by natural surfaces, or wet, i.e. the indirect transfer of substances from the atmosphere in or on the surface of hydrometeors, i.e. rain, snow, hail, etc. It has been estimated that 60–70 per cent of the wet deposition of sulphur and nitrogen is effected by cloud condensation. The impact of wet deposition is dependent on the ecological nature of the receiving surface. On a terrestrial vegetation cover the significant variables are the roughness and the hydrophotic quality of the vegetation surface and the age of the constituent plants. Trees, and particularly evergreen trees, are very efficient filters of pollutants from cloud and mist. The resulting high precipitation acidity can leach metabolites directly from their leaves at a rate that cannot be made good by uptake from the often nutrient-deficient soils on which they grow. The correlation between tree damage and natural (at high altitude) or anthropogenically induced (power stations, etc.) concentration of acid deposition was early established.

Acid rain and forest decline. Decline, particularly of mature trees, in forests and woodland as a result of either competition, disease, insects, climatic stress or pollution is a long-recognised common occurrence throughout the world. Visible and usually progressive symptoms include leaf yellowing (as a result of chlorosis or breakdown of chloroplasts) and browning; premature leaf fall (necrosis); die-back of outer branches; and crown thinning with the tree eventually dying from the top downwards. The more recent, highly publicised forest decline in both north-western Europe and north-eastern USA has been both more rapid and more widespread than previously (See Table 18.2). This so-called *novel decline* or

Table 18.2 Diffusion of novel forest decline in Europe (compiled from data in Treshaw and Anderson 1989)

Date	*Species*	*Location*
Early 1970s	Silver fir (*Abies alba*)	Germany (old trees – 60 years about 600 m)
1976	Norway spruce (*Picea abies*)	Germany, Norway, Sweden
1980s	Scots pine (*Pinus sylvestris*)	Central Europe (France, Poland, Czechoslovakia)
	Beech (*Fagus sylvatica*)	Western Europe
	Oak (*Querus robur* and *Q. petracea*)	Western Europe
1986	Birch (*Betula* spp.)	Western Europe
	Maple (*Acer* spp.)	Western Europe
	Ash (*Fraxinus* spp.)	Western Europe
	Alder (*Alnus* spp.)	Western Europe

Neuer Waldschaden (Wellburn 1988) has been distinguished by the suddenness of its occurrence in the early 1970s and the rapidity of its spread over an extensive and geographically varied area. It has affected young as well as mature coniferous and deciduous trees, and has been characterised by considerable variation in decline symptoms in different species or species combinations in particular but not mutually exclusive physical sites varying in altitude, oceanicity and soil parent material. A common feature, however, has been the decrease in the radial growth of all affected trees before the appearance of visible symptoms of decline. The amount of growth reduction and the period when it occurred have been verified by dendrological analyses.

The causes of the novel forest decline still remain unresolved. Although it has often been linked to the impact of acid rain, there is in fact no clear evidence of a direct causal relationship. Precipitation tends to be naturally slightly acid (pH 5.6), consequent on the absorption of atmospheric carbon dioxide and the production of a weak carbonic acid (H_2CO_3). As already noted, anthropogenic acidification of the atmosphere commenced at least 200 years ago. However, although the areas affected have increased in both Europe and North America (see Figs 18.1 and 18.2), the large-scale acidification of German forests had been recorded from the

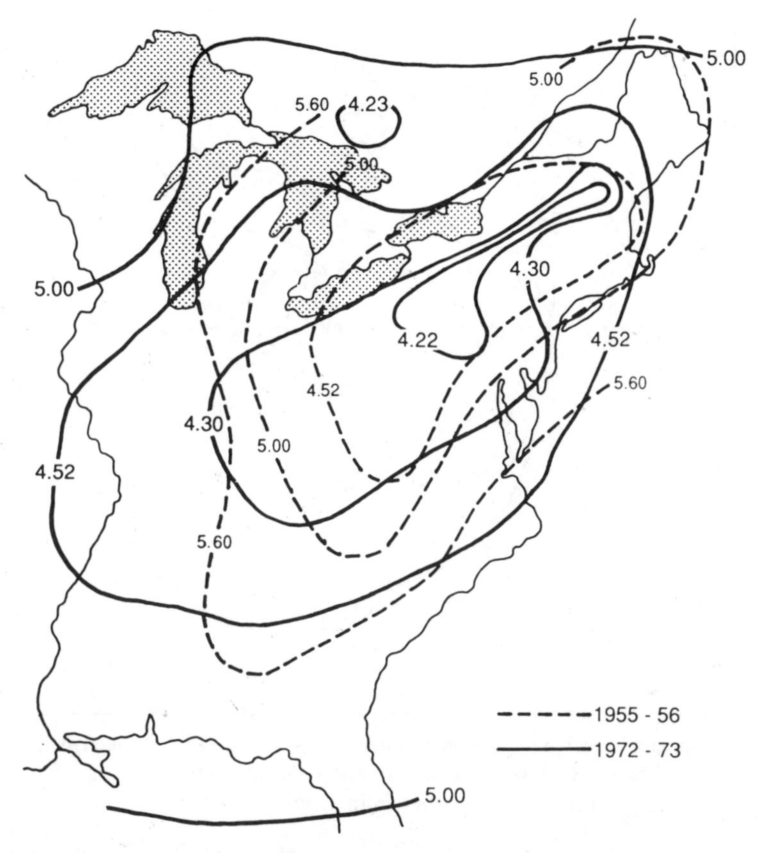

Fig. 18.1 Increase in acidity of precipitation over the eastern United States from 1955/56 (broken line) to 1972/73 (solid line); numbers = pH values (after Likens 1976)

1920s and by 1940 the point of *acid saturation* (pH 4.2) had been reached. The efficient filtering capacity of forests has resulted in the impact of wet deposition being greatest on forest soils of naturally low pH and with a limited capacity to buffer pH changes. Soil acidification results in the chemical breakdown of complex minerals and the release and mobilisation of toxic aluminium ions. This is known to cause a drastic reduction of the fine roots of trees with the impaired uptake of calcium and magnesium which results in the eventual decline and death of affected trees. There is, however, no real correlation between aluminium toxicity and the novel forest decline, which can be equally severe on calcareous soils (pH 6.5) low in aluminium. The possible effects of other chemicals

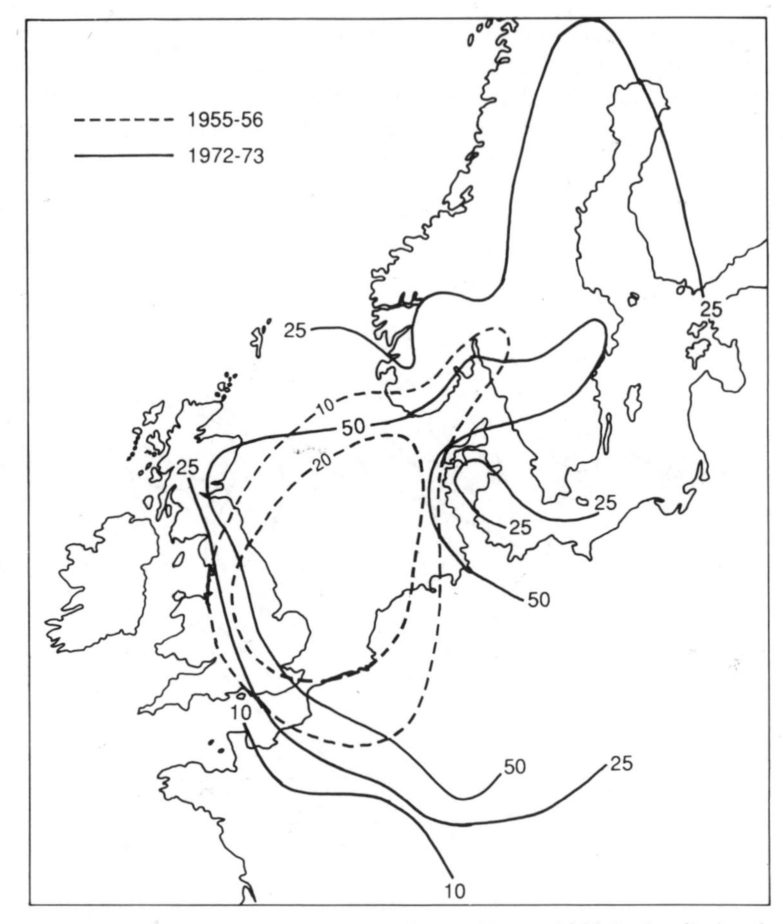

Fig. 18.2 Increase in acid precipitation over north-west Europe 1956 (broken line) and 1974 (solid line) based in volume-weighted average concentration of H^+ ions in microequivalents per litre of water (adapted from Likens *et al.* 1979)

associated with acid rain such as the heavy metals, hydrogen peroxide and ozone have received less attention.

Other acidity-independent variables now considered to be directly or indirectly involved in the recent forest decline are nutrient deficiencies. Other predisposing natural factors that have been suggested are climatic stress, pathogens and silvicultural practices. There is, in fact, a close correlation between reduced tree growth as reflected in annual tree-ring widths and weather conditions such as the summer drought in the late 1970s and early 1980s in Europe and the exceptionally severe frosts in the late winter and early spring of the years 1981–83 inclusive. Unfavourable soil and climate conditions may have made trees more susceptible to pathogenic infections. Despite the continued popular association of forest decline and acid rain, it is now clear that the causes of the novel forest decline are complex, the result of a number of interacting natural and novel factors, the combination and relative importance of which probably vary from one area to another but each of which reflects the total impact of air pollution in a particular environmental context.

Water acidification. There is now a large and still growing body of evidence of increasing acidity in freshwater bodies in areas in Europe and North America with similar naturally acid, little buffered substrata. Records of temporal acidity increase have been obtained by analysis of the known pH tolerance of species of diatoms in lake sediments of varying depths in Scandinavia and Scotland. In southern Sweden and western Norway increasing acidity would appear to have started early (in the late 1920s and early 1930s) consequent on the growth of metallurgical industries. Low pH values often vary seasonally, particularly low values being recorded after heavy autumn rain following a dry summer and after rapid snow-melt. As in the soil, acidity is often accompanied by a high concentration of aluminium and other metallic ions. Three processes are thought to be involved in the acidification of water: direct input of acid precipitation; changes in soil due to acid precipitation; and the increased concentration of cations (including H^+) resulting from an increase in sulphate (SO_4^{2-}) concentrations in surface runoff (Overrein *et al.* 1980).

Acidification has had a considerable impact on the freshwater ecosystems affected. All have experienced a significant decline in fish populations due mainly to the high mortality of eggs and fry in acid water. Adult fish can be killed, as in many fish farms, by a rapid influx of acid water as a result of snow-melt, a particular pollutant emission, or stress from toxic combinations of acid water and aluminium. It has been noted that more than half the fish population in Norway was lost during the period 1940–80 (Overrein *et al.* 1980) and many lakes in the south of the country are practically devoid of fish. However, tolerance of acidification varies between (see Table 18.3) and within species (dependent on size, age, etc.).

Stressful conditions as a result of acidity and aluminium toxicity have been exacerbated in many freshwater systems by the deposition of other substances associated with acid rain. The most harmful are the persistent and fat-soluble polychlorinated biphenyls (PCBs) and mercury. With increasingly acid conditions

Table 18.3 Ranking of some important Norwegian fish according to acidification sensitivity – 1. highest; – 9 lowest (from Overrein *et al.* 1980)

Rainbow trout	(*Satmo gairdneri*)
Salmon	(*S. salar*)
Sea trout	(*S. truttax*)
Brown trout	(*S. truttax*)
Perch	(*Perca fluviatalis*)
Char	(*Salvelinus alpinus*)
Brook trout	(*S. fontanalis*)
Pike	(*Esox lucins*)
Eel	(*Anguilla rughilla*)

decomposition of organic matter is retarded as a result of the decline in bacterial decomposers and an increase in the less efficient fungi. Reduced cycling of nutrients (particularly of phosphate which becomes immobilised as aluminium phosphate) is reflected in lower productivity and a reduction in the diversity of the invertebrate fauna (i.e. zooplankton, insects, crustaceans, snails and bivalves).

Chlorofluorocarbons and the 'ozone hole'

In 1974 a family of complex synthetic chemical compounds called chlorofluorocarbons (CFCs) were added to the growing list of air pollutants. Known under the trade name of Freons they were widely used in refrigerators, air conditioners, aerosol sprays, etc. Inert, non-flammable and non-toxic, they can, however, remain unchanged in the atmosphere for up to 50 years (CF-11) or even a 100 (CF-12) (Allaby 1986). They can also be carried up by warm ascending air into the stratosphere where exposure to UVR causes their disintegration. The chlorine atoms produced in this process act as catalysts in the conversion of ozone (O_3) to oxygen (O_2) and as a result contribute to the thinning and eventual destruction of the ozone layer. However, the CFCs account for less than half the chlorine in the atmosphere. Methyl chloride produced by vegetation and wood-burning and chloromethane in fungal decomposition are chemically similar to CFCs and are important sources of chlorine. As Allaby notes (1986), they have been produced ever since the fungal attack on wood began. It is now known that because ozone is formed by solar radiation and removed by chemical reactions, the ozone layer always thins over the Poles during the winter period. The Antarctic *ozone hole* was first detected in September–November 1987. Although it closed in the succeeding summer it reappeared in the following years. A similar but smaller 'hole' has been observed over the Arctic. The initial assumption that CFCs were responsible for the depletion of the ozone layer, which would expose the earth's surface to biologically disastrous levels of UVR, was widely publicised by the media. Public concern generated political pressure to protect the ozone layer by restricting or even banning the use of CFCs. By the time the Montreal Protocol to the UN Convention had been signed in 1988, the leading manufacturers were already developing new, less stable CFCs. Other substances that could be used as alternatives (e.g. carbon dioxide, vinyl chloride, hydrocarbons and ammonia) are, however, less environmentally favourable! Chlorofluorocarbons pollutants make a more significant contribution to the greenhouse gases than to the destruction of ozone.

The greenhouse gases

The CFCs like the naturally occurring water vapour and gases carbon dioxide (CO_2), ozone (O_3), methane (CH_4) and ethylene (C_2H_4) can absorb short-wave solar radiation. In doing so they become warm and transmit heat in the form of long-wave radiation to other gases and into the atmosphere where it is radiated in all directions. The process whereby these gases trap heat that might otherwise be lost by reflection or outgoing radiation and keep the atmosphere warmer than it would otherwise be is comparable to that in a greenhouse from which the name has been derived. As shown in Table 18.4, the *greenhouse* (or *radiation-forcing*) *gases* vary in that part of the spectrum where they can absorb solar radiation and in their absorptive strength. While carbon dioxide is present in the highest concentration, it does not have the highest absorptive strength, being surpassed in this respect by sulphur dioxide and particularly by the recently emitted CFCs. Nitrous oxide and methane absorb radiation more strongly at wavelengths 700–1300 nm than carbon dioxide or water vapour; the CFCs absorb where previously radiation could escape from the atmosphere.

Table 18.4 Some infra-red trapping gases in the atmosphere with approximate wavelength (μm = micrometre wavelength), and absorption bands and their relative absorption strengths (from Treshaw and Anderson 1989)

Gases	*Absorption band (μm)*	*Relative absorption strength*
Ethylene (C_2H_4)	98–10.4	420
Sulphur dioxide (SO_2)	7.2–13.2	750
	8.5–9.0	100
Ozone (O_3)	9.4–9.8	330
	12.4–12.8	320
	5.7–9.1	12
Carbon dioxide (CO_2)	13.2–16+	240
Nitrous oxide (N_2O)	7.9–8.4	200
Methane (CH_4)	7.3–9.8	200
CFC-12 (Cl_2F_2)	10.8–11.3	1500
	9.1–9.6	1400
CFC-11 (Cl_3F_2)	8.3–8.8	700
CH_3Cl_3	12.3–13.2	1100
	7.8–8.7	140
Carbon tetrachloride (CCl_4)	12.4–13.2	1400

It has now been established that the concentration of carbon dioxide and other greenhouse gases has been slowly increasing at a rate of *c.* 0.2 per cent year^{-1}. Measurements of carbon dioxide trapped in polar ice-sheets have dated the increase from the beginning of this century – from 270 ppm to 354 ppm in 1984. There are indications that the northern hemisphere is slowly becoming warmer, though the rate of this trend is difficult to assess. Increases in average temperatures of between 1.5 and 4.5 °C by the end of the century have been predicted. The consequences of a given amount of warming for the future climatic regimes of the earth are even less predictable. Increased concentration of carbon dioxide could increase potential primary biological productivity. Higher temperatures could result in less rainfall in some areas, but more in others, with implications for the ranges of wild and domestic plants and animals alike. In the words of Treshaw and

Anderson (1989: 179): 'no air pollutant however phytotoxic, widely distributed or health-threatening, compares in destructive potential to this global disaster'.

WATER POLLUTION

A very high proportion of the anthropogenic pollutants are discharged either directly or indirectly into surface, coastal and marine waters. The largest contribution comes from human sewage (treated or untreated), agricultural waste (i.e. liquid waste, farmyard slurry, silage effluent, fertiliser and pesticide residues) and industrial waste composed of varying amounts and proportions of water, suspended solids, organic solvents, oils and dissolved organic and inorganic chemicals. Among the most ecologically serious effects of such pollutants are:

1. An increase to the *biological oxygen demand* (BOD) (i.e. the amount of dissolved oxygen used in chemical and microbiological action when a sample of water is incubated for 5 days in the dark (Dix 1981);
2. Nutrient enrichment or *eutrophication* of the water;
3. *Toxification* of the water.

Biological oxygen demand (BOD)

The addition of biodegradable wastes increases the BOD by an amount dependent on the organic content of the effluent. Silage and some chemicals have BOD values of 50 000 and 30 000 mg l^{-1} respectively. Some types of industrial waste (from paper pulp, pharmaceutical manufacture and coal carbonisation) produce effluent with values of up to 10 000–25 000 mg l^{-1}; farmyard waste and sewage produce less than *c.* 2000 mg l^{-1} (Dix 1981). As a result the oxygen content of the water into which these pollutants are discharged is rapidly depleted and an anaerobic environment produced (Fig. 18.3). Decomposition and nutrient cycling are retarded and organic matter accumulates with the production of methane and ammonium sulphides. In addition, effluent with toxic substances such as heavy metals, cyanides and sulphides can kill off aerobic bacteria and hence retard decomposition even though sufficient oxygen may be present, while surface films of oil and hard detergents, which are more resistant to biodegradation, can prevent the reoxygenation of the water from the surface. The main agents of decomposition become the slimy fungi – the so-called *sewage-fungi.* Many aquatic species, particularly fish, are intolerant of oxygen deficiency and in the most severely polluted water practically all aerobic organisms are eliminated apart from sludge worms (Turbificidae) and bloodworms (*Chironomous* spp.).

The impact of pollutants with a high BOD or which create anaerobic conditions is dependent on the volume, concentration, composition and the duration of the effluent discharge, on the one hand, and on the rate at which the water is reoxygenated on the other. The latter is a function of the volume and rate of water flow which varies both temporally and spatially. Until the development of efficient sewage-treatment plants and the initation of other water-purification schemes, an increasing volume of effluent discharge in those rivers draining large urban/ industrial catchments in North America and Europe resulted in the oxygen sag moving progressively upstream from the end of the nineteenth century to the

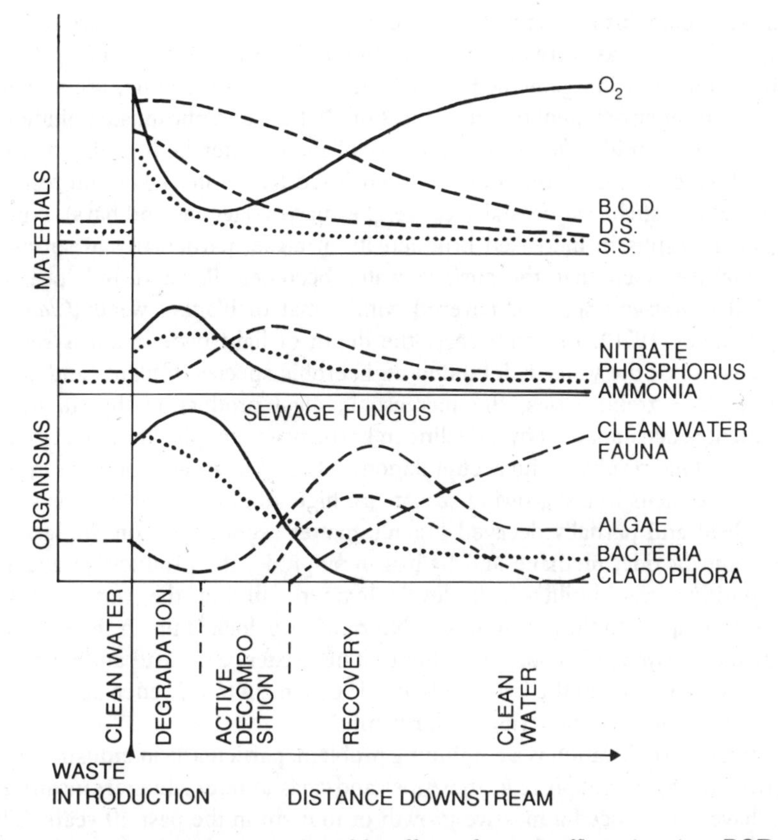

Fig. 18.3 Diagrammatic representation of the effects of organic effluent in a river: BOD – biological oxygen demand; DS = dissolved salts; SS = suspended salts (from Hynes 1963)

1950s. However, some rivers (e.g. Thames, Clyde) have been 'revived' to the extent that a reduction in BOD has been accompanied by a recovery of fish species numbers and the return of the particularly pollutant-sensitive migrating salmon.

Eutrophication (or nutrient enrichment)

Some types of domestic, industrial and agricultural waste products contain significant amounts of two of the most important plant nutrients, nitrogen and phosphorus, in the form of nitrates and phosphates. The discharge of both has increased since the 1950s. In the first case, intensification of farming in the Western world has been accompanied by an increasing production of liquid manure (*slurry*) and the use of inorganic fertilisers. Up to 40 per cent of the highly soluble nitrates applied may be lost in surface water runoff and by soil leaching. Phosphates are less mobile and tend to become fixed and protected from leaching in agricultural soils. However, increase in phosphate pollution has resulted largely from the increase in the use of detergents during the 1950s which subsequent changes from non-biodegradable to biodegradable materials has in fact not reduced.

All water contains a certain amount of these two nutrients which are necessary for aquatic as well as terrestrial primary biological productivity. Ageing lakes in which organic and inorganic sediment is gradually accumulating, are subject to natural nutrient enrichment or eutrophication. Nitrate and phosphate pollution has resulted in the rapid *cultural eutrophication* of many water bodies, the ecological effects of which are most pronounced in enclosed freshwater lakes and reservoirs and in shallow, sheltered coastal areas (e.g. estuaries, lagoons and bays). Nutrient enrichment results in increased productivity of algae, particularly in spring and early summer, such that the surface water becomes like a turbid 'pea soup' (unicellular *Monodus* spp.) or covered with a mat of blanket weed (*Cladophora* sp.) (Mellanby 1970). In either case, the depth of light penetration is reduced, reoxygenation of the water is inhibited and aerobic species of plants and animals decrease. In extreme cases, the increase in algal production due to nutrient enrichment is accompanied by a decline in herbivorous zooplankton due to oxygen deficiency. This results in the accumulation of DOM in an environment in which the oxygen deficits in the growing season are high and nutrient cycling is slowing down. Dead and partially decayed organic matter accumulates on the lake floor and carbon dioxide, methane and hydrogen sulphide, the products of anaerobic decomposition, can build up to levels toxic to all but the very specialised organisms adapted to these conditions. Nutrients are 'locked up' in the DOM and the aquatic ecosystem becomes moribund. The existence of dead and dying lakes was widely publicised in the late 1950s and 1960s in central Sweden and the lower Great Lakes (Ontario and Erie) of North America.

Cultural eutrophication is a continuing problem, particularly in industrial/urban estuaries and in mountain (e.g. the Alps) and coastal (e.g. Mediterranean) areas which have experienced a massive growth of tourism in the past 40 years. Those water bodies most susceptible to the deleterious impact of cultural eutrophication are inland or coastal wetlands, small to medium-sized lakes, large deep-water bodies in which disturbance and turnover of bottom water are negligible and in estuarine and shallow coastal waters where tidal flushing is not strong enough to counteract the input of nutrients from the land.

ENVIRONMENTAL TOXIFICATION

Much anthropogenic waste contains potentially toxic substances which can effect the gradual or sudden injury or death of organisms. In so far as toxic pollutants can be discharged directly or indirectly into land, water or air they will be, for convenience, grouped together. Three important types of toxicants can be distinguished: the heavy metals and PCBs, pesticides and radioactive waste.

Heavy metals and PCBs

The heavy metals include iron, zinc, copper, nickel, lead, mercury and cadmium. They may be directly discharged into air, soil or water in gaseous, liquid or solid industrial wastes or in sewage sludge on to farmland. Although the first four are trace minerals (or micronutrients) essential in very small concentrations for the health of organisms, all are potentially toxic above a normally very low threshold concentration (i.e. *threshold limiting value* (TLV)) dependent on the species involved. In contrast, the PCBs are complex synthetic organic compounds chemically produced in the manufacture of plastics, rubber, paper etc. They are stable, and insoluble in water. If ingested by animals they can accumulate in the body tissues, break down very slowly and have proved lethal to fish, shellfish (particularly mussels) and fish-eating birds in particular.

Pesticides

Pesticides are, as the term suggests, substances used to control animal and plant (weed) pests. Before the Second World War most of the chemicals used for this purpose were either naturally derived and ecologically safe or, if chemical toxins, were used in small amounts. The development and increased application of synthetic pesticides were stimulated by the chemical research and the intensification of agriculture during and after the Second World War (see Fig 18.4). Public concern about the pollutant effect of pesticides was generated by the publication of Rachel Carson's book *Silent Spring* (1964) which drew attention to the effects of the wholesale indiscriminate use of pesticides and, particularly at that time, of DDT, the first multipurpose biocide to be used not least in clearing mosquito-infested tropical wetlands and delousing soldiers during the Second World War. Since then the number and range of pesticides have escalated, that of approved agrochemicals alone has risen more than a hundred-fold since 1944.

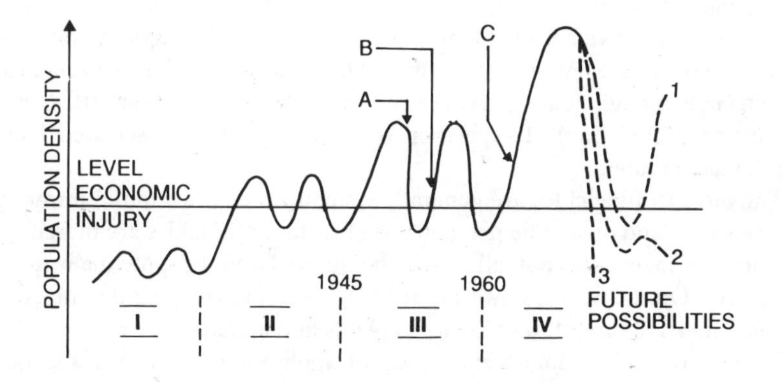

Fig. 18.4 History of pests and pesticides: (I) pre-agriculture; (II) agriculture without pesticides; (III) agricultural with pesticides; (IV) agriculture with pesticides-resistant strains. A = direct density-independent effect of pesticides; B = repression of density-dependent controlling factors; C = natural selection of strains resistant to pesticides: (1) use of new pesticide; (2) partial return and use of complexity; (3) increased simplification by control of crop environment (after Moore 1967)

The three main groups are the insecticides, the herbicides and the fungicides. The ecological effect of an insecticide is, in part, a function of the type of chemical used and how it enters the environment. The routes by which an insecticide may enter and leave an ecosystem can be direct or indirect; normal or the result of misuse. The effect of insecticide pollution on *non-target species* is dependent on the specific toxicity, the stability and persistence of the particular chemical in the environment and in organic tissues of the affected organism. The *organo-chlorines* (or chlorinated hydrocarbons) are more toxic than the *organo-phosphate* pesticides. The former are non-biodegradable and hence are very persistent in soil, air and water, from where they can be absorbed by plants and animals and in which they can effect many physiological and metabolic processes. The organo-chlorines are also easily soluble in fats and fatty tissues and hence can be rapidly absorbed by organisms, particularly insects whose cuticle is more easily penetrated than the mammalian skin. Finally, they are very mobile, capable of being volatilised and of being carried great distances as aerosols; low levels of both DDT and DDD have been detected in some fish and fish-eating birds such as penguins in the Antarctic. By the 1960s it was established that the organo-chlorine pesticides were affecting the health and reproductive ability of birds and fishes at the end of the food chain as well as many beneficial pollinating insects. They have now been largely replaced in farming by the unstable, biodegradable and short-lived organo-phosphates and carbonates. These do not accumulate in organic tissues and are quick-acting. However, some such as dioxin are so highly toxic as to be a hazard to those who apply them.

Herbicides are used more extensively and in greater amounts than insecticides and, it is maintained, are less ecologically damaging (Mellanby 1970). Some are selective in the largely broad-leaved crop weeds they destroy; some are residual, i.e. applied to the soil in order to retard or inhibit weed growth in a germinating crop; some are systemic or translocated chemicals which penetrate the plant's system. Non-selective herbicides are normally used on non-agricultural sites. The more recently developed organo-hormonal herbicides are the growth regulators. If used in large enough quantities these can (as in the Vietnam War) effect complete destruction (defoliation) of vegetation and, as a result, its associated fauna and inhibit regeneration.

The modern fungicides are generally systemic in contrast to the formerly used surface sprays and dust. The presently used surface fungicides are mercury-based and are, in many but not all cases, being replaced by potentially less toxic substances. On the whole, however, in order to avoid damaging the infected plant tissues, fungicides are of low toxicity to plants and animals.

Because of their mobility and mode of application, pesticides and pesticide residues can affect aquatic and terrestrial ecosystems other than those to which they have been deliberately applied. They frequently kill non-target species which may, in fact, be predators of the target species or food plants of non-harmful – indeed beneficial – insects such as pollinating butterflies. Many, particularly insect pests, can build up a resistance to biocides relatively rapidly and hence necessitate their replacement by newer and more toxic substances. The behaviour of many pesticides in the soil is still not well understood. However, it is known that some (such as aldicarbs used on sugar-beet) may produce metabolites and degradation products which may be more toxic than the original pesticide (OECD 1986). And all too little is known about the possible synergistic effect of pesticide mixtures or of the combination of a pesticide and other compounds including solvents and carrier substances with which they are mixed.

Table 18.5 Some persistent pollutant nuclides: source and half-life (from Dix 1981)

Source	*Nuclide*	*Half-life*
Nuclear weapon testing	Strontium-89	53 days
	Strontium-90	28 days
	Caesium-137	28 years
	Carbon-14	5570 years
Nuclear reactors and airborne discharges	Krypton-85	10.8 years
	Tritium-3	12.3 years
	Carbon-14	5570 years
Fuel reprocessing and airborne discharges	Krypton-85	10.8 years
	Tritium-3	12.3 years

Radioactive fall-out

Testing of nuclear weapons (on the ground, in the air and under the sea), nuclear reactors and nuclear fuel-processing plants are sources of the waste product of nuclear fission. These nuclides (Table 18.5) are deposited as radioactive fall-out on land and water. Some break down very rapidly; others can persist in the environment for long periods of time and can be absorbed directly by aquatic animals or indirectly from the vegetation by herbivores. The contamination of hill sheep in Britain and of reindeer in Scandinavia following the Chernobyl disaster revealed that mountain areas were particularly susceptible to fall-out from mist and rain and that the nucleides were very effectively absorbed by wet, particularly bog moss (*Sphagnum* sp.), vegetation.

Biological concentration

The biological hazards of biocide pollutants are exacerbated by the process of *biological concentration* whereby a particular chemical becomes concentrated in animal tissues at levels far in excess of those in which it occurs in the environment from which it was absorbed. This can occur in two ways. In the first, some organisms, particularly aquatic detrital feeders (e.g. clams, mussels, oysters), are capable of concentrating biocides by *c.* 70 000–100 000 times that of the water in which they live and in which pollutant concentration may be less than 1 ppb. This degree of concentration is more likely to be toxic for the animals (including humans) that eat them than for those in which biocides are concentrated.

The second type of biological concentration is sometimes referred to as *food-chain concentration* or *magnification* (see Fig. 18.5). Stable biocides which are ingested by animals from plants, soil or water are not excreted. As a result they accumulate in the body at rates roughly proportional to food composition and intake. Because of the low rate of energy conversion the concentration of the biocides increases by a factor of 10 as it is passed from a lower to a higher trophic level in the food webs with the result that the small amounts initially absorbed can attain lethal levels in animals at the end of food chains such as fish, predatory birds and humans. The degree of concentration is dependent on the particular

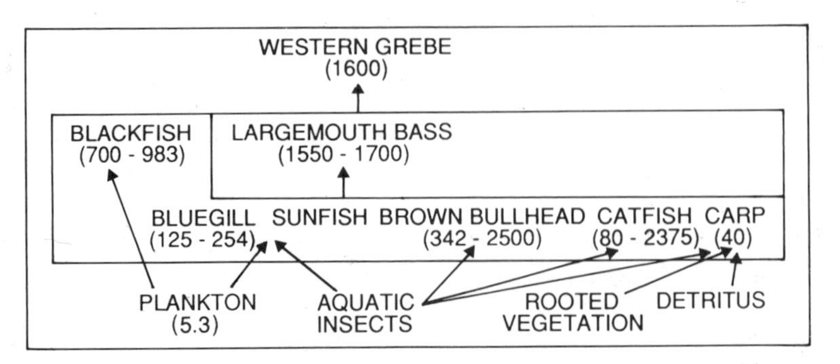

Fig. 18.5 Trophic web concentration of DDD (= stable derivative of DDT). The estimated concentrations are for visceral fat of each type. An initial application of 0.002 ppm DDD is assumed. Degree of concentration with species (from Foin 1976 based on Rudd 1964)

animal tissues in which the pollutant is stored and the number of trophic levels through which it passes. Strontium-90, for instance, is stored mainly in bone, hence its tendency to become concentrated is much less than that of caesium-137 which is stored in the flesh of herbivores and concentrated in that of their predators. In Alaska during its passage from lichens via caribou to man the concentration of the caesium-137 doubled; in wolves and foxes which ate the flesh of the caribou concentration could be trebled. Woodwell (1971) demonstrated that 20 years after DDT had been used to control mosquitoes in a marsh ecosystem in New York there were residues of *c.* 32 kg ha^{-1} in the upper layer of the mud. The phytoplankton contained 0.04 ppm, minnows 1 ppm and one species of predatory bird 75 ppm – a concentration by a factor of more than 1000 over that of the primary plant producers.

As Pimm (1984: 329) so cogently remarks: 'Most of our planet is dominated neither by pristine ecological communities, nor by species on the brink of extinction. Rather, most ecological communities we observe are fragmented, harvested, and polluted, stressed in various ways by humans and their technology.'

Chapter

19

Managed ecosystems

While most of the biosphere's existing ecosystems have been, and continue to be, directly or indirectly modified or completely altered by human activity, a high proportion are deliberately managed in order to facilitate the exploitation of a particular organic resource, i.e. a stock of material of use or of value to man. The value and management of organic resources are dependent on two basic biological characteristics:

1. *The renewability of organisms*, i.e. their capability of self-perpetuation, in contrast to inorganic resources which, although they may originally have been derived from organic matter, are formed so slowly that, from the human point of view, their limits of supply can be regarded as finite.
2. *The biotic potential* (or productivity) of organisms according to Watt (1968) is a function of their reproductive capacity (or fecundity) versus the age curves (life expectancy) for each species. Ricklefs (1973) defines it as the inherent capacity for increase as determined by the rate at which each individual contributes to future populations. However, as he notes, as population density increases so the resources available per individual decreases so that the theoretical maximum biotic potential cannot be attained. Species with a high biotic potential produce more young than the environmental resources can support. Hence intra-species competition is initially high, survival rates are low and reproductive waste is high compared to those with a lower biotic potential.

YIELD OR HARVEST

The principal objective of organic resource management is to obtain the maximum production (i.e. yield or harvest) without impairing the renewability of the resource. This is possible because individual or biomass productivity can be increased by judicious harvesting which will ensure as high a rate of stock turnover as possible. This can be effected by either removing (i.e. thinning) individuals from a young, high-density stock or from an older slow or non-growth, mature stock, thereby making more resources available to the remainder and hence accelerating their rate of growth. The level or intensity of exploitation will depend on the biotic potential of the species, and the higher the potential the larger the harvest that can be taken without endangering the renewability of the resource. However, increase in productivity of a population will occur only if growth is normally regulated by density-dependent factors and if the population size is limited, at the time of exploitation, by the available environmental resources (Ricklefs 1973; Putnam and Wratten 1984). While the maximum intensity of exploitation results in maximum productivity, it does not necessarily produce the *optimum yield.* This is because though productivity may be high the size (or *biomass*) may be reduced below its potential maximum. The optimum yield or *maximum sustainable yield* per unit time will theoretically be that beyond which increased intensity of exploitation results in a decrease in the stock of reproductive individuals or biomass. Where population size is regulated by density-dependent factors optimum yield can be expressed according to Watt (1968) as

$$\max (P_b) = \max [B_t + I(\mathbf{X})] - B_t$$

where P_b is the biomass productivity from time (t) to time ($t + i$), B_t the biomass at t and $B_t + I(\mathbf{X})$ the biomass at $t + 1$ as a function of vector $\mathbf{X}$ of variables X_1, where $i = I \ldots n$, which governs biomass production over t to $t + I$.

In the case of populations regulated by extrinsic non-density-dependent factors such as weather, optimum yield will be related more to minimising wastage due to natural mortality because of unfavourable environmental conditions, i.e.

$$\max (Y) = B_t - \min (R_t)$$

where Y is the yield at time t, B_t the biomass present at time t and $\min (R_t)$ the minimum number of reproducing individuals left at t in order to ensure replacement of Y by $t + I$ (Watt 1968). Laboratory-based experiments to assess optimum rates of exploitation of small animals establish levels (see Table 19.1) that are related to biotic potential. The response of an aquarium population of guppies (*Bebistes reticulatus*) is illustrated in Fig. 19.1 which shows that the maximum exploitation rate possible is just under 50 per cent per 3 weeks at which an adult population of *c.* 30 individuals could maintain a yield of 33.

Table 19.1 Optimum rates of exploitation for several laboratory populations of small organisms (Watt 1968); (60) per cent population/three weeks (from data in Putnam and Wratten 1984)

Species	*Exploitation rates (% adults/day)*
Blow fly	99 (60)
Daphnia	23 —
Algae	13 (45)
Flour beetle	3 (40)
Guppies	2 (40)

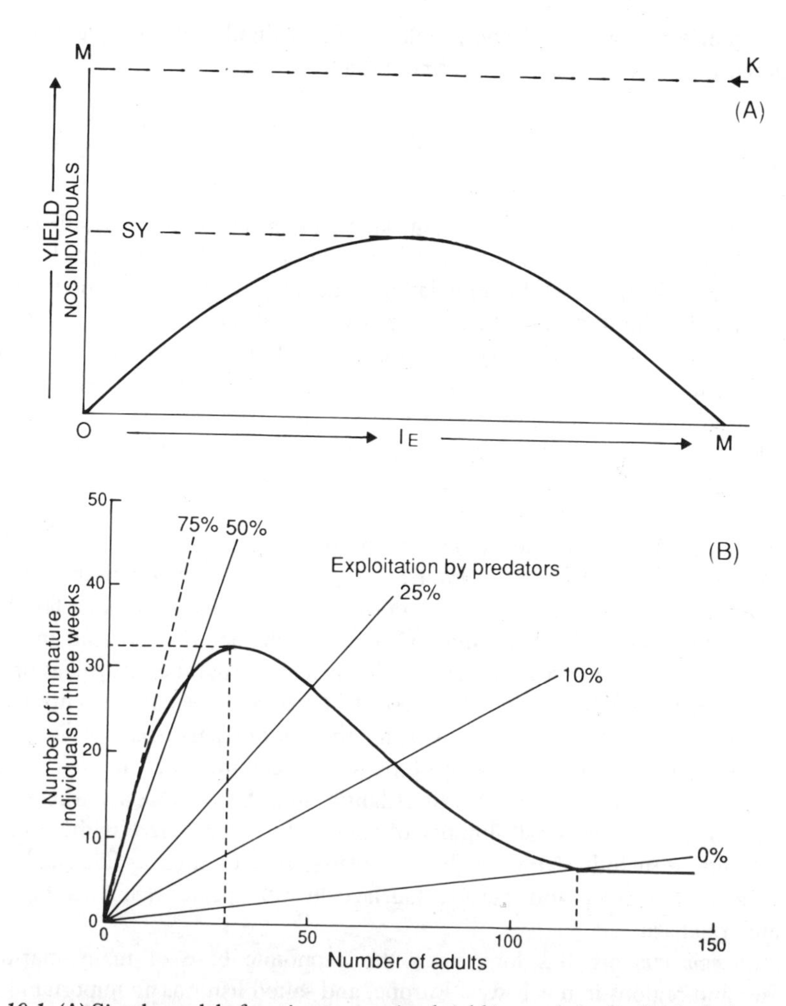

Fig. 19.1 (A) Simple model of maximum sustained yield in number of individuals from a given stock. K = carrying capacity of the environment; I_E = intensity of exploitation; M = maximum; O = no yield; SY = sustained yield; (B) Recruitment rate curve (R) and hypothetical rates for aquarium populations of guppies. In the absence of predation (0%) the natural mortality would stabilise the population at about 120 individuals. The maximum exploitation rate possible is just under 50% per 3 weeks at which point an adult population of *c.* 30 and a yield of 33 would be maintained (after Gulland 1964)

The concept of the maximum sustained yield expresses an ideal goal which is difficult to quantify and hence attain in reality. It is, in fact, a relative and variable concept dependent on a combination of behavioural as well as demographic characteristics of the species concerned, the quantity or quality of the resource or resources for which it is being exploited at a particular time and on the availability of a long-term data base about population or biomass responses to varying intensities (techniques and strategies) of exploitation and environmental conditions.

Ecosystem exploitation and the problems of sustained biological yield will be considered in relation to fishing, forestry and agriculture.

FISHING

Fish constitute not only the one important source of protein food that is still hunted but also the only important food animal that has not been domesticated. Fish are totally wild and the successful manipulation of aquatic ecosystems in the interests of exploitation is limited to particular species in small areas of fresh or marine water. The production of fish in specially constructed nutrient-enriched fishponds is an ancient practice still employed throughout much of Asia which finds expression in the larger modern intensive fish farms in which all stages of the fish life cycle are controlled and managed to provide eggs for small hatcheries, fry to stock recreational fishing areas and/or commercial fresh or processed fish and fish products.

Exploitation of the marine ecosystem, however, has concentrated on those animals whose size and population numbers, in particular areas, permit the greatest ease of catching and which, because of custom or prejudice have been favoured more than others as food. This is more particularly characteristic of cold and cool temperate than of warmer tropical seas. In the former the number of species of fish and other marine animals is small, but they most frequently occur concentrated in such numbers (shoals) as to make the exploitation of one particular species easier and, hence, more profitable. The bulk of the world's commercial fish catch is taken from the colder waters of the North Atlantic and Pacific Oceans where it is composed of a relatively small number of species for which there is the greatest demand. These include demersal (bottom-living) species such as cod, haddock, hake, plaice and sole, and pelagic (surface-living) species such as herring, mackerel, pilchard.

Commercial fishing has long been the economic base of many maritime countries and regions in north-west Europe, and salted fish was an important item of trade between northern and the Mediterranean areas in medieval times. At the end of the nineteenth century, however, demand from a rapidly increasing population and the substitution of steam-engines for sailing ships witnessed a rapid increase in fish landings. Larger, faster boats and more efficient techniques for locating and catching fish resulted in an increasing intensity of fishing in terms of catch per boat or man-hour. Within the last 50 years faster diesel-powered vessels, combined with facilities for refrigeration and processing, have allowed wider fishing ranges than ever before. Between 1948 and 1970 world fish landings trebled, since when the volume has levelled off, fluctuating around an average of 60 million tonnes per annum (Tait 1981). Intensity of fishing, however, did not decline and there has been growing concern about the extent of overexploitation and the consequent collapse of fish stocks since the 1950s. Commercial fishing continues to maximise immediate or short-term harvests and increasingly to threaten the viability of certain species and fishing grounds. The problem is most acute, with the exception of whaling, in the cool North Atlantic and Pacific areas.

The most intensively exploited is the North Sea, the focus of six coastal European countries plus others. Since the end of the Second World War stocks which had built up in two preceding interwar periods have been fished so intensively that the collapse of its cod and haddock stocks has been recently reported.

WHALING

Also since the end of the 1970s, the survival of whale stocks has become an urgent, though internationally contentious, issue. Whales, the largest of the marine mammals, constitute a valuable multipurpose resource for bone, meat and oil. With a low fecundity and long growth curve they are particularly susceptible to overexploitation. The development of larger wider-ranging vessels, and more efficient locating and catching methods (e.g. the harpoon gun and harpoon explosives) at the beginning of this century resulted in a rapid intensification of commercial whaling and an increasing exploitation of the Antarctic *krill* feeding grounds of the baleen filter-feeder whales.

The subsequent history of whaling has been one of the progressive depletion of the species (see Fig. 19.2). Initial logistic harvesting models indicated that

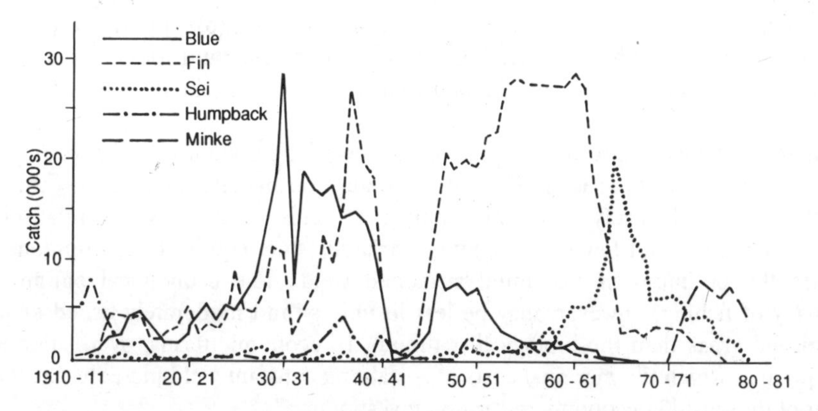

Fig. 19.2 Catches of baleen whales in the southern hemisphere 1910–77. The usual lengths of whales in the commercial catches were: blue 21–30 m; fin 17–26 m; sei 14–16 m; humpback 11–15 m; minke 7–10 m (from Allen 1980)

maximum yield could be sustained at 80 per cent of the equilibrium density. It has, however, now been realised that single-species models based on logistic curves are too simple since they depend on the single species and fail to take subspecies and species interaction into consideration (Ricklefs 1973). There are ten major species of large commercially fished whales, including one toothed (the sperm whale) and nine baleen filter-feeders, all of which have interacting subspecies. Another problem is that whale population change is slow and, even if a moratorium on whaling was universally accepted, it would take some time before sufficient data on their recovery response by which to estimate maximum sustained yield became available. However, without this, all whales may be doomed to extinction.

OVERFISHING

Under natural conditions, the size and productivity or biomass of a particular species depend on the amount of food available, and the balance between reproduction and growth on the one hand and mortality from natural causes on the other. This may be explained as follows:

$$S_2 = S_1 + (A + S) - M$$

where S_2 is the weight of stock in second year, S_1 the weight of stock in first year, A the annual increment to weight of stock by reproductive recruitment, G the annual increment to weight of stock by growth of recruits and $A + G - M$ the natural yield which if removed would theoretically represent the *equilibrium catch* which could be taken without depleting the stock. Overfishing occurs when fish are extracted at a rate greater than they can be replaced and the size of the stock starts to decline. Tait (1981) distinguishes two types of overfishing. The first is growth related; catches decline because of the number of fish caught before they attain their optimum growth rate. The second is recruitment related, and the stock declines because of a reduction in the number of spawners and hence in the number of new recruits coming into the stock. The symptoms of overfishing reveal themselves when, although the intensity and efficiency of fishing do not change there is (a) a diminution in the average size of fish caught, (b) an ever-increasing proportion of the weight of the catch is composed of smaller and younger fish than of larger, more mature individuals, and finally (c) a decline in yield in terms of weight caught.

Between the extremes of underfishing and overfishing, both of which result in low natural yields, there is considered to be an intermediate intensity of fishing which is the biologically optimum fishing rate. This should, given a relatively stable population and low environmental variability (i.e. equilibrium conditions), ensure the optimum or maximum sustained yield. The economical optimum intensity of fishing, however, may be less in under- and moderately fished or in overfished areas than the biological optimum. In economic terms, overfishing is that point at which the marginal costs of sustaining constant yield increase, i.e. the point of diminishing economic return on investment.

SUSTAINED YIELD

While the concept of sustained yield is relatively simple, its application to the exploitation of fish stocks is particularly difficult because of the variable nature of the aquatic environment, the mobility of aquatic animals, and the lack of data on which to base predictive calculations, although there are reasonably long-term fishing records. Information about the world's fish stocks is limited to the weight of annual landings of commercial species and, in some cases, the age distribution of the catch. Mathematical models of fish populations, although used in formulating management strategies, tend to be based on selected demographic characteristics of single-species populations, and do not always take into account all the factors which are necessary for a close estimation of the maximum sustained yield in the real-life situation. Among the important variables affecting yield are environmental conditions and the breeding characteristics of the species. As Clapham (1981) points out, sustained yield can only be considered in relation to a specific species in a specific environment.

Environmental variation, particularly of weather, may be reflected in either a population's mortality and or its fertility at all ages in the same or different seasons – the more variable the environment the more vulnerable will a species be to overexploitation at a given intensity of fishing. Also, overfishing may be the result of one of many fishing strategies (Clapham 1981) – such as the age class exploited or the time of fishing. Given a high degree of environmental variability, population size will tend to fluctuate randomly between high and low, and a constant minimum population is necessary for renewability.

However, the response of populations to environmental fluctuations varies between species. This has been demonstrated by comparisons of naturally stable and unstable species in the relatively stable environment of the North Sea with its small range of annual temperature variation, shallow depth and high plankton production. As indicated in Fig. 19.3, the variation in catches of plaice and sole up to 1960 was small. This high degree of stability has been attributed to very specific and tightly circumscribed spawning grounds and a high dependence on a particular copepod for food. Unlike the majority of fish, its population growth is density dependent and is regulated by efficient feedback mechanisms. In contrast, haddock

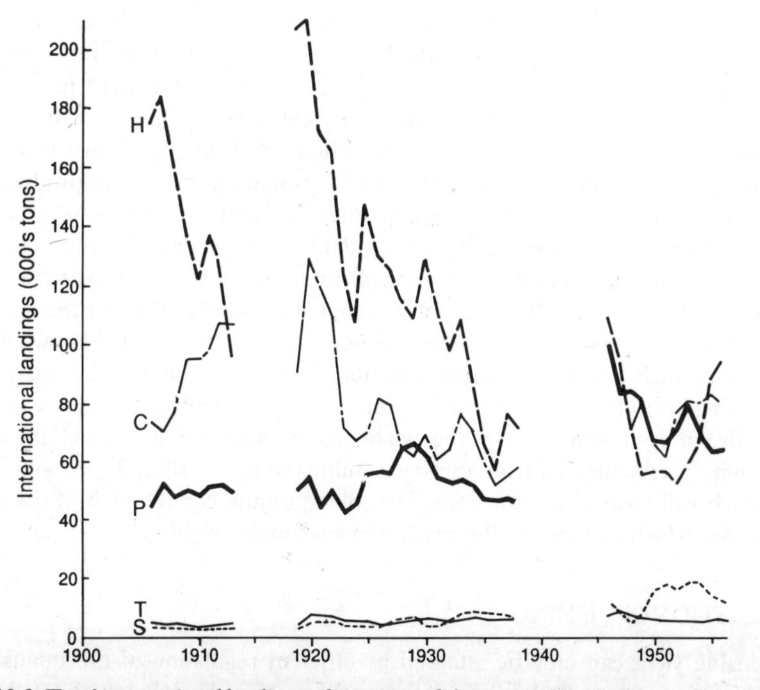

Fig. 19.3 Total international landings of important fish species: H = haddock; C = cod; P = plaice; T = turbot; S = sole (from Krebs 1985 redrawn from Beverton 1962)

and other catches in the same environment fluctuated widely over the same period of time, due to large variations in annual reproduction rates.

In the much less stable environment of the eastern Pacific Ocean, the highly unstable sardine fishery collapsed at a period when fishing intensity of older individuals was high, which coincided with exceptionally cold water conditions along the coast of California (Murphy 1966). In Peruvian waters the anchovy occupies a similar niche to that of the sardine. The anchovy has a naturally more stable population. Like the North Sea plaice, high fertility compensates for high mortality and the density of juvenile fish is regulated by food supply. However, since the anchovy is short-lived (3 years) it is very susceptible to overfishing and its maximum sustainable yield is much lower than less fertile and longer-lived species. Susceptibility to overfishing is exacerbated by the unstable environment as a result of periodic shifts (El Niño) in the balance of warm and cold currents off shore. During longer periods of warm water the anchovy's niche is restricted and its reproductive capacity is reduced. This happened in the later 1960s at a time when a growing demand for fishmeal in Peru resulted in increased fishing intensity and the eventual collapse of the anchovy industry in 1971 (Paulik 1971).

It is not, as yet, possible to manage on any but a very small scale the natural (physical and biological) factors affecting the yield of commercial fisheries. Hence the only feasible way of achieving sustained yields is by regulating the methods and intensity of fishing. However, as Tait (1981) notes, the relationships between fishing intensity, stock yields and profitability are complex; the same weight of catch can be composed of a wide range of fish size classes each of which will have a different impact on the size of the stock and the size and sustainability of the yield. The two main concepts on which fishing regulations have for long been based are those of surplus production and the analytical (or yield per recruit) respectively. In the former, the equilibrium yield (which represents the natural increase) is that catch which reduces the whole stock to a level just below the carrying capacity of the environment, thereby promoting maximum productivity. Catch size can be regulated by fishing intensity, and the effects of changing intensity can be predicted from the data available. In the latter case, the yield from a given intensity is predicted in *recruitment units* (i.e. the annual year classes of which a stock is composed) from each or one particular year class in the life span of a particular species. This approach is dependent on more detailed knowledge of fish biology than that of surplus production. However, it allows the analysis of both fishing effort (boat or man-hours), catching techniques (i.e. mesh size of nets) on the size or age composition of the catch; and the mathematical prediction of the maximum sustainable yield for varying combinations of fishing effort and mesh size which will ensure *eumetric fishing*, i.e. the optimum combination of these two parameters which will ensure the maximum sustainable yield.

REGULATION OF FISHING

Sustainable yield can only be attained by efficient regulation of the intensity of fishing. This is particularly difficult given the state of knowledge about the the marine ecosystem and a mobile stock unconstrained by national boundaries.

Considerable efforts have, however, been made at both national and international levels to control fishing and minimise the risks of overfishing. Imposition of closed seasons, regulation of mesh size and restrictions on the size of fish landed were among the early recommendations of International Fishery Councils. A further possibility, which has been the focus of much research, is that of restocking depleted fishing grounds with young fry raised in specialised hatcheries. Although the establishment of a shad fishing industry off the coast of California in the 1960s, by the transference of fry from hatcheries on the east coast of North America, was successful, it is a costly procedure involving the risk of excessive loss of transferred fry which may jeopardise the establishment of a minimum viable population.

It has frequently been noted that the stock of potential food from the sea is very considerable. In the first place there are many species which are unfished or underfished because of cultural preferences. Also, the amount of economically exploitable food – in the form of fish and crustaceans (shellfish) – represents a relatively small proportion of the total marine organic resource. A large proportion of the biological productivity is consumed by other organisms of little human food value. Competition is greatest in the shallow water over the continental shelf where productivity of plant and animal plankton is high and only a small percentage (1–2 per cent) is consumed by fish. The rest sustains a large volume of invertebrate animals. It has been calculated that, if only a quarter of such competitors as sea-urchins, starfish and crabs were eliminated, at least ten times the existing weight of fish could be supported (Hardy 1965). This author invoked a Wellsian vision of a time when the floor of the continental shelf might be 'weeded and harrowed' in order to increase the productivity of demersal fish in particular. He speculated even further about the possibility of submarine ranching or fish herding. However, the vastness of the sea, the extreme mobility and relative invisibility of its life and the particular nature of its pastures are problems peculiar to the rational exploitation of the marine ecosystem yet to be solved. As Clapham (1981: 104) notes: 'the management of fisheries will ultimately require social institutions attuned to the reality of fish populations as well as to managers who must make significant decisions outside the direct control of any higher authority'. They should be based on fishing areas rather than political limits. However, this would, as in the case of any form of marine ecosystem management, be dependent on a degree of international political co-operation which is difficult to envisage at present. At the national level, many countries have established 200-mile territorial limits within which fishing can hopefully be more easily regulated than on a multinational basis.

FORESTRY

Although forests, particularly in Europe, have been managed since at least medieval times, their level of management is still relatively low compared with agriculture. Even the most intensively developed forests still retain many of the basic characteristics of natural, albeit simplified, forest ecosystems.

One of the main forestry management problems is the length of a tree's life cycle and its precarious regeneration (see p. 184). The principal economic forest product is the stem (bole) wood or timber and a tree attains *economic maturity* or *maximum economic age* when the annual rate of increase in volume of wood (i.e. current annual increment (CAI)) begins to decline (Cousens 1974; Packham and Harding 1982). This, in economic terms, is the point of diminishing returns beyond which increase in volume fails to compensate for the cost of continued management (see Fig. 19.4). However, economic maturity will depend on a

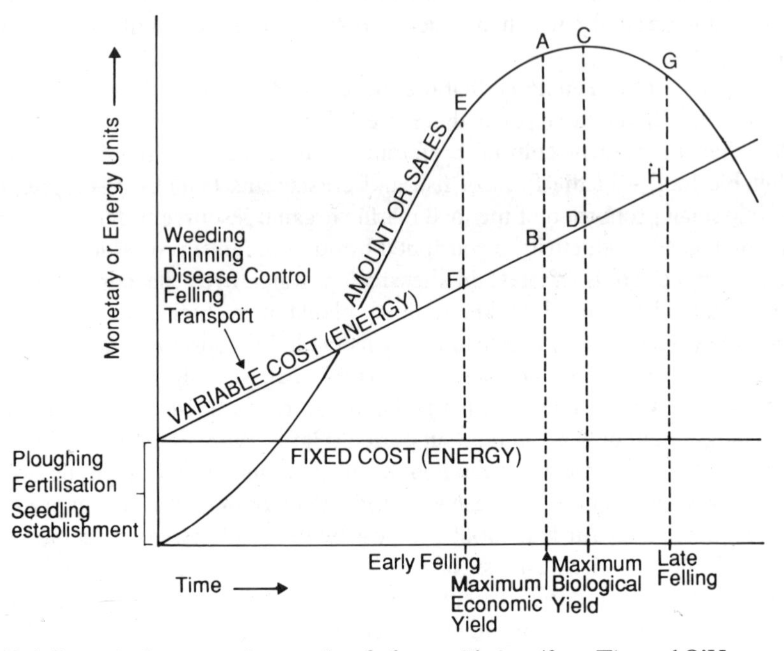

Fig. 19.4 Stages in the economic maturity of a forest with time (from Tivy and O'Hare 1981)

complex of variables, including the purpose and hence the size and quality for which a particular species is grown; the environmental or site conditions; and the returns on investment which will be dependent on the cost of production, transport from site to market and the current demand for the product.

A particular advantage of tree longevity is that the successional development of a forest stand is accompanied by increasing variation in the types of product and, hence, the potential for multiple use of the ecosystem. Management during growth aims to protect seedlings and saplings from inter- and intra-species competition. The latter is achieved by progressive thinning of the dominant trees and of the substorey strata combined with the removal of lower and/or dead branches. In the more intensively managed commercial forests of Europe, herbicides, fertilisers and pesticides may be applied during the initial growth phase.

The most prevalent forest pests are the *phytofagus* (plant-eating) insects. They are, in fact, the most abundant herbivores in natural forest ecosystems in which it has been estimated that they normally consume *c.* 5–10 per cent of the annual NPP in temperate forests (Gosz *et al* 1978). Insect populations are subject to natural fluctuations. However, particularly in single-species forests devastatingly destructive infestations can occur resulting in complete (100 per cent) defoliation and subsequent death of the forest stand. Among some of the most virulent pests of commercial forests are the spruce bud worm, the bark beetle, the tent caterpillar and the gypsy-moth caterpillar. It has been suggested that between the two extremes, a moderate insect infestation may be advantageous. Generally older trees are more prone to attack than the younger, understorey saplings of their mature parents. As a result, decline and death of the latter may, because of increase in environmental resources, stimulate growth of saplings and increase the productivity of the stand (Mattson and Addy 1975). Also, as has previously been noted (see p. 298), fire can perform a similar function in killing older trees, clearing wood litter and recycling nutrients.

COPPICING

Crucial to all types of forest management is the harvesting and regeneration of the particular crop. In Europe forests have been managed for many centuries and pure stands (groves) of single species were produced by the selection of favoured trees (e.g. beech, oak, fir) dependent on the region, while the unwanted trees were weeded out (Edlin 1976). Initially these 'domestic' and often communal forests were valued more for their fruit and for forage for pigs than for their wood. One of the oldest and simplest methods of management for wood production was that of coppicing (Pennington 1969; Peterken 1981). Coppicing involves the cutting at regular intervals of young branches and stems produced from the bole(s) of a tree. It is dependent on the natural ability of certain, mainly hardwood, trees to produce branches (or *suckers*) from either the bole or the stumps (or stools) of felled trees. In the latter case regular felling of the bole stimulates sucker growth and results in a multi-stemmed shrub growth. The stems can then be harvested on a rotation the length of which is dependent on the species and size of wood required. This can vary from 8 (willow, hazel) to 30 years (oak). Coppicing is undoubtedly one of the most effective methods of rapid high production of 'small wood' for a variety of purposes, e.g. fuelwood, fencing and hurdles, charcoal and bark tannin. In time, however, regenerative vigour and yield decline with coppicing, not least because it constitutes a continual drain on resources due to animal browsing of leaves and a consequent reduction in the volume of the nutrient cycle. Akin to coppicing is that of *pollarding*, whereby the tree crown is removed in order to promote a bushy growth of suckers at a height beyond the reach of browsing animals. A variant of the simple coppice was that of the *coppice-with-standards* in which a number of coppiced trees (standards – often of oak or ash) were allowed to reach maturity either for large constructional timber and/or a seed crop; hence the term 'mother trees'.

Coppicing was the most common method of forest woodland management throughout medieval Europe and up until the end of the eighteenth century when the decline in demand for small wood was superseded by the increase in that for large constructional lumber and for pulpwood from the soft- rather than hardwoods. Although still practised on a limited, specialised scale in Europe, evidence of former coppice management had left its mark on the form of many scattered remnants of old now unmanaged woodlands (Peterken 1981).

However, within the last four decades or so the advantages of coppicing have attracted increasing attention for the rapid, rotational production of the following:

1. *Biomass*, particularly on land of low agricultural capability as an alternative energy source or for low-grade pulpwood; development for this purpose has already been pioneered on bog-land areas in Scandinavia and Ireland.
2. *Domestic* fuel in developing countries where fuelwood demand is increasing, but the remaining forest resources are dwindling at an even faster rate. This has stimulated the search for and planting of fast-growing, usually secondary, tropical species – to provide fuelwood, pulpwood and to re-establish the forest cover on cleared and degraded land. Rapid growth species such as *Gemelina arborea* cultivated in Brazil and West Africa can grow at rates of 5–10 m $year^{-1}$ and can be coppiced on a 7–10-year rotation.

TREE HARVESTING

Increasing demand for large constructional lumber and pulpwood has been accompanied by increasingly sophisticated methods of silviculture, i.e. the raising, tending and regeneration of a wood crop. The methods, to a certain extent, depend on whether the forest is a long-standing natural or near-natural ecosystem or is a recent plantation. In the first case, the emphasis will be on methods of harvesting designed to ensure natural regeneration especially in mixed, uneven-aged stands. These include *selection cutting* of trees identified either because of age and/or species. This used to be the common method of harvesting widely dispersed high-value hardwood species (teak, ebony, mahogany) required for particular purposes in tropical forests. It requires individual tree extraction often under difficult conditions but, if performed skilfully, creates a gap in the mature forest which allows rapid regeneration by stimulating the growth of seedlings which can exist in the shade of the dominant tree canopy. *Shelter-wood cutting* is better suited to trees whose seedlings require a certain amount of shade for successful regeneration (oak, yellow poplar, hickory). In this case the mature trees are felled in stages, often in strips or blocks of a size and shape which will maintain conditions suitable for seed regeneration and provide the saplings with adequate shelter.

Selective harvesting can, if skilfully performed, initiate and speed up the process of natural forest regeneration. It maintains the perceived aesthetic values of a mixed uneven-aged forest in which the resources are more fully exploited and the total phytobiomass is greater than in monocultural plantations. The latter, however, give a higher economic return when the demand is for large-scale specialised commercial production of large wood.

Clear-felling is now the commonest method of felling commercial forest stands. In the early stages of forest exploitation in the New World countries of North America, Australia and New Zealand forests were indiscriminately felled, all the larger trees being removed and the cut-over land abandoned to the successional development of a species-impoverished growth of secondary forest or woodland which subsequently was often further depleted by culling for domestic fuel, livestock and game grazing and browsing. From the end of the eighteenth century, and particularly in the first half of the twentieth century, the continued decline of forest acreages, at a time when wartime demands were increasing, engendered concern and stimulated extensive afforestion (or reafforestation) of areas from which forest growth had long disappeared or which had more recently been abandoned from agricultural (usually pastoral) use.

Plantation forests
Single-species commercial plantations are now established, after preliminary land clearing, drainage and surface ploughing, by planting out nursery-grown seedlings produced from seed provenance selected to give the maximum production on a particular site. The recognition of ecotypes within many of particularly the softwood species of pine, larch, spruce and fir have facilitated the introduction of exotic species which give high yields of the required quality from a species-rich to a species-poor area, e.g. the Sitka spruce (*Picea stitchensis*) from north-west America to north-west Europe and the *Eucalptyus* spp. from Australia to Mediterranean Europe. Areas are planted sequentially in single species (with or without a nurse tree crop), even-aged blocks or compartments which are clear felled when economic age is reached and replanted on a rotation which varies dependent on site conditions and type of product required. Forest plantations are the most intensively managed silvicultural systems. High yields per hectare and per man-hour of labour have been attained at the ever-increasing expense of high energy costs for mechanisation; fertilisers, herbicides and pesticides often applied by aerial spraying; seedling production; and of environmental costs (externalities that are less easily identified or quantified).

SUSTAINED YIELD

The principle of sustained yield has long been implicit in all types of forest and woodland management. In simple terms this means ensuring that the harvest does not exceed growth. The establishment of an optimum sustained yield from such a long-run resource is complex because it involves not only a knowledge of the annual growth rate of single and/or groups of species throughout the economic life of the tree(s) but also the establishment of the amount and time of cutting at various growth stages if more than one size of wood is required. The *allowable cut* is the maximum volume of timber that can be harvested in a given time period in order to sustain yield from a given forest area, region or political unit. To achieve this aim the forest should be managed so that there is a sequence of age classes in the growing stock. The felling cycle may vary in length dependent on species, environmental conditions and demand for type and size of wood. The harvest should then be not more than the rate of growth recruitment into the economically mature age class.

It is unlikely that optimum forest yield can be maintained without an increasing level of management. Forest growth results in the long-term immobilisation of a large proportion of the ecosystem's nutrient capital (particularly of calcium and potassium) in the wood. Nutrient depletion with harvesting can constitute as high a drain on the soil fertility as can an agricultural crop, and this must be made good if long-term yields are to be sustained.

AGRICULTURE

Agriculture can be defined as the cultivation and production of crop plants and/or livestock products (Tivy 1991). It involves the management of both the organisms and the environment (or habitat) in which they are produced. The former aims, through selection and breeding, to produce a crop or domestic animal that will give the highest yield and/or quality in a specific habitat. The latter endeavours to minimise the biophysical constraints of the habitat by cultivation. This comprises a number of processes of which the two most important are as follows:

1. *Propagation*, i.e. the preparation of the soil by some form of tillage in order to ensure the best conditions possible for crop germination and growth;
2. *Protection*, i.e. from competition for primary resources by weeds and from the direct or indirect reduction in yield potential of crops and livestock by pests.

ENVIRONMENTAL MANAGEMENT

That part of the biophysical environment most amenable to management is the soil which, from the farmer's point of view, can be defined as the medium in which most crops are grown. Soil management or cultivation involves a range of processes of which the most important are tillage, drainage, irrigation and fertilisation.

Tillage

Tillage is the process of opening up, disturbing and turning over (inverting) the soil. In some less technically advanced societies in the humid tropics, this involves the use of a simple 'digging stick' or hoe which does little more than scratch and expose the soil surface. The plough and harrow have long been the almost universally employed tillage implements. The heavy ox-drawn wooden plough, which furrows and breaks up, but does not invert the soil, was in use 5000 years ago in the Middle East and is still used in parts of India. The prototype of the modern plough which opens up and inverts the soil was introduced into Europe in the medieval period. Today there is a vast range of tractor driven ploughs, harrows and mechanical cultivators designed for particular soil conditions and types of tillage. Traditional tillage in humid temperate regions involves primary ploughing to *c.* 20 cm soil depth and harrowing to *c.* 10 cm. It buries pre-existing surface organic matter and thereby reduces competition from weeds, speeds up the decomposition and mineralisation of DOM and SOM, and facilitates the growth of young roots and shoots. The main aim of tillage is to produce a *soil tilth* (or structure) that will provide as good a seed (or plant) bed as possible for the penetration of seedling roots and shoots and for subsequent plant growth.

SOIL STRUCTURE

Soil structure is an expression of the way in which the individual soil mineral particles are aggregated into clumps or peds of varying size, shape and stability. The agronomically ideal tilth is that in which the majority of peds are in the form of porous but stable crumbs 2–5 mm diameter. A crumb structure combines free drainage and aeration, provided by the inter-crumb spaces (*macropores*), with the high nutrient and water retention of the *micropores* and on the large surface area provided by the fine-grained crumb.

The structure produced by tillage, however, is a function of the initial soil texture, the organic matter content, methods of tillage and weather conditions. A sandy loam is the optimum texture for the development of a crumb structure. The proportion of clay particles is such as to ensure optimum water and nutrient retention and to aid crumb aggregation. While the latter is enhanced, aggregation into stable porous crumbs is, however, dependent on the presence of well-decomposed mull humus which can combine with the clay fraction to form a colloidal complex. In addition to the humus content, living soil organisms make an important contribution to structural development. Many, particularly worms, which live and feed in the soil serve to mix, invert and bind the mineral and humus particles together. In addition, a dense network of fine grass roots which, on decay, contribute organic matter during growth, facilitates the separation and aggregation of peds and is also associated with the development of a crumb structure.

The soil structure finally produced is, however, ultimately dependent on the methods of tillage and the weather conditions before, during and after tillage. Cultivation when the soil is too dry and hard or too wet and sticky can result in the production of large hard 'clods' which are resistant to breakdown by mechanical processes or by weathering (e.g. alternate freeze/thaw or drying and wetting). Also, continuous arable cultivation can result in over-tillage and the deterioration of soil structure, impeded drainage either because of soil compaction or of the development of a compacted impermeable plough-pan, and/or of accelerated soil erosion. The exception, however, is cultivation of wet rice, a 'swamp' crop whose seed-bed must be an impermeable saturated soil submerged in a depth of water dependent on the particular variety. In this case the soil is tilled and mixed up when wet in order to impede too rapid drainage.

Soils – particularly the infertile latosols – in extremely humid tropical regions are very susceptible to overcultivation as a result of the use of modern methods of mechanised tillage developed in temperate environments. Under the prevailing climatic conditions the exposure of the soil causes rapid decomposition of organic matter, intensive soil leaching and alternating saturation and hardening of the mineral fraction. The latter is dominated by iron and aluminium resulting in a sandy soil structure with low nutrient and water-holding properties.

SOIL-WATER MANAGEMENT

To achieve its maximum yield potential, a crop requires sufficient water to maintain maximum rate of photosynthesis during its growth period. This occurs when the soil is at or just below *field capacity*, i.e. the maximum amount of water that can be retained by capillary tension after natural drainage. Water in excess of field capacity causes anaerobic conditions which retard both organic decomposition and root growth. Tillage helps to mitigate too rapid downward percolation of water in light sandy soils and too slow percolation in heavy clayey soils, particularly during the growing season. However, climatic and landform conditions may be such that because of either an excess of precipitation over evapotranspiration or a permanently high water-table in coastal or riverine lowlands there is a soil-water surplus for the whole or part of the thermal growing season, and artificial drainage becomes necessary to achieve a satisfactory crop yield. Techniques vary from the oldest and simplest, whereby cultivation mounds or the ridges and furrows produced by tillage facilitate the evacuation of surface water, to the more sophisticated methods of underground, lined or unlined drainage channels and pumping where the aim is to lower the level of the water-table below the depth required for cultivation.

In contrast, where for part or the whole of the growing season the potential evapotranspiration exceeds precipitation, the soil-water deficit may be such as to check or halt growth or, in arid climates, to inhibit cultivation completely. The practice of irrigation whereby water is transferred from a source of supply (river, lake, reservoir or ground-water) to an area otherwise suitable for cultivation, is as old as that of tillage, particularly in the arid and semi-arid areas of the Middle East – one of the early 'cradles' of sedentary agriculture. Irrigation techniques range from the simple, manually dependent, short-distance transference of water from river or small pond or dam to the large-scale, long-distance transport from permanent exotic rivers or large-scale storage reservoirs. Effective irrigation, however, is dependent on the maintenance of a low water-table and the application of sufficient water to allow downward drainage as well as surface soil-water retention in order to counteract the development of excessive soil salinity consequent on the precipitation of salts near or on the soil surface from a raised water-table.

FERTILITY MANAGEMENT

Agriculture inevitably disrupts the theoretical natural nutrient cycle in which the loss of soil nutrients by erosion and leaching is balanced by two gains from rock weathering, volcanic dust and the atmosphere. Cropping, whether of plant or animal produce, constitutes a drain on the soil nutrient pool at a greater rate than can be made good by natural processes and, in consequence, yields will decrease unless made good by some means of restoring or maintaining soil fertility. Indeed, the management of soil fertility, which entails not only nutrient return but conditions which will minimise loss by leaching and erosion, maximise retention and ensure nutrient uptake in quantity and at a rate that will allow a crop to realise its maximum yield potential, is central to cultivation and is dependent upon both methods of soil tillage and of soil-water management.

Shifting agriculture

The earliest method of fertility management which is still common in many areas of subsistence or near-subsistence agriculture in the humid tropics is that of *shifting cultivation*. A small area is cleared of its vegetation cover by cutting and/or burning. Nutrient-rich ashes fertilise the soil which is then cropped continuously for 2 or 3 years, by which time yields decline because of loss of fertility and the area is abandoned. Given time the natural vegetation is re-established. The length of time (the rotation period) before the area is cultivated again should, theoretically, be not less than that required for the pre-existing ecosystem to reach maturity – which in the tropical rain forest may be 100 or more years. In practice, because of rapid tree growth, the rotation period can be 15–30 years. However, in many areas population growth has resulted in a reduction to less than 5 years, a period insufficient for adequate nutrient return. In other tropical and temperate areas where inherent soil fertility was greater and its decline under cultivation has not been so rapid, variants of shifting and sedentary cultivation have developed. The most fertile soils are cropped continuously or are cultivated on a short crop–fallow rotation. Livestock grazed on the fallow return a high proportion of nutrients in their dung. The less fertile soils may be cultivated on a long-term fallow rotation and/or used for extensive livestock grazing.

Crop rotation

The traditional method of fertility management which developed from the sixteenth century onwards in north-west Europe and became fundamental to methods of agriculture in all the temperate countries of the world was that based on the rotation of annual crops of varying nutrient demands with a 'break' or 'rest' crop – usually a grass/legume mixture sown down and maintained for 1–6 or more years. The latter promoted the build-up of nitrogen-rich organic matter and, on loamy soils, a well-developed crumb structure. This *grass (or ley)* crop, combined with an arable fodder crop supported a greater number of livestock than had previously been possible, and farmyard manure, often supplemented by urban sewage, was the main means of fertilising the cultivated soil. In highly populated areas of the Far East human waste (*night soil*) had long been used to fertilise padi rice soils.

From the latter half of the nineteenth century inorganic (so-called artificial) fertilisers derived from mineral deposits or, as in the case of phosphates, as by-products of industrial processes became available and supplemented those from farmyard manure. After the Second World War the development of the chemical industries and methods of synthesising nitrogen fertiliser, the use of highly concentrated, soluble inorganic nutrients all but replaced organic sources particularly in areas of intense arable farming in the developed countries of the world.

CROP PROTECTION

The second, through equally important, aim of agricultural management is to protect the crop from the direct or indirect depressive effects of weeds, pests and pathogens. These organisms can seriously reduce the amount or quality of crop or animal yields. They do so by competing for environmental resources of light, water and nutrients in particular and by reducing vigour and making both crops and livestock more susceptible to disease by direct consumption of crops and parasitism of animals. Originally, in less technically advanced types of agriculture, crop mixtures and crop rotation combined with labour-intensive weeding helped to control levels of infestation. Increasing crop and livestock specialisation, and large-scale monoculture of genetically uniform organisms which resulted in an increase in pest populations, together with the availability and use of chemical herbicides and pesticides since the 1950s, have tended to exacerbate management problems. The rapid evolution of pesticide resistance of organisms necessitated the breeding of even more pest- and disease-resistant crops which in turn have stimulated the evolution of increasingly virulent pests and pathogens. This trend, combined with the problems of environmental contamination by pesticides, has witnessed a revival of formerly more popular methods of biological protection. The latter include the use of suitable pest-predator species or the depression of pest populations by induced sterility of the particular species.

AGRO-ECOSYSTEMS

Agriculture creates a particular type of ecosystem – an *agro-ecosystem* – whose composition, functioning and persistence are entirely dependent on human manipulation of its biotic and abiotic components. Agro-ecosystems differ from unmanaged (wild) and other less managed ecosystems in having a simpler biotic composition and structure. The number of species and associated life forms is smaller and the dominance of one or two crops and/or livestock is common. Not only are agro-ecosystems species-poor but the populations of domestic plants and animals tend to be more genetically uniform and, therefore, less resistant to environmental stresses than their wild ancestors or near relatives.

Simplification of agro-ecosystem composition is accompanied by that of structure and function. Energy flow through the agro-ecosystem is channelled along fewer and shorter food chains, and a high proportion of the available light energy is hence made available to the crop plants and from them either directly or indirectly via domesticated livestock to humans. The number of trophic levels, in this much simplified food web, is two or at most three. Also, the biomass of large herbivores, in the form of domestic livestock, can be much greater than in other ecosystems in ecologically equivalent environments. In such circumstances, a smaller proportion of the primary biological production enters the detrital or decomposing route. A higher proportion is exported from the system as in the harvested plant crop or in the livestock which consumed it. In consequence the energy pool of dead and decaying organic matter in the soil is, in general, less than that in other ecosystems in similar environments. Simplification of the energy flow is accompanied by modification of that of the nutrient cycle. In the unmanaged ecosystem, most of the nutrients taken up from the environment are returned and recycled by the decomposition of organic matter. The loss of nutrients from agro-ecosystems is accelerated by export of the crops and livestock and by the increased rate of decomposition and nutrient leaching as a result of soil tillage.

The production of short-life (annual) crops and livestock speeds up the rate of nutrient cycling and the labile nutrient pool tends to be more variable in volume and more highly concentrated in the animal (livestock) than the plant (crop) biomass than in other terrestrial ecosystems. Also, nutrient loss takes place at a rate greater than can be made good by natural gains. Unless loss by export is compensated by the input of fertilisers the agro-ecosystem will inevitably 'run down', productivity will decline and biomass production decrease. However, management not only endeavours to maintain the nutrient balance and ecosystem productivity but, by increasing nutrient input and reducing pest/pathogen competition, to increase productivity and, as a result, the yield harvested from a particular crop or animal.

Agro-ecosystems are frequently described as ecologically unstable when compared with wild or less managed ecosystems. They lack the species diversity and trophic complexity which is considered fundamental to homeostasis (a steady state maintained by negative feedback mechanism) in the face of normal environmental disturbance. The comparison, however, lacks validity because the stability of an agro-ecosystem is not dependent on natural regulatory mechanism but on how successfully the system is managed in the face of normal and abnormal environmental variation. Also, the agro-ecosystem is a much more open system with a greater number and larger volume of inputs and outputs than in wild or less managed ecosystems. Agro-ecosystems are maintained and are dependent on a high level of either direct or indirect energy subsidies to compensate for the continuous drain of energy out of the system. The number and amount of outputs from an agro-ecosystem greatly exceed those from others. As well as the harvested crop or livestock they produce more waste. This is because the harvested crop normally represents a small, though variable, proportion of the total plant or livestock biomass produced and because of the amount of excreta from livestock. And because of the import of animal feedstuff a given agro-ecosystem can produce more animal waste than can be recycled within the system. Some organic waste is recycled; some, however, together with surplus chemicals, can leak from the agro-ecosystem and become a polluting input into other wild or managed ecosystems (see p. 340).

TYPES OF AGRO-ECOSYSTEMS

Agro-ecosystems vary in the extent to which they differ from wild or less managed ecosystems dependent on the methods and intensity of management. Smith and Hill (1975) identify a continuum between unmanaged and little-managed ecosystems on the one hand and intensively managed agro-ecosystems on the other, on the basis of four main parameters: biological diversity, net energy

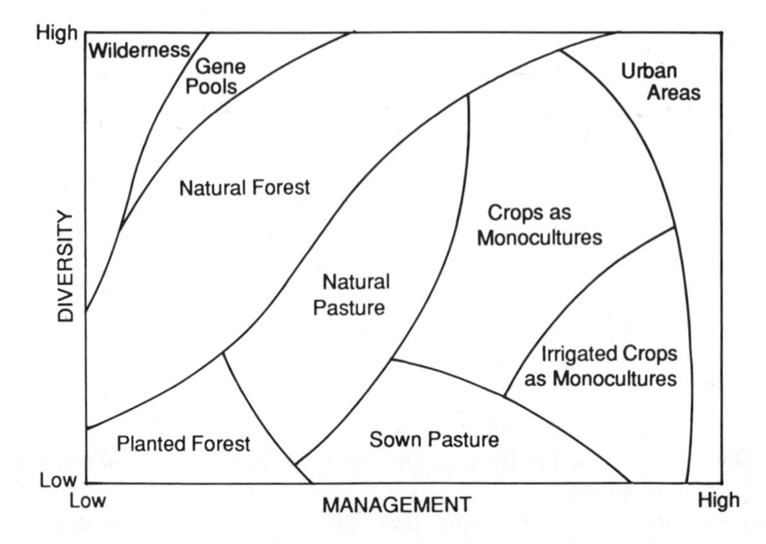

Fig. 19.5 The relationship between agro-ecosystems in terms of degree of management (from Smith and Hill 1975)

balance, intensity of management and management responsibility (see Fig. 19.5). Intensity of management can be expressed in terms of the number and amount of energy subsidies and the net energy balance as the difference between the energy equivalents of the inputs to and those of the outputs from the system.

LOW-INTENSITY AGRO-ECOSYSTEMS

Low-intensity agro-ecosystems (see Table 19.2) as exemplified by shifting agriculture (swidden, lalang) are commonly practised throughout the humid tropical rain forest environment of South America, Africa and South-east Asia. It is characterised by a low level of management, the main aim of which is to produce as large and as reliable a yield of food as possible. To this end an extremely diverse mixture of native crops of varying height, life form, nutrient requirements and growing period is grown together – as many as eighty varieties of thirty species of food crop have been noted in Zaire (Okigbo and Greenland 1976). They provide a year-round food supply, a protective soil cover on easily erodible soils and, because of varying intra- and interspecific variations, minimise the risk of total crop failure due to climate, pests and/or disease. Competition from weeds is reduced by the crop cover and weeding of all non-crop plants except seedlings or saplings of 'mother trees'. The latter contribute to the rapid regrowth of forest when the plot is abandoned, because of declining fertility after about 3 years. In the diversity of its composition and in the variation in the growth height of the crops this agro-ecosystem replicates some of the basic characteristics of the rain forest (Ruthenberg 1976), while the management of the environment is on a scale and intensity that results in an ecosystem modification not much greater than that in a naturally produced gap in the forest cover.

Table 19.2 The energy input in (thousand kcal ha^{-1}) for various items used in corn production from 1700 to 1985 (from Pimentel *et al* 1990)

Years	*1920*	*1945*	*1954*	*1964*	*1975*	*1980*	*1985*
Labour	65	31	23	15	10	7	6
Machinery	278	407	648	907	925	1 018	1 018
Draft animals	886	0	0	0	0	0	0
Fuel							
Petrol	0	1 200	1 500	1 250	600	500	400
Diesel	0	228	342	741	912	878	878
Nitrogen	168	630	1 555	2 331	2 940	2 940	3 192
Phosphorus	0	50	82	227	353	409	365
Potassium	0	15	50	70	155	195	187
Lime	3	46	39	64	69	134	134
Seeds	44	161	421	520	520	520	520
Insecticides	0	0	13	27	50	60	60
Herbicides	0	0	7	40	300	300	350
Irrigation	n.d.	125	250	625	2 000	2 125	2 250
Drying	0	9	15	145	458	640	760
Electricity	1	8	24	60	90	100	100
Transport	25	44	67	89	82	90	89
Total input	1 302	2 492	4 111	5 935	8 855	9 916	10 309
Total ratio $I : Y$	5.8	3.4	2.5	2.9	2.3	2.6	2.9
Total yield corn	7 520	8 528	10 288	17 060	20 575	26 000	26 000

MODERATE-INTENSITY AGRO-ECOSYSTEMS

Mixed farming systems in which the production of livestock products (meat, milk) is combined with that of arable fodder and food crops, was the traditional and formerly more widespread type of agriculture in temperate regions of the world. It is characterised by a moderate, though relatively high, intensity of management. Compared with labour-intensive shifting agriculture, mechanisation during and after the Second World War resulted in a drastic reduction in farm workers and draught animals and a concomitant increase in output per man-hour (Table 19.3). However, until relatively recently the nutrient store in the soil was maintained by the use of a grass/legume crop of varying duration which fixed atmospheric nitrogen, built up organic matter and rehabilitated soil structure, all of which declined during the preceding arable cereal and root cultivation. Nutrient inputs were replenished by deposits of livestock excrement on the grazed grass crop and a high proportion of these nutrients, particularly of phosphorus and potassium consumed, were returned by the application of farmyard manure and other organic wastes produced to the arable cropland. These organic inputs were supplemented by a moderate to low input of inorganic fertiliser, particularly phosphate, potassium and calcium. Nutrient losses by denitrification and leaching were limited. In its ideal form this type of agro-ecosystem approaches an almost closed, self-sustaining, nutrient- and energy-conserving system.

Table 19.3 Energy inputs and outputs in rice production with varying intensities of management (from Pimentel 1979)

	Energy (000 kcal ha^{-1})		
	Borneo	*Japan*	*California*
Inputs			
Direct energy			
Labour	0.626	0.84	0.008
Axe and hoe	0.016	—	—
Machinery	—	0.189	0.360
Diesel	—	—	3.264
Petrol	—	0.910	0.657
Gas	—	—	0.354
Indirect energy			
Nitrogen	—	2.088	4.116
Phosphorus	—	0.225	0.201
Seeds	0.392	0.813	1.140
Irrigation	—	0.910	1.299
Insecticides	—	0.348	0.191
Herbicides	—	0.699	1.119
Drying	—	—	1.217
Electricity	—	0.007	0.380
Transport	—	0.051	0.121
Outputs			
Rice yield	7.318	17.598	22.3698
(protein yield)	(141 kg)	(364 kg)	(462 kg)
Energy efficiency	7.08	2.45	1.55

HIGH-INTENSITY AGRO-ECOSYSTEMS

The modern high-intensity agro-ecosystems are those in which increasingly large yields are achieved by commensurably large energy subsidies needed for the intensive management of both the physical habitat and the biotic components (see Tables 19.2 and 19.3). The recent intensification of agriculture has been accompanied by increasing specialisation of production; the spatial separation of crop and livestock production; and an increase in the size of farm and field units to facilitate increased machine size.

Intensive breeding programmes have produced genetically pure, high-yielding varieties of crops whose growth potential can only be realised by a high input of nutrients and whose growth form is suitable for mechanisation. Also of livestock which require high-energy (protein-rich) feedstuff for maximum production. High productivity has, however, been achieved at the expense of a loss of resistance of crops and livestock to stress. Dense planting and continuous cultivation without a break crop or rest of genetically uniform crops has increased the risk of yield depression by weeds, pests and disease and has necessitated an ever-increasing application of herbicides and insecticides combined with a continual search for increasingly disease-resistant crops.

An ever higher input of inorganic fertilisers, particularly of nitrogen, has all but replaced the use of farmyard manure in arable cultivation. In livestock farming, excreta tend increasingly to be returned to the land in a liquid (slurry) form. The latter and the soluble inorganic fertilisers are susceptible to loss by soil leaching, particularly if applied in excess of crop needs or at the wrong time. Together with other waste chemicals they become pollutants in other non-agricultural ecosystems.

Except in the case of some soft fruit, there are few cultivation processes that have not been completely mechanised, and intensive agriculture subjects the soil to a high frequency of disturbance and particularly by mechanical impact. And in the absence of an annual supply of manure or a grass/legume rotation the soil organic matter content tends to decrease and the soil loses its structural stability becoming susceptible to accelerated erosion by wind and/or precipitation. The removal of hedges and associated field verges, copses and shelter-belts together with the drainage of wetland completes the creation of the simplest type of ecosystem. Not only is it dependent on increasingly sophisticated management which aims to achieve the maximum yield-producing fit between the organic and inorganic components but in doing so it becomes increasingly difficult to sustain the increasingly high yields. For some crops on particular soils in certain areas, there are already indications that the point of diminishing returns has been reached, i.e. the ratio between the cost of inputs to that of outputs has decreased. And in those situations where the system cannot be economically sustained it is abandoned and replaced by another agro-ecosystem or non-agro-ecosystem managed at a low level of intensity for a lower but hopefully more sustainable yield. Indeed, there is already some doubt as to whether it is ecologically possible to maintain the production of increasingly high agricultural yields indefinitely or even over a moderately long time-scale without a biotechnical breakthrough in genetic engineering which succeeds in producing more biologically efficient crops and livestock from the same or smaller resource base.

SUSTAINABLE AGRICULTURE

While it is possible to estimate the maximum sustainable yield of a single animal species population quantitatively on the basis of its previous response to exploitation, such calculations do not allow for the complications of interactions between the exploited and unexploited species or for the extent to which increased competition and predation following stock reduction by harvesting may affect sustainability. Maintenance of a level of sustainability must also depend on maintenance of environmental and economic conditions. Even in this unlikely situation, genetic drift resulting in the reduction in the vigour of a continually harvested stock might well limit long-term sustainability.

The sustainability of agricultural production (i.e. optimum yields of a plant crop) is much less easy to assess. In this case the exploited stock is the soil (the nutrient, water and oxygen store). The *nutrient status* can give a comparative measure of relative fertility, but the yield potential will be dependent on nutrient availability which is a function of the climate, soil chemistry and physical structure. An alternative index of crop sustainability is that of *soil capability*, i.e. the capacity of a soil to produce a given crop for an indefinite period without exhaustion, waste or degradation. Several methods of assessing soil capability have been employed at different times and places (Tivy 1991). Those most commonly used are based on the presence or absence of those soil and land characteristics which are considered most important determinants of agricultural potential. Land is rated according to the particular type and severity of its principal inherent physical limitations of local climate, relief and soil (depth, texture, wetness, erodibility, etc.) under a given level of management and assuming relatively constant climatic and economic conditions. Those with no limitations should, theoretically, be capable of sustaining higher yields with greater energy efficiency for longer than those with severe limitations. However, what is difficult to assess is the optimum yield that a given soil can maintain from a minimum and or moderate intensity of management. Given the continual export of crops and animals it is unlikely that any soil could maintain a constant level of fertility without management which ensures maintenance of its organic content on which its structure and hence other resource characteristics depend. Although technology is a resource that can make good deficiencies in, or be substituted for, other resources in a soil-less agriculture (e.g. as in the case of hydrophonics or a synthetic soil material) none has yet been developed on a scale necessary to maintain the world's increasing demand for food.

Chapter

20

The urban ecosystem

The built environment (i.e. that associated with the interior and exterior of buildings and with land covered by man-made materials) attains its most complex expression in large, densely built-up urban areas. The increasingly high concentration of population in the large conurbations of the developed world and the very rapid rate of urbanisation in the developing countries have, particularly since the Second World War, focused attention on urban problems and stimulated the growth of urban studies. The initial analytical approach to the 'form and function' of towns and cities has, particularly within the last two decades, been supplemented by a synthetical approach. The latter attempts to explain the city in terms of a dynamic system, analogous to a biological system, in which the flow of energy and the circulation of materials are the two most important integrators. A closely related, more recent trend, has been the growing interest in urban ecology and the urban ecosystem.

THE CONCEPT OF THE URBAN ECOSYSTEM

The concept of the urban ecosystem is, however, despite the wide usage and acceptance of the term, still vague and undefined. At present, as Gilbert (1989) notes, it is used to describe three approaches to the study of the organic components in the built environment. The first is that of human ecology as exemplified by the Boyden *et al* (1981) study of Hong Kong which analyses the energy-flow relationships between the human population and their urban (built) habitat. The second is that of landscape ecology with an emphasis on land-use diversity and pattern in cities, often from an aesthetic point of view. The focus of the third approach is on the characteristics of nature (or wildlife) in the urban habitat. The latter is the theme of this chapter. The wildlife (flora and fauna) of urban areas is attracting increasing interest related to its scientific, ecological and desirable aesthetic values on the one hand, and concern about associated problems of pollution and human health on the other.

Within the last 15–20 years there has been an increasing amount of published data on types of wildlife found in towns and cities, much of which is descriptive of particular species or, less commonly, of communities. All too few inter- or intraregional comparisons of wildlife in cities are as yet available by which to test the urban ecosystem concept. The aim of this chapter is, in the light of available data, to analyse the biological characteristics of the urban habitat, the features common to the urban flora and fauna and the extent to which they vary within the urban and between it and the non-urban (countryside) habitat.

THE URBAN HABITAT

BIOCLIMATE

The urban habitat is incontrovertibly the most unnatural one. A high proportion of the area is either covered by man-made material or drastically altered by present or past urban uses, and the condition of the ambient atmosphere is greatly affected by the spatial distribution of urban structures. The distinctive features of the urban climate and associated hydrology have received more attention than any other attribute. Table 20.1 summarises the degree of modification of the main climatic elements in built-up areas. Among the most bioclimatically important effects are,

Table 20.1 Average changes in climatic factors caused by urbanisation (from Sukopp and Werner 1983)

Factor		*Comparison with rural environs*
Radiation	Global	2–10% less
	Ultraviolet, winter	30% less
	Ultraviolet, summer	5% less
	Sunshine duration	5–15% less
Temperature	Annual mean	1–2 °C less
	Sunshine days	2–6 °C more
	Greatest differences at night	11 °C more
Wind-speed	Annual mean	10–20% less
	Calms	5–20% more
Relative humidity	Winter	2 °C less
	Summer	8–10% less
Precipitation	Total	5–30% more
Cloudiness	Cover	5–10% more
	Fog, winter	100% more
	Fog, summer	30% more
	Condensation nuclei	10 times more
	Gaseous admixtures	5–25 times more

first, the higher average air temperatures and air pollution levels than in the non-urban environs, and second, extreme variations in the hydrological regime and modification of airflows. The *urban heat island* is a universally recognised phenomenon which is reflected in the extension polewards of the range of warmth-requiring species of plants (see Fig. 20.1) and animals (e.g. cockroaches). The Chelsea Physic Garden, London, a walled garden established some 125–150 years ago, has an olive tree (*Olea europea*) which sometimes produces ripe fruit. More visibly evident, however, are the biological effects of the greater intensity of air pollution over built-up areas. Particulate matter (smoke), sulphur dioxide (SO_2) nitrous oxide (N_2O), nitrogen dioxide (NO_2) and the mixture of chemicals subsumed under the term 'acid rain' are in general inimical for plant growth. They can act indirectly by reducing the amount of sunlight and directly through the deposition of pollutants on leaves and roots. Among the biological effects are those noted or recorded on urban trees such as stunted growth, deciduousness of normally perennial species of which the conifers appear to be the most vulnerable, early leaf fall of deciduous species. At very high pollution levels development of all but the fastest-growing species (e.g. birch, willow, poplar) is suppressed. In addition, the existence of lichen deserts is indicative of high levels of pollution in city centres (O'Hare 1974). A number of studies have demonstrated the varying susceptibility of different species of lichens to air pollution; the number, vitality and density of lichens on a wide range of substrates decrease with increasing pollution levels from the non-urban city periphery to the city centre. Air pollution has also been responsible for the evolution of melanism in certain groups of animals (more particularly moths) whose natural habitat has been affected either directly by the deposition of soot or indirectly by the reduction of a former light-coloured lichen cover. Under these circumstances selection operates in favour of dark-coloured moth varieties which are less susceptible to predation than the originally light-coloured individuals. In contrast, city parks and gardens can benefit from air pollution which suppresses common pests such as aphids.

Because of the high proportion of hard, impermeable surfaces many with steep or vertical surfaces, runoff of surface precipitation is rapid, often at a rate greater than that at which it can be evacuated by the surface and underground drainage system. Areas of rapid runoff alternate with those at ground level which are susceptible to flooding; the latter is intensified by the detrital load (particularly leaves and other types of litter) which clog drains. In addition, surface evaporation is high in heavily built-up areas and alternating arid and flood conditions create a particularly stressful and unstable environment for organisms.

ENERGY FLOW

In face of the reduction in the vegetated surface primary biological production in urban areas is, to a lesser or greater extent, severely curtailed. This, however, is compensated by the availability of secondary or detrital energy. Cities are predominantly detrital systems in that they produce more organic and inorganic waste material than can be biologically recycled and which has to be evacuated by one means or another. A high proportion of urban waste is organic material in the

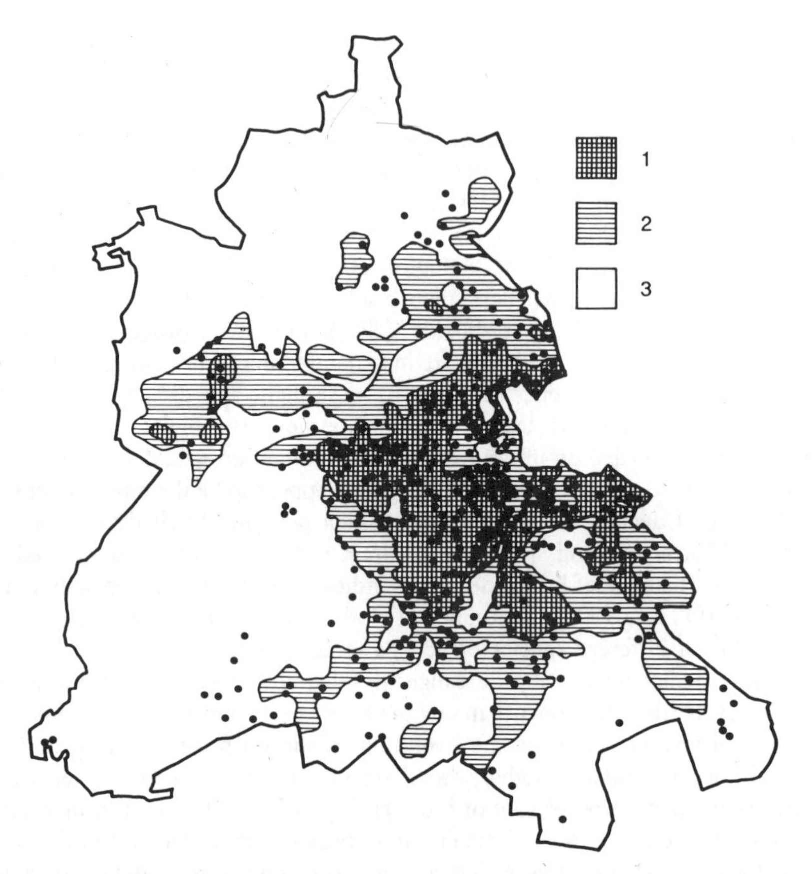

Fig. 20.1 Map of West Berlin showing spontaneous regeneration of the thermophilous Tree of Heaven (*Ailanthas altissima*). The heaviest shading indicates the warmest temperature zones. 1–3 = temperature zones from warmest (1) to coolest (3) (from Sukopp and Werner 1986)

form of sewage and foodstuff. As a result wildlife is characterised by many scavenging animals which have been able to take advantage of this rich detrital source of food and also of the voluntary feeding provided by urban dwellers. Indeed, it has been estimated that in some urban areas this secondary energy source can be at least 25–50 per cent that of insolation.

HABITAT DIVERSITY

Despite these general characteristics urban areas are structurally complex with a high diversity of meso- and microhabitats. This heterogeneity is a function of two factors: the first is the variety of land uses, each with a different association and density of buildings and a differing ratio of built to unbuilt surfaces found within the urban area. Habitat diversity is increased further by the considerable development of edges (or fringes) between microhabitats and by the diverse microrelief associated with man-made surfaces of varying area height and gradient. The second factor is habitat instability as a result of severe human disturbance in urban areas, e.g. movement, noise, cleaning operations, and a continually changing land use accompanied by the destruction and the construction of buildings.

Urban flora and fauna

The habitat diversity of urban areas is **reflected** in a greater number of wildlife species than in a comparably sized area of more homogeneous non-urban use. The urban biota, however, is characterised by a relative paucity of indigenous species in comparison to the numbers of recently introduced colonists. Many of the exotics are garden escapees like the common *Rhodendron ponticum*, the rose-bay willow herb, Michaelmas daisy, golden rod, evening primrose, Japanese knot-weed, etc. or aliens like the feral pigeon, cockroaches and the ship rat (*Rattus rattus*). Multifarious means of transport and routeways associated with the continual movement of people, materials and vehicles have facilitated their immigration into the city. As Gilbert (1989) notes, urban areas show enhanced permeability to immigrants due to increased opportunities for dispersal and reduced competition in open habitats with many new unoccupied niches. The most successful immigrants are those originally non-urban plants and animals that have been able, by only minor changes in form or habit, to become adapted to the urban habitat – some because they are scavengers, and opportunistic generalists, others because they are highly specialised in some way. On the one hand, birds for whom the urban habitat provides a multitude of safe nesting and roosting sites and, on the other, omnivorous, surface-feeding, nocturnal and crepuscular animals have an advantage over others. In addition, many urban organisms are weeds and pests of other managed or modified non-urban ecosystems. The populations of other colonists have attained weed or pest status in urban areas. Finally, urban species diversity is enhanced by successful hybridisation and deliberate introduction.

VARIATIONS IN THE URBAN HABITAT

Any consideration of the urban ecosystem must inevitably raise the question of the scale, in terms of the size (area) and degree of habitat diversity, at which a distinct urban biota develops. Sukopp and Werner (1983) identify nine types of urban land use together with their consequences for climate, soil, and plant and animal life:

1. Residential closed and dispersed (with gardens);
2. Industrial sites and technical utility installations;
3. Inner-city vacant land;
4. Green areas and recreational facilities;
5. Cemeteries;
6. Traffic – streets, railways and waterways;
7. Waste disposal areas;
8. Forests;
9. Fields;
10. Water bodies (reservoirs and recreational lakes).

These are all characteristic of many large Western cities. The size, form and land-use pattern of urban areas, however, vary with the age, site, function and the physical and cultural environment in which the town or city has developed. With relatively few exceptions most work to date on urban ecology has been in the modern West European and American cities; that on the large and fast-growing metropolises of the Third World or the ancient but still viable walled cities of Europe and Asia is relatively sparse. For this reason the following analysis of ecological variations will be considered in relation to three main urban habitats characteristic of the Western city: the city centre, the residential suburbs and what Gilbert (1989) terms the commons (waste and derelict land).

THE CITY CENTRE

The city centre in this context is that part of the town or city dominated by the built environment, covered almost entirely by hard man-made surfaces and used primarily for commercial, industrial and, in some instances, residential purposes. It is a particularly severe and stressful biological environment whose surface relief is frequently compared with that of an intensely dissected plateau with deep narrow canyons, high steep cliffs and isolated buttes and mesas. Lacking in potential soil parent material the hard surfaces are subjected to extreme spatial and temporal variations of temperature, of humidity, of shade and light, and of shelter from or exposure to high wind-force exacerbated by altitude, wind funnelling and turbulence as well as by a continuous movement of people and traffic. Open 'green' spaces are relatively few and, apart from heavily used central parks, are small, e.g. window boxes, roof gardens and urban trees such as the London plane (*Platanus hybrida*), sycamore (*Acer pseudoplatanus*) and white beam (*Sorbus aucuparia*). Spontaneous plant colonisation is confined to gutters, corners, ledges, etc. where sufficient dust and organic debris can collect and provide a rooting medium for one or two individuals of different species (see Table 20.2).

The most successful animals are those species of bird for whom the built environment of the city centre provides habitat opportunities comparable to those to which they are adapted elsewhere. The nooks and crannies of urban structures

Table 20.2 The most frequent wild flora of the London W1 postal district (from Gilbert 1989)

Commonest species	*Slightly less frequent species*
Chenopodium album	*Achillea millefolium*
Epilobium angustifolium	*Bullis perennis*
Plantago major	*Buddleia davidii*
Poannua	*Cirsium arvense*
Sagina procumbers	*Cenyza canadersis*
Senecio squalidus	*Dryopteris filix-mas*
Sonchus olevaceus	*Epilobium ciliatum*
	Pteridium aquilinum
	Rumix cebtusifolius

offer safe nesting sites for gulls, pigeons, starlings, house sparrows, swifts, hawks (kestrel, black kite) for whom the whole of the urban area is a prolific source of food. The kestrel populations expanded in London at the beginning of the 1970s with a change in diet from voles and insects to small urban birds, mammals, insects and earthworms. Gulls and pigeons are the most ubiquitous. Feral pigeons (*Columbia livia*) are common to cities world-wide (Coghlan 1990) and those in most of the city centres of Europe are descendants of the rock dove. The latter probably ranged from India, the Middle East and North Africa to north-west Britain and may have been captured and kept for food since the Neolithic period. The pigeon was introduced to North America and South Africa in the first half of the seventeenth century and to Australia in 1788. It is a voracious feeder eating a wide variety of vegetable materials and whose diet is supplemented by 'voluntary feeding'. Pigeons have reached pest status in most cities, polluting buildings and streets and are generally suspected of being a human health hazard. Gulls are equally widespread. Monaghan (1980) notes that fifteen species of seabird have been recorded as nesting in occupied buildings in a wide variety of localities in Britain. Seven are species of gull of which the herring gull (*Larus argentatus*) and lesser black-backed gull (*L. fuscus*) are the most common. The growth in the urban-nesting herring gull throughout much of Europe is thought to reflect the saturation of their more traditional colonial sites consequent on the increased availability of food from waste dumps. In city parks where large trees are available the carrion crow, jay and magpie are now numerous. Other birds including starlings, wagtails and several species of gull are autumn/winter roosters in urban areas.

Mammals, with the exception of the brown rat, house mouse, and the feral cat and dog, and the grey squirrel are less commonly adapted to the built environment. The rat, however, has long been one of man's most common camp-followers. The black ship-rat (*R. rattus*) – the carrier of bubonic plague across Europe in the fourteenth century – frequents dockside areas. Its numbers, however, have been declining with changes in port technology and in types of ships. In contrast, the urban sewage network has provided the subterranean brown rat (*R. norvegicus*) with an ideal breeding and feeding habitat.

The interiors of urban buildings are perennial or seasonal breeding and feeding niches for a vast array of invertebrates, particularly insects which either feed and/or breed on organic matter; and many are pests and/or human parasites. Some of the most prolific such as species of ants, crickets and cockroaches are tropical aliens which have been able to colonise heated restaurants, warehouses and factories. Ten of the most common species of cockroaches originated in Africa, the oriental species (*Blattus orientalis*) is now four times more prevalent than the northern German (*B. germanicus*) species. They are now the most resilient and highly urbanised insect in city centres where high-rise residential buildings have (as in parts of London) become subject to major infestations.

RESIDENTIAL SUBURBS

In many modern Western cities the density of buildings and hard surfaced areas decreases from the centre towards the periphery of the urbanised area while the amount of open green space increases. The latter comprise domestic gardens and allotments, public gardens and parks of one kind or another, cemeteries and small water reservoirs. All are planted and managed areas in which a wide diversity of lawn grass and trees and shrubs are the most common types of habitat. Habitat diversity, a large varied and well-distributed natural and voluntary food supply combined with ample cover, water, and a variety of nesting and roosting sites is reflected in the number of species of birds, including year-long breeders and transient roosters and feeders. The suburban area supports a richer bird community than either the city centre (see Table 20.3) or the adjacent non-urban area. Gilbert

Table 20.3 The relationship between numbers of breeding species of bird and the distance from the centre of London (both approximations). Data compiled from Hounsome 1979 (from recent editions of the *London Bird Report*, the *London Naturalist*, *Bird Life in the Royal Parks*, and from Simms (1975)

Site	*Number of breeding species*	*Distance from Charing Cross (km)*
1. St James/Green Park	22	1.5
2. Hyde Park/Kensington Gardens	29	3.0
3. Regent's Park/Primrose Hill	34	3.5
4. Holland Park	24	5.5
5. Clapham Common	17	5.75
6. Bishop's Park	16	7.25
7. Wandsworth Common	18	10.0
8. Wimbledon Common	46	11.25
9. Brent Reservoir	44	12.5
10. Richmond Park	56	12.5
11. Kew Gardens	39	12.75
12. Greenwich Park	27	14.0
13. Ham House	38	14.5
14. Osterly Park	46	15.5
15. Hampton Court	42	18.0
16. Bushey Park	47	18.0

(1989) notes that data from the Garden Bird Survey being undertaken by the British Trust for Ornithology indicates that the fifteen most frequently recorded birds in British gardens during autumn/winter 1987/88 are, in order of frequency, blue-tit, blackbird, robin, house sparrow, starling, great-tit, dunnock, chaffinch, greenfinch, wren, song thrush, collared dove, coal-tit, magpie and black-headed gull.

The suburban area also supports several omnivorous mammals; apart from the domestic and feral dogs and cats, squirrels, hedgehogs and foxes, racoons and skunks in North America and possums in Australia, and monkeys (such as the macaque in Hong Kong). All are non-urban natives which have moved into and adapted to suburban areas, have increased in numbers and have in some instances penetrated into the city centre inner suburbs. The grey squirrel is now ubiquitous throughout urban areas in Britain and North America, while the urban fox population has grown steadily in British towns particularly during the past 40 years (MacDonald and Newdick 1980) probably as a result of the decimation of rabbits (which was a major food source in its non-urban habitat) by myxomatosis as well as of a reduction in cover in the intensively farmed countryside. It tends to favour large, long gardens with plenty of hedge cover. The suburban biota has recently been enriched by an increasing number of escapees (frequently exotic aliens) from zoos, pet shops and homes such as foxes, hamsters, monkeys, cockatoos and parrots, mina birds, etc. most of which do not have sufficient breeding potential to establish viable populations independently of human care. Along the east coast of the USA the white-tailed deer, which was practically extinct at the beginning of the century, has become a real pest. It is reported (*The Independent*, 25.9.91) that Philadelphia, the fifth largest city in the USA, has a resident population of 1500; the large open-plan suburban gardens provide a rich source of forage (particularly yew, rhododendron, azalea) adjacent to dense woodland.

COMMONS

As already noted, the commons include unused waste land and formerly used but now abandoned derelict land. The amount and distribution vary from one urban area to another. It can be found as plots of varying size from the centre to the periphery of the city. It is characterised by a wide diversity of microrelief and a substratum which may still be sealed by a hard surface, or be composed of one or more materials varying in physical structure (building rubble, mine or quarry waste, sand, salt or clay), and chemical composition (from nutrient rich to extremely acid, alkaline or frequently toxic). Accessible common sites are frequently used as informal playgrounds, transit areas or rubbish dumps, and hence tend to be extremely and constantly disturbed by trampling. Few are permanent with their length of existence dependent on the rate of land-use development and redevelopment in the urban area concerned.

The commons are the main urban habitats to be actively colonised by plants from the surrounding countryside. Particularly common are the ruderal species initially adapted to the harsh open, exposed and continually disturbed habitats of seashores, river banks, steep unstable slopes and cultivated land. Dependent on the age of the site, its location and human accessibility the commons can exhibit a wide range of vegetation types which illustrate all stages in succession from pioneer to, in some instances, scrub and woodland. Pioneer grassland is usually the most common community, followed by areas on which a dense thicket of bramble, hawthorn and elder scrub has become established – or fast growing pioneer trees such as birch and sycamore. Many of the commons support mixed plant assemblages of native, alien and garden escapees unlike any community in the countryside and which can be regarded as true urban associations. In addition, the commons have become the preferred sites for the planting of urban woods or forests either for amenity and/or recreational purposes. Finally, an uneven land surface, due to previous dumping or quarrying or to mine subsidence, has left on many commons a legacy of ill-defined marshy land or water bodies of varying size and depth in which a relatively undisturbed species-rich flora and fauna with a high conservation value may have become established.

URBAN ISLANDS

Urban ecosystems exhibit many of the ecological characteristics of islands. In the first place cities are islands of built habitat surrounded and isolated one from the other by unurbanised countryside. Their biota have initially been established by colonisation, not only by natural agents from the adjacent countryside but from a global range consequent upon the deliberate or accidental transport by humans. It is, however, doubtful if the equilibrium concept of island biogeography is applicable to the urban island. In the first place many cities are continuing to grow in area and, with ever-changing forms, to provide an increasing diversity of habitats in what were suburban or peri-urban areas. Second, urban expansion in many countries reduces isolation of formerly more widely and clearly separate cities between which countryside remnants become the distinctive islands. Third, because of the continual presumed high mortality of animals consequent on predation by pets and feral carnivores, death (particularly by vehicles) and a rapid microhabitat change results in a high turnover of populations which inhibits the attainment of equilibrium. However, as Gilbert notes, constant site disturbance and redevelopment facilitate the regeneration of the urban ecosystem by creating gaps in a development cycle which has varied with time, in different types of cities.

The urban island owes its species diversity (in total much greater than that of the often uniform agro-ecosystems or forest and woodland which surround it) to its habitat diversity. This has also favoured the establishment of characteristic urban populations of plants and animals, some species of which have already developed distinct urban races or ecotypes (e.g. Michaelmas daisy, evening primrose, dandelions) or successful hybrids, e.g. coy dog (dog × coyote) in America and the jackal × wolf in Greece.

URBAN LANDSCAPE PLANNING

The biological components have played a significant role in urban development and urban planning. Eighteenth- and nineteenth-century urbanisation saw the development of the formal city parks in which aesthetic considerations dominated their design and use. The new garden cities of the first half of the twentieth century, while initially influenced by the emphasis on urban forms and physical planning, witnessed the beginning of what has been called 'the greening of the city' after the Second World War. This has been accompanied by a broader approach to the functions of the urban biota and their significance for urban planning and management in terms not only of its aesthetic but of its educational/recreational value, and also of its implications for the urban environmental quality. This has been reflected in the changing concept of the urban park. As a result of the increasing ecological simplification of the countryside and the mounting pressure for recreational land, the role of the urban park has been or is being reassessed in many developed Western countries. Its value as a type of multipurpose land use in which wildlife conservation, recreation and education can be profitably integrated has been recognised. The image of the urban park as an open space where biotic diversity and use were subordinated to the achievement of tidiness and cleanliness has been replaced by that of a semi-natural transition – a countryside enclave with greater biotic diversity than that of the rural countryside. However, despite urban redevelopment in many of the older industrial cities and the decentralisation of industry, cities are the most polluted of all the man-dominated ecosystems and the most important sources of pollution of others.

Chapter

21

Conservation

As the rate of exploitation of the biosphere has accelerated so too have the problems of the management of its renewable organic resources multiplied. Central to all aspects of environmental management is that of conservation. In common with many other ecological terms, conservation is a common or general word that has acquired a more or less specific scientific meaning. In general it can mean either preservation or maintenance of the status quo, or 'wise use'. It was originally defined by Pinchot, the first Director of the US Forest Service – the father of the conservation movement – as resource use designed to ensure the greatest good for the greatest number for the longest time, or in the words of the late J. F. Kennedy, as the prevention of waste and despoilment while preserving improving and renewing the quality and usefulness of our resources. It can encompass the protection and preservation of plant and animal species at one extreme and the management of the environment at the other. Indeed, conservation implies an attitude of mind as much as a process – a philosophy as well as a technique.

Inherent in the concept of the conservation of organic resources are two views which, literally interpreted, might appear to be so contradictory as to defy reconciliation. The first is that (considered in Ch. 20) of *sustained yield* of those products for which there is the greatest demand and which would give the maximum economic return on capital invested. The second is that of the protection and preservation of what is perhaps misleadingly called 'nature' or 'wildlife' hence *nature conservation* which is the theme of this concluding chapter.

DEVELOPMENT OF THE CONSERVATION MOVEMENT

The concepts, aims and methods of conservation developed from a concern for the way in which humans were misusing and depleting their natural resources by overexploitation, leading to the simplification and impoverishment of plant and animal life accompanied by reduction of biological production and, in extreme cases, the destruction of organic resources. Neither this process nor the concept of conservation are recent. Literary sources record the existence in China in the third century BC of official forest and watershed conservators whose responsibility it was to regulate timber extraction and ensure the maintenance of the last remaining stands of mountain forests. Medieval laws in Europe designated areas in which hunting and the exploitation of forest products were strictly conserved for their exclusive use and to ensure the protection of a viable stock of game animals for hunting for food and for sport.

The wasteful exploitation of the recently colonised 'virgin lands', particularly in North America and other temperate regions in the southern hemisphere, was stimulated by an abundance of land and expanding world markets for agricultural products. At the same time, however, knowledge about biological principles and a greater understanding of the ecological implications of organic resource use were beginning to emerge. To this end the publication of George Marsh's now classic book, *Man and Nature: Physical Geography as Modified by Human Action*, in 1864, had a profound influence, not least in the USA. What was to become known as the 'Modern Conservation Movement' was born of the realisation of the extent to which ruthless, uncontrolled clear-felling was destroying the last great stands of virgin forest in the western part of the USA. The Forest Services and National Parks were established in 1905 and 1916 respectively, on the public domain. Drought and agricultural depression in the 1930s had an even more dramatic and universal impact. They revealed the appalling degree and extent to which soils had been ruthlessly overcultivated. The 'Dust Bowl' of Texas and Oklahoma and the gully-ravaged hillsides of the Tennessee Valley became the classic case-studies of the disease of soil erosion. These served to focus attention on the extent of the problem in other parts of the world. The establishment of the US Soil Conservation Service in 1934 and the Wildlife Service in 1935 were major landmarks in the history of the Conservation Movement.

Until the Second World War conservation policies and techniques had progressed more rapidly in the USA than elsewhere and were mainly concerned with problems of forest and soil use. Since the Second World War, increasing pressure on land in Western urban/industrial countries plus the growing populations and needs of the underdeveloped tropical countries have intensified the problems attendant on the exploitation of organic resources. Attention has become increasingly focused on the need to control and regulate the use of organic resources; the concept of conservation as a method of 'applied ecology' or 'resource-use planning' is now universally accepted, if not always implemented.

This change in the perception of conservation was reflected in the designation in 1956 of the International Union for the Conservation of Nature (IUCN) in place of the earlier more restricted Union for the Preservation of Nature (IUPN) founded in 1951. The former recognised the need to combine preservation and sustained use of renewable natural resources for the benefit of mankind. By 1980 the World Conservation Strategy (IUCN/UNEP/WWF) was firmly based on the basis of the reciprocal interrelationships between conservation and development, while the World Conservation and Economic Development Conference in Ottawa in 1987 clearly recognised that sustainable (i.e. long-term) economic development was, particularly in the tropical lands, dependent on management of natural resources for a variety of purposes. The concept of conservation has, since its late nineteenth-century inception, evolved from that of nature protection for its own sake to that for the benefit of human beings.

NATURE CONSERVATION

Since the initiation of the Conservation Movement the principles and policies have evolved as the problems of what to conserve, how to conserve and for what purpose or what end should nature conservation be pursued have become more complex as human demands on the biosphere have increased. However, the fundamental aim has always been implicitly or explicitly the preservation of 'wild' species and habitats and, in so doing, the maintenance of maximum biodiversity at the local, regional and global scale. This has been justified on five separate grounds:

1. *Ethical* – it is morally wrong that what can be considered the common heritage of future generations should be squandered by the present generation.
2. *Scientific* – preservation in the interests of either pure or applied research since knowledge of organisms and their ecological relationships is essential to the efficient management of organic resources. This can only be sucessfully pursued in actual habitats, and the more diverse and varied that are available for empirical observation and experiment the better.
3. *Aesthetic* reasons for the preservation of wildlife cannot be ignored, in fact they have been given increasingly serious consideration in conservation management. Species diversity and habitat variety are generally accepted as intellectually more satisfying and emotionally more pleasing than monotonous uniformity of scenery. Some of the most attractive rural landscapes in many parts of the world, in which a variety of wild and managed habitats are intermixed, were deliberately created for the aesthetic satisfaction (as well as social prestige!) of previous landowners. Scenic diversity and variety also enhance the value of land for recreational use. And public concern for wildlife conservation has been stimulated by the increasing number of people who pursue outdoor activities and have a greater knowledge and interest in natural history than ever before.
4. *Economic* – because of the financial spin-off from recreation and tourism, conservation has acquired a more readily acceptable and comprehensible justification. There are also clear indications that species and habitat diversity are not only ecologically desirable but, in the long term, economically desirable and profitable. Although the positive relationship between species diversity and stability has come under considerable criticism, the conservation of spatial diversity and ecologically healthy and pleasing landscape is still strongly supported.
5. *Potential organic resources* – only a very small proportion of extant species

have been exploited for food, industrial materials and, not least, medicinal drugs. As the depletion of existing organic stocks are reduced so the need to preserve as diverse a gene pool as possible of both wild species and cultigens becomes more urgent.

SPECIES CONSERVATION

Conservation management may be species, habitat or ecosystem oriented. Some of the earliest conservation strategies practised, and for which legislation was framed, was to protect and conserve species either of economic value or of particular ecological interest. It has involved the outlawing of the collection of certain birds' eggs and the picking of rare plants; and the establishment of closed seasons for hunting and fishing. In this respect, the ornithological lobby has been the most powerful and successful. One of the earliest pieces of conservation legislation was an Act (1869) passed to protect designated sea birds in Britain.

Species conservation, however, is concerned mainly with known endangered and threatened species of plants and animals. The former are those such as the white rhinoceros, the Siberian tiger, the blue whale, the panda and narrow endemic plant species (see Table 21.1) in immediate danger of extinction; the latter are

Table 21.1 Status of endemic plant vascular taxa on selected oceanic islands (from IUCN)

Island	*Total*	*Percentage rare, threatened or extinct*
Ascension Island	11	91
Azores	55	55
Canary Islands	569	67
Galapagos	229	66
Juan Fernandez	118	81
Lord Howe Island	75	97
Madeira	131	66
Mauritius	280	61
Norfolk Island	48	96
Rodrigues	48	96
St Helena	49	96
Seychelles	90	81
Socotra	215	61

those which, though still abundant in parts of their range, are known, like the grizzly bear, the chimpanzee and the African elephant, to have recently declined sharply in numbers (Shafer 1990). The designated endangered or threatened species, however, tend to be those rare and/or large mammals and rare species of flowering plants which have attracted much attention. The actual number must be far greater particularly in face of the rapid decimation of the tropical rain forest which it is estimated, contains about 50 per cent of the earth's terrestrial species, the majority of which are represented by small habitat-specific populations. It is now realised that the conservation of certain keystone species is essential to that of many others, which can quickly become endangered if the former declines or dies out in a particular habitat. Clapham (1981), for instance, quotes the example of the alligator in the Everglade swamps of Florida which digs water-holes in the mud in which during periods of low water many fish species can survive.

Effective species conservation, however, depends on a knowledge of the size and structure (age and sex ratio) of the minimum viable population below which conservation becomes increasingly difficult or impossible because of inbreeding depression and sterility in a small and decreasing gene pool. However, some species experience what is called a *population bottleneck* (i.e. a very rapid reduction in size and genetic diversity below the demographic minimum viable level) and yet manage to maintain themselves or even recover. Such is the case of the Northern African elephant in about twenty-four locations (Shafer 1990). Apart from protection of existing populations, methods of saving endangered species from extinction include: expanding their range by introduction to other similar habitats where the species has become extinct; breeding in captivity and releasing in the wild; and finally preservation in zoos and botantic gardens. In the first instance, species protection frequently results in a population increase beyond the carrying capacity of the available habitat, in over-predation or overgrazing and in numbers to the level of a pest in other managed ecosystems. This is reflected in conflicts between fishermen and seals, between arable farmers and red deer in the Highlands of Scotland and between householders and possums in Australia. In the second instance methods are expensive and uncertain. Successful breeding of rare animals in captivity is variable and the survival of those reintroduced to the wild difficult.

SPECIES REINTRODUCTION

The oryx (Price 1989) is perhaps one of the outstanding examples of a successful reintroduction. Exterminated by hunting in Saudi Arabia in 1972, it had been bred in zoos without difficulty. Its successful reintroduction to Oman in 1980 has been attributed to its highly developed adaptation to a harsh environment in which there are few competitors and no predators other than humans. In contrast (Price 1989 reported by Trudge 1992) the adaptation of 100 young orang-utan released in the tropical rain forest of Sumatra has proved more difficult because of the very complex environment with numerous potential competitors and predators and the lengthy learning period (normally parent dependent) necessary for survival. This is also a problem for the golden lion tamarind presently being introduced into Brazil.

The ultimate success of introductions will depend on whether the initially small populations (because of expense) can avoid becoming genetic bottlenecks. Species, such as the oryx which breed in one-stud 'harems' and the orang-utan (the least fecund of all primates) do not become reproductive until they are 13–15 years old and then breed only at 6-year intervals. In contrast, the reintroduction of predatory and scavenging birds such as sea-eagle (west coast of Scotland), the Californian condor (Death Valley) and the griffon vulture (France) with a higher fertility and a balanced sex ratio have a greater chance of rapidly building up genetically viable and variable populations.

WILDLIFE MANAGEMENT

Successful species conservation must, obviously, be linked to efficient management. This, however, is dependent on an understanding of the species' demographic habits, and particularly, habitats for which sufficient data are rarely available and which was not always appreciated by early animal preservationists. Wildlife management, in the interests of game animals hunted for food and/or sport, generally has long been practised and has been most successful when based on long familiarity and inherited experience, as among the hunting communities of the African savanna, and of the deer, boar, moose, caribou and the red grouse hunters in Europe and North America. As in the case of fishing, hunting must maintain the sex : age ratios of the population if a viable breeding stock is to be sustained, and ensure a range of habitats of sufficient extent to satisfy seasonal grazing regimes for not just one species population but for groups of species, as in the case of the wild game herds of the African savanna. In this context, the most advanced habitat management is that of controlled rotational burning of vegetation designed to stimulate new and more nutritous growth and in some cases to provide a range of successional vegetation types for cover and feeding for one or more species.

GENE POOL CONSERVATION

It is now recognised that there is an increasingly urgent need to conserve intra-species as well as species diversity in both wild and domesticated plants and animals. This is particularly relevant in the tropics where a very large, unknown and as yet untapped pharmaceutical resource is fast disappearing. Two years ago a National Bio-diversity Institute was set up in Costa Rica (Heredia) with financial aid from the USA. The Merck Chemical Company was to undertake chemical prospecting in the tropical rain forest all of which (= 50 per cent total land surface) is now in parks and reserves. An agreement has been signed allowing Merck to screen any plant extracts collected and thereby support the Institute's scheme to inventory, within a 10-year period, the forest's entire flora and fauna (Joyce 1991). Also, in view of the increasing genetic homogeneity of domesticated species there is a growing need to conserve wild ancestors of old but now little or unused varieties of present-day crops and breeds of livestock. It is estimated that 50 years ago 80 per cent of the wheat grown was composed of old varieties, 95 per cent of which have now disappeared (Holdgate 1991). In response to this trend an International Board for Plant Genetic Resources, funded by several UN agencies, has been set up to establish registered seed banks for collecting and maintaining traditional crop strains.

NATURE RESERVES

Nature reserves include all public or private land areas that have been officially designated for the protection of their biological resources. They range from those in which the protection of wildlife is the sole aim to those in which this is combined with the protection of the economic, educational, scenic and tourist values of its biological resources as in the case of national and other parks. Some 4 per cent of the earth's land surface is contained within nature reserves of one kind or another. Although initially intended to be representative samples of habitats characteristic of a local, regional or global area, their distribution and location tend to be haphazard, dependent on the availability of land at the time of formation. Also, they are more often than not part of a habitat than a distinct ecological unit. A very high percentage are less than 100 km^2 and only about 2 per cent are over 10 000 km^2. Few, if any, provide the entire home range particularly necessary for large migratory mammals such as buffalo, zebra, wildebeest and gazelles in Africa and of caribou and elk in the Boreal zone of North America.

Reserves of what ever size can create as many conservation problems as they solve. All are subject to both intrinsic and extrinsic variables. Few reserves enclose climax plant communities when initially established; more include successional stages effected by burning and grazing regimes which need to be maintained to ensure their conservation. Removal of domestic grazing animals can result in a change in vegetation composition, sometimes with reduced diversity, as pre-climax or climax dominants become established. Similarly, animal protection can, as in the case of the African elephant, cause serious overgrazing and deterioration of the biological resources concerned.

Shafer (1990) records that in 1988 the US National Parks Service considered that 60 per cent of their parks were threatened by past and present extrinsically generated impacts. These included: the inmigration of alien species of plants and animals not native to the park environment; legal and illegal (poaching) hunting; intensive recreational use; remnant scars of previous land uses; degradation of air and water; urban development along park boundaries. Even wildlife refuges, in which access and use are restricted, are subjected to alien inmigration, environmental pollution and poaching. And impacts increase when buffer zones (of similar but less protected habitat type) are absent or have been reduced in width or completely eliminated.

Spellberg (1991) notes that the setting up of nature reserves can lead to fragmentation and isolation of once continuous communities, especially when the biological character of the areas surrounding and separating them is damaged or altered. Reserves, then, become islands and their reduced animal populations become vulnerable to genetic drift and faunal collapse. And at present serious consideration is being given to the creation of a European network of special biogenetic reserves designed to counteract the rapid rate of fragmentation and isolation of natural communities.

CONSERVATION EVALUATION

The selection of areas to be set aside as nature reserves has to be done within the context of multiple and competing land uses. This has necessitated the selection of criteria by which their relative ecological value can be assessed and their establishment defended in face of the economic use of the same biological and/or land resource use. The most commonly used criteria (Usher 1986) in developed Western countries are as follows:

1. *Species diversity* expressed as either total number (richness); as relative (percentage) numbers; or as the complexity of the trophic structure (or food web).
2. *Size or area* on the assumption that the larger the reserve the greater will be its species diversity and the more likely it is to be at or above the minimum viable size for the maintenance of the community or the particular species for whose conservation it has been selected.

(While diversity and size can be expressed quantitatively other important criteria can only be assessed subjectively. They are given below.)

3. *Rarity of species* (or species community) as has been previously noted, rare species are normally those with small and more widely dispersed populations than common ones. They could, however, be a small but highly localised population related to a particular uncommon habitat such as on serpentive rocks. Rarity, however, can only be expressed in relation to a continuum from rare to common. If species distribution has been surveyed on the basis of a spatial network (vice-counties or grid squares) then its rarity can be ranked in terms of a percentage of the areas in which it occurs. However, rarity is also relative to the scale of the area surveyed. A species may be locally (county) rare but regionally or nationally common, so that its rarity value will, all other criteria being equal, be less than in the case of regional and, particularly, national rarity.
4. *Naturalness* is an expression of the extent to which the site has been modified by human activities. Given the range and varying intensity of possible human impacts in the past and for which a historical record of some kind may or may not be available, this is a particularly difficult subjective criterion to apply consistently. In the event it tends to be closely correlated with diversity on the perhaps questionable assumption that a more natural area would be more species rich.

(Finally, two criteria particularly important in the selection of nature reserves are typicalness and representativeness.)

5. *Typicalness* expresses the extent to which a reserve is typical of the ecosystem of which it is an isolate. Simberloff (1986) maintains that in this context rarity and typicalness are mutually exclusive.
6. *Representativeness* is an expression of the extent to which a reserve is a good

example of a community or ecosystem type characteristic at a given regional, national or global scale. Its assessment, however, is dependent on knowing the range of communities or ecosystems in a area of a particular scale – as, for example, in the case of the range of variation, of associations (facies) within the heathlands of north-West Europe.

7. *Replaceability* (Netherlands) is the ability and ease with which a community or ecosystem can regenerate after disturbance, i.e. its resilience (see p. 295).

Other criteria, some of which are attributes of the biological resources considered, include fragility (i.e. population or ecosystem response to disturbance) and uniqueness (or community rarity). Some are not attributes of the organic resources but of those extrinsic conditions which may influence the value of one reserve *vis-à-vis* another, e.g. vulnerability to impact or degree to which endangered, accessibility and scientific educational scenic and/or tourist value.

Effective evaluation depends on as comprehensive a survey of the attributes of the species or community as possible, and the replicability of the methods by which the criterion is 'measured' and the final evaluation calculated. While site evaluation should, theoretically, be undertaken before selection and designation, this depends on the availability of land use, land capability and land planning surveys – as in the Netherlands and Scandinavian countries. In others, not least Britain, the rate of development and depletion of wildlife resources is such that selection may have to be based on a coarse evaluation, which is, hopefully, refined after designation. This involves the extent to which criteria are weighted or not. The final stage in conservation evaluation is, as in Britain, a hierarchical grading of sites in terms of their relative national, regional and global scale. In this case rarity, size and representativeness are more important classificatory criteria. However, because of economic, social and political constraints on the selection of nature reserves and the pressure of particular conservation interests, reserves within a particular country or region are frequently unrepresentative or incompletely representative of the actual range of ecosystems present. Some (often wetlands) are over-represented, others under-represented (woodland) and, as was previously noted, small reserves are much more numerous than large ones.

There has been considerable discussion, but very little consensus, regarding the applicability of island biogeography theory to the design (size, form and distribution) of nature reserves in order to ensure maximum species diversity. In developed countries, particularly Europe, the discussion, irrespective of the validity or value of the theory, is largely academic because opportunities to redesign existing reserves and to establish many new ones is limited by land availability. Three aspects of reserve design that have received serious consideration and have been applied in both Europe and North America are reserve zoning, conservation corridors and wilderness character. Reserve zoning is by no means new; it was applied to national parks in India and Sri Lanka when they were under British rule. The model was that of delineating concentric buffer zones based on the carrying capacity for wildlife. The central core is strictly protected from all uses other than for scientific and educational purposes. The role of the surrounding zones is to shield (i.e. buffer) the core from alien impacts and to provide space for wide-ranging species and for widely dispersed rare species which are not included in the core. In developed countries today the concept of the buffer zone is considered an important element in multiple-use reserves and parks which protect the core particularly from intensive recreational use and associated developments.

The nature and ecological significance of *habitat corridors* for conservation are analysed by Forman and Godorn (1986). These are mainly linear landscape features such as hedges, shelter-belts, streams, railway and power lines, etc. Their potential conservation value as summarised by Shafer (1990) depends on their width, continuity and reserve connectivity which allow a higher immigration rate, provide increased foraging area for wide-ranging species, cover from predators, diversity of habitats, refugia from disturbance, and maintain links between seasonal ranges where the access from one to another has been disrupted by developments such as routeways, oil pipelines, etc. However, these must be weighed against the potential disadvantage of corridors for conservation, i.e. such as high immigration rates which are likely to allow the spread of disease, pests and weeds and exotic aliens, to facilitate the distribution of feral animals and to provide cover for predators all of which could result in the conservation costs outweighing the benefits.

WILDERNESS AREAS

The overriding concern of the early American conservationists was to save the great virgin forest stands of the USA from decimation by the lumbermen. To this end forest preserves were designated which, it was intended, should be maintained in their pristine state – 'for ever wild' (Tivy 1972). While a few such preserves still remain (as in the Adirondack Park) the National Forest and the National Parks Service adopted a management policy based on multipurpose, sustained use. However, after the Second World War the rapid increase in tourism in both areas generated concern about the impact of recreational activities and the need to preserve some of the – by now fast disappearing – wilderness. With the passage of the US Wilderness Act in 1964 *wilderness* acquired statutory recognition and was legally defined in the USA as an area of undeveloped Federal land retaining its primeval character and influence, without permanent improvements or human habitation which is protected and so managed as to preserve its natural condition. To this end access is confined to those on foot, horseback or canoe and only shooting and fishing of wildlife necessary to sustain biological resources are permitted.

Since then the wilderness concept has been universally accepted, but its definition and delineation are less easy to apply in small, highly populated countries with much less public land and no really untouched natural territory as in the USA. Nevertheless, scenically wild remote areas could justify designation as wilderness and protection from human intrusion other than by walkers. However, wilderness is a relative concept which poses problems for both its spatial identification and its conservation. Wilderness is a resource designed not only to protect wild areas but to provide people with a 'wilderness experience'. However, the statutory designation of these areas has already created problems in the USA by attracting an increasing number of visitors with consequent deleterious impacts. Wilderness carrying capacity (i.e. the level of use beyond which the quality of the wilderness experience is reduced) is very low. As a result controlled access and management are often necessary to protect the wilderness from, rather than for, people (Tivy and O'Hare 1981) in many national parks today.

NON-RESERVE CONSERVATION

The creation of nature and wildlife reserves and parks of one kind or another has resulted not only in fragmentation of wild habitats but an increasing dichotomy between protected and unprotected areas. While the former can create more problems than they solve, the conservation value of the latter has, until fairly recently, been neglected. Among those non-reserve habitats which can provide refugia for plants and animals are derelict and urban areas, farmland, woodland and forest, as well as abandoned ponds, reservoirs and canals. Derelict land (i.e. formerly used but now temporarily or permanently abandoned) is a characteristic landscape element in all developed countries. It can be found within and continguous to urban areas or in the countryside. It encompasses an extremely diverse range of habitats from the sterile (e.g. toxic waste) to the relatively productive (e.g. water bodies). The biota is more often than not a ruderal one whose composition reflects that of the surrounding land uses. Some of the less accessible sites, such as abandoned quarries and water bodies, can form species-rich islands with a relatively high conservation value in an otherwise impoverished matrix. Urban areas provide a diversity of habitats which are either used by non-urban transients or have attracted what are now urban-adapted species to the extent that the abundance of some taxa is well above that in the adjacent non-urban land (see Ch. 20).

Although farming and forestry have been for long (and indeed still are) regarded as anathema by conservation ecologists, they nevertheless provide habitats with their own characteristic communities of wild species, not all of which are pests. In those quite considerable areas in Europe and North America where very intensive farming systems have not obliterated the traditional farm hedges, copses and shelter-belts, uncultivated field edges, wetland and barns and the rotation of grassland with two or more arable crops, these provide permanent or temporary habitats for a wide range of species many of which have become adapted to the seasonal and annual cultivation cycles and are farm endemics. On those areas where modern intensive, particularly arable, farms have resulted in a loss of many, often rare, farm species there is a growing pressure to encourage a greater interest in, and commitment to, conservation farming. The most extreme example at present is probably that of organic farming which aims to produce healthy food without the use of pesticides or inorganic fertilisers, but which also results in a pest-rich habitat.

Throughout the world, forestry (see Ch. 17) is now dominated by monocultural plantations of quick-growing and often exotic species such as conifers, eucalypts and hybrid species (e.g. poplar, cypress). Frequently these replace a formerly more diverse non-conifer woodland community or a species-rich secondary habitat such as grass or heathland. In some instances conifer afforestation is accompanied by the reintroduction of animal species which were native to the pre-existing forest cover, as for instance the crossbill, the pine marten and the capercaillie in Scotland. In others, efforts have been made particularly in tourist areas of high scenic value to mitigate the undesirable visual effects of conifer culture by a broad-leaved screen on routeways and/or planting of mixed stands of deciduous and coniferous species.

SOIL CONSERVATION

Soil is the basic organic resource and an integral component of terrestrial ecosystems. On a human time-scale it is a finite resource. Soil erosion is a natural process which involves the relatively slow movement of its material from one place to another by gravity, water and/or wind. Theoretically this loss is compensated by weathering of parent material and soil development so that an equilibrium state between loss and gain is maintained. This, however, can be disrupted by a disturbance of the protective stabilising vegetation cover which exposes the mineral substratum to accelerated (or cultural) soil erosion. As has already been noted (see p. 328) this is not only one of the first, but is still one of the most serious and widespread effects of human impact on the environment. It is now a particularly serious problem in semi-arid and humid tropical regions.

Eroded soil may be transported from one land area to another. But ultimately much will find its way by surface flow and wind into water bodies and eventually into the oceans where it is lost from the terrestrial ecosystem. Concern about the rate, volume and spatial scale of accelerated soil erosion in the Dust Bowl era resulted in the establishment of the US Soil Conservation Service in 1934 to advise farmers about methods of soil management that would minimise the dangers of 'clean cultivation' and downslope tillage. It has been estimated that soil erosion losses from agricultural land in the USA is 5 times that removed by natural erosion and could reach 10 000 times in extreme circumstances (Warren 1974). The Conservation Service attempts to predict the rate of soil erosion according to farm type on the basis of precipitation potential, soil erodibility, slope length and gradient, cropping regimes and conservation practices with the aim of restricting the average maximum loss to 11.2 t ha^{-1} $year^{-1}$ (range 1.5–13.0).

Soil management has a long history, particularly in the once or still densely settled agricultural areas of the Mediterranean, the Middle and the Far East as reflected in terracing of hillslopes, bunding land to control water flow and the application of night soil to padi-rice fields. The modern methods of soil conservation used in many parts of the developing world retain the early developed basic terracing techniques in the following:

1. Contour cultivation involving the alternation of crop-strips and grassed water-control strips planted parallel to the contours of the ground;

2. Strip cropping in which strips of two or more different crops are contour or field planted;
3. Particularly in semi-arid areas, tillage methods which do not produce too fine a texture; the protection of dry-season fallow soil with organic mulches; the establishment of grass : arable crop rotations; the planting of shelter-belts characteristic of the prairie and steppe wheat-growing areas of the world.

Fundamental, however, to the conservation of the soil fabric and its potential productivity (or fertility) is the maintenance of an amount and type of organic matter (humus) which will minimise structural collapse and accelerated erosion.

CONSERVATION IN DEVELOPING COUNTRIES

Conservation issues in the developing – particularly tropical – countries of the world differ from those in the developed countries by reason not only of environmental and biological contrasts but of the rate of population growth and exploitation of organic resources. As a result the countries involved are caught between the horns of the dilemma of the urgent need for economic development on the one hand, and on the other the need to protect the biological resources on which economic development depends before they are completely decimated. Since the beginning of this century the reduction of the original humid tropical forest and savanna woodland and associated loss of species has been increasing at an exponential rate. Table 21.2 illustrates the relative loss of habitats in terms of

Table 21.2 Loss of natural habitat in tropical Asia (from IUCN)

Country	*Original wildlife habitat (km²)*	*Remaining wildlife habitat (km²)*	*Habitat loss (%)*
Bangladesh	142 100	13 300	91
India	3 017 000	615 100	80
Indonesia	1 446 400	746 900	48
Malaysia	356 200	210 100	41
Nepal	117 100	53 900	54
Pakistan	165 900	39 800	76
Thailand	507 300	130 000	74
Vietnam	332 100	66 400	80
Total	6 084 100	1 875 500	69

natural vegetation in tropical Asia. This, as Pellew (1991) notes, can be translated into habitat loss for forest-dwelling primates with extensive home ranges (see Table 21.3). The area of primary tropical rain forest in Africa and South America is continuing to contract at a rate which, unless curtailed, could see its complete destruction, together with the loss of a very high percentage of the world's terrestrial plants and animals (many of which have not yet been identified) by the

Table 21.3 Range loss and habitat protection for some primate species in South-east Asia (from IUCN)

Species	*Original range (km^2)*	*Remaining range (km^2)*	*Loss (%)*	*Range protected (%)*
Orang-utan	553 000	207 000	63	2.1
Siamang	465 100	169 800	63	6.8
Agile gibbon	532 300	184 300	65	3.7
Bornean gibbon	395 000	253 000	36	5.1
Javan gibbon	43 300	1 600	96	1.3
Stump-tailed macaque	1 547 000	556 500	64	3.7
Rhesus macaque	1 732 000	568 600	67	2.8
Proboscis monkey	29 500	17 800	40	4.1
Doue langur	296 000	72 300	76	3.1
Banded leaf monkey	450 800	168 200	63	7.5
Phayre's leaf monkey	708 600	193 200	73	3.8
Average of 27 species			65	5.0

end of the 1990s. It has, for instance, been estimated that over half the world's species are found in just over seven countries – Brazil, Mexico, Colombia, Zaire, Madagascar, Indonesia and Australia (Holdgate 1991).

In view of the socio-economic problems and level of technology of the developing countries it is perhaps ironic that they have a history of conservation that predates the Modern Conservation Movement in the Western world and that eleven of the poorest countries have set aside over 10 per cent of their respective land (30 per cent in Botswana) for wildlife conservation. The conservation ethos is deeply entrenched in the culture of many of those communities whose subsistence is dependent on natural resources and their cyclic regeneration. Taboos associated with the indiscriminate slaughter of reproductive female animals, together with the social value and/or sacredness of others, have probably evolved from the initial need to conserve. The colonial powers introduced modern methods of conservation in the late nineteenth century in the form of game and forest preserves, particularly in Africa and Indo-Asia. Few, however, were large enough to provide the size and habitat diversity of home range required by large migratory herbivores and their predators. Protection, particularly of elephants, without management (what is called *benign neglect*) resulted in a population increase beyond the carrying capacity of the reserve and severe habitat degradation by overgrazing.

Within this century the number of wildlife reserves and national parks in the developing countries has increased. So too have the problems of their management. Overprotection has either been exacerbated or been replaced by sharp decreases in species populations during the past 25 years as a result of overexploitation and, more seriously, poaching, of for example, elephant and rhinoceros for ivory and horn. The total world population of the latter species is now estimated at 11 000 – half that considered safe for survival (Trudge 1992). Of the five rhinoceros species, the Java with a population of 50 is now the rarest large mammal, while the African black rhino (now 3500) has experienced a 95–98 per cent decrease in the last 20 years. However, the elephant and the rhinoceros are relatively resilient, slow reproducers with a long generation time and they can recover from very low levels without genetic drift. There are three conservation options:

1. Zoos which presently contain breeding populations;
2. Strict protection combined with prohibition of trade in ivory or horn as favoured by Kenya;
3. Sustainable production of rhinoceroses and rhinoceros products within the local economy as advocated in Zimbabwe.

On the one hand, the latter strategy would make Kenya rhinoceroses (and elephants) more attractive to poachers since at present it is not possible to distinguish ivory from different sources. On the other, trade in wildlife to zoos and private owners is so lucrative that a change in land use from livestock to wildlife farming is taking place. Also, where the wildlife is an integral part of the local economy protection from poaching has proved most effective. Predictions of faunal collapse within the next 5000 years in the existing small East African parks have been postulated (Soulé and Wilcox 1980) because of isolation and lack of management (benign neglect). This viewpoint has, however, been rejected because according to Shafer (1990) not all African reserves have become as ecologically isolated as those in many temperate countries and some tropical regions such as for example, Costa Rica. However, in time many reserves will probably become partly or wholly isolated and may, for the larger sedentary animals, have already become so – and their populations may no longer be viable.

Whatever their size, one of the major factors in the recent faunal decline in tropical reserves and parks has been human impact within and from outwith the protected area. In the developing countries the majority of the large and fast-growing populations are still directly dependent on the biophysical resource base for food and trade; in pastoral communities the socio-economic structure is related to numbers rather than quality of livestock held. Also, many reserves and parks impinge on and partially or wholly overlap migratory or sedentary tribal territories in which large herbivores and livestock (cattle, sheep and goats) share and compete to a greater or lesser extent for the same resources. For example, in the Amboseli National Park in Kenya (488 km^2) the same land is used by wildlife and domestic livestock. In the wet season migratory mammals disperse over a home range of *c.* 5000 km^2. However, in the dry season this contracts to 600 km^2 where grazing and water are available and where some 80 per cent of the migrant wildlife use the same land as the Masai cattle.

Methods of wildlife conservation in the developing countries have, to a considerable extent, been influenced by those of the Western world. Many of the national parks and other types of reserves were originally established, and some continue to be directed and managed, by European or North American expertise. In the past half-century they have provided for ecological research. They vary in the emphasis of their conservation aims from those in which wildlife exploitation is prohibited or restricted, to those with a highly protected core and a multiple land-use buffer zone to those in which tourism and wildlife conservation are interdependent. Indeed, tourism (with wildlife as the main attraction) accounts for 30 per cent of Kenya's gross national product. These reserves, as Pellew (1991) comments, have to a greater extent than in the developed countries, exacerbated the economic gulf between those who use them and those who live in and around them, and also exacerbated the conflicts between conservation and other economic types of land use. There is a growing awareness of the need to integrate the protected with the surrounding non-protected areas to their mutual benefit. Greater tourist development within and on the periphery of reserves could generate even more income than it does at present. This could be used, on the one hand, to employ the number of staff needed to deal effectively with the poaching problem and, on the other, to compensate other land users for restrictions on land-use practices (including hunting), livestock reduction and livestock losses due to predation. Shafer (1990) maintains that the conservation benefits would more than compensate for these costs and that revenue from tourism could be eighteen times that from a fully developed beef enterprise on a similar area.

THE FUTURE

The particular conservation problems of the developing countries combined with those posed by the ecological implications of rapidly increasing environment pollution have contributed to a return in conservationism to that of the sustainable use of all natural resources. This found expression in the World Conservation Strategy formulated by the IUCN, the WWF and UNEP in 1980 and to which the majority of countries are signatories. It embodies three aims:

1. To maintain essential ecological processes and life-support systems (e.g. soil regeneration and protection) by the recycling of nutrients and the cleansing of waters on which human survival and development depend.
2. To preserve genetic diversity (the range of genetic material found in the world's organisms) on which depend many of the above processes and life-support systems, the breeding programmes necessary for the protection and improvement of cultivated plants, domesticated organisms and micro-organisms, as well as much scientific and medical advance, technical innovation and the security of the many industries that use living resources.
3. To ensure the sustainable utilisation of species and ecosystems (notably fish and other wildlife, forests and grazing lands) which support millions of rural communities as well as major industries.

Among the difficulties of achieving these aims, the lack of capacity to conserve must, in this writer's view, be one of, if not the most, serious. This is, first, because of the lack of basic data on the quantity and distribution of remaining organic resources of the biosphere in terms of species, communities and ecosystems; second, because of a lack of adequate techniques with which to deal with organic conservation at varying scales and under varying environmental conditions. In the former instance, although the World Conservation Monitoring

Centre was established in 1982 by the IUCN, its activities had to be curtailed in 1988 to managing three types of data on species, areas and utilisation and trade in wildlife products and derivatives. In the latter, the problem, according to Pellew (1991), arises directly from the difference between the science of ecology and the data it presently generates and the practice of nature conservation and the data it requires. And in this respect ecologists on whose science the successful manipulation of organic materials is dependent might be thought of as 'fiddling while Rome burns'. However, nature conservation entails more than the management of biota and their habitats; it is a type of land- and water-use management which as such must share and compete for a finite resource with a variety of uses, all of whose demands for land and water are increasing as population and urban/industrial development increases. Although 'nature' cannot be given an economic price tag, conservation of natural resources must nevertheless increasingly be justified not only on the basis of the socio-economic opportunity costs or benefits it incurs in a particular time–space location but on a long-term global basis. Sustaining maximum global biological diversity and sustaining the renewability of the biosphere will in the long run be dependent on maintaining the quality of the environment of the atmosphere, the hydrosphere and the lithosphere.

Assignments

A2.1 First law of thermodynamics: 'Energy is neither created nor destroyed but may be transformed from one form to another.'
Second law of thermodynamics: 'Energy transformation is from a highly organised and concentrated form to a less organised and dispersed form.'
Explain how energy input to and output from the biosphere conforms to these two laws.

A2.2 Assess the significance of the hydrological cycle for the biosphere.

A2.3 Compare and contrast the role and circulation of carbon and nitrogen in the biosphere.

A2.4 Analyse the factors affecting the routes and rates of geochemical cycling.

A3.1 Explain the factors affecting inter- and intraregional variations in the length of the growing season and the accumulated temperatures.

A3.2 Discuss the concepts of habitat, niche and competition with reference to (a) animals and (b) plants. Consider the way in which these concepts are interrelated.

A4.1 Assess the validity of explanations that have been proposed for disjunct ranges of biotic taxa.

A4.2 Discuss the statement: 'Endemism is an important key to the understanding of existing biotic distributions.'

A4.3 Critically review methods of analysing environmental changes in the Quaternary period.

A4.4 Environment: the evolutionary template. Discuss.

A5.1 Why is it difficult to distinguish between density-dependent and density-independent factors affecting population growth?

A5.2 Compare and contrast fire and herbivores as ecological variables.

A5.3 To what extent are the community and continuum concepts of vegetation mutually exclusive?

A6.1 Consider the ways in which the biological efficiency of environmental resource use might be expressed. Discuss the problems of quantifying the expressions identified.

A6.2 Analyse Table 6.1 graphically and explain the principal intra- and inter-ecosystem relationships between absolute and relative primary productivity, and biomass.

A6.3 Explain why crop yield does not necessarily reflect potential net primary biological productivity.

A7.1 Table A7.1 shows the rate of litter fall and disappearance (gm^2) in two types of vegetation. Analyse graphically and explain the differences shown.

Table A7.1

Months		$N<D$	$A-S$	*ND*	*AS*	*ND*	*AS*
Old litter	A	80	40	10	50	40	150
Fresh litter	A	420	—	320	—	300	—
Old litter	B	220	200	190	280	226	400
Fresh litter	B	660	—	500	—	630	—
Years			(1)		(2)		(3)

A7.2 Contrast the roles of mull and mor humus in nutrient cycling.

A7.3 Assess the effect of human activities of cycling of (a) phosphorus and (b) nitrogen.

A8.1 Assess the relative role of biotic and abiotic factors in the course of plant succession.

A8.2 Explain the factors that may affect the nature and rate of vegetation succession on a coal bing (spoil heap), an abandoned railway, an old sand and gravel working, and a clear-felled forest site.

A8.3 Explain the changes in the associated flora, fauna and soil that might be expected to result from the replacement of a deciduous woodland with a coniferous plantation.

A8.4 Analyse the relationship between climate and soil formation.

A9.1 Mean annual (MAP) and current annual production (CAP) of a Scots pine stand. Plot the data shown in Table A9.1 to show the difference between MAP and CAP. At what age does this stand attain (a) ecological maturity, (b) maturity for timber production? Consider the factors that might affect these two values.

A9.2 Analyse and discuss the variables which may account for differences in the biomass characteristics of the selected woodlands given in Table A9.2.

A9.3 Graph the data in Table A9.3 and comment on and suggest explanations for the inter- and intra-species variations shown.

A9.4 Compare and contrast the impact of human activities on the tropical rain forest and temperate forest.

Table A9.1

Age (*years*)	*MAP*	*CAP*
	($t\ ha^{-1}\ year^{-1}$)	
10	0.5	1.0
20	2.0	4.0
30	5.0	14.0
40	8.0	16.0
50	9.5	14.0
60	10.0	12.0
70	10.0	10.0
80	10.0	7.0
90	9.7	6.0
100	9.5	5.0
120	9.4	4.0
130	9.3	4.0
140	9.3	3.5
150	8.1	3.0
160	7.9	3.0

Table A9.2 Difference between mean monthly air temperatures within forest and those above (+) and below (−) in the open (°C)

	J	*F*	*M*	*A*	*M*	*J*	*J*	*A*	*S*	*O*	*N*	*D*
Scots pine	+1	0	0	0	−1	−2	−2	−2	−1	0	+1	+2
Beech	+1	0	+1	+1	−2	−4	−5	−4	−3	0	0	+1
Norway spruce	+3	+2	−1.5	−3	−2	−2	−3	−2	−2	0	0	+2

A10.1 Contrast the ecological significance of snow and permafrost in the taiga and tundra ecosystems.

A10.2 Compare and contrast nutrient cycling in the taiga and tundra ecosystems.

A10.3 Why is the tundra ecosystem so vulnerable to the impact of human activities?

A11.1 Analyse the ecological role and status of grasses in (a) sand-dune succession, (b) rough grazing in Britain, (c) agro-ecosystems.

A11.2 Compare and contrast the relative importance of 'predisposing causal, resulting and maintaining factors' in affecting the character and establishment of temperate and tropical grasslands.

A11.3 Discuss the validity of the climax concept in relation to the tropical savanna biome.

A12.1 Variability of precipitation – the controlling factor in desert ecosystems. Discuss.

A12.2 Analyse the differences in adaptive strategies of evader and tolerator organisms in desert environments.

Table A9.3

Trees	*Pinus nigra*	*Pinus sylvestris*	*Betula verrucosa*	*Quercus borealis*	*Picea abies*	*Nothofagus truncata*	*Pseudotsuga taxifolia*	*Evergreen gallery forest*
Location	NE Scotland	E England	Moscow	Minnestoa USA	Sweden	New Zealand	Washington State, USA	Thailand
Status	Plantation	Plantation	Natural	Natural	Natural	Natural	Natural	Natural
Age (years)	48	55	67	57	58	110	52	—
Height (m)	14	16	26	17	17	21	17	19
Number trees per ha	1112	760	—	800	924	490	1157	16 200
Biomass								
Tree leaves	5.6	7.2	2.6	3.5	9.1	2.7	12.0	19.0
Tree branches	11.2	12.3	11.3	49.5	14.3	42.0	17.9	50.0
Tree trunks	95.1	96.7	156.7	111.9	85.2	224.8	174.3	225.2
Shrubs and herbs	7.0	2.6	2.0	0.6	1.0	0	0.1	0.2
Roots	34.0	34.1	43.1	15.0	60.0	39.2	12.3	88.5
Dead branches on trees	10.0	10.0	2.0	21.9	2.6	1.1	11.2	—
Organic matter on ground	22.0	45.0	3.0	36.7	78.0	16.7	117.3	3.0

A12.3 Assess the attributes of desert ecosystems that might make them more or less vulnerable to the impact of human activities.

A13.1 (a) Analyse factors other than size and remoteness for which there could be a close correlation with the species diversity of islands.
(b) Why might 'island equilibrium' be an unattainable goal?

A13.2 Consider the relationship between island biogeography and continental drift.

A13.3 Analyse the advantages and disadvantages of one large nature reserve and an equal area of several small reserves.

A13.4 From an available map of appropriate size, plot the distribution and size of existing nature reserves and assess their relative vulnerability to external impacts.

A14.1 Compare and contrast the island characteristics of oceanic islands and mountains.

A14.2 Plant species that live in alpine zones do so because they can. Discuss.

A14.3 With reference to mountains, explain why the study of ecological boundaries is so important.

A14.4 Formulate a hypothesis with regard to mountain ecosystems to be field tested in a week, a month or a year. Choose one of these periods and suggest a methodology.

A15.1 The phytoplankton are often described as 'the pastures of the sea'. Discuss.

A15.2 Consider the criteria that might be used to delimit marine (flora and fauna) biogeographical regions.

A15.3 (a) Analyse the factors affecting the role of nutrient cycling in the sea.
(b) Contrast the marine with terrestrial nutrient cycling in terms of 'pools', reservoirs and sinks.

A15.4 What parameters might you choose to identify important commercial fishing grounds?

A16.1 Assess the relative value of species richness and species equitability as expressions of diversity.

A16.2 Consider why the concepts of ecosystem stability and equilibrium are untenable at present or in the past.

A16.3 Explain the differences in the relative resilience of the tundra with the taiga and of tropical rain forest with the savanna.

A16.4 (a) Fragility can only be a relative concept. Discuss.
(b) Rank, with justification, the relative fragility of sand dune, marsh, peat-bog, birch wood and bracken ecosystems.

A17.1 Analyse the differences between wild and domesticated plants.

A17.2 Assess the role of anthropogenic factors in organic evolution. To what extent could it be said that the production of cultivars compensate for the extinction of wild species?

A18.1 Consider the problems of assessing the ecological effects of pollution.

A18.2 Compare and contrast the ecological impact of pollution of terrestrial with that of aquatic ecosystems.

A19.1 Analyse and explain the basic differences between managed and unmanaged ecosystems.

A19.2 (a) Construct a graph to illustrate the most important trends indicated in Table 19.2 and comment on their significance.
(b) With relation to Table 19.3, discuss the possible expression of productivity and efficiency of rice production that could be formulated.

A19.3 Assess the validity and value of the concept of sustained yield of biotic resources.

A20.1 Compare and contrast the possible ecosystem characteristics of a modern Western city with those of a Mediterranean and a Third World city.

A20.2 Analyse the factors that determine the characteristics of urban biota.

A20.3 Discuss the validity of the concept of the urban ecosystem.

A21.1 Home range; bottleneck; buffer zone; corridors; keystone species; endangered species. Discuss the above terms and explain their relevance for wildlife conservation.

A21.2 How far are the aims and methods of 'game' management compatible with those for wildlife conservation?

A21.3 Compare and contrast the problems of establishing and managing nature reserves in the developed and lesser developed countries.

A21.4 Conservation without management and compromise defeats its own ends. Discuss.

Glossary

abiotic non-living

acid rain precipitation with pH less than 5.0 containing high amounts of SO_2, NO_2 and other acidifying pollutants

aerobic oxygen requiring

albedo ratio between insolation (short-wave) and outgoing (long-wave) radiation from earth's surface

allelopathy inhibition of one type of organism by substances secreted or excreted by another

allogenic affected by factors external to organism or community

anion a negatively charged particle (ion)

aspection seasonal appearance of a stand of vegetation

association an actual, or type of, vegetation unit distinguished from others by its species composition

autecology the study of a species population in relation to its habitat

autogenic affected by factors intrinsic to organism or community

autotrophic organism which obtains its food energy from inorganic elements (self-feeding)

base status percentage cation exchange sites in soil colloids occupied by metallic cations (Na, Ca, Mg, NH, S)

benthos all the organisms living on a subaquatic surface

biodegradable capable of being decomposed by living organisms e.g. bacteria and fungi

biological concentration accumulation of substances in organic tissues: sometimes synonymous with magnification

biological cycle the circulation of elements from an inorganic form to an organic form by the process of photosynthesis and respiration (decomposition)

biological magnification magnification or increase in concentration of a substance in organic tissues with passage along a food chain

biological (or biochemical oxygen demand (BOD) a measure of the oxygen required to degrade organic matter in a water body; also used as an approximate measure of the amount of organic matter

biomass amount of living material (expressed in weight or energy equivalent) in a given site or condition

biome a major biotic region characterised by a distinctive flora and fauna (biota) usually associated with a distinctive climatic regime
biotic potential theoretical maximum rate of species population growth in the absence of any extrinsic limiting factors
bottleneck a point at which a specific population is reduced to numbers so low as to reduce genetic variety and/or endanger its continued existence
bulk density the mass (weight) of a unit volume of dry soil

capillary soil water water retained in soil micropores and which can move, usually upwards, as a result of capillarity (i.e. attraction of surface-water molecules in contact with a solid surface)
carrying capacity the maximum number, density or volume of organisms that can be maintained by a given environmental resource base
cation positively charged particle (ion)
catena (soil) a group of soils varying continuously in relation to an environmental gradient
cauliflory production of flowers on tree trunks
CFC chlorofluorocarbons – a group of inert gases used as aerosols, refrigerator gases, solvent cleaners, etc. which, above a certain level, can cause destruction of the ozone layer
chelation the formation of mobile, reversible, complex molecules of metallic and organic molecules
cheluviation movement down soil profile of chelates in solution
clay fraction that proportion of soil mineral matter composed of particles less than 2 μm diameter
climax vegetation (or ecosystem) assumed stage of maximum development (height and/or biomass) under prevailing environmental conditions
colloid a mixture in which very fine particles of one substance remain suspended in the other; the soil colloidal complex is formed of clay particles and humic acids
community a distinctive grouping of plant and/or animal species populations usually associated with a particular habitat
compensation point light intensity at which the rate of photosynthesis equals that of respiration
cryophyte plant adapted to grow on/in ice or snow
cryoturbation disturbance of soil by differential freezing and thawing

deme local inbreeding species population
demersal organisms living at or near the bottom of deep seas
denitrification the breakdown in the soil of nitrogen compounds into gaseous nitrogen
density (biological) numbers of individual organisms occupying a given environmental space
density-dependent factor any variable (usually biotic) whose rate is positively or negatively correlated with rate of change of population density
density-independent factor any variable whose rate is not correlated with rate of change in population density
desertification spread of arid environmental conditions as a result of climatic change or human activities
dioecious with male and female reproductive organs on the same plant
DOM dead organic matter
dominance condition when or one or more species by reason of size, number, etc. exert an influence or control on habitat and hence on type of associated species

ecosystem any group of interdependent, interacting species populations and their environment
ecotone zone of transition between two different communities
ecotype subspecies (variety or race) adapted to a particular habitat
ectothermic (see poikilotherm)
edaphic pertaining to the soil
eluviation removal of materials in suspension or solution ('leaching') from soil surface horizon downwards
endemic of organisms found in particular, and usually limited, geographical area
endemism the proportion of species found only in a particular biotic region
endotherm an organism able to increase its 'core temperature' above that of its environment
energy capacity of doing work inherent in a body or system (see kinetic energy)
epiphyte a plant which uses another for support
eutrophic nutrient rich (usually applied to water)
eutrophication nutrient enrichment by natural or human activities (i.e. pollution)
evapotranspiration loss of water from combined soil surface and vegetation cover by plant transpiration and atmospheric evaporation

faculative organism able to exist in more than one habitat or on more than one prey
fallow land left uncultivated after a period, or between periods, of cultivation
fauna the assemblage of animal taxa on a given area
fecundity the number of young produced by an organism during its life
fertility capability of reproduction by an organism or of biological production by a soil
field capacity (soil) the amount of water held by a soil after drainage of all the gravity water
flora the assemblage of plant taxa in a given area
flux rate the rate of flow of energy or matter from one state to another
food chain a sequence of groups of food-dependent organisms
food web a set of food chains linking species at different trophic levels

garrigue low, open, shrubby Mediterranean vegetation, often associated with limestone
gene physical units of heredity that carry information from one cell/generation to another
gene pool sum total of all genes in a breeding population at a particular time
geophyte perennial plant with underground storage organs
gley poorly drained soil in which iron occurs in reduced ferrous state imparting a bluish colour
greenhouse gas gases such as CO_2, CH_4 and water vapour which trap short-wave isolation and reradiate it as heat to the earth's surface and atmosphere
guano dry accumulation of excrement rich in nitrogen (particularly uric acid) produced by birds and reptiles

halophyte a plant tolerant of high salt concentration in soil, air or water
heliophyte plant (with high light compensation point) adapted to high intensity of sunlight
heterotroph organism dependent on organic food (other feeding)
heterozygosity pertaining to an organism which contains two different forms of the same gene in all diploid cells (i.e. with two pairs of male and female chromosomes)
homeostasis maintenance, within limits, of a relatively constant condition or function by an organism, a population or a community in face of changing external environmental conditions as a result of negative feedback mechanisms

humus amorphous colloidal organic substance produced as the end-product or organic decomposition in the soil
hydrophyte plant adapted to growing in water
hygrochasy inhibition of seed germination of one plant species by the chemical action of another

illuviation process whereby substances are deposited from a higher to a lower soil horizon
insolation incoming (short-wave, infra-red) solar radiation
iron pan a thin iron-indurated soil horizon

keystone species organism (usually animal) which exerts a particularly powerful influence in a community
kinetic energy (E_k) or 'free energy' continually dissipated as heat, expressed as $E_k = \frac{1}{2}MV^2$, where M = mass and V = velocity of moving body
krill the main food of whalebone whales in Antarctic seas; composed largely of small pelagic shrimp-like euphausids
krumholtz dense thicket of small recumbent or prostrate trees (knee-pine) at the mountain tree-line
K-selected organisms in which selection is for conservative efficient use of resources, usually perennial plants and long-generation animals

laterite residual product of humid tropical weathering formed of Fe and Al which hardens on exposure to form a duricrust
latosol deep intensely weathered and leached soils rich in Fe, Al and Mn, characteristic of humid tropics
leaching removal in downwards suspension or solution of materials from the upper soil horizon
leaf area index (LAI) ratio of area of leaf surface to underlying ground area
legume member of pea family Leguminosae (peas, beans, lentils, etc.); most have nodulated roots
ley meadow or temporary sown grass crop
loess fine, wind-deposited silt

marsupial mammals with a pouch into which young are born in an undeveloped state
maquis dense evergreen shrub vegetation in Mediterranean regions
melanism dark skin caused by excess of dark-brown pigment (melanin)
meristem area of a plant in which cell division (hence growth) occurs
microclimate climate 'near ground' where atmospheric conditions are modified by natural or man-made structures
mineralomass the total amounts (weight) of mineral elements in a given amount of soil
moder decaying organic matter intermediate in form and composition to mor and mull
mor acid, nutrient-deficient, partially decomposed organic matter formed on soil surface
mull mild, nutrient-rich, well-decomposed, well-distributed soil organic matter
mycorrhiza symbiotic association between a fungus and the roots of another higher plant

necrosis localised death of plant or animal tissue
necton free-swimming animals in the upper (pelagic) layers of the oceans
niche the multidimensional environmental space occupied and used by a species within an ecosystem
nitratophile plants with high nitrogen requirements

nitrification conversion of organic nitrogen into inorganic nitrates by soil bacteria
nitrogen fixation the use of gaseous nitrogen by free-living or symbiotic soil bacteria to synthesise organic nitrogen compounds

obligate organism specific to a particular habitat or prey
oligotrophic nutrient deficient
oligotrophication process whereby a medium becomes nutrient deficient
optimum yield maximum amount of biomass or number of organisms that can be cropped without reducing the breeding capacity of the population
oxidation addition of oxygen to the molecule of a substance, or more generally any reaction during which electrons are lost from an atom
ozone (O_3) triatomic form of oxygen formed by ultraviolet radiation of O_2
ozone layer that area in the upper atmosphere (ozonosphere) where there is a sufficient concentration of O_3 to absorb incoming ultraviolet radiation

paludification (from paludal or marshy) creation of marshy ground conditions
PAN peroxyl-acetyl-nitrate, secondary pollutant produced by photochemical reactions of primary pollutants in atmosphere
parasite an organism which lives on another to the latter's detriment
peat organic deposit usually over 50 cm thick formed as a result of partial decay of organic matter in a saturated site
pelagic of the upper zone of open sea
permafrost ground permanently frozen for over 2 years
pest any organism detrimental to human comfort, health, economic activities
photoperiodism reaction of organisms to periodic variation in length of exposure to light
phraeophyte organisms, usually plant, dependent on or capable of reaching ground-water
phytofagus plant eating
placental animals whose young form from placental tissues in the mother's body
plagioclimax (or deflected climax) end-point of organic succession under influence of factors other than climate (i.e. fire, grazing, etc.)
plankton organisms whose movement is dependent on water in which they live; phyto – plants; zoo – animals
plough pan compacted, relatively impermeable soil horizon as a result of continuously ploughing at a constant depth
polyploid organisms whose cells contain more than 2N chromosomes
podzol soil profile characterised by whitish-grey intensely podzolised and leached A horizon and an iron/aluminium humus-enriched B horizon
podsolisation the formation in organic acids of soluble chelates (organic-metallic complexes of Fe, Al), and the dissociation of clay humus colloids, susceptible to leaching
production sometimes used synonymously with productivity or for total biomass produced in a growing season or year
productivity (biological) increase in amount of biomass or energy equivalent produced by an organism or a community per unit time per unit area
prokaryote (or procaryote) unicellular organism (neither bacteria or blue-green algae) whose cells lack an enclosed nucleus
putak lattice of ice crystals 3–10 mm long formed at base of snow layer
pyrophyte plant adapted to survive, and/or whose reproduction is dependent on, burning

range the geographical area within which an organism is present

ranker (soil) young soil characterised by an A and C horizon
reduction (chemical) process whereby oxygen is lost from a compound and hydrogen or other anion gained
r-selected species organisms in which selection is for opportunistic strategies in which a high rate of increase is a key characteristic; annual plants and short-generation animals
ruderal plant living on 'waste' or open ground: often weeds

saprophyte an organism that feeds on dead/decaying organic matter
sciaphyte shade-tolerant plant
selva tropical rain forest
sere stage in the development of vegetation
solifluction 'soil flow', i.e. movement of saturated soil downslope under influence of gravity
SOM soil organic matter
stand a discrete unit of vegetation which is distinguishable from others by the homogeneity of form or composition
stom(ata) openings in epidermis of leaves and some stems through which gaseous and water exchange takes place
structure (soil) tilth; the size and shape of the compound mineral particles or aggregates which comprise the soil fabric
succession sequence of stages (seres) in the development of vegetation from colonisation to climax
symbosis the co-habitation of two or more organs to their mutual benefit (e.g. mutualism)
syneocology study of communities (usually plant)

taiga (Fin. swamp forest) name given to Boreal forest
texture (soil) feel of a soil dependent on relative proportion of mineral particles of a particular size class, e.g. sandy, silty, clayey
thermocline the usually sharp boundary between warm upper and lower colder water
thermokarst ground surface depressions caused by variable thawing of permafrost
thermoperiodism response of organisms to periodic temperature variations
tiller a shoot formed at the base of a plant; common in grasses and cereals
trace element (micronutrient) an element necessary in a very small amount for any biological process
transpiration loss of water from green parts of plants via stomata
trophic level that stage at which food energy passes from one group of organisms to another, e.g. producers–herbivores–carnivores
tropophyte plant that does not respond to changes in climate conditions

viviparous (plants) seeds germinating in fruiting body while still attached to plant

water-table upper level of ground-water

xerophyte plant adapted to grow in dry areas
xylopodium woody tuberous growth (lignotuber) produced at base of bole of some trees

References

Allaby M (1986) *Green Facts: The Greenhouse Effect and Other Key Issues*. London, Hamlyn

Allaby M (1989) *Guide to Gaia*. London, MacDonald Optima

Allaby M, Lovelock J (1983) *The Great Extinction*. London, Secker & Warburg

Allen K R (1980) *Conservation and Management of Whales*. Seattle, University of Washington Press

Anderson E (1967) *Plants, Man and Life*. Berkeley & Los Angeles, University of California Press

Anderson J M, MacFayden A (eds) (1976) *The Role of Terrestrial and Aquatic Organisms in Decomposition Processes*. Oxford, London, Melbourne, Edinburgh, Blackwell Scientific Publications

Anderson M S (1951) *A Geography of Living Things*. London, English Universities Press

Anderson R M, May R M (eds) (1982) *Population Biology and Infectious Diseases*. Berlin, New York, Springer-Verlag

Atkins Research and Development (n.d.) *A Monograph on Water Pollution – its Dispersion and Effects*. Epsom, Surrey

Aubreville A (1938) La forêt coloniale: les forêts de l'Afrique occidentale française. *Annales Academie des Sciences Coloniales. Paris* **9**: 1–245

Auclair D (1983) 3.2 Natural mixed forests and artificial monospecific forests. In: Moon H A and Gordon M (eds) *Disturbance and Ecosystems – Components of a Response*. Berlin, Springer-Verlag: 71–82

Austin M P (1985) Continuum concept, ordination methods and niche theory. *Annual Review of Ecology and Systematics* **16**: 39–61

Bagnouls G, Gaussen H (1957) Les climats biologiques et leur classification. *Annales de Géographie* **66**: 193–220

Baker H G (1974) The evolution of weeds. *Annual Review of Ecology and Systematics* **5**: 1–24

Bandry J (1988) Hedgerows and hedgerow-networks as wild life habitats in Europe. In: Park J R (ed) *Environmental Management in Agriculture*. London, Belhaven Press: 111–24

Bannister P (1976) *Introduction to Physiological Plant Ecology*. Oxford, Blackwell Scientific Publications

Barnes R K, Mann K H (1980) *Fundamentals of Aquatic Ecosystems*. Oxford, Blackwell Scientific Publications

Barnes R S K (1984) *Estuarine Biology*. Studies in Biology 49. London, Edward Arnold Biology

Barry R G, L van Wie C (1974) Topo- and microclimatology in alpine areas. In: Barry R G, Ives J D (eds) *Arctic and Alpine Environments*. London, Methuen: 72–83

Barton H N (1988) 7. Speciation. In: Myers A A, Gillen P S (eds) *Analytical Biogeography*. London and New York, Chapman & Hall: 185–218

Bates M (1964) *Man in Nature* (2nd edn). Englewood Cliffs NJ, Prentice-Hall

Battarbee R W *et al.* (1988) *Lake Acidification in the United Kingdom 1800–1986, Evidence from Analysis of Lake Sediments*. Prepared for the Department of the Environment/Palaeoecology Research Unit, Department of Geography, University of London. London, Ennis Publishing

Bazzaz F A (1983) Characteristics of populations in relation to disturbance in natural and man-modified ecosystems. In: Mooney H A and Gordon M (eds) *Disturbance and Ecosystems – Components of a Response*. Ecological Studies 44. Berlin, Springer-Verlag: 259–76

Beadle N W C (1951) The misuse of climate as an indicator of vegetation and soils. *Ecology* **32**: 343–45

Beatley J C (1974a) Phenological events and their environmental triggers in Mojave Desert ecosystems. *Ecology* **55**: 856–63

Beatley J C (1974b) Effects of rainfall and temperatures on the distribution and behaviour of *Larrea tridentia* (creosote bush) in the Mojave Desert of Nevada. *Ecology* **55**: 245–61

Bell M K (1974) Decomposition of herbaceous plants. In: Dickinson C H, Pugh C J F (eds) *Biology of Plant Litter Decomposition*. New York, Academic Press: 37–68

Benecke N, Davis M R (1980) *Mountain Environments and Sub-alpine Tree Growth*. Christchurch, New Zealand Forest Service, Forest Research Institute

Beverton R S H (1962) Long-term dynamics of certain North Sea fish populations. In: Le Cren E D, Holdgate M W (eds) *The Exploitation of National Populations*. Oxford, Blackwell Scientific Publications: 242–59

Billings W D (1974) Arctic and alpine vegetation: plant adaptations to cold summer climates. In: Barry R C, Ives, J D (eds) *Arctic and Alpine Environments*. London, Methuen: 403–43

Birch D (1981) Dominance in marine ecosystems. *American Naturalist* **118**(2): 262–83

Birks H J B, West R G (eds) (1973) *Quaternary Plant Ecology*. Oxford, Blackwell Scientific Publications

Birot P (1965) *Les Formations Végétales du Globe*. Paris, SEDES

Bjorkmann O (1973) Comparative studies of photosynthesis in higher plants. In: Giese A C (ed) *Photophysiology* Vol. 8. New York, Academic Press: 7–63

Bjorkman O, Berry J (1973) High efficiency photosynthesis. *Scientific American* **229**(4): 180–93

Bliss L C (1975) Arctic tundra ecosystems. *Annual Review of Ecology and Systematics* **4**: 359–400

Bliss L C (1981) 1.2 North American and Scandinavian tundras and polar deserts. In: Bliss L C, Heal O W, Moore J J (eds) *Tundra Ecosystems: A Comparative Analysis*. Cambridge, Cambridge University Press: 8–24

Bliss L C, Heal O W, Moore J (1981) *Tundra Ecosystems: A Comparative Analysis*. Cambridge, Cambridge University Press

Bolin B (1970) The carbon cycle. *Scientific American* **223**(8): 124–30

Bolin B, Cook R B (eds) (1983) *The major biogeochemical cycles and their interactions*. SCOPE 21. Chichester, John Wiley & Sons

Bolin B, Döös B R, Jäger J N, Warwick R A (eds) (1986) *The Greenhouse Effect, Climate change and Ecosystems*. Chichester, John Wiley & Sons

Boney A D (1989) *Phytoplankton* (2nd edn). New Studies in Biology. London, Melbourne, Auckland, New York, Edward Arnold

Bormann F H, Likens G E (1979) *Pattern and Process in Forested Ecosystems*. New York, Springer-Verlag

Bornkamm R (1981) Rates of change in vegetation during secondary succession. *Vegetatio* **47**: 213–20

Boyden S, Millar S, Newcombe K, O'Neill B (1981) *The Ecology of a City and its People: The Case of Hong Kong*. Canberra, Australian National University Press

Bradshaw A D, Goode D A, Thorp E (1986) *Ecology and Landscape Design*. British Ecological Society Symposium Vol. No. 14. Oxford, Blackwell Scientific Publications

Brady N C (1974) *The Nature and Properties of Soils*. (8th edn). New York, Macmillan

Braun-Blanquet J (1932) *Plant Sociology* (trans. revised and ed. by Fuller G D and Conrad H S). New York, MacGraw-Hill

Bray R J, Curtis J T (1957) An ordination of upland forest communities of Southern Wisconsin. *Ecological Monographs* **27**: 325–49

Bridges E M (1978) Soil – the vital skin of the earth. *Geography* **63**: 254–61

Briggs D, Walters S M (1984) *Plant Variation and Evolution* (2nd edn). Cambridge, Cambridge University Press

Brown J, Miller P C, Tieszen L L, Bunnell F L (eds) (1980) *An Arctic Ecosystem*. Stroudsburg, Pa, Dowden, Hutchinson & Ross

Brown J H, Kodric-Brown A (1977) Turn-over rates in insular biogeography: effect of immigration on extinction *Ecology* **58**: 445–9

Brown V K (1985) Insect herbivores and plant succession. *Oikos* **44**: 17–22

Brown V K, Southwood T R F (1987) Secondary succession: patterns and strategies. In: Gray A J, Crawley M J, Edwards P J (eds) *Colonisation, Succession and Stability*. Oxford, Blackwell Scientific Publications: 315–38

Budyko M (1963) *Weather and climate*. New York, McGraw-Hill

Bunnell F L (1981) Ecosystem synthesis – a 'fairytale'. In: Bliss L C, Heal O W, Moore J J (eds) *Tundra Ecosystems: A Comparative Analysis*. Cambridge, Cambridge University Press: 635–6

Bunney S (1990) Mammoth killers could have done it with stone. *New Scientist* **125**(1702): 35

Bunt J S (1975) Primary productivity of marine ecosystems. In: *Primary Productivity of World Ecosystems*. Washington DC, National Academy of Sciences: 169–202

Bunting A H (1975) Time, phenology and the yield of crops. *Weather* **30**: 312–25

Burgess R L, Sharpe D M (eds) (1988) *Forest Island Dynamics in Man Dominated Landscapes*. New York, Springer-Verlag

Burnett J D (ed) (1964) *The Vegetation of Scotland*. Edinburgh, Oliver & Boyd

Butzer K W (1974) *Accelerated Soil Erosion – A Problem of Man-Land Relations*. Perspectives on Environment. Washington, American Association of Geographers

Cain R (1974) *The Geography of Flowering Plants* (4th edn). Harlow, Longman

Cairns Jr J (1980) *The Recovery Process in Damaged Ecosystems*. Michigan, Ann Arbor Science Publishers

Caldwell M M (1975) Primary production of grazing lands. In: Cooper J P (ed) *Photosynthesis and Productivity in Different Environments*. Cambridge University Press: 41–75

Caldwell M M (1981) Plant responses to ultra-violet radiation. In: Lange O L, Nobel P S, Osmund C B, Zeigler H (eds) *Physiological Plant Ecology I.* Berlin, Springer-Verlag: 169–97

Carlquist S (1965) *Island Life*. New York, American Museum of National History

Carlquist S (1974) *Island Biology*. New York, Columbia University Press

Carson R (1963) *Silent Spring*. London, Hamish Hamilton

Cassola F (1979) Shooting in Italy: the present situation and future perspectives. *Biological Conservation* **12**: 122–34

Cherritt J M (ed) (1989) *Ecological Concepts*. Oxford, Blackwell Scientific Publications

Cicerone R J (1987) Changes in stratospheric ozone. *Science* **237**: 35–42

Clapham Jr W B (1973) *Natural Ecosystems*. London, Collier-Macmillan

Clapham Jr W B (1981) *Human Ecosystems*. New York, Macmillan Publishing Co, London, Collier-Macmillan

Clark, W C, Munn R E (eds) (1986) *Sustainable Development of the Biosphere.* Cambridge, Cambridge University Press, International Institute for Applied Systems Analysis, Luxembourg, Austria

Clements F E (1904) *The Development and Structure of Vegetation*. Botanical Survey of Nebraska 7. The Botanical Seminar, Lincoln, Nebraska

Clements F E (1916) *Plant Succession.* Washington, Carnegie Institute of Washington

Clements F E (1936) The nature and structure of the climax: Ecology **24**: 253–48

Cloud P, Gibor A (1970) The oxygen cycle. *Scientific American* **223**(3): 111–23

Clout H, Wood P (1986) *Wild Life and the City*. Harlow, Longman

Clutton-Brock J (1980) *Domesticated Animals from Early Times*. London, British Museum and Heinemann

Clutton-Brock J (1992) How wild beasts were tamed. *New Scientist* **133**(1808): 41–5

Cody M L, Diamond J M (eds) (1975) *Ecology and the Evolution of Communities*. Cambridge Mass, Harvard University Press

Coghlan A (1990) Pigeons, pests and people. *New Scientist* 1 Dec: 48–51

Cole B J (1980) Trophic structure of a grassland community. *Nature* **288**: 76–7

Cole Lamont C (1958) The ecosphere. *Scientific American* **198**: 83–92

Cole M M (1986) *The Savnnas Biogeography and Geobotany*. London, Academic Press

Colemann D C (1976) A review of root production processes and their influence on soil biota in temperate ecosystems. In: McAnderson J M, McFayden A (eds) *The Role of Terrestrial and Aquatic Organisms in the Decay Process.* Oxford, Blackwell Scientific Publications

Coleman D C, Andrews R, Ellis J E, Singh J B (1976) Energy flow and partitioning in selected man-managed and natural ecosystems. *Agro-ecosystems* **3**: 45–54

Connell J H, Slayter R D (1977) Mechanisms of succession in natural communities and their role in community stability and organisation. *American Naturalist* **111**: 1119–44

Connor E F, Simberloff D W (1979) The assembly of species communities: chance or competition. *Ecology* **60**: 1132–40

Cooper J (1975) *Photosynthesis and Productivity in Different Environments.* Cambridge, Cambridge University Press

Cooper W S (1931) A third expedition to Glacier Bay, Alaska. *Ecology* **12**: 61–95

Cooper W S (1939) A fourth expedition to glacier bay, Alaska. *Ecology* **20**: 130–55

Cottam G (1981) Patterns of succession in different forest ecosystems. In: West D C, Slingart H H, Botkin D B (eds) *Forest Succession Concepts and Applications*. New York, Springer-Verlag: 178–84

Coughtrey P J, Martin M H, Unsworth M I (1987) *Pollutant Transport and Fate in Ecosystems*. Special Publication No. 6 of the British Ecological Society. Oxford, Blackwell Scientific Publications

Coupland R T (1979) *Grasslands of the World*. International Biological Programme 18. Cambridge, Cambridge University Press

Cousens J (1974) *An Introduction to Woodland Ecology*. Edinburgh, Oliver & Boyd

Cowles H C (1901) The physiographic ecology of Chicago and vicinity: a study of the origin, development and classification of plant communities. *Botanical Gazette* **27**: 95–11; 167–202; 281–308; 361–91

Cox C B (1974) Vertebrate palaeodistributional patterns and continental drift. *Journal of Biogeography* **1**: 75–94

Cox C B, Healy I N, Moore P D (1973, 1976) *Biogeography: An Ecological and Evolutionary Approach*. Oxford, Blackwell Scientific Publications

Crawford A K, Liddle M D (1977) The effect of trampling on neutral grassland. *Biological Conservation* **12**: 135–42

Crawley M J (1983) *Herbivory: The Dynamics of Animal–Plant Interactions*. Oxford, Blackwell Scientific Publications

Crawley M J (ed) (1986) *Plant Ecology*. Oxford, Blackwell Scientific Publications

Crisp D J (1975) Secondary productivity in the sea. In: *Productivity of World Ecosystems*. Washington DC, National Academy of Science: 71–89

Crocker K L, Mayor J (1955) Soil development in relation to vegetation and surface age at Glacier Bay, Alaska. *Journal of Ecology* **43**: 427–8

Croizat L (1952) *Manual of Phytogeography*. The Hague, Junk

Cromack K (1981) Below ground processes in forest succession. In: West B E, Shugart A H, Botkin D B (eds) *Forest Succession: Concepts and Applications*. New York, Springer-Verlag: 161–73

Crowther J G (1969) Von Humboldt: exploror of a new world. *New Scientist* **42**: 534–6

Cruickshank, J G (1972) *Soil Geography*. Newton Abbot, David and Charles

Cumming D D M (1982) The influence of large herbivores on savanna structure in Africa. In: Huntley B J, Walker B H (eds) *Ecology of Tropical Savannahs*. Berlin, Springer-Verlag: 217–45

Curtis J T (1956) The modification of mid-latitude grasslands and forests by man. In: Thomas W L (ed) *Man's Role in Changing the Face of the Earth*. Chicago, University of Chicago Press: 721–36

Curtis J T (1957) *The Vegetation of Wisconsin: An Ordination of Plant Communities*. Madison, Wisconsin, University of Wisconsin Press

Curtis L F, Courtney R M, Trudgill S (1976) *Soils in the British Isles*. Harlow, Longman

Cushing D H (1980) *Fisheries Biology: A Study in Population Dynamics* (2nd edn). Madison, Wisconsin, Madison University Press

Darlington P J (1957) *Zoogeography: The Geographical Distribution of Animals*. New York, Wiley

Darlington P J (1963) *Chromosome Botany and the Origins of Cultivated Plants* (2nd edn). London, Allen & Unwin

Daubenmire R F (1968) Ecology of fire in grasslands. *Advances in Ecological Research* **57**: 209–66

Davies D B, Eagle D J, Finney J B (1971) *Soil Management*. Ipswich, Suffolk, Farming Press

Davis, M B (1981) Quaternary history and the stability of forest communities. In: West D C, Shugart H H, Botkin B D (eds) *Forest Succession Concepts and Applications*. New York, Springer-Verlag: 132–53

Davis W M (1899) The geographical cycle. *Geographical Journal* **14**: 481–504

Dawkins R (1988) *The Blind Watchmaker*. London, Penguin Books Ltd

De Angelis D L (1980) Energy flow, nutrient cycling and ecosystem resilience. *Ecology* **61**: 764–7

De Angelis D L, Post III W M (1990) *Nutrient Cycling and Food Web Dynamics*. New York, Chapman & Hall

De Candolle A (1885) *Géographie Botanique Raisonnée en Exposition des Faits Principaux et des Lois Concernant la Distribution des Plantes d'Epoque Actuelle*. Paris, Masson et Cie

Deevy E S (1970) Mineral cycles. *Scientific American* **223**(3): 148–59

Deevy E S (1971) The human population. In: *Man and the Ecosphere. Readings from Scientific American*. San Francisco, W M Freeman: 49–56

Delcourt P A, Delcourt H R (1987) *Long Term Forest Dynamics of the Temperate Zone*. Ecological Studies Vol. 63. New York, Springer-Verlag

Dell B, Hopkins A J M, Lamont B B (1986) *Resilience of Mediterranean-type Ecosystems*. Dordrecht, Junk

Delwiche C C (1974) The Nitrogen Cycle. In: *Ecology, Evolution and Population Biology. Readings from Scientific American*. San Francisco, W H Freeman: 237–40

Denno R F, McClure S (1983) *Variable Plants and Herbivores in Natural and Managed Systems*. New York, Academic Press

Detling J K (1988) 7. Grasslands and savannas: regulation of energy flow and nutrient cycling in herbivores. In: Pomeroy L R, Alberts J J (eds) *Concepts of Ecosystem Ecology*. New York, Academic Press: 33–148

Diamond J M (1974) Colonisation of exploded volcanic islands by birds: the supertramp strategy. *Science* **184**: 803–6

Diamond J M (1975) The island dilemma: lessons of modern biogeographic studies for the design of nature reserves. *Biological Conservation* **7**(2): 129–46

Diamond J M (1984) Biogeographic mosaics in the Pacific. In: Rodovsky F J, Raven P H, Shomers S H (eds) *Biogeography of the Tropical Pacific*. Lawrence, Kansas, Association of Systematic Collections.

Diamond J M (1987) Human use of world resources. *Nature* **528**: 479–80

Diamond J M (1988) Urban extinction of birds. *Nature* **333**: 393–4

Diamond J, Case J T (eds) (1986) *Community Ecology*. New York, Harper & Row

Diamond J M, May R M (1976) Theoretical ecology. In: May R M (ed) Oxford, Blackwell Scientific Publications: 163–87

Di Castri F, Goodall D W, Specht R C (eds) (1981) *Mediterranean-type Shrublands: Ecosystems of the World* Vol. II. Amsterdam, Elsevier

Dickinson C H, Pugh G J F (eds) (1974) *Biology of Plant Litter Decomposition*. New York, Academic Press

Dickman C R (1987) Habitat fragmentation and vertebrate species richness in urban environments. *Journal of Applied Ecology* **24**: 347–51

Dimbleby G W (1962a) The development of British heathlands and their soils. *Oxford Forestry Memoirs* No. **23**: 1–121

Dimbleby G W (1962b) Post-glacial changes in soil profile. *Proceedings of the Royal Society London* B **161**: 355–62

Dix H M (1981) *Environmental Pollution*. Chichester, John Wiley and Sons

Dix R L :1964) A history of biotic and climatic changes within the North American grassland. In: Crisp D S (ed) *Grazing Terrestrial and Marine Environment*. Oxford, Blackwell Scientific Publication: 71–90

Dorney R S, McLellan P W (1984) The urban ecosystem: its spatial structure, its scale relationships and its sub-system attributes. *Environments* **16**(1): 9–20

Douglas I (1983) *The Urban Environment*. London, Edward Arnold

Dowding P, Chapin III F S, Wielgolaski F E, Kilfeather P (1981) Nutrients in tundra ecosystems. In: Bliss L C, Heal O W and Moore J J (eds) *Tundra Ecosystems*. Cambridge, Cambridge University Press: 647–84

Drude D (1897) *Manual de Géographie Botanique*. Paris, Klincksieck
Drury W H, Nisbet I C T (1973) Succession. *Journal of the Arnold Arboretum* **53**: 331–68
Dryness C T, Viereck L A, van Cleve K (1986) Fire in taiga communities of interior Alaska. In: van Cleve K (ed) *Forest Ecosystems in the Alaskan Taiga*. New York, Springer-Verlag: 74–88
Duchafour P (1965) *Précis de Pédologie*. Paris, Masson et Cie
Duchafour P (1976) Dynamics of organic matter in soils of temperate regions and its action on pedogenesis. *Geoderma* **15**: 31–40
Dunbar M J (1973) Stability and fragility in Arctic ecosystems. *Arctic* **26**(3): 179–85
Duvigneaud P, Denaeyer-de Smet S (1970) Biological cycling of minerals in temperate deciduous forests. In: Reichle D (ed) *Analysis of Temperate Forest Ecosystems*. London, Chapman and Hall; Berlin, Springer-Verlag: 199–229
East R (1981) Area requirements and conservation status of large African animals. *Nyala* **7**: 3–20
Eden M J (1974) Palaeoclimatic influences and the development of the savannas in southern Venezuela. *Journal of Biogeography* **1**: 95–109
Edlin H L (1976) *The Natural History of Trees*. London, Weidenfeld & Nicolson
Edson M M, Foin C T, Knapp C M (1981) Emergent properties and ecological research. *American Naturalist* **118**: 593–6
Edwards P J (1989) Insect herbivory and plant defence theory. In: Grubb P J, Whittaker B (eds) *Toward a More Exact Ecology*. Oxford, Blackwell Scientific Publications: 275–99
Ellenberg H (1971a) Introductory survey. In: Ellenberg H (ed) *Integrated Experimental Ecology*. New York, Chapman Hall: 1–15
Ellenberg H (1971b) Nitrogen content, mineralisation and cycling. In: Duvigneaud P (ed) *Productivity of Forest Ecosystems*. Proceedings Brussels Symposium 1969. Paris, UNESCO: 509–14
Ellenberg H (1979) Man's influence on tropical mountain ecosystems in South America. *Journal of Ecology* **67**: 401–16
Elliot-Fisk D L (1983) The stability of the northern Canadian tree limit. *Annals of the Association of American Geographers* **73**: 560–76
Elton C S (1958) *The Ecology of Invasions by Plants and Animals*. London, Methuen
Elton C S (1966) *The Pattern of Animal Communities*. London, Methuen
Evans F C (1956) Ecosystem as the basic unit in ecology. *Science* **123**: 1227–8
Evans F S, Dahl F (1955) The vegetation structure of an abandoned field in South-East Michigan in relation to environmental factors. *Ecology* **36**: 685–706
Evenari M (1985a) The desert environment. In: Evenari M, Noy-Meir I, Goodall D (eds) *Hot Deserts and Arid Shrublands. Ecosystems of the World* Vol. 12A. Amsterdam, Elsevier Scientific Publications: 1–22
Evenari M (1985b) Adaptations of plants and animals to the desert environment. In: Evenari M, Noy-Meir I, Goodall D (eds) *Hot Deserts and Arid Shrublands. Ecosystems of the World* Vol. 12A. Amsterdam, Elsevier Scientific Publications: 79–92
Evenari M, Noy-Meir I, Goodall D W (1985) *Hot Deserts and Arid Shrublands A and B. Ecosystems of the World* 12A/B. Amsterdam, Elsevier Scientific Publications
Ewel J J, Berish C, Brown B, Price N, Raich J (1981) Slash and burn impacts in a Costa Rican wet forest site. *Ecology* **62**: 816–29
Eyre S R (1968) *World Vegetation and Soils* (2nd edn). London, Edward Arnold
FAO (1978) Agro-ecological zones. Vol. 1. *Methodology and Results for Africa*. Rome, FAO
FAO/UNEP (1975) *Pilot Study on Conservation of Animal Genetic Resources*. Rome, FAO
Farnworth E G, Golley F B (eds) (1974) *Fragile Ecosystems: Evaluation of Research and Applications in the Neotropics*. New York, Springer-Verlag
Fenner M (1985) *Seed Ecology*. London, Chapman & Hall

Finegan B (1984) Forest Succession. *Nature* (London) **312**: 109–14
Fisher R A, Turner N C (1978) Plant productivity in the arid or semi-arid zones. *Annual Review of Plant Physiology* **29**: 277–317
Flenley J R (1979) *The Equatorial Rain Forest: A Geological History*. London, Butterworth
Flessa K W (1980) Biological effects of plate tectonics and continental drift. *Bio Science* **30**: 518–83
Flohn H (1969) Local wind systems. In: Landsberg H E (ed) *General Climatology* Vol. 12. Amsterdam, Elsevier Scientific Publications: 139–72
Foin T C Jr (1976) *Ecological Systems and the Environment*. Boston, Houghton Mifflin Company
Forman R T T, Godorn M (1986) *Landscape Ecology*. New York, John Wiley & Sons
Fosberg F R (1974) Phytogeography of atolls and other coral islands. *Proceedings 2nd International Coral Reef Symposium* **1**: 389–96
Frankel O H, Bennett E (eds) (1980) *Genetic Resources of Plants – their Exploitation and Conservation*. Oxford, Blackwell Scientific Publications
French N R (1979) *Perspectives in Grassland Ecology*. New York, Springer-Verlag
Frenkel R E, Harrison C M (1974) An assessment of the usefulness of phytosociological and numerical classificatory methods for the community biogeographea. *Journal of Biogeography* **1**: 27–58
Frenzel B (1968) The Pleistocene vegetation of northern Eurasia. *Science* **161**: 637–49
Fridriksson S (1975) *Surtsey – Evolution of Life on a Volcanic Island*. London, Butterworth
Fry G L A (1991) Conservation in agricultural ecosystems. In: Spellberg I F (ed) *The Scientific Management of Temperate Communities for Conservation*. Oxford, Blackwell Scientific Publications: 415–43
Fry G L A, Cooke A S (1984) *Acid Deposition and its Implications for Nature Conservation in Britain*. London, Nature Conservancy Council
Furley P A, Newey W W (1982) *Geography of the Biosphere. An Introduction to the Nature, Distribution and Evolution of the World's Life Zones*. London, Butterworth
Futuyma D J (1973) Community structure and stability in constant environments. *American Naturalist* **107**: 443–4
Gause G F (1932) Experimental studies on the struggle for existence 1. Mixed populations of two yeast species. *Journal of Experimental Biology* **9**: 389–402
Gause G F (1934) *The Struggle for Existence*. New York, Macmillan (Hafner) Press (reprinted 1964)
Gaussen H (1954) *Géographie de Plantes* (2nd edn). Paris, Armand Colin
Gee J H R, Giller P S (1987) *Organisation of Communities Past and Present*. British Ecology Society Symposium 27. Oxford, Blackwell Scientific Publications
Gehlbach F R (1981) *Mountain Islands and Desert Seas*. College Station Texas, Texas A M University Press
Geiger R (1969) Topoclimate. In: Flohn H (ed) *General Climatology*. Vol. 2 of Landsberg H E (ed) *World Survey of Climatology*. Amsterdam, Elsevier Scientific Publications: 105–17
Gentry A H, Lopex-Parodi J (1980) Deforestation and increased flooding of the Upper Amazon. *Science* **210**: 1354–6
Gerlach S A (1981) *Marine Pollution: Diagnostics and Therapy*. Berlin, Springer-Verlag
Gerrard J (1991) Mountains under pressure. *Scottish Geographical Magazine* **107**(2): 75–83
Gersmehl P J C (1976) An alternative biogeography. *Annals Association American Geographers* **66**: 223–41
Gilbert F S (1980) The equilibrium theory of island biogeography: fact or fiction. *Journal of Biogeography* **7**: 209–35

Gilbert O L (1983) Wildlife of Britain's wasteland. *New Scientist* **97**: 824–9
Gilbert O L (1989) *The Ecology of Urban Habitats*. London, Chapman & Hall
Gilbertson D D, Kent M, Pyatt F B (1985) *Practical Ecology for Geography and Biology*. London, Hutchinson
Gill A M, Groves R A, Noble I R (eds) (1981) *Fiore and the Australian Biota*. Canberra, Australian Academy of Science
Giller P S (1984) *Community Structure and the Niche*. London, Chapman & Hall
Gilpin M E, Soulé M E (1986) Minimum viable populations: processes of species extinction. In: Soulé M E (ed) *Conservation Biology: The Science of Scarcity and Diversity*. Sunderland, Mass, Sinauer Associates: 19–34
Gleason H A (1926) The individualistic concept of the plant association. *Bulletin of the Torrey Botanical Club* **53**: 7–26
Glenn-Lewin D (1990) *Succession*. London, Chapman & Hall
Gliessman S R (ed) (1990) *Agroecology: Researching the Ecology for Sustainable Agriculture*. Berlin, Springer-Verlag
Glinka K P (1927) *The Great Soil Groups of the World and their Development*. (trans. from the German by Marbut C F). University of Michigan, Ann Arbor
Godwin H (1975) *The History of the British Flora* (2nd edn). Cambridge, Cambridge University Press
Goldammer J G (ed) (1990) *Fire in the Tropical Biota*. Ecological Studies 84. Oxford, Blackwell Scientific Publications
Golley F B, Medina E (eds) (1975) *Tropical Ecological Systems*. New York, Springer-Verlag
Good R (1974) *The Geography of Flowering Plants* (3rd edn). Harlow, Longman
Goodall D W, Perry R W (eds) (1979) *Arid-Land Ecosystems: Structures, Functioning and Management*. Cambridge, Cambridge University Press
Goreau T J, de Mello W Z (1988) Tropical deforestation: some effects on atmospheric chemistry. *Ambio* **17**: 275–81
Gorham E, Vitousek P M, Reiners W A (1979) The regulation of chemical budgets over the course of terrestrial ecosystem succession. *Annual Review Ecology and Systematics* **10**: 53–84
Gorman M (1979) *Island Ecology*. London, Chapman & Hall; New York, A Halsted Press Brook, John Wiley & Sons
Gosz S R *et al* (1978) The flow of energy in a forest ecosystem. *Scientific American* **238**: 93–102
Goudie A (1981) *The Human Impact: Man's Role in the Environmental Change.* London, Blackwell
Gould R (1991) Pests and pollution join forces. *New Scientist* **1786**: 13
Gourou P (1961) *The Tropical World* (3rd edn). Harlow, Longman
Grace J, Ford E D, Gravis P G (eds) (1981) *Plants and their Atmospheric Environment*. British Ecological Society Symposium 21. Oxford, Blackwell Scientific Publication
Graham S A (1941) Climax forests of the upper penin of Michigan. *Ecology* **22**: 355–62
Gray A J, Crawley M J, Edwards P J (1987) *Colonisation, Succession and Stability*. The 26th Symposium of the British Ecological Society held jointly with the Linnean Society, London. Oxford, Blackwell Scientific Publications, Oxford
Greenslade P J M (1968) Island patterns in the Solomon Islands bird fauna. *Evolution* **22**: 751–61
Greenslade P J M (1983) Adversity selection and habitat templet. *American Naturalist* **122**: 352–65
Greig-Smith P (1964) *Quantitative Plant Ecology* (2nd edn). London, Butterworth

Grime J P (1979) *Plant Strategies and Vegetation Processes*. Chichester, John Wiley & Sons

Grime J P (1987) 20. Dominant and subordinate components of plant communities: implications for succession, stability and diversity. In: Gray A J, Crawley M J, Edwards P J (eds) *Colonisation, Succession and Stability*. Oxford, Blackwell Scientific Publications: 413–28

Grisebach A (1877–78) *La Végétation du Globe Après la Disposition Suivant Les Climats*. Paris, Ballure

Grubb P J (1986) The ecology of establishment. In: Bradshaw A D, Goode D A, Thorp E (eds) *Ecology and Landscape Design*. Oxford, Blackwell Scientific Publications: 83–97

Grubb P J (1987) Some generalising ideas about colonisation and succession in green plants and fungi. In: Gray A J, Crawley M H, Edwards P J (eds) *Colonisation, Succession and Stability*. Oxford, Blackwell Scientific Publications: 81–102

Grubb P J, Whittaker J B (1989) *Towards a More Exact Ecology*. Oxford, Blackwell Scientific Publications

Gulland J A (1964) Application of mathematical models to fish populations. In: Cren Le (ed) *The Exploitation of Natural Animal Populations*. Oxford, Blackwell Scientific Publications

Hadley N E (1972) *Desert species and adaptation. American Scientist* **60**: 338–47

Hadley N E (ed) (1975) *Environmental Physiology of Desert Organisms*. Stroudsberg, Pa, Dowden, Hutchinson & Ross

Hadley N E, Szarek S R (1981) Productivity of desert ecosystems. *Bioscience* **32**: 747–53

Haeck J, Woldendorp J W (eds) (1985) *Structure and Functioning of Plant Populations*. Amsterdam, North-Holland

Hafez E S E (1969) *Adaptation of Domestic Animals*. Philadelphia, Lea & Febiger

Hall C A S, De Angelis D L (1985) Models in ecology: paradigms found or paradigms lost? *Bulletin of the Ecological Society of America* **66**(3): 339–45

Hallam A (1972) Continental drift and the fossil record. *Scientific American* **226**(4): 42–52

Hallam A (ed) (1977a) *Atlas of Palaeobiology*. Amsterdam, Elsevier Scientific Publications

Hallam A (ed) (1977b) *Patterns of Evolution as Illustrated by the Fossil Record*. Amsterdam, Elsevier Scientific Publications

Hallé F, Oldeman R A A, Tomlinson P B (eds) (1978) *Tropical Trees and Forests: An Architectural Analysis*. Berlin, Springer-Verlag

Hämet-Ahti L (1981) The boreal zone and its biotic subdivisions. *Fennica* **159**(1): 69–75

Hardy A C (1965) *The Open Sea – The Natural History*. London, Collins

Harlan J R (1971) Agricultural origins: centres and non-centres. *Science* **174**: 468–74

Harlan J R (1975) *Crops and Man*. Madison, Wisconsin, American Society of Agronomy

Harley J L, Lewis C H (eds) (1985) *The Flora and Vegetation of Britain: Origins and Changes – The Facts and their Interpretation*. New York, Academic Press

Harper J L (1977) *Population Biology of Plants*. New York, Academic Press

Harris D R (1976) The ecology of swidden cultivation in the upper Orinoco rain forest, Venezuela. *Geographical Review* **614**: 75–95

Harris D R (1979) Tropical vegetation: an outline of some misconceptions. *Geography* **59**: 240–50

Harris D R (1980) *Human Ecology in Savanna Environments*. New York, Academic Press

Harris D R, Hillman G C (1989) *Foraging and Farming – the Evolution of Plant Exploitation*. London, Unwin Hyman

Harris S, Rayner J M V (1986) Urban foxes (*Vulpes vulpes*): population estimates and habitat requirements in several British cities. *Journal of Animal Ecology* **55**: 571–91

Harrison C, Goldsmith F B, Morton A (1986) Description and analysis of vegetation. In: Chapman J B, Moore P D (eds) *Methods in Plant Ecology*. Oxford, Blackwell: 437–524

Hassell M P, Anderson R M (1989) Predator–prey and host–pathogen interactions. In: Cherritt J M (ed) *Ecological Concepts. The Contribution of Ecology to an Understanding of the Natural World*. Oxford, Blackwell Scientific Publications for British Ecological Society: 147–98

Hayes A J (1979) A microbiology of plant litter decomposition. *Science Progress* **66**: 25–42

Heal O W, MacLean S F (1975) Comparative productivity in ecosystems. In: van Dobben W H, Lowe-MacConnell R H (eds) *Unifying Concepts in Ecology*. The Hague: 67–88

Heinselman Miron L (1981) Fire and succession in the coniferous forest of North America. In: West D C, Shugart A H, Botkin D B (eds) *Forest Succession*. New York, Springer-Verlag: 374–405

Heiser Jnr C B (1981) *Seed to Civilisation – the Story of Food* (2nd edn). San Francisco, W H Freeman & Co

Helliwell D R (1976a) The effects of size and isolation on the conservation value of wooded sites in Britain. *Journal of Biogeography* **3**: 407–16

Helliwell D R (1976b) The extent and location of nature conservation areas. *Environmental Conservation* **3**: 255–8

Hemming J (ed) (1985) *Change in the Amazon Basin: Man's Impact on Forests and Rivers*. Manchester, Manchester University Press

Hengereld R (1989) *Dynamics of Biological Invasion*. New York, Chapman & Hall

Herbertson A J (1905) The major natural regions: an essay in systematic geography. *Geographic Journal* **25**: 300–12

Heslop-Harrison J (1973) The plant kingdom: an inexhaustible resource? *Transactions Botanical Society (Edinburgh)* **42**: 1–15

Heslop-Harrison J (1974) Genetic resource conservation: the end and the means. *Journal Royal Society of Arts* **1974**: 157–69

Heywood V H, Moore D M (eds) (1984) *Current Concepts in Plant Taxonomy*. New York, Academic Press

Hill A R (1975) Ecosystem stability in relation to stresses caused by human activities. *Canadian Geography* **XIX (3)**: 206–19

Hill T A (1977) *The Biology of Weeds*. Studies in Biology No. 79, Institute of Biology. London, Edward Arnold

Hills T L (1960) Savanna: a review of a major research problems in tropical geography. *Canadian Geographer* **4**: 216–78

Hogg A J (1978) Island biogeography theory and the nature reserve design. *Journal of Biogeography* **1**: 117–20

Holdgate M W (1991) Conservation in a world context. In: Spellberg I F (ed) *The Scientific Management of Temperate Communities for Conservation*. Oxford, Blackwell Scientific Publications: 1–26

Holdgate M W, Woodman M J (1978) *The Breakdown and Restoration of Ecosystems*. New York, Plenum

Holland P G (1978) An evolutionary biogeography of the genus *Aloë*. *Journal of Biogeography* **5**: 213–16

Holling C S (1973) Resilience and stability of ecological systems. *Annual Review of Ecology and Systematics* **4**: 1–23

Holzner W, Werger J A, Iknsima I (1983) *Man's Impact on Vegetation*. The Hague, Junk

Hood A F M (1982) Fertiliser trends in relation to biological productivity within the United Kingdom. *Philosophical Transactions of the Royal Society* **B296**: 315–28

Horn H S (1975) Forest succession. *Scientific American* **232**: 90–8

Houghton R A *et al.* (1985) Net flux of carbon dioxide from tropical forests in 1981. *Nature (London)* **316**: 1617–20
Hounsome M (1979) Bird life in the city. In: Laurie C *Nature in Cities.* Chichester, John Wiley & Son: 179–201
Hubbell T H (1968) The biology of islands. *Proceedings National Academy of Science USA* **60**: 22–32
Hudson N (1971) *Soil Conservation.* London, Batsford
Huntley B J (1982) South African savannahs. In: Huntley B J, Walker B D (eds) *Ecology of Tropical Savannahs.* Berlin, Springer-Verlag: 101–19
Huntley B J, Walker B H (eds) (1982) *Ecology of Tropical Savannahs.* Berlin, Springer-Verlag
Hustich I (1979) Ecological concepts and biogeographical zonation in the North: the need for a generally accepted terminology. *Holarctic Ecology* **2**: 208–17
Hutchinson G E (1957) Concluding remarks. Cold Spring Harbour Symposium on Quantitative Biology **22**: 415–27
Hutchinson G E (1965) The niche; an abstractly inhabited hyper-volume. In: *The Ecological Theatre and the Evolutionary Plan.* New Haven, Conn., Yale University Press: 26–78
Hutchinson J (1926) *The Families of Flowering Plants 1 Dicotyledons.* London, MacMillan
Huttel C (1978) Root distribution and biomass in a central amazonian rain forest. In: Tomlinson P B, Zimmerman H H (eds) *Tropical Trees as Living Systems.* Cambridge, Cambridge University Press
Hynes H B N (1963) *The Biology of Polluted Waters.* Liverpool, Liverpool University Press
Illies J (1974) *Introduction to Zoogeography.* London, Macmillan
Isaac E (1970) *The Geography of Domestication.* New Jersey, Prentice-Hall
IUCN (International Union for Conservation of Nature) (1975) *Red Data Book.* Gland, Switzerland, IUCN
IUCN (1980) *World Conservation Strategy.* Gland, Switzerland, IUCN
IUCN (1988) 40 Years in Conservation. *IUCN/UICN Bulletin* (Special Issue) **19** (7112)
IUCN/UNEP/WWF (International Union for Conservation of Nature, United Nations Environmental Programme, World Wildlife Fund) (1986) *The World Conservation Strategies for Sustainable Development.* Gland and Cambridge, IUCN
IUCN/WWF (1987) *Centres of Plant Diversity. A Guide and Strategy for their Conservation.* Richmond, Surrey, IUCN Threatened Plants Unit
Ives J D (1974a) Permafrost. In: Ives J D, Barry R G (eds) *Arctic and Alpine Environments.* London, Methuen: 159–94
Ives J D (1974b) Biological refugia and the nunatuk hypothesis. In: Barry R G, Ives J D (eds) *Arctic and Alpine Environments.* London: Methuen: 605–68
Ives J D, Barry R G (eds) (1974) *Arctic and Alpine Environments.* London, Methuen
Ives J, Pitt D C (eds) (1988) *Deforestation: Social Dynamics in Watersheds and Mountain Ecosystems.* London, Routledge
Jackson J B C (1981) Interspecific competition and species' distribution: the ghost of theories and data past. *American Zoologist* **21**: 889–901
Jacobs J (1975) Diversity, stability and maturity in ecosystems influenced by human activities. In: van Dobben W H, Lowe-McConnell R H (eds) *Unifying Concepts in Ecology.* The Hague, Junk: 187–207
Jacobs J (1990) *Tropical Rain Forests.* Berlin, Springer-Verlag
Jantsch E (1980) *The Self-Organising Universe.* Oxford, Pergamon Press
Janzen D (1975) *Ecology of Plants in the Tropics.* Studies in Biology 58. London, Edward Arnold

Jarvis P G, Leverenz J W (1983) Productivity of temperate, deciduous and evergreen forests. In: Lange O L, Nobel P S, Osmond C B, Zugler H (eds) *Physiological Plant Ecology iv. Ecosystem Processes: Cycling Productivity and Man's Influence. Encylopedia of Plant Physiology*, New Series Vol. 12D. Berlin, Springer-Verlag: 233–80

Jarvis P G, Monteith J L, Shuttleworth W J, Unsworth M H (1989) *Forests, Weather and Climate.* London, The Royal Society

Jensen V (1974) Decomposition of angiosperm tree leaf litter. In: Dickinson C H, Pugh C J F (eds) *Litter Decomposition* Vol. I. New York, Academic Press: 69–104

Jewell P A (1974) 12. Managing animal populations. In: Warren A, Goldsmith B (eds) *Conservation in Practice*. Chichester, John Wiley & Sons: 185–98

Jones H G, Flowers T J, Jones M B (1989) *Plants under Stress*. Society for Experimental Biology Seminar Series 39. Cambridge, Cambridge University Press

Jordan C F (1987) *Amazonian Rain Forests? Ecosystem Disturbance and Recovery*. Ecological Studies Vol. 67. New York, Springer-Verlag

Jordan C F, Kline J R (1972) Mineral cycling. Some basic concepts and their application in the tropical rain forest. *Annual Review of Ecology and Systematics* **3**: 33–50

Jordan C F, Todd R L, Escalante G (1979) Nitrogen conservation in a tropical rain forest. *Oecologia* **39**: 123–8

Joyce C (1991) Prospectors for tropical medicines *New Scientist* **132**(1791): 36–40

Justic C O, Townshend J R G, Holben B N, Tucker C J (1985) Analysis of the phenology of global vegetation using meteorological satellite data. *International Journal of Remote Sensing* **6**: 1274–318

Karl D M, Wirsen C O, Jannasch H W (1980) Deep sea primary production at the Galapagos hydro-thermal vents. *Science* **207**: 1345–7

Keast D (1971) Continental drift and the biota of the southern continents. *Quarterly Review Biology* **46**: 335–78

Kennish M L (1986) *Ecology of Estuaries* Vol. 1. New York, Academic Press

Kent M (1987) Island biogeography and habitat conservation. *Progress in Physical Geography* **11**(1): 91–102

Kershaw A P (1983) Considerations nouvelles sur la flore et la végétation australiennes. *Espace Géographique* **3**: 185–94

Kimmins J P, Wein R W (1986) Introduction. In: van Cleeve K *et al.* (eds) *Forest Ecosystems in the Alaska Taiga*. Berlin, Springer-Verlag: 2–8

King J, Nicholson I A (1964) In: Burnett J (ed) *The Vegetation of Scotland*. Edinburgh, Oliver & Boyd: 168–206

Knox G A (1984) The key role of krill in the ecosystem of the southern ocean with special reference to the contention in the conservation of Antatctic marine resources. *Ocean Management* **9**: 113–56

Kononova M M (1961) *Soil Organic Matter* (2nd English edn). Oxford, Pergamon

Köppen W (1918) Klassification der Klimate nach Temperatur, Niederschlag und Jahreslauf. *Petermanns Mitt.* **64**: 133–203; 243–8

Kozlowski T T, Ahlgren R C (1974) *Fire and Ecosystems*. New York, Academic Press

Krebs C J (1985) (3rd edn) *Ecology. The Experimental Analysis of Population Distribution and Abundance*. New York, Harper & Row

Kruger F J, Mitchell D T, Jarvis J U M (1983) *Mediterranean-type Ecosystems: The Role of Nutrients*. Berlin, Springer-Verlag

Kruk Reinke *et al.* (eds) (1988) *The Tropical Rain Forest*. London, Springer-Verlag

Küchler A W (1949) *Vegetation Mapping*. New York, Ronald Press

Lacey C J, Walker J, Noble I R (1982) Fire in Australian tropical savannas. In: Huntley B J, Walker B H (eds) *Ecology of Tropical Savannahs*. Berlin, Springer-Verlag: 246–72

Lack D (1943) The age of some more British birds. *British Birds* **36**: 214–21
Lack D (1970) Island birds. *Biotropica* **2**: 29–31
La March Jr V C, Graybill D A, Fritts H C, Rose M R (1984) Increasing atmospheric carbon dioxide. Tree ring evidence for growth enhancement in natural vegetation. *Science* **225**: 1019–21
Lamb H F (1985) Palynological evidence for post-glacial change in the position of tree limits in Labrador. *Ecological Monographs* **55**(2): 241–58
Larcher W (1975) *Physiological Plant Ecology*. Berlin, Springer-Verlag
Larson F (1940) The role of the bison in maintaining the short grass plains. *Ecology* **21**: 113–21
Larsen J A (1980) *The Boreal Ecosystem*. New York, Academic Press
Larsen J A (1982) *Ecology of Northern Lowland Bogs and Conifer Forests*. New York, Academic Press
Larsen J A (1989) *The Northern Forest Border in Canada and Alaska Biotic Communities and Ecological Relationships*. Ecological Studies Vol. 70. New York, Springer-Verlag
Law R, Watkinson A R (1989) Competition. In: Cherritt J M (ed) *Ecological Concepts*. Oxford, Blackwell Scientific Publications: 243–84
Lee J A, McNeill S, Rorison I H (1983) *Nitrogen as an Ecological Factor*. Symposium British Ecological Society 22. Oxford, Blackwell Scientific Publications
Lee J A, Seaward M R D (eds) (1982) *Urban Ecology*. Oxford, Blackwell Scientific Publications
Lee K E, Wood T G (1971) *Termites and Soils*. New York, Academic Press
Le Houérou H N (1979) Long-term dynamics in arid-land vegetation and ecosystems of North Africa. In: Goodall D W, Perry R W (eds) *Arid-Land Ecosystems: Structures, Functioning and Management*. Cambridge, Cambridge University Press: 357–80
Leith H (1975) Primary productivity in ecosystems comparative analysis of global patterns. In: van Dobben W H, Lowe-McConnell R H (eds) *Unifying Concepts in Ecology*. The Hague, Junk: 67–88
Leith H, Whittaker R H (1975a) *Primary Production of the Major Vegetation Units of the World. Primary Productivity of World Ecosystems*. Washington DC, National Academy of Sciences
Leith H, Whittaker R H (eds) (1975b) *Primary Productivity of the Biosphere*. Ecological Studies 14. New York, Berlin, Heidelberg, Springer-Verlag
Lemée M G (1967) *Précis de Biogéographie*. Paris, Masson et Cie
Leopold A C, Kriedemann P E (1975) *Plant Growth and Development* (2nd edn). New York, McGraw-Hill
Levitt J (1972) *Responses of Plants to Environmental Stresses*. New York, Academic Press
Lewis M R (1989) The variegated ocean: a view from space. *New Scientist* **1685**: 37–40
Lewis O A M (1986) *Plants and Nitrogen*. Studies in Biology 166. London, Edward Arnold Biology
Liddle M J (1975a) A selective review of the ecological effects of trampling on natural ecosystems. *Biological Conservation* **7**: 17–35
Liddle M J (1975b) A theoretical relationship between the primary productivity of the vegetation and its ability to tolerate trampling. *Biological Conservation* **8**: 251–6
Likens G E (1972) *Nutrients and Eutrophication*. Special Symposium Vol. 1. American Society of Limnology and Oceanography. Kansas, Lawrence
Likens G E (1975) *Primary Productivity in Aquatic Ecosystems. Productivity of World Ecosystems*. Washington DC, National Academy of Sciences: 185–202
Likens G E (1976) Acid precipitation. Chemical Engineering News **54**: 29–37
Likens G E *et al* (1979) Acid rain. *Scientific American* **241**: 39–47

Lindemann R (1942) The trophic–dynamic aspect of ecology. *Ecology* **23**: 399–418
Ling K A, Ashmore M R (1987) *Acid Rain and Trees*. London, NCC (Commissioned from Department of Pure and Applied Biology of Imperial College of Science and Technology)
Livingstone D A (1975) Late Quaternary climatic change in Africa. *Annual Review of Ecology and Systematics* **6**: 249–80
Löffler H (1984) The importance of mountains for animals distribution, species, speciation and faunistic evolution (with special attention to inland waters). *Mountain Research and Development* **4**: 229–304
Longman K A, Jenk J (1987) *Tropical Forest and its Environment* (2nd edn). Harlow, Longman
Louw G, Seely M (1982) *Ecology of Desert Organisms*. Harlow, Longman
Löve A, Löve D (1974) Origin and evolution of the arctic and alpine flora. In: Ives J D, Barry R G (eds) *Arctic and Alpine Environments*. London, Methuen: 571–604
Lovelock J E (1972) Gaia seen through the atmosphere. *Atmospheric Environment* **6**: 579
Lovelock J E (1979) *Gaia. A New Look at Life on Earth*. London, Oxford University Press
Lovelock J E (1986) Gaia: the world as a living organism. *New Scientist* **112**: 25–8
Lovelock J E, Epton S R (1975) The quest for Gaia. *New Scientist* **6**, Feb: 304–6
Lovelock J E, Margulis L (1973) Atmospheric homeostasis by and for the biosphere: the Gaia hypothesis. *Tellus* **26**: 2
Lowrance R, Stinner B R, House G S (1984) *Agricultural Ecosystems*. New York, Wiley Inter-Science
Lucas G, Synge H (1978) *The IUCN Plant Red Data Book*. Switzerland, IUCN Morges
Lynch J D (1988) Refugia. In: Myers A A, Giller P S (eds) *Analytical Biogeography*. New York, Chapman & Hall: 311–41
Mabberley D J (1983) *Tropical Rain Forest Ecology*. Glasgow, Blackie
Mabbuth J A (1984) A global assessment of the status and trends of desertification. *Environmental Conservation* **11**: 103B
MacArthur R H, Wilson E O (1967) *The Theory of Island Biogeography*. Princeton University Press, NJ
MacArthur R H (1972) *Geographical Ecology*. New York, Harper & Row
McCluskey D S (1981) *The Estuarine Ecosystem*. Tertiary Level Biology. Glasgow and London, Blackie
MacDonald D W, Newdick M T (1980) The distribution and ecology of foxes, *Vulpes vulpes* (L.) in Urban Areas. In: Bornkham R, Lee J A, Seaward M R D (eds) *Urban Ecology*. Oxford, Blackwell: 111–21
McIntosh R P (1967) The continuum concept of vegetation. *Botanical Review* **33**: 130–87
McIntosh R P (1970) Community competition and adaptation. *Quarterly Review of Biology* **45**: 259–80
McIntosh R P (1980) The relation between succession and the recovery process in ecosystems. In: Cairns J (ed) *The Recovery Process in Damaged Ecosystems*. Michigan, Ann Arbor Science Publications: 11–62
MacIntosh R P (1981) Succession and ecological theory. In: West D C, Shugart H H, Botkin D B (eds) *Forest Succession: Concepts and Application*. New York, Springer-Verlag: 10–23
Mackay D W, Taylor W K, Henderson A R (1978) The recovery of the polluted Clyde. *Proceedings of the Royal Society of Edinburgh* **76B**: 135–52
McLuskey D S (1981) *The Estuarine Ecosystem*. Glasgow and London, Blackie
MacMahon J (1981) Successional processes: comparison among biomes with special reference to probable roles of and influences of animals. In: West D C, Shugart H H,

Botkin B D (eds) *Forest Succession: Concepts and Application*. New York, Springer-Verlag: 278–304
McVean D N, Ratcliffe D A (1962) *Plant Communities in the Scottish Highlands*. Monograph of the Nature Conservancy No. 1. London, HMSO
Mangelsdorf P C (1953) Wheat. *Scientific American* **189**: 50–9
Mann K H (1957) The breeding, growth and age structure of a population of the leech (*Helobdella stagmalis*). *Journal of Animal Ecology* **26**: 171–7
Mann K H (1969) The dynamics of aquatic ecosystems. *Advancements in Ecological Research* **6**: 1–81
Margulis L, Guesso R (1991) Kingdoms in turmoil. *New Scientist* **23**, March: 46–50
Margulis L, Lovelock J E (1974) Biological modulation of the Earth's atmosphere. *Icarus* **21**: 474
Marshall L G (1988) Extinction. In: Myers A, Giller P S (eds) *Analytical biogeography*. New York, Chapman & Hall: 217–54
Martin A H (1978) Evolution of the Australian flora and vegetation through the Tertiary: evidence from pollen. *Alcheringa* **12**: 181–202
Mason C F (1977) *Decomposition*. Institute of Biology, Studies in Biology No. 74. London, Edward Arnold
Mason I L (1984) *Evolution of Domestic Animals*. Harlow, Longman
Mather J R, Toshioka G A (1968) The role of climate in the distribution of vegetation. *Annals of the Association of American Geographers* **58**: 29–41
Matthews J D (1989) *Silvicultural Systems*. London, Oxford University Press
Matthews J R (1955) *The Origin and Distribution of the British Flora*. London, Hutchinson's University Library
Mattson H H, Addy N D (1975) *Phytofagus* insects as regulators of forest primary production. *Science* **190**: 515–22
May R M (1982) Introduction. In: Anderson R M, May R M (eds) *Population Biology and Infectious Diseases*. Berlin and New York, Springer-Verlag: 1–12
May R M (1989) *Levels of Organisation in Ecology in Ecological Concepts*. Oxford, Blackwell Scientific Publications: 339–68
Mayr E (1970) *Population, Species and Evolution*. Cambridge, Mass, Harvard University Press, Oxford University Press
Mayr E (1976) *Evolution and the Diversity of Life*. Cambridge, Mass, Harvard University Press
Meadows P S, Campbell J T (1988) *An Introduction to Marine Science* (2nd edn) Glasgow and London, Blackie
Meentemeyer V (1984) The geography of organic decomposition rates. *Annals Association American Geographers* **74**: 551–60
Meentemeyer V, Box E O, Thompson K (1982) World patterns and amounts of plant litter production. *Bio Science* **32**: 125–8
Meigs P (1953) World distribution of arid and semi-arid homoclimates. In: *Arid Zone Programme 1 Reviews of Research in Arid Zone Hydrology*. Paris, Unesco: 203–10
Meijden E van der (1989) Mechanisms in plant population control. In: Grubb D, Whittaker J J (eds) *Towards a More Exact Ecology*. British Ecological Society, Blackwell Scientific Publications: 163–84
Mellanby K (1970) *Pesticides and Pollution*. Oxford, Collins
Merriam C H (1894) Laws of temperature control of the geographic distribution of terrestrial animals and plants. *National Geographic Magazine* **6**: 229–38
Messerli B, Ives J D (1984) Mountain ecosystems: stability and instability. *Mountain Research*, Special Publication **3**(2): 77–175; **3**(3): 263–97; **4**(1): 1–71

Metello J M, Aber J D, Muratone J F (1984) Nitrogen and lignin control of hardwood leaf litter decomposition. *Ecology* **63**: 621–6

Micklin P (1988) Desiccation of the Aral Sea: a water management disaster in the Soviet Union. *Science* **241**: 1170–6

Miles J (1979) *Vegetation Dynamics*. London, Chapman & Hall

Miles J (1985) The pedogenic effects of different species and vegetation types and the implications of succession. *Journal Soil Science* **36**: 571–84

Miles J (1987) Vegetation and succession: past and present perceptions. In: Gray A J, Crawley M J, Edwards P J (eds) *Colonisation, Succession and Stability*. Oxford, Blackwell Scientific Publications: 1–30

Miller A A (1950) Climatic requirements of some major vegetation formations. London, *Advancement of Science* **185**: 90–140

Miller C S (1974) Decomposition of leaf litter. In: Dickinson D J, Pugh G F (eds) *Litter Decomposition* Vol. 1. New York, Academic Press: 105–78

Mitchell J (1973) The effect of bracken distribution on moorland vegetation and soils. PhD thesis, University of Glasgow

Molchanov A A (1963) *The Hydrological Role of Forests* (trans. from Russian). Jerusalem, Israel Programme for Scientific Translation

Monaghan P (1980) The breeding ecology of urban nesting gulls. In: Bornkamm R, Lee J A, Seaward M R D (eds) *Urban Ecology*. Oxford, Blackwell Scientific Publications: 111–21

Monod T (1973) *Les Déserts*. Paris, Horizons de France

Mooney H A (1986) Photosynthesis. In: Crawley M (ed) *Plant Ecology*. Oxford, Blackwell Scientific Publications: 345–75

Mooney H A, Drake J A (1986) *Ecology of Biological Invasions of North America and Hawaii.* Ecological Studies Vol. 58. New York, Springer-Verlag

Mooney H A, Gordon M (eds) (1983) *Disturbance and Ecosystems: Components of a Response*. Ecological Studies 44. Berlin, Springer-Verlag

Mooney H A, Gulman S L (1982) Constraints on leaf structure and function in reference to herbivory. *Bioscience* **32**: 198–206

Moore J W (1967) *The Changing Environment*. New York, Springer-Verlag

Moran J, Morgan M D, Wiersma J H (1980) *Introduction to Environmental Science*. San Francisco, W H Freeman

Morey P E (1973) *How Trees Grow*. Studies in Biology No. 39. London, Edward Arnold

Morgan N C (1980) Secondary production. In: Le Cren E D, Lowe-McDonnell R H (eds) *The Functioning of Freshwater Ecosystems*. Cambridge, University Press: 247–340

Morgan R P C (1979) *Soil Erosion.* Harlow, Longman

Morgan R P C (1986) *Soil Erosion and Conservation*. Harlow, Longman

Morisset P, Payette S (eds) (1983) Tree line ecology. *Nordicana* No. 47. University of Laval, Quebec

Moss B (1980) *Ecology of Fresh Waters*. Oxford, Blackwell Scientific Publications

Mueller-Dumbois D, Bridges K W, Carson H J (1981) *Island Ecosystems*. Hutchinson Ross, Stroudsberg, Pa

Murie A (1944) The wives of Mt McKinley. *Fauna National Parks* V5 No. 5

Murphy D D (1989) Conservation and confusion: wrong species, wrong scale, wrong conclusions. *Conservation Biology* **3**: 82–4

Murphy G I (1966) Population-biology of the Pacific sardine (*Sardinops caerulea*) *Proceedings of the Californian Academy of Science* 4th Series **34**(1): 1–84

Myers A A, Giller P S (eds) (1988) *Analytical Biogeography*. New York, Chapman & Hall

National Academy of Sciences (1975) *Productivity of World Ecosystems*. Washington DC

Naveh Z (1982) Landscape ecology as an emerging branch of human ecosystem science. *Advances in Ecological Research* **12**: 189–237

Naveh Z, Lieberman A S (1984) *Landscape Ecology: Theory and Application*. New York, Springer-Verlag

Neilson S, Duinker P (1987) The extent of forest decline in Europe. *Environment* **29**, 4–10; 30–31

Nelson-Smith A (1972) Effects of the oil industry on shore life in estuaries. *Proceedings Royal Society of London* **180B**: 487–96

Newbigin M (1948) *Plant and Animal Geography* (2nd edn). London, Metheun

Newcombe K (1977) Nutrient flow in a major urban settlement: Hong Kong. *Human Ecology* **1**: 179–208

Newcombe K (1978) The metabolism of a city: the case of Hong Kong. *Ambio* **7**: 3–15

Noss R F (1983) A regional landscape approach to maintain diversity. *Bio Science* **33**: 700–6

Noy-Meir I (1973) Desert ecosystems: environment and producers. *Annual Review Ecology and Systematics* **4**: 25–52

Noy-Meir I (1974) Stability in arid ecosystems and the effects of man on it. *Proceedings 1st International Congress on Ecology*. The Hague

Noy-Meir I (1980) Structure and function of desert ecosystems. *Israel Journal of Botany* **28**: 1–19

Noy-Meir I (1985) *Desert Ecosystem: Structure and Function in Hot Deserts and Shrublands*. Evanari M, Noy-Meir I, Goodall D W (eds) *Ecosystems of the World* Vol. 12A. Amsterdam, Elsevier Scientific Publications: 93–104

Nye P H (1960) The soil under shifting cultivation. *Technical Commonwealth Bureau Soils* **51**: 1–156

Nye P H (1961) Organic matter and nutrient cycles under moist tropical forest. *Plant and Soil* **8**: 333–46

Oberlamder T M (1979) Characterisation of arid climates according to combined water-balance parameters. *Arid Environment Journal* **2**: 219–41

Odum E P (1969) The strategy of ecosystem development. *Science* **164**: 262–70

Odum E P (1971) *Fundamentals of Ecology* (3rd edn). Philadelphia, W B Saunders

Odum E P (1983) *Basic Ecology*. New York, Saunders College Publishing

OECD (Organisation for Economic Co-operation and Development) (1986) *Water Pollution by Fertilisers and Pesticides*. Paris, OECD

Oechel W C, van Cleve K (1986) The role of bryophytes in nutrient cycling in the taiga. In: van Cleve K, Chapin III F S, Viereck L A, Dryness C T, Flanagan P W (eds) *Forest Ecosystems in the Alaskan Taiga*. New York, Springer-Verlag: 121–54

O'Hare G P (1974) Lichens and bark acidification as indicators of pollution in west central Scotland. *Journal of Biogeography* **2**: 135–46

Okigbo B N, Greenland D J (1976) Inter-cropping systems in tropical Africa. *American Society of Agronomy* Special Publication No. 27: 63–120

Oldeman R A A (1978) Architecture and energy exchange of dicotyledon trees in the forest. In: *Tropical Trees in Living Systems*. Cambridge University Press: 353–60

Olson S L, James H E (1982) Fossil birds of the Hawaiian Islands: evidence for wholesale extinction by man before Western contact. *Science* **217**: 633–5

O'Neill R V, DeAngelis D L, Waide J B, Allen T F H (1986) *A Hierarchical Concept of Ecosystems*. Princeton NJ, Princeton University Press

Open University (1975) *Ecology* (Course 5323). Unit 2 Primary Production in Ecosystems

Otterman, J (1974) Baring high-albedo soils by over-grazing. A hypothesized desertification mechanism. *Science* **186**: 531–3

Overrein L N, Hans Martin S, Arne T (1980) *Acid Precipitation and Effects in Forest and Fish*. Oslo, Final Report of the SNSP Project 1972–1980

Owen J, Owen D F (1975) Suburban gardens: England's most important nature reserves. *Environmental Conservation* **2**: 53–9

Owen-Smith, N (1982) Factors influencing the consumption of plant products by large herbivores. In: Huntley B J, Walker B H (eds) *Ecology in Tropical Savannahs*. Berlin, Springer-Verlag: 359–405

Packham J R, Harding J L (1982) *Ecology of Woodland Processes*. London, Edward Arnold

Paine R T (1969) A note on trophic complexity and community stability. *American Naturalist* **103**: 91–3

Paine R T (1974) Inter-tidal community structure. *Oecologia* **15**: 93–120

Paine R T (1980) Food web: linkage, interaction, strength, community. *J. Animal Ecol.* **49**: 667–86

Paulik G J (1971) Anchovies, birds and fishermen in the Peru current. In: Murdoch W W (ed) *Environment.* Sunderland, Mass, Sinauer Association, 156–8

Park J R (1988) *Environmental Management in Agriculture*. London, Belhaven Press

Partin S A (1982) *Pollution and the Biological Resources of the Ocean*. London, Butterworth Scientific Publications

Paul E A (1989) Soils as components and controllers of ecosystems. In: Cherritt J M (ed) *Processes in Ecological Concepts*. Oxford, Blackwell Scientific Publications: 153–74

Pearce D, Barber E, Markandyn A (1989) *Sustainable Development: Economics and Environment in the Third World*. London, Edward Elgar Publishers

Pearce R B, Brown R H, Blaser R F (1967) Photosynthesis in plant communities as influenced by leaf angle. *Crop Science* **7**: 321–4

Pears N (1985) *Basic Biogeography* (2nd edn). Harlow, Longman

Pears N V (1968) Some recent trends in classification and description of vegetation. Stockholm, Geografisker Annaler. **50A**: 162–72

Pearsall W H (1950) *Mountains and Moorlands*. London, Collins New Naturalist Series

Pellew R A (1991) Data management for conservation. In: Spellberg I F *et al.* (eds) *Scientific Management of Temperate Communities for Conservation.* Oxford, Blackwell Scientific Publications: 505–22

Pellew R A, Harrison J D (1988) A global data base on the status of biological diversity. In: Mounsey H, Tomlinson R F (eds) *Building Databases for Global Science*. London, Taylor & Francis: 330–9

Penman H L (1963) *Vegetation and Hydrology Technical Communication No. 53.* Farnham, Commonwealth Bureau of Soil, Harpenden, Royal Commonwealth Agricultural Bureau

Pennington W (1969) *The History of British Vegetation*. London, English Universities Press

Pereira H C (1973) *Land Use and Water Resources*. Cambridge University Press

Perkins E J (1974) *The Biology of Estuaries and Coastal Waters*. London, Academic Press

Persson T (ed) (1980) *Structure and Function of Northern Coniferous Forests*. Stockholm, Swedish National Sciences Research Council Ecological Bulletin 32: 607

Peterken G F (1981) *Woodland Conservation and Management in the UK*. London, Chapman & Hall

Petrusewicz K, Grodzinski W L (1975) *The Role of Herbivores as Consumers in Various Ecosystems*. Productivity of World Ecosystems. Washington DC, National Academy of Sciences; 64–70

Petrusewiez K, Macfayden A (1970) *Productivity of Terrestrial Animals: Principles and Methods*. IBP Handbook No. 13. Oxford, Blackwell Scientific Publications

Phillips D L, MacMahon J A (1981) Competition and spacing patterns in desert shrubs. *Journal of Ecology* **69**: 97–115

Pianka E R (1970) On r- and K-selection. *American Naturalist* **104**: 592–7
Pianka E R (1988) *Evolutionary Ecology* (4th edn). New York, Harper & Row
Pickett S T A, White P S (eds) (1985) *The Ecology of Natural Disturbance and Patch Dynamics*. New York, Academic Press
Pielou E C (1979) *Biogeography*. New York, Wiley
Pimentel D, Dazhong W, Grampietro M (1990) Technological changes in energy use in US agricultural production. In: Gliess S R (ed) *Agroecology*, Ecological Studies 78. New York, Springer-Verlag: 305–19
Pimentel O, Pimentel M (eds) (1979) *Food, Energy and Society*. London, Edward Arnold
Pimm S L (1982) *Food Webs*. London, Chapman & Hall
Pimm S L (1984) The complexity and stability of ecosystems. *Nature* **307**: 321–6
Pimm S L (1986) Community stability and structure. In: Soulé M E (ed) *Conservation Biology: A Science of Scarcity and Diversity*. Sunderland Mass, Sinquer Association: 309–29
Pimm S L (1988) Energy flow and trophic structure. In: Pomeroy L R, Alberts J (eds) *Concepts of Ecosystem Ecology*. New York, Springer-Verlag: 263–78
Pimm S L, Lawton J H (1977) The number of trophic levels in ecological communities. *Nature* **268**: 329–31
Pitty A F (1979) *Geography and Soil Properties*. London, Methuen
Polunin N (ed) (1986) *Ecosystems: Theory and Application*. Chichester, Wiley
Polunin N, Burnett J (eds) (1990) *Maintenance of the Biosphere*. Edinburgh, Edinburgh University Press
Pomeroy L R, Alberts J J (eds) (1988) *Concepts of Ecosystem Ecology*. Ecological Studies Vol. 67. New York, Springer-Verlag
Poore M E D (1955–56) The use of phytosociological methods in ecological investigations. *Journal of Ecology* **43**: 226–69, 606–51; **44**: 28–50
Poore M E D (1962) The method of successive approximation in descriptive ecology. *Advances in Ecological Research* **1**: 35–66
Potts G R (1986) *The Partridge: Pesticides, Predation and Conservation*. London, Collins
Potts G R, Aebischer N J (1989) Control of population size in birds: the grey partridge as a case study. In: Grubb P D, Whittaker J J (eds) *Towards a More Exact Ecology*. British Biological Society, Blackwell Scientific Publications: 141–62
Prance G T (1984) Completing the inventory. In: Heywood V H, Moore D M (eds) *Current Concepts in Plant Taxonomy*. New York, Academic Press: 365–96
Price L W (1981) *Mountains and Man*. Berkeley: University of California Press
Price P W (1984) The concept of the ecosystem. In: Hufaker C B, Rabo R L (eds) *Ecological Entomology*. New York, John Wiley: 19–50
Price S (1989) *Animal Re-introductions: The Arabian Oryx in Oman*. Cambridge, Cambridge University Press
Proctor J, Woodwell S R J (1975) The ecology of serpentine soils. *Advances in Ecological Research* **9**: 255–366
Pruitt W D Jnr (1978) *Boreal Ecology*. Institute of Biology Studies in Biology No. 91. London, Edward Arnold
Putnam R J, Wratten S O (1984) *Principles of Ecology*. London, Chapman & Hall
Ragotzkie R A (ed) (1983) *Man and the Marine Environment*. Boca Raton, Fla CRC Press
Ramade R (1984) *Ecology of Natural Resources*. Chichester, John Wiley and Sons
Raunkiaer C (1934) *The Life Form of Plants and Statistical Plant Geography*. Oxford, Clarendon Press
Raven P, Axelrod D I (1972) Plate tectonics and Australasian pala-biography. *Science* **176**: 1379–86
Raven P, Axelrod D I (1974) Angiosperm biogeography and past continental movements. *Annals Missouri Botanic Garden* **61**: 539–67

Reichle D E (ed) (1970) *Analysis of Temperate Forest Ecosystems*. Ecological Studies No. 1. London, Chapman & Hall

Reichle D E (ed) (1981) *Dynamic Properties of Forest Ecosystems*. Cambridge, Cambridge University Press

Reichle D E, O'Neill R V, Harris W F (1975) Principles of energy and material exchange in ecosystems. In: van Dobben W H, Lowe-McConnell R H (eds) *Unifying Concepts in Ecology*. The Hague, Junk: 27–43

Reiners W A (1983) *Disturbance and Basic Properties of Ecosystem Energetics. Disturbance and Ecosystems: Components of a Response.* Ecological Studies 44. Berlin, Springer-Verlag: 83–98

Remmert H (1980) *Arctic Animal Ecology*. Berlin, Springer-Verlag

Reuss J O, Johnson D W (1986) *Acid Deposition and the Acidification of Soils and Waters: An Analysis.* Ecological Studies Vol. 59. Berlin, Springer-Verlag

Richards B N (1974) *Introduction to the Soil Ecosystem*. Harlow, Longman

Richards P W (1973) The tropical rain forest. *Scientific American* **229**: 58–67

Richards P W (1979) *The Tropical Rain Forest*. Cambridge University Press

Ricklefs R E (1973) *Ecology*. London, Nelson

Ricklefs R E (1987) Community diversity: relative role of local and regional processes. *Science* **235**: 167–71

Ricklefs R E, Cox G W (1972) Taxon cycles in the West Indian avifauna. *American Naturalist* **106**: 195–219

Riley G A (1972) Patterns of production in marine ecosystems. In: Wiens J A (ed) *Ecosystem Structure and Function*. Oregon, University Annual Biology Colloquia **31**: 91–112

Risser P G (1988) Abiotic controls in North American grasslands. In: Pomeroy L R, Alberto J J (eds) *Concepts of Ecosystem Ecology*. New York, Springer-Verlag: 115–29

Ritchie J C (1988) *Post-glacial Vegetation of Canada*. New York, Cambridge University Press

Roberts N (1989) *The Holocene: An Environmental History*. London, Basil Blackwell

Rodin L E, Bazilevich N I (1967) *Production and Mineral cycling in Terrestrial Vegetation*. Edinburgh, Oliver Boyd

Ross C A (ed) (1976) *Palaeobiogeography*. Stroudsburg, Pa, Dowden, Hutchinson & Ross

Rosswall T (1976) (2nd edn 1979) The internal nitrogen cycle between micro-organisms, vegetation and soil. In: Svensson B H, Soderlind R (eds) *Nitrogen, Phosphorus and Sulphur Global Cycles*. Stockholm, SCOPE Report of Ecological Bulletin, 22, published in collaboration with Royal Swedish Academy of Science, SCOPE and UNEP: 157–67

Rougierr G (1962) *Biogéographie des montagues*. Cours de l'université de Besançon. Paris V, Centre de Documentation Universitaire

Roughgarden J (1983) Competition and theory in community ecology. *American Naturalist* **122**: 583–601

Rowe J S, Scotter G W (1973) Fire in the boreal forest. *Quaternary Research* **3**: 444–64

Royal Commission on Environmental Pollution (1972) *Pollution in Some British Estuaries and Coastal waters*. CMND 5054. London, HMSO

Rudd R L (1964) *Pesticides and the Living Landscape*. Madison, University of Wisconsin

Ruthenberg H (1976) *Farming Systems in the Tropics* (2nd edn). Oxford, Oxford University Press

Rydén B E (1981) 4. Hydrology of the northern tundra. In: Bliss L C, Heal O W, Moore J J (eds) *Tuneta Ecosystems: A Comparative Analysis*. Cambridge, Cambridge University Press: 79–89

Salisbury E J (1929) The biological equipment of species in relation to competition. *Journal of Ecology* **17**: 197–222

Sanford W, Wangari E (1985) Tropical grasslands: dynamics and utilisation. *Nature and Resources* **21**(3): 12–34

Satchell J (1974a) Introduction. In: Dickinson C I, Pugh G F (eds) *Litter Decomposition* Vol. I. New York, Academic Press: i–xxvi

Satchell J E (1974b) Interface of animate/inanimate matter. In: Dickinson C J, Pugh G F (eds) *Litter Decomposition* Vol. I. New York, Academic Press: xiii

Sauder J D (1969) Oceanic islands and biogeographic theory: a review. *Geographical Review* **59**: 582–93

Sauer C O (1950) Grassland climax, fire and man. *Journal of Range Management* **3**: 1–6

Sauer C O (1952) *Agricultural Origin and Dispersions*. American Geographical Society of New York (reprinted 1969 MIT Press)

Saunders D A, Arnold G W, Burdige A A, Hopkins A J M (1987) *Nature Conservation: The Role of Remnants of Native Vegetation*. Surrey, Beatty

Saunders D, Hobbs R (1989) Corridors for conservation. *New Scientist* **122**: 63–8

Schimper A W F (1903) *Plant Geography on a Physiological Basis* (trans. by Fisher W R, revised and ed. by Groom P, Balfour I B). Oxford, Clarendon Press

Schoener T W (1989) The ecological niche. In: Cherritt J M (ed) *Processes in Ecological Concepts*. Oxford, Blackwell Scientific Publications: 79–115

Schonewald-Cox C M, Bayless J W (1986) The boundary model: a geographical analysis of design and conservation of nature reserves. *Biological Conservation* **38**: 305–22

Schuster R (1972) Continental movements, Wallace's line and Indo-Malaysian Australian land plants. *Botanical Review* **38**: 3–87

Scott-Russell R (1977) *Plant Root Systems: Their Function and Interaction with the Soil*. New York, McGraw-Hill

Scurlock J, Hall D (1992) The carbon cycle. *'Inside Science'* **51**: 1–4; *New Scientist* Nov: 1793

Seddon G (1983) Biological pollution in Australia. *Resource Management and Optimisation* **2**: 243–58

Seely M K (1979) Ecology of a living desert. *South African Journal of Science* **75**: 298–303

Sestak Z (ed) (1985) *Photosynthesis during leaf development* T: V511. The Hague, Junk

Shafer C L (1990) *Nature Reserves: Island Theory and Conservation Practice*. Washington and London, Smithsonian Institute

Sheridan D (1981) *Desertification of the United States*. Washington, DC Council on Environmental Quality, US Government Printing Office

Shimwell D W (1972) *The Description and Classification of Vegetation*. Seattle, University of Washington Press

Shmida A (1985) Biogeography of the desert flora. In: Evenari M, Noy-Meir I, Goodall D (eds) *Hot Deserts and Arid Shrubland*. Amsterdam, Elsevier Scientific Publications: 23–78

Shonocks B (ed) (1984) *Evolutionary Ecology*. London, Blackwell

Shugart H H, Urban D L (1988) Scale, synthesis and ecosystem dynamics. In: Pomeroy L W, Alberts J J (eds) *Concepts of Ecosystem Ecology*. New York, Springer-Verlag: 277–89

Shugart H H, Urban D L (1989) Factors affecting the relative abundance of forest tree species. In: Grubb P J, Whittaker J B (eds) *Towards a More Exact Ecology*. Oxford, Blackwell Scientific Publications: 249–74

Silverton J W (1987) *Introduction to Plant Population Ecology* (2nd edn). London, Longman

Simberloff D S (1969) Experimental zoogeography of islands: a model for insular colonization. *Ecology* **50**: 296–314

Simberloff D S (1974) Equilibrium theory of island biogeography and ecology. *Annual Review of Ecology and Systematics* **5**: 161–87

Simberloff D S (1986) Design of nature reserves. In: Usher M B (ed) *Wildlife Conservation Evaluation.* London, Chapman & Hall: 315–69

Simberloff D S, Abele L G (1976) Island biogeography theory and conservation practice. *Science* **191**: 285–6

Simberloff D W, Abele L G (1982) Refuge design and island biogeographic theory: effects of fragmentation. *American Naturalist* **120**: 121–7

Simberloff, D S, Wilson E D (1969) Experimental zoo geography of islands: the colonisation of empty islands. *Ecology* **50**: 278–96

Simberloff D S, Wilson E D (1970) Experimental zoo geography of islands: a two year record of colonisation. *Ecology* **51**: 934–7

Simmonds N W (ed) (1976) *Evolution of Crop Plants*. London, Longman

Simmons I G (1974) *The Ecology of Natural Resources*. London, Edward Arnold

Simmons I G (1979) *Biogeography: Natural and Cultural.* London, Edward Arnold

Simms E (1975) *Birds of Town and Suburb*. London, Collins

Sims P L *et al.* (1978) The structure and function of ten western North American grasslands. *Journal of Ecology* **66**: 251–85, 573–97

Slatter R J (1978) Ecological effects of trampling on sand dune vegetation. *Journal of Biological Education* **12**: 81–96

Smathers G A, Mueller-Dombois D (1974) *Invasion and Recovery of Vegetation after a Volcanic Eruption in Hawaii.* Washington DC, National Park Service Science Monograph No. 5

Smil V (1983) Deforestation in China. *Ambio* **12**(5): 226–31

Smith D F, Hill D M (1975) Natural agricultural ecosystems. *Journal of Environmental Quality* **4**(2): 143–5

Smyth B (1987) *City Wildspace*. London, Hilary Shipham

Solomon M E (1969) *Population Dynamics*. London, Edward Arnold

Soulé M (ed) (1986) *Conservation Biology. The Science of Scarcity and Diversity.* Sunderland Mass, Sinauer Associates

Soulé M, Wilcox A (eds) (1980) *Conservation Biology. An Evolutionary Ecological Perspective.* Sunderland Mass, Sinauer Associates

Sousa W P (1984) The role of disturbance in natural communities. *Annual Review of Ecology and Systematics* **15**: 353–91

Spedding C W R (1975) *Biology and Agricultural Systems.* New York, Academic Press

Spellerberg I F (1991) Biological basis of conservation. In: Spellerberg I F, Goldsmith F B, Morris F (eds) *Scientific Management of Temperate Communities for Conservation.* Oxford, Blackwell Scientific Publications: 293–322

Spellerberg I F, Goldsmith F B, Morris M G (1991) *The Scientific Management of Temperate Communities for Conservation*. Oxford, Blackwell Scientific Publications

Sprent J I (1987) *The Ecology of the Nitrogen Cycle.* Cambridge Studies in Ecology. Cambridge, Cambridge University Press

Spurr S H, Barnes B V (1980) *Forest Ecology* (3rd edn). New York, Wiley

Stace C A (1989) *Plant Taxonomy and Biosystematics* (2nd edn). London, Edward Arnold

Stark N M, Jordan C F (1978) Nutrient retention by the root mat of an Amazonian rain forest. *Ecology* **59**: 434–7

Stearns F, Montag T (eds) *The Urban Ecosystem: A Holostic Approach.* Stroudsberg, Dowden, Hutchinson & Ross

Steele J H (1974) *The Structure of Marine Ecosystems*. Oxford, Blackwell Scientific Publications

Steele J H (1985) A comparison of terrestrial and marine ecological systems. *Nature* **313**: 355–8

Stehli F G (1968) Taxonomic gradients in pole location, the recent model. In Drake E T (ed) *Evolution and Environment.* New Haven, Yale University Press: 163–227

Steinberg O'R H (1987) Aggravation of floods in the Amazonian river as a consequence of deforestation. *Geografiska Annaler* Set A **69**: 201–19

Stewart W N (1983) *Palaeobotany and the Evolution of Plants*. London, Cambridge University Press

Stoddart D R (1969) Islands as ecological laboratories. *New Scientist* **41**: 20–3

Stolarski R S (1988) The Antarctic ozone hole. *Scientific American* **255**: 30–7

Stork N, Gaston K (1990) Counting species one by one. *New Scientist* **11** August: 43–6

Stott P (1981) *Historical Plant Geography*. London, George Allen & Unwin

Stout J D, Tate K R, Molloy L F (1976) Decomposition processes in New Zealand soils with particular respect to rates and pathways of plant degradation. In: Anderson J M, Macfayden A (eds) *The Role of Terrestrial and Aquatic Organisms in Decomposition Processes*. Oxford, Blackwell Scientific Publications: 97–144

Strang R M, Johnston A H (1981) Fire and climax spruce forest in central Yukon. *Arctic* **34**(1): 60–1

Street H E (1987) British endemics. In: Walters S M (ed) *Essays in Plant Taxonomy*. London and New York, Academic Press: 263–76

Strong D R Jr, Simberloff D, Abele L G, Thistle A B (eds) (1984) *Ecological Communities: Conceptual Issues and the Evidence*. Princeton NJ, Princeton University Press

Sukopp H, Werner P (1983) Urban environments and vegetation. In: Holzmer W, Werger M J A, Ikusima J (eds) *Man's Impact on Vegetation*. The Hague, Junk: 247–76

Sukopp H, Werner P (1986) Biotype mapping in urban areas of the Federal Republic of Germany. *Landschaff und Stadt* **18**: 25–8

Swift M J, Heal O W, Anderson J M (1979) *Decomposition in Terrestrial Ecosystems*. Oxford, Blackwell Scientific Publications

Swingland I R, Greenwood P J (1983) *The Ecology of Animal Movement*. Oxford, Clarendon Press

Tait R V (1972, 1981) *The Elements of Marine Ecology*. (2nd edn). London, Butterworth

Talbot F H, Russell B C, Anderson G R V (1978) Coral reef fish communities: unstable, high-diversity systems. *Ecological Monographs* **48**: 425–40

Tamm C O (1991) *Nitrogen in Terrestrial Ecosystems*. Ecological Studies 81. New York, Springer-Verlag

Tamm C O, Halbächen T (1988) Changes in soil acidity in two forest areas with different acid deposition: 1920s–1980s. *Ambio* **17**: 56–61

Tangi M (1977) Tourism and the environment. *Ambio* **6**: 336–41

Tansley A G (1935) The use and abuse of vegetational concepts. *Ecology* **16**: 284–307

Tansley A G (1954) *An Introduction to Plant Ecology* (3rd edn). London, Allen & Unwin

Thompson J D (1990) *Spartinia anglica*, characteristic feature or invasive weed of coastal salt marshes. *Biologist* **37**(1): 9–12

Thorne R F (1972) Major disjunctions in the geographic ranges of seed plants. *Quarterly Review of Biology* **47**: 365–411

Thornthwaite C W (1933) The climates of the earth. *Geographical Review* **23**: 433–40

Thornthwaite C W (1948) An approach towards a rational classification of climate. *Geographical Review* **38**: 55–94

Tieszen L L (1978) *Vegetation and Production Ecology of an Alaskan Arctic Tundra*. New York, Springer-Verlag

Tilman D (1982) *Resource Competition and Community Structure.* New Jersey, Princeton University Press

Tilman D, Kilham S S, Kilham P (1982) Phytoplankton community ecology: the role of limiting nutrients. *Annual Review of Ecology and Systematics* **13**: 349–72

Tivy J (1971) Biogeography: A study of plants in the ecosystem (1st edn). Edinburgh, Oliver Boyd

Tivy J (1973a) *The Organic Resources of Scotland.* Edinburgh, Oliver & Boyd

Tivy J (1973b) *The Concept and Determination of Carrying Capacity of Recreational Land in the USA: A Review.* Perth, Countryside Commission for Scotland

Tivy J (1975) Environmental impact of cultivation. In: Lenihan J, Fletcher W W (eds) *Food, Agriculture and the Environment.* Glasgow, Blackie: 21–47

Tivy J (1979) Recreation and its relevance for shore erosion. In: Huxley T (ed) *Shore Erosion Around Loch Lomond.* Perth, Countryside Commission for Scotland: 1–10

Tivy J (1980) *Recreational Impact on Freshwater Lochsides in Scotland.* Perth, Countryside Commission for Scotland

Tivy J (1991) *Agricultural Ecology.* Harlow, Longman

Tivy J, O'Hare G (1981) *Human Impact on the Ecosystem.* Edinburgh, Oliver & Boyd

Tivy J, Rees J (1978) Recreational impact on Scottish lochshore wetlands. *Journal of Biogeography* **5**: 93–108

Tomaselli R (1977) The degradation of the Mediterranean maquis. *Ambio* **6**(6): 256–66

Torrey J G (1978) Nitrogen fixation by actinomycetes in nodulated ecosystems: *Bioscience* **25**: 586–92

Towns D, Atkinson I (1991) New Zealand's restoration ecology. *New Scientist* 20 April: 36–9

Townsend C R (1980) *The Ecology of Streams and Rivers* (2nd edn). Studies in Biology No. 122. London, Edward Arnold Biology

Townsend W N (1972) *An Introduction to the Scientific Study of Soil* (5th edn). London, Edward Arnold

Tranquilli W (1979) *Physiological Ecology of the Alpine Timberline.* Ecological Studies No. 31. Berlin, Springer-Verlag

Treshaw M, Anderson F K (1989) *Plant Stress from Air Pollution.* Chichester, John Wiley & Sons

Trewartha G T (1954) *An Introduction to Climate.* London, McGraw-Hill

Troll C (ed) (1972) *Geoecology of the High Mountain Regions of Eurasia.* Erdwiss Forsch 4. Wiesbaden, Franz Steiner Verlag: 1–3

Troll C (1973a) High mountain belts between the polar caps and the equator: their definition and lower limit. *Arctic Alpine Research* Part 2, **5**(3): A19–A28

Troll C (1973b) The upper timberlines in different climatic zones. *Arctic Alpine Research* Part 2, **5**(3): A3–A18

Trudge C (1992) The wild way to save threatened species. *The Independent* 2.5.1992

Trudgill S T (1977, 1988) *Soil and Vegetation Systems.* (2nd edn) Oxford, Clarendon Press

Turner M G (ed) (1987) *Landscape Heterogeneity and Disturbance.* Ecological Studies No. 64. New York, Berlin, Springer-Verlag

Tuza W J (ed) (1972) *Continents Adrift. Readings from Scientific American.* San Francisco, W H Freeman

Tyler-Miller G (1989) *Resource Conservation and Management.* New York, Chapman & Hall

Uko P J, Dimbleby G W (eds) (1969) *The Domestication of Plants and Animals.* London, Duckworth

Ulanowicz R E (1989) Energy flow and productivity in the oceans. In: Grubb P J, Whittaker

J B (eds) *Towards a More Exact Ecology*. Oxford, Blackwell Scientific Publications: 327–52

UNEP (United Nations Environment Programme) (1978a) *African Elephant Database Project: Final Report.* GRID Case Study Series No 2. Nairobi, UNEP

UNEP (1978b) *Uganda Case Study: A Sampler Atlas of Environmental Resource Datasets within GRID.* Grid Case Study Series No. 1. Nairobi, UNEP

Unesco (United Nations Educational Scientific and Cultural Organisation) (1979) *Map of the World Distribution of Arid Regions.* MAB Technical Notes No. 7. Paris, Unesco

Unesco (MAB) (1973) *No. 11 Ecology and Rational Use of Island Ecosystems.* Paris, Unesco

Unesco (MAB) (1974) *No. 14 Impact of Human Activities on Mountain Ecosystems.* Paris, Unesco

Unesco (1978) *Tropical Forest Ecosystems.* Paris, Unesco

Usher M B (ed) (1986) *Wildlife Conservation Evaluation.* London, Chapman & Hall

Vale T R, Vale G R (1976) Suburban bird populations in western California. *Journal of Biogeography* **3**: 157–65

Valentine D H (ed) (1972) *Taxonomy, Phytogeography and Evolution.* New York, Academic Press

Valentine J W (1978) The evolution of multicellular plants and animals. *Scientific American* **219**(3):105–18

Valiela I (1984) *Marine Ecological Processes.* New York, Berlin, Heidelberg, Tokyo, Springer-Verlag

Van Cleve K, Chapin III F W, Viereck L A, Dyrness C T, Flanagan P W (1986) *Forest Ecosystems in the Alaskan Taiga: a Synthesis of Structure and Function.* Ecological Studies 57. New York, Springer-Verlag

Van der Hammen T (1974) The Pleistocene changes of vegetation and climate in tropical South America. *Journal of Biogeography* **1**: 3–26

Van Dobben W H, Lowe-McConnell R H (1975) *Unifying Concepts in Ecology.* The Hague, Junk

Van Dyme G M (1975) *An Overview of the Ecology of the Great Plains Grasslands with Special Reference to Climate and its Impact.* US/IBP Grassland Biome Technical Review No. 290. Fort Collins Colorado State University

Van Hook R I, Johnson D W, West D C, Mann L K (1982) Environmental effects of harvesting forests for energy. *Forestry Ecology and Management* **4**: 79–94

Van Hulst R (1978) On the dynamics of vegetation patterns: environmental and vegetational change. *Vegetatio* **38**: 65–75

Van Hulst R (1980) Vegetation dynamics or ecosystem dynamics: dynamic sufficiency in succession theory. *Vegetatio* **43**: 147–51

Van Wieren (1991) The management of large mammals. In: Spellberg I F *et al* (eds) *Scientific Management of Temperate Communities for Conservation.* Blackwood Scientific Publications: 103–28

Vavilov I N (1949–50) The origin, variation, immunity and breeding of cultivated plants. *Chronica Botanica* **13**: 1–366

Vickery P J (1972) Grazing and net primary production of a temperate grassland. *Journal of Applied Ecology* **9**: 307–14

Viereck L A (1966) Plant succession and soil development on gravel outwash of the Muldow Glacier, Alaska. *Ecological Monographs* **36**: 181–90

Viereck L A, van Cleve K, Dryness C T (1986) Forest ecosystem distribution in the taiga environment. In: van Cleve K *et al* (eds) *Forest Ecosystems in the Alaskan Taiga.* New York, Springer-Verlag

Vitousek P (1982) Nutrient cycling and nutrient use efficiency. *American Naturalist* **119**: 553–72

Vitousek P M (1983) The effect of deforestation on air, soil and water. In: Bolin B, Cooke R B (eds) *The Major Biogeochemical Cycles and Their Interactions* (Scope 21). Chichester, John Wiley & Sons: 223–45

Vitousek P M, Reiners W A (1975) Ecosystem succession and nutrient retention: a hypothesis. *BioScience* **25**: 376–81

Vitousek P M, White P S (1981) Process studies in succession. In: West D C, Shugart H H, Botkin B D (eds) *Forest Succession: Concepts and Applications*. New York, Springer-Verlag: 267–76

Vogl R J (1980) Perturbation-dependent ecosystems. In: Cairns J Jr (ed) *Recovery Process in Damaged Ecosystems*. Michigan, Ann Arbor Publications 63–94

Wace N (1978) *The Character of Oceanic Island Resources and the Problem of their Rational Use and Conservation. The Use of High Mountains of the World.* Wellington, New Zealand, IUCN

Walker B H, Noy-Meir I (1982) Aspects of the stability and resilience of savanna ecosystems. In: Huntley B H, Walker B H (eds) *Ecology of Tropical Savannahs.* Berlin, New York, Springer-Verlag: 556–90

Walker D (1970) Direction and rate in some British post-glacial hydroseres. In: Walker D, West R G (eds) *Studies in the Vegetational History of the British Isles*. Cambridge, Cambridge University Press: 117–39

Walker D (1982) The development of resilience in burned vegetation. In: Newman E I (ed) *The Plant Community as a Working Mechanism.* Special Publication of the British Ecological Society London, Blackwell Publications: 27–43

Walker D (1989) Diversity and stability. In: Cherritt J M (ed) *Ecological Concepts.* Oxford, Blackwell Scientific Publications: 115–46

Walker L R, Zabada J C, Chapman F S III (1986) The role of life history processes in primary succession on an Alaskan flood plain. *Ecology* **67**: 1243–53

Walter H (1971) Ecology of Tropical and Sub-tropical Vegetation. In: Burnett J H (ed) (trans. by Mueller-Dombois D). Edinburgh, Oliver & Boyd

Walter H (1973) *Vegetation of the Earth in Relation to Climate and Ecophysiological Conditions*. New York, Springer-Verlag

Walters S M, Street H E (eds) (1978) *Essays in Plant Taxonomy*. London and New York, Academic Press

Wardle P (1974) Alpine timber lines. In: Ives J D, Barry R G (eds) *Arctic and Alpine Environments*. London, Methuen: 371–402

Waring R H (1989) Ecosystems: fluxes of matter and energy. In: Cherritt J M (ed) *Ecological Concepts*. Oxford, Blackwell Scientific Publications 17–42

Waring R H, Schlesinger W H (1985) *Forest Ecosystems: Concepts and Management.* London, Academic Press

Warren A, Goldsmith F B (1983) *Conservation in Perspective.* Chichester, John Wiley & Sons

Watkinson A R (1981) *The Biological Aspects of Rare Plant Conservation.* Chichester, John Wiley & Sons

Watkinson A R (1986) Plant population dynamics. In: Crawley M J (ed) *Plant Ecology*. Oxford, Blackwell Scientific Publications

Watson A (1991) Gaia 'Inside Science'. *New Scientist* **48** (6 July): 1–4

Watt A S (1964) The community and the individual. *Journal of Ecology* **52**: 203–11

Watt A S (1981) A comparison of grazed and ungrazed grassland in an East Anglican breck. *Journal of Ecology* **69**: 499–508

Watt K E F (1968) *Ecology and Natural Resource Management.* New York, McGraw-Hill
Watt K E F (1974) *The Titanic Effect.* Stamford, Conn, Sinauer & Associates
Watts D (1978) The new biogeography and its niche in physical geography. *Geography* **63**: 324–37
Watts D (1971) *Principles of Biogeography.* London, McGraw-Hill
Webb N R, Le Haskins L E (1980) An ecological survey of heathlands in Poole Basin, Dorset, England. *Biological Conservation* **17**: 281–96
Webb R H, Wilshire H G (eds) (1980) *Environmental Effects of Off-Road Vehicles: Impacts and Management in Arid Regions.* New York, Springer-Verlag
Webb W L, Lavenroth W K, Szarek S R, Kinerson R S (1983) Primary production and abiotic controls in forests, grasslands and desert ecosystems in the United States. *Ecology* **64**: 134–51
Webber P J (1974) Tundra primary productivity. In: Ives V D, Barry R G (eds) *Arctic and Alpine Environments.* London, Methuen: 445–73
Weigert R G, Owen D F (1971) Trophic structure, available resources and population densities in terrestrial ecosystems versus aquatic ecosystems. *Journal of Theoretical Biology* **30**: 69–81
Wein R, MacLean D A (eds) (1983) *The Role of Fire in Northern Circumpolar Ecosystems.* Toronto, John Wiley & Sons
Wein R W, Riewe R R, Methuen I R (eds) (1983) *Resources and Dynamics of the Boreal Zone.* Ottawa, Association of Canadian Universities for Northern Studies
Wellburn A (1988) *Air Pollution and Acid Rain: The Biological Impact.* Harlow, Longman
Wells P V (1965) Scarp woodland, transported grassland soils and the concept of grassland climate in the Great Plains Region. *Science* **148**: 246–9
West D C, Shugart H H, Botkin D B (eds) (1981) *Forest Succession, Concepts and Applications.* New York, Springer-Verlag
West N A (ed) (1983) *Temperate Deserts and Semi-deserts.* Ecosystems of the World 5. Amsterdam, Elsevier
West N E (1979) Nutrient cycling in desert ecosystems. In: Goodall D W, Perry R A, Howe K M W (eds) *Arid-land Ecosystems: Structure, Functioning and Management* Vol. 2. Cambridge, Cambridge University Press: 301–24
West N E, Skujins J (1977) The nitrogen cycle in North American cold winter and semi-desert ecosystems. *Oecologia Plantarum* **12**: 45–53
West R G (1968) *Pleistocene Geology and Biology.* Harlow, Longman
Western D, Pearl M (eds) (1989) *Conservation for the Twenty-First Century.* New York, Oxford University Press
White J (ed) (1985) *The Population Structure of Vegetation.* Dordrecht, Junk
White R G,, Bunnell F L, Gaare E, Skogland T *et al* (1981) Ungulates on arctic ranges. In: Bliss L C, Heal O W, Moore J J (eds) *Tundra Ecosystems: a comparative study.* Cambridge, Cambridge University Press: 397–484
White R W (1979) *Introduction to the Principles and Practice of Soil Science.* Oxford, Blackwell Scientific Publications
Whitmore T C (1988) *Tropical Rain Forests of the Far East* (2nd edn). Oxford, Clarendon Press
Whitmore T C, Prance G T (1987) *Biogeography and Quaternary History in Tropical America.* Oxford, Clarendon Press
Whittaker R H (1951) A criticism of the plant association and climax concepts. *Northwest Science* **25**: 17–31
Whittaker R H (1953) A consideration of climax theory: the climax as a population and pattern. *Ecological Monographs* **23**: 41–78

Whittaker R H (ed) (1973) Ordination and classification of communities. In: *Handbook of Vegetation Science* Vol. 5. The Hague, Junk: 1–737

Whittaker R H (1975) *Communities and Ecosystems.* New York, Macmillan; London, Collier-Macmillan

Whittaker R H (1977) Evolution of species diversity in land communities. *Evolutionary Biology* **10**: 1–67

Whittaker R H *et al* (1974) The Hubbard Brook ecosystem study, forst biomass and production. *Ecological Monographs* **44**: 233–54

Whittaker R H, Feeny P P (1971) Allelochemics: chemical interactions between species. *Science* **171**: 757–70

Whittaker R H, Levin S A (1977) The role of mosaic phenomena in natural communities. *Theoretical Population Biology* **12**: 117–39

Whittaker R H, Levin S A, Root L B (1975) On the reasons for distinguishing niche habitat and ecotope. *American Naturalist* **109**: 479–82

Wiegolaski F E, Bliss L C, Svoboda J, Doyle G (1981) Primary production in the tundra. In: Bliss L C, Heal O W, Moore J J (eds) *Tunda Ecosystems: a comparative study*. Cambridge, Cambridge University Press: 187–226

Wilcox B A (1980) Insular ecology and conservation. In: Soulé E, Wilcox B A (eds) *Conservation Ecology.* Sunderland Mass, Sinauer Associates: 95–118

Wilcox B A (1988) Tropical deforestation and extinction. *Red List of Threatened Animals.* Cambridge and Gland, Switzerland, IUCN: v–x

Williamson M (1981) *Island Populations.* Oxford, Oxford University Press

Williamson P, Gribbin J (1991) How plankton change the climate. *New Scientist* **129(1760)**: 48–52

Wilson E O, Peter F M (eds) (1988) *Biodiversity.* National Academy Press

Wilson E O, Willis E O (1975) Applied biogeography. In: Cody M J, Diamond J M (eds) *Ecology and Evolution of Communities.* Harvard Unversity Cambridge, Mass, Belknap Press: 522–36

Witkamp M (1971) Soils as components of ecosystems. *Annual Review of Ecology and Systematics* **2**: 85–110

Witkamp M, Ausmus B S (1976) Processes in decomposition and nutrient transfer in forest systems. In: Anderson J M, MacFayden A (eds) *The Role of Terrestrial and Aquatic Organisms in Decomposition Processes.* Oxford, Blackwell Scientific Publications: 37–396

Wood T G (1976) The role of termites in decomposition processes. In: Anderson J M, MacFayden A (eds) *The Role of Terrestrial and Aquatic Organisms in Decomposition Processes.* Oxford, Blackwell Scientific Publications: 145–68

Woodman R G, Dodd J L, Bowman R A, Clark F E, Dickinson C E (1978) Nitrogen budget of a short-grass prairie ecosystem. *Oecologia* **34**: 363–76

Woodward F I, Sheehy J E (1983) *Principles and Measurements in Environmental Biology.* Cambridge, Butterworths

Woodward F L (1987) *Climate and Plant Distribution.* Cambridge University Press

Woodwell G M (1970) Effects of pollution on the structure and physiology of the ecosystem. *Science* **168**: 429–33

Woodwell G M (1971) Toxic substances and ecological cycles. In: *Man and the Ecosphere* Readings from Scientific American. San Francisco, W H Freeman Co: 137–218

Woodwell G M (1978) The carbon dioxide question. *Scientific American* **238**: 34–43 (reprint)

Woodwell G M (1983) The blue planet: of wholes and parts and man. In Mooney H A, Gordon M (eds) *Disturbance and Ecosystems: Components of a Response.* Ecological Studies 44. Berlin, Springer-Verlag: 2–10

World Resources (1988–89) *World Resources Report. An Assessment of the Resource Base that Supports the Global Economy.* New York, World Resources Institute. International Institute for Environment and Development and UNEP. Basic Books

Wright H E (1971) Late Quaternary vegetation history of North America. In: Inrekian K K (ed) *The Late Cenozoic Glacial Ages*. New Haven, Conn, Yale University Press: 425–65

Wynne-Edwards V C (1962) *Animal Dispersion in Relation to Social Behaviour.* London, Edward Hafner

Yalden D W (1980) Urban small mammals. *Journal of Applied Zoology (London)* **191**: 408–6

Young S (1990) Gardeners of the underworld. *New Scientist* **1728**: 50–3

Zeuner F E (1963) *A History of Domesticated Animals*. London, Hutchison

Zeven A C, Zhukovosky P M (1975) *Dictionary of Cultivated Plants and their Centres of Diversity.* Wageningen, Netherlands, Centre for Agricultural Publishing and Documentation

Zisweiler V (1967) *Extinct and Vanishing Animals.* New York, Longman/Springer-Verlag

Index